Integrated Methods for Optimization

Recent titles in the **INTERNATIONAL SERIES IN OPERATIONS RESEARCH & MANAGEMENT SCIENCE**
Frederick S. Hillier, Series Editor, *Stanford University*

Gass & Assad/ *AN ANNOTATED TIMELINE OF OPERATIONS RESEARCH: An Informal History*
Greenberg/ *TUTORIALS ON EMERGING METHODOLOGIES AND APPLICATIONS IN OPERATIONS RESEARCH*
Weber/ *UNCERTAINTY IN THE ELECTRIC POWER INDUSTRY: Methods and Models for Decision Support*
Figueira, Greco & Ehrgott/ *MULTIPLE CRITERIA DECISION ANALYSIS: State of the Art Surveys*
Reveliotis/ *REAL-TIME MANAGEMENT OF RESOURCE ALLOCATIONS SYSTEMS: A Discrete Event Systems Approach*
Kall & Mayer/ *STOCHASTIC LINEAR PROGRAMMING: Models, Theory, and Computation*
Sethi, Yan & Zhang/ *INVENTORY AND SUPPLY CHAIN MANAGEMENT WITH FORECAST UPDATES*
Cox/ *QUANTITATIVE HEALTH RISK ANALYSIS METHODS: Modeling the Human Health Impacts of Antibiotics Used in Food Animals*
Ching & Ng/ *MARKOV CHAINS: Models, Algorithms and Applications*
Li & Sun/ *NONLINEAR INTEGER PROGRAMMING*
Kaliszewski/ *SOFT COMPUTING FOR COMPLEX MULTIPLE CRITERIA DECISION MAKING*
Bouyssou et al/ *EVALUATION AND DECISION MODELS WITH MULTIPLE CRITERIA: Stepping stones for the analyst*
Blecker & Friedrich/ *MASS CUSTOMIZATION: Challenges and Solutions*
Appa, Pitsoulis & Williams/ *HANDBOOK ON MODELLING FOR DISCRETE OPTIMIZATION*
Herrmann/ *HANDBOOK OF PRODUCTION SCHEDULING*
Axsäter/ *INVENTORY CONTROL, 2nd Ed.*
Hall/ *PATIENT FLOW: Reducing Delay in Healthcare Delivery*
Józefowska & Węglarz/ *PERSPECTIVES IN MODERN PROJECT SCHEDULING*
Tian & Zhang/ *VACATION QUEUEING MODELS: Theory and Applications*
Yan, Yin & Zhang/ *STOCHASTIC PROCESSES, OPTIMIZATION, AND CONTROL THEORY APPLICATIONS IN FINANCIAL ENGINEERING, QUEUEING NETWORKS, AND MANUFACTURING SYSTEMS*
Saaty & Vargas/ *DECISION MAKING WITH THE ANALYTIC NETWORK PROCESS: Economic, Political, Social & Technological Applications w. Benefits, Opportunities, Costs & Risks*
Yu/ *TECHNOLOGY PORTFOLIO PLANNING AND MANAGEMENT: Practical Concepts and Tools*
Kandiller/ *PRINCIPLES OF MATHEMATICS IN OPERATIONS RESEARCH*
Lee & Lee/ *BUILDING SUPPLY CHAIN EXCELLENCE IN EMERGING ECONOMIES*
Weintraub/ *MANAGEMENT OF NATURAL RESOURCES: A Handbook of Operations Research Models, Algorithms, and Implementations*

* *A list of the early publications in the series is at the end of the book* *

INTEGRATED METHODS FOR OPTIMIZATION

by
JOHN N. HOOKER

John N. Hooker
Carnegie Mellon University
Pittsburgh, PA, USA

Library of Congress Control Number:

ISBN-10: 0-387-38272-0 (HB)   ISBN-10: 0-387-38274-7 (e-book)
ISBN-13: 978-0-387-38272-2 (HB) ISBN-13: 978-0-387-38274-6 (e-book)

Printed on acid-free paper.

© 2007 by Springer Science+Business Media, LLC
All rights reserved. This work may not be translated or copied in whole or in part without the written permission of the publisher (Springer Science + Business Media, LLC, 233 Spring Street, New York, NY 10013, USA), except for brief excerpts in connection with reviews or scholarly analysis. Use in connection with any form of information storage and retrieval, electronic adaptation, computer software, or by similar or dissimilar methodology now know or hereafter developed is forbidden.
The use in this publication of trade names, trademarks, service marks and similar terms, even if the are not identified as such, is not to be taken as an expression of opinion as to whether or not they are subject to proprietary rights.

9 8 7 6 5 4 3 2 1

springer.com

Contents

Preface		xiii
1. INTRODUCTION		1
1.1	A Unifying Framework	3
1.2	Modeling to Reveal Problem Structure	5
1.3	The Role of Duality	7
1.4	Advantages of Integrated Methods	9
1.5	Some Applications	11
1.6	Plan of the Book	12
1.7	Bibliographic Notes	13
2. SEARCH		15
2.1	The Solution Process	16
	2.1.1 Search	17
	2.1.2 Inference	18
	2.1.3 Relaxation	19
	2.1.4 Exercises	20
2.2	Branching Search	21
	2.2.1 Branch and Infer	21
	2.2.2 Branch and Relax	22
	2.2.3 Example: Freight Transfer	24
	2.2.4 Example: Production Planning	30
	2.2.5 Example: Employee Scheduling	33
	2.2.6 Example: Continuous Global Optimization	41
	2.2.7 Example: Product Configuration	45
	2.2.8 Branch and Price	51
	2.2.9 Example: Airline Crew Scheduling	53

		2.2.10 Exercises	59
	2.3	Constraint-Directed Search	63
		2.3.1 Constraint-Directed Branching	65
		2.3.2 Example: Propositional Satisfiability	67
		2.3.3 Partial-Order Dynamic Backtracking	72
		2.3.4 Example: Propositional Satisfiability	74
		2.3.5 Relaxation in Constraint-Directed Search	75
		2.3.6 Logic-Based Benders Decomposition	76
		2.3.7 Example: Machine Scheduling	78
		2.3.8 Exercises	86
	2.4	Local Search	88
		2.4.1 Some Popular Metaheuristics	89
		2.4.2 Local Search Conceived as Branching	90
		2.4.3 Relaxation	93
		2.4.4 Constraint-Directed Local Search	94
		2.4.5 Example: Single-Vehicle Routing	95
		2.4.6 Exercises	100
	2.5	Bibliographic Notes	103
3.	INFERENCE		105
	3.1	Completeness	106
		3.1.1 Basic Definitions	107
		3.1.2 Domain Completeness	108
		3.1.3 Bounds Completeness	110
		3.1.4 k-Completeness	110
		3.1.5 k-Consistency	112
		3.1.6 Backtracking and Width	112
		3.1.7 Exercises	114
	3.2	Inference Duality	115
		3.2.1 Strong Duality and Completeness	116
		3.2.2 Certificates and Problem Complexity	117
		3.2.3 Sensitivity Analysis	118
		3.2.4 Duality and Constraint-Directed Search	119
		3.2.5 Exercises	121
	3.3	Linear Inequalities	121
		3.3.1 A Complete Inference Method	122
		3.3.2 Domain and Bounds Completeness	124
		3.3.3 k-Completeness	125
		3.3.4 Linear Programming Duality	127

	3.3.5	Sensitivity Analysis	130
	3.3.6	Basic Solutions	131
	3.3.7	More Sensitivity Analysis	133
	3.3.8	Domain Reduction with Dual Multipliers	135
	3.3.9	Classical Benders Cuts	136
	3.3.10	Exercises	138
3.4	General Inequality Constraints		140
	3.4.1	The Surrogate Dual	141
	3.4.2	The Lagrangean Dual	143
	3.4.3	Properties of the Lagrangean Dual	144
	3.4.4	Domain Reduction with Lagrange Multipliers	146
	3.4.5	Exercises	147
3.5	Propositional Logic		148
	3.5.1	Logical Clauses	149
	3.5.2	A Complete Inference Method	150
	3.5.3	Unit Resolution and Horn Clauses	152
	3.5.4	Domain Completeness and k-Completeness	152
	3.5.5	Strong k-Consistency	154
	3.5.6	Completeness of Parallel Resolution	155
	3.5.7	Exercises	157
3.6	0-1 Linear Inequalities		158
	3.6.1	Implication between Inequalities	159
	3.6.2	Implication of Logical Clauses	161
	3.6.3	Implication of Cardinality Clauses	163
	3.6.4	0-1 Resolution	165
	3.6.5	k-Completeness	167
	3.6.6	Strong k-Consistency	168
	3.6.7	Exercises	169
3.7	Integer Linear Inequalities		170
	3.7.1	The Subadditive Dual	171
	3.7.2	The Branching Dual	175
	3.7.3	Benders Cuts	178
	3.7.4	Exercises	181
3.8	The Element Constraint		182
	3.8.1	Domain Completeness	183
	3.8.2	Bounds Completeness	185
	3.8.3	Exercises	187
3.9	The All-Different Constraint		187

	3.9.1 Bipartite Matching	188
	3.9.2 Domain Completeness	189
	3.9.3 Bounds Completeness	191
	3.9.4 Exercises	193
3.10	The Cardinality and Nvalues Constraints	194
	3.10.1 The Cardinality Constraint	194
	3.10.2 Network Flow Model	195
	3.10.3 Domain Completeness for Cardinality	197
	3.10.4 The Nvalues Constraint	198
	3.10.5 Exercises	198
3.11	The Circuit Constraint	199
	3.11.1 Modeling with Circuit	200
	3.11.2 Elementary Filtering Methods	201
	3.11.3 Filtering Based on Separators	202
	3.11.4 Network Flow Model	204
	3.11.5 Exercises	206
3.12	The Stretch Constraint	207
	3.12.1 Dynamic Programming Model	208
	3.12.2 Domain Completeness	210
	3.12.3 Exercises	211
3.13	Disjunctive Scheduling	212
	3.13.1 Edge Finding	213
	3.13.2 Not-First, Not-Last Rules	217
	3.13.3 Benders Cuts	222
	3.13.4 Exercises	228
3.14	Cumulative Scheduling	230
	3.14.1 Edge Finding	231
	3.14.2 Extended Edge Finding	236
	3.14.3 Not-First, Not-Last Rules	238
	3.14.4 Energetic Reasoning	239
	3.14.5 Benders Cuts	241
	3.14.6 Exercises	244
3.15	Bibliographic Notes	245
4. RELAXATION		249
4.1	Relaxation Duality	251
4.2	Linear Inequalities	252
	4.2.1 Linear Optimization	252

	4.2.2	Relaxation Dual	255
	4.2.3	Exercises	257
4.3	Semicontinuous Piecewise Linear Functions		259
	4.3.1	Convex Hull Relaxation	259
	4.3.2	Exercises	260
4.4	0-1 Linear Inequalities		261
	4.4.1	Chvátal-Gomory Cuts	262
	4.4.2	0-1 Knapsack Cuts	266
	4.4.3	Sequential Lifting	266
	4.4.4	Sequence-Independent Lifting	269
	4.4.5	Set Packing Inequalities	271
	4.4.6	Exercises	273
4.5	Integer Linear Inequalities		275
	4.5.1	Chvátal-Gomory Cuts	276
	4.5.2	Gomory Cuts	278
	4.5.3	Mixed Integer Rounding Cuts	282
	4.5.4	Separating Mixed Integer Rounding Cuts	285
	4.5.5	Integral Polyhedra	286
	4.5.6	Exercises	290
4.6	Lagrangean and Surrogate Relaxations		291
	4.6.1	Surrogate Relaxation and Duality	292
	4.6.2	Lagrangean Relaxation and Duality	292
	4.6.3	Lagrangean Relaxation for Linear Programming	293
	4.6.4	Example: Generalized Assignment Problem	295
	4.6.5	Solving the Lagrangean Dual	296
	4.6.6	Exercises	297
4.7	Disjunctions of Linear Systems		298
	4.7.1	Convex Hull Relaxation	298
	4.7.2	Big-M Relaxation	300
	4.7.3	Disjunctions of Linear Inequalities	303
	4.7.4	Disjunctions of Linear Equations	306
	4.7.5	Separating Disjunctive Cuts	307
	4.7.6	Exercises	311
4.8	Disjunctions of Nonlinear Systems		313
	4.8.1	Convex Hull Relaxation	313
	4.8.2	Big-M Relaxation	316
	4.8.3	Exercises	318
4.9	MILP Modeling		318

- 4.9.1 MILP Representability 319
- 4.9.2 Example: Fixed-Charge Function 321
- 4.9.3 Disjunctive Models 324
- 4.9.4 Knapsack Models 329
- 4.9.5 Exercises 333
- 4.10 Propositional Logic 335
  - 4.10.1 Common Logical Formulas 335
  - 4.10.2 Resolution as a Tightening Technique . . . 340
  - 4.10.3 Refutation by Linear Relaxation 343
  - 4.10.4 Input Resolution and Rank 1 Cuts 344
  - 4.10.5 Separating Resolvents 348
  - 4.10.6 Exercises 350
- 4.11 The Element Constraint 352
  - 4.11.1 Convex Hull Relaxations 353
  - 4.11.2 Big-M Relaxations 357
  - 4.11.3 Vector-Valued Element 358
  - 4.11.4 Exercises 361
- 4.12 The All-Different Constraint 362
  - 4.12.1 Convex Hull Relaxation 362
  - 4.12.2 Convex Hull MILP Formulation 368
  - 4.12.3 Modeling Costs with Alldiff 369
  - 4.12.4 Example: Quadratic Assignment Problem . 372
  - 4.12.5 Exercises 374
- 4.13 The Cardinality Constraint 375
  - 4.13.1 Convex Hull Relaxation 376
  - 4.13.2 Convex Hull MILP Formulation 378
  - 4.13.3 Exercises 378
- 4.14 The Circuit Constraint 379
  - 4.14.1 0-1 Programming Model 379
  - 4.14.2 Continuous Relaxations 380
  - 4.14.3 Comb Inequalities 382
  - 4.14.4 Exercises 385
- 4.15 Disjunctive Scheduling 385
  - 4.15.1 Disjunctive Relaxations 386
  - 4.15.2 MILP Relaxations 388
  - 4.15.3 A Class of Valid Inequalities 390
  - 4.15.4 Exercises 392
- 4.16 Cumulative Scheduling 393

Contents   xi

        4.16.1 MILP Models   394
        4.16.2 A Class of Valid Inequalities   398
        4.16.3 Relaxation of Benders Subproblems   400
        4.16.4 Exercises   410
   4.17 Bibliographic Notes   411

5. DICTIONARY OF CONSTRAINTS   415
   0-1 linear   416
   All-different   417
   Among   418
   Bin packing   419
   Cardinality   420
   Cardinality clause   421
   Cardinality conditional   422
   Change   422
   Circuit   423
   Clique   424
   Conditional   424
   Cumulative scheduling   425
   Cutset   426
   Cycle   426
   Diffn   427
   Disjunctive scheduling   428
   Element   429
   Flow   430
   Indexed linear   431
   Integer linear   432
   Lex-greater   433
   Linear disjunction   433
   Logic   434
   MILP   435
   Min-n 436
   Network design   437
   Nonlinear disjunction   438
   Nvalues   438

Path	439
Piecewise linear	440
Same	441
Set covering	441
Set packing	442
Soft alldiff	443
Stretch	444
Sum	444
Symmetric alldiff	445
Symmetric cardinality	446
Value precedence	446
References	449
Index	475

Preface

Optimization has become a versatile tool in a wide array of application areas, ranging from manufacturing and information technology to the social sciences. Methods for solving optimization problems are equally numerous and provide a large reservoir of problem-solving technology. In fact, there is such a variety of methods that it is difficult to take full advantage of them. They are described in different technical languages and are implemented in different software packages. Many are not implemented at all. It is hard to tell which one is best for a given problem, and there is too seldom an opportunity to combine techniques that have complementary strengths.

The ideal would be to bring these methods under one roof, so that they and their combinations are all available to solve a problem. As it turns out, many of them share, at some level, a common problem-solving strategy. This opens the door to integration—to the design of a modeling and algorithmic framework within which different techniques can work together in a principled way.

This book undertakes such a project. It deals primarily with the unification of mathematical programming and constraint programming, since this has been the focus of most recent research on integrated methods. Mathematical programming brings to the table its sophisticated relaxation techniques and concepts of duality. Constraint programming contributes its inference and propagation methods, along with a powerful modeling approach. It is possible to have all of these advantages at once, rather than being forced to choose between them. Continuous global optimization and heuristic methods can also be brought into the framework.

The book is intended for those who wish to learn about optimization from an integrated point of view, including researchers, software developers, and practitioners. It is also for postgraduate students interested in a unified treatment of the field. It is written as an advanced textbook, with exercises, that develops optimization concepts from the ground up. It takes an interdisciplinary approach that presupposes mathematical sophistication but no specific knowledge of either mathematical programming or constraint programming.

The choice of topics is guided by what is relevant to understanding the principles behind popular linear, mixed integer, and constraint programming solvers—and more importantly, integrated solvers of the present and foreseeable future. On the mathematical programming side it presents the basic theory of linear and integer programming, cutting planes, Lagrangean and other types of duality, mixed integer modeling, and polyhedral relaxations for a wide range of combinatorial constraints. On the constraint programming side it discusses constraint propagation, domain filtering, consistency (reconceived as completeness), global constraints, and modeling techniques. The material ranges from the classical to the very recent, with some results presented here for the first time.

The ideas are tied together by a search-infer-and-relax algorithmic framework, an underlying theory of inference and relaxation duality, and the use of metaconstraints (a generalization of global constraints) for modeling. One chapter takes the reader through a number of examples to develop the algorithmic ideas and illustrate modeling techniques. Two subsequent chapters focus, respectively, on inference methods from constraint programming and relaxation methods from mathematical programming. The final chapter is a dictionary of constraint types and is intended as a sourcebook for modeling.

I would like to acknowledge the many collaborators and former students from whom I have learned much about integrated problem solving. They include Kim Allen Andersen, Ionuț Aron, Alexander Bockmayr, Endre Boros, Srinivas Bollapragada, Jonathan Cagan, Vijay Chandru, Milind Dawande, Giorgio Gallo, Omar Ghattas, Ignacio Grossmann, Peter Hammer, Latife Genç Kaya, Hak-Jin Kim, Maria Auxilio Osorio, Greger Ottosson, Gabriella Rago, Ramesh Raman, Erlendur Thorsteinsson, Paul Williams, Hong Yan, and Tallys Yunes. I am also indebted to Tarik Hadzic, Samid Hoda, Gil Jones, Matt Streeter, John Turner, and Tallys Yunes for correcting a number of mistakes in the manuscript, although I of course take responsibility for any remaining errors.

Chapter 1

INTRODUCTION

Optimization by nature requires a multi-faceted approach. Some classes of optimization problems, such as linear programming models, can be solved by an all-purpose method. But most problems need individual attention. Typically, the secret to solving a problem is to take advantage of its particular structure.

The result is a proliferation of optimization methods. Thousands of journal articles address narrowly defined problems, as they must if the problems are to be solved in a reasonable amount of time. Not only this, but the articles are distributed across several literatures that speak different technical languages. Chief among these are the literatures of mathematical programming, constraint programming, continuous global optimization, and heuristic methods. This imposes an obvious burden on anyone who seeks the right method and software to solve a given problem.

There has been significant progress toward developing robust, general-purpose solvers within some of the individual disciplines. Each generation of mathematical programming software solves a wider range of problems, and similarly for constraint programming and global optimization software. Yet even these solvers fail to exploit most problem-specific methods—with the partial exception of constraint programming, whose concept of a global constraint provides a clue to how to overcome this weakness. Also, the four disciplines continue to move in largely separate orbits. This not only imposes the inconvenience of becoming familiar with multiple solvers, but it passes up the advantages of integrated problem solving.

Recent research has shown that there is much to be gained by exploiting the complementary strengths of different approaches to optimization. Mathematical programmers are expert at relaxation techniques and polyhedral analysis. Constraint programming is distinguished by its inference techniques and modeling power. Continuous global optimization is known for its convexification methods, and heuristic methods for their search strategies. Rather than choose between these, one would like to have them all available to attack a given problem. Some problems submit to a single approach, but others benefit from the more flexible modeling and orders-of-magnitude speedup in computation that can occur when ideas from different fields are combined.

The advantages of integrated methods are being demonstrated in a growing literature but, as a result, they themselves are multiplying. If there are many solution methods that might be combined, there are even more ways to combine them. The problem of proliferation seems only to be compounded by efforts to integrate.

A change of perspective can bring order into this chaos. Rather than look for ways to combine methods, one can look for what the methods already have in common. Perhaps there is a general problem-solving strategy that the various communities have arrived at independently, albeit from different directions and with different emphases, because it is a strategy that works.

This book takes such a perspective. It develops an algorithmic framework in which the different optimization methods, and more importantly their many combinations, are variations on a theme. It proposes a modeling practice that can bring problem-specific methods into the scheme in a natural way. In short, it seeks an underlying unity in optimization methods.

The book emphasizes the integration of mathematical programming and constraint programming in particular, since this is where most of the research on integrated methods has been focused to date. Nonetheless, some attempt is made to show how global optimization and heuristic methods fit into the same framework.

Unification is good for learning as well as practice. Students of operations research or computer science who confine themselves to their own field miss the insights of the other, as well as an overarching perspective on optimization that helps make sense of it all. This book is therefore designed as a graduate-level optimization text that belongs to neither field but tries to construct a coherent body of material from both.

Neither this nor any other text covers all concepts related to optimization. There is nothing here about stochastic optimization, multiobjective programming, semidefinite programming, or approximation methods, and only a little about nonlinear programming and dynamic programming. Some important ideas of combinatorial analysis are left out, as are whole areas of constraint programming. A broad selection of topics is nonetheless presented, guided primarily by what is relevant to the major general-purpose solvers of linear and mixed integer programming, constraint programming, and to some extent global optimization. The emphasis, of course, is on how these topics form more than just a miscellaneous collection but are parts of an integrated approach to optimization.

Although not comprehensive, the material presented here is more than adequate to provide a substantial grounding in optimization. It is a starting point from which the student can explore other subdisciplines from an ecumenical point of view.

1.1 A Unifying Framework

Optimization methods tend to employ three strategies that interact in specific ways: *search*, *inference*, and *relaxation*. Search is the enumeration of problem restrictions, while inference and relaxation make the search more intelligent. Inference draws out implicit information about where the solution might lie, so that less search is necessary. Relaxation replaces the problem with a simpler one whose solution may point the way to a solution of the original one.

Search is fundamental because the optimal solution of a problem lies somewhere in a solution space, and one must find it. The solution space is the set of solutions that satisfy the constraints of the problem. Some solutions are more desirable than others, and the objective is to find the best one, or at least a good one. In many practical problems, the solution space tends to be huge and multidimensional. Its boundaries may be highly irregular, and if there are discrete as well as continuous variables, it is full of gaps and holes.

Whether searching for lost keys or a fugitive from the law, a common strategy is to divide up the territory and focus on one region at a time. Optimization methods almost invariably do the same. A region is the solution set for a *restriction* of the problem, or a problem to which additional constraints have been added. A sequence of restrictions are solved, and the best solution found is selected. If the search is exhaustive, meaning that the entire search space is covered, the best solution found is optimal.

The most prevalent scheme for exhaustive search is *branching*: splitting the territory (say) in half, splitting it again if either half is still too large to manage, and so forth. Another basic strategy is *constraint-directed search*: whenever a region has been searched, a constraint or *nogood* is created that excludes that part of the search space, and perhaps other parts that are unpromising for the same reason. The next examined region must satisfy the constraints so far accumulated.

Branching methods include the popular branch-and-cut methods of mixed integer programming and branch-and-infer methods of constraint programming, on which all the major commercial solvers are based. Constraint-directed methods include Benders decomposition in mathematical programming and such nogood-based methods as branching with clause learning for propositional satisfiability problems, and partial order dynamic backtracking. Continuous global optimizers use a branching scheme to divide space into multidimensional boxes. Local search or heuristic methods likewise enumerate a sequence of problem restrictions, represented by a sequence of neighborhoods. The distinction of branching and constraint-directed search carries over to heuristic methods: a greedy adaptive search procedure, for example, is an incomplete form of branching search, and tabu search is an incomplete constraint-directed search.

Inference is a way of learning more about the search space, so as not to waste time looking in the wrong places. Police might deduce from a street map that a suspect would never frequent certain neighborhoods, just as a problem solver might deduce from a constraint that certain variables would never take certain values in an optimal solution. This not only shrinks the region to be searched but, sometimes more importantly, reduces its dimensionality. Mathematical programming systems use inference methods in the form of cutting planes and preprocessing techniques, while constraint programming systems rely heavily on domain filtering and bounds propagation.

Relaxation enlarges the search space in a way that makes it easier to examine. A common strategy is to replace the current problem restriction with a continuous, linear relaxation. This replaces the corresponding region of the search space with a polyhedron that contains it, thus smoothing out the boundary and filling in the holes, and simplifying the search for an optimal solution. Relaxation can help in several ways. The solution of the relaxation may happen to lie inside the original search space, in which case it solves the current problem restriction. If not, the optimal solution of the relaxation may be no better than the best feasible solution found so far, in which case one can immediately move on to another region. Even if this does not occur, the optimal solution of the relaxation may provide a clue as to where the optimal solution of the original problem lies.

Search, inference, and relaxation reinforce each other. Restricting the problem in a search process allows one to draw more inferences and formulate relaxations that are closer to the original. Inference accelerates the search by excluding part of the search space, as when filtering reduces domains in constraint programming, when logic-based Benders cuts (a form of nogood) are generated in Benders decomposition, or when items are added to the tabu list in local search. It deduces constraints that can strengthen the relaxation, such as cutting planes in integer programming.

Relaxation abbreviates the search by providing bounds, as in the branch-and-relax and branch-and-cut methods of integer programming. The solution of the relaxation, even when infeasible in the original problem, can provide information about which problem restriction to examine next. In branch-and-cut methods, for example, one branches on a variable that has a fractional value in the solution of the relaxation. In constraint-directed search, the set of accumulated nogoods is, in effect, a relaxation whose solution defines the next problem restriction to be solved. Less obviously, the solution of the relaxation can direct inference, since one can give priority to deducing constraints that exclude this solution. The separating cuts of integer programming provide an example of this.

This search-infer-and-relax scheme provides a framework within which one can mix elements from different traditions. As a search scheme one might select branching or nogood-based search. For inference one might apply domain filtering to some constraints, generate cutting planes for others, or for others use some of the interval propagation and variable fixing methods characteristic of global optimization. Some constraints might be given relaxations based on integer programming models, others given relaxations that are tailor-made for global constraints, others relaxed with the factorization methods used in continuous global solvers, and still others left out of the relaxation altogether.

1.2 Modeling to Reveal Problem Structure

Search-infer-and-relax methods can succeed only if they exploit problem-specific structure. Experience teaches that inference and relaxation can be blunt instruments unless sharpened with specific knowledge of the problem class being solved. Yet, it is impractical to invent specialized techniques for every new type of problem that comes along.

The answer is to analyze constraints rather than problems. Although every problem is different, certain patterns tend to recur in the constraints. Many scheduling problems require, for example, that jobs run sequentially without overlapping. Other problems require that employees be assigned to work a specified number of days in a row. It is not hard to identify structured subsets of constraints that keep coming up. Each subset can be represented by a single *metaconstraint*, and specialized inference and relaxation methods can be designed for each metaconstraint. The modeler's choice of metaconstraints can then communicate much about the problem structure to the solver. In particular, it dictates which inference and relaxation techniques are used in the solution process.

This might be called *constraint-based control*, which can extend to the search procedure as well as the choice of inference and relaxation methods. The user begins by choosing the overall search algorithm, perhaps branching or constraint-directed search, and perhaps an exhaustive or heuristic version of it. The choice of metaconstraints determines the rest. In a branching framework, for example, the search branches when the solution of the current relaxation violates one or more constraints. A priority list designates the violated constraint on which to branch. The constraint "knows" how to branch when it is violated, and the search proceeds accordingly.

This scheme can work only if the modeler uses constraints that are rich enough to capture substantial islands of structure in the problem. This requires a change from the traditional practice of mathematical programming, which is to build models with a small vocabulary of primitive constraints such as inequalities and equations. It recommends something closer to the constraint programmer's practice of using *global constraints*, so named because each constraint stands for a collection of more elementary constraints whose global structure is exploited by the solver.

This book therefore advocates modeling with metaconstraints, which generalize the idea of global constraints. A metaconstraint may consist of a set of inequalities of a certain type, a set of constraints to be activated under certain conditions, or a global constraint familiar to the constraint programming world, to mention only a few possibilities. The advantages of metaconstraints are twofold. They not only reveal the problem structure to the solver, which may lead to faster solution, but they allow one to write more concise models that are easier to follow and easier to debug.

Modeling with metaconstraints immediately raises the issue of what to do when the metaconstraints begin to proliferate, much as the special-purpose algorithms that were mentioned earlier. New problems often require new metaconstraints to capture a substructure that did not occur in previous problems. Yet, this is not the stumbling block that it may seem to be.

The lexicon of metaconstraints can grow large, but there are still many fewer constraints than problems, just as there are many fewer words than sentences. In any field of endeavor, people tend to settle on a limited number of terms that prove adequate over time for expressing the key ideas. There is no alternative, since most of us can master only a limited vocabulary. This is true of technical domains in particular, since a limited number of technical terms tend to evolve and prove adequate for most situations. Sailors must know about halyards, booms, mizzens, and much else, but the nautical vocabulary is finite and learnable.

The same applies to modeling. In any given domain, practitioners are likely to develop a limited stock of metaconstraints that frequently arise. There might be one stock for project management, one for process scheduling, one for supply chains, and so forth, with much overlap between them. In fact, this has already happened in some domains for which specialized software has developed, such as project scheduling. There will be many metaconstraints overall, just as there are many technical terms in the world. But no one is obliged to know more than a small fraction of them.

Computer-based modeling systems can ease the task further. There is no need to write models in a formal modeling language in which one must get the syntax right or generate error messages. An intelligent user interface can provide menus of constraints, conveniently organized by application domain or along other dimensions. Selecting a constraint activates a window that allows one to import data, set parameters, and choose options for inference, relaxation, and search. The window contains links to related constraints that may be more suitable. The system prompts the user with checklists or queries to guide the modeling process. The solver keeps updating the solution of a small problem as the modeling proceeds, so that the modeler can see when the solution begins to look reasonable.

1.3 The Role of Duality

Duality is a perennial theme of optimization. It occurs in such forms as linear programming duality and its special cases, Lagrangean duality, surrogate duality, and superadditive duality. It is also a unifying theme for this book for two reasons. These various duals turn out to be closely related, a fact that helps to unify optimization theory. They can all be classified as *inference duals* or *relaxation duals*, and in most cases as both. Secondly, the two types of duals help elucidate how search, inference, and relaxation relate to each other: inference duality is a duality of search and inference, while relaxation duality is a duality of search and relaxation.

Inference duality arises as follows. An optimization problem can be seen as the problem of finding a set of values for the problem variables that minimize the objective function. But it can also be seen as the problem of inferring from the constraint set the tightest possible lower bound on the value of the objective function. In the first case, one searches over values of the variables and in the second, one searches over proofs. The problem of finding the proof that yields the best bound is the inference dual.

The precise nature of the inference dual depends on what inference method one uses to derive bounds. If the inference method is complete for the problem class in question, the inference dual has the same optimal value as the original problem. One particular inference method, nonnegative linear combination, yields the classical linear programming dual for linear problems and the surrogate dual for general inequality-constrained problems. A slightly different inference method gives rise to the all-important Lagrangean dual. In fact, the close connection between surrogate and Lagrangean duals, which are superficially unrelated, becomes evident when one regards them as inference duals. Still another inference method yields the superadditive dual, which arises in integer programming.

Inference duality is a unifying concept because, first of all, it can be defined for any optimization problem, not just the inequality-constrained problems traditionally studied in mathematical programming. Secondly, it can serve as a general basis for sensitivity analysis, which examines the sensitivity of the optimal solution to perturbations in the problem data, thus revealing which data must be accurate to get a meaningful solution. Most importantly, the proof that solves the inference dual can yield nogoods or logic-based Benders cuts for a constraint-directed search method. Methods as disparate as Benders decomposition, a classic technique of operations research, and the Davis-Putnam-Loveland method with clause learning, which rapidly solves propositional satisfiability problems, are closely related because they both generate nogoods by solving an inference dual. In principle, any inference dual can give rise to a nogood-based algorithm. In planning and scheduling problems, for example, the use of new inference duals has resulted in computational speedups of several orders of magnitude.

Relaxation duality is a duality of search and relaxation, or more precisely a duality of restriction and relaxation. A motivation for solving a sequence of restrictions is that the restrictions are easier to solve than the original. Since relaxations are also designed to be easier than the original, one might ask whether a problem can be addressed by solving a sequence of relaxations. It can, if the relaxations are parameterized by *dual variables*, which allow one to search the space of relaxations by enumerating values of the dual variables. The solution of each relaxation provides a bound on the optimal value, and the problem of finding the best bound is the relaxation dual. In general, an enumeration of relaxations does not solve the problem, as does an enumeration of restrictions, because the best bound may not be equal to the optimal value of the original problem. The bound may nonetheless be useful, as in the surrogate and particularly Lagrangean duals, which were originally conceived as relaxation duals rather than inference duals.

1.4 Advantages of Integrated Methods

The academic literature tends to emphasize computation speed when evaluating a new approach to problem-solving, perhaps because it is easily measured. Practitioners know, however, that model development time is often at least as important as solution time. This argues for the convenience of having all the modeling and algorithmic resources available in a single integrated system. One can try several approaches to a problem without having to learn several systems and port data between them. Metaconstraints can also be a significant time saver, as they lead to simpler models that are easier to build and maintain.

Nonetheless, the computational advantages are there as well. Certainly one need never pay a computational price for using an integrated system, since the traditional techniques can always be available as one of the options. However, experience confirms that a more broad-based strategy can substantially speed computation. This is borne out by a sampling of results from the literature. The focus here is on methods that integrate constraint programming (CP) and mixed integer/linear programming (MILP), which can be classified roughly by the type of integration they use.

Many integrated methods combine CP with linear relaxations developed in MILP. One study, for example, combined a CP algorithm with an assignment problem relaxation (Section 4.14.2) and reduced-cost variable fixing (Section 3.3.8) to solve lesson timetabling problems two to fifty times faster than CP [127]. Another study [279] combined CP with convex hull relaxations to solve problems with piecewise linear costs two to two hundred times faster than MILP (Section 4.3). A third study [315] solved product configuration problems thirty to forty times faster than MILP (which was faster than CP) by combining convex hull relaxations for variable indices with constraint propagation (a similar problem is discussed in Section 2.2.7). Linear relaxations combined with logic processing (Section 3.5) solved a boat party scheduling problem in five minutes that MILP could not solve in twelve hours, and solved flow shop instances four times faster than MILP [198].

Experiments have been conducted with other kinds of relaxations as well. A combination of CP and Lagrangean relaxation (Sections 3.4.2, 4.6) solved automatic digital recording problems one to ten times faster than MILP, which was faster than CP [300]. CP assisted by semidefinite programming relaxations obtained significantly better suboptimal solutions of stable set problems than CP alone in a fraction of the time [177]. Logic processing and linear quasi-relaxations solved nonlinear structural design problems up to 600 times faster than MILP and solved two of the problems in less than six minutes when MILP could not solve them in twenty hours [60].

Branch-and-price integer programming methods (Section 2.2.8) have also been combined with CP processing, particularly in the area of airline and transit crew scheduling (such a problem is discussed in Section 2.2.9). In one study [344], this approach solved urban transit crew management problems that involved up to 210 trips, while traditional branch and price could accommodate only 120 trips. A CP-based branch-and-price method was the first to solve the eight-team traveling tournament problem [114].

Perhaps the greatest speedups have been achieved by integrating CP and MILP through generalized forms of Benders decomposition (Section 2.3.6). One study [203] solved minimum-cost machine allocation and scheduling problems 20 to 1000 times faster than CP or MILP. A subsequent study [314] improved upon these results by an additional factor of ten. More recent work has extended the applicability of logic-based Benders methods. One industrial implementation [316] solved, in ten minutes, polypropylene batch scheduling problems at BASF that were previously insoluble. A CP/MILP hybrid solved twice as many call center scheduling problems as traditional Benders [44]. A different CP/MILP hybrid solved planning and scheduling problems, with resource-constrained scheduling, 100 to 1000 times faster than CP or MILP when minimizing cost or makespan [189], 10 to 1000 times faster when minimizing the number of late jobs, and about ten times faster (with much better solutions when optimality was not obtained) when minimizing total tardiness [191] (Section 2.3.7 shows how to solve a simplified minimum-makespan problem with a logic-based Benders technique). Finally, a hybrid Benders approach was applied [278] to obtain speedups of several orders of magnitude relative to the state of the art in sports scheduling.

It is important to bear in mind that none of these results were achieved with the full resources of integration. They are also preliminary results obtained with experimental codes. Integrated solution software will doubtless improve over time. The chief advantage of integrated methods, however, may be that they encourage a broader perspective on problem solving. This may inspire developments that would not have been possible inside individual disciplines.

1.5 Some Applications

Integrated methods have been successfully applied in a wide variety of contexts. The literature can again be roughly organized according to the type of integration used.

Applications that combine CP with cutting planes (Sections 4.4 and 4.2) include the orthogonal Latin squares problem [9], truss structure design [60], processing network design [161, 198], single-vehicle routing [292], resource-constrained scheduling [106], multiple machine scheduling [58], boat party scheduling [198], and the multidimensional knapsack problem [258]. Cutting planes for disjunctions of linear systems (Section 4.7.5) have been applied to factory retrofit planning, strip packing, and zero-wait job shop scheduling [294].

Convex hull relaxations for disjunctions of linear and nonlinear systems (Sections 4.7.1 and 4.8.1) have been used to solve several chemical process engineering problems [223, 224, 277, 295, 326]. Convex hull relaxations of piecewise linear constraints have been used in a CP context to solve fixed charge problems and transportation problems with piecewise linear costs [279], as well as production planning problems with semicontinuous piecewise linear costs [259, 260].

Applications that combine CP with reduced-cost variable fixing (Sections 3.3.2 and 3.3.2) include the traveling salesman problem with time windows [239], product configuration [239], fixed charge network flows [213], and lesson timetabling [127].

CP-based Lagrangean methods (Sections 3.4.2, 4.6) have been applied to network design [95], automatic digital recording [300], traveling tournament problems [45], the resource-constrained shortest path problem [140], and the general problem of filtering domains [212].

The most popular application of CP-based branch-and-price methods is to airline crew assignment and crew rostering [76, 120, 209, 216, 301]. Other applications include transit bus crew scheduling [344], aircraft scheduling [160], vehicle routing [291], network design [77], employee timetabling [107], physician scheduling [141], and the traveling tournament problem [113, 114].

Benders methods that combine MILP with CP or logic based methods (Sections 2.3.6–2.3.7) have been developed for circuit verification problems [201], integer programming [80, 199] and the propositional satisfiability problem [185, 199]. A series of papers have described applications to planning and scheduling [81, 185, 169, 189, 191, 203]. Other applications include dispatching of automated guided vehicles [92], steel production scheduling [168], batch scheduling in a chemical plant [229], and polypropylene batch scheduling in particular [316]. CP-based Benders methods have also been applied to real-time scheduling of computer processors [67] and traffic diversion problems [339].

1.6 Plan of the Book

The three main chapters of the book reflect the three parts of the search-infer-and-relax framework. Chapter 2, on search, begins with an overview of the algorithmic framework, and then successively takes up branching algorithms, constraint-directed search, and heuristic methods. This chapter serves the equally important function of introducing the main ideas of the book by example. Nine problems are used to illustrate how models are constructed and how search, inference, and relaxation interact to solve them. One can get a very good idea of what integrated problem solving is all about by reading this chapter alone.

The rest of the book develops the ideas of Chapter 2 in a more systematic way—except for those relating to continuous global optimization and local search, which are discussed only in Chapter 2. Chapter 3, on inference, begins with some basic concepts from the constraints literature. It develops a theory of inference for inequality-constrained problems and propositional logic. It covers four types of inference duality (linear programming, surrogate, Lagrangean, and subadditive), and from inference duality it derives sensitivity analysis, some domain reduction techniques, and Benders decomposition. The remainder of the chapter presents filtering methods for some popular global constraints. It concludes with bounds reduction algorithms for disjunctive and cumulative scheduling, which have contributed much to the success of constraint programming.

Chapter 4, on relaxation, has a stronger flavor of mathematical programming due to its discussion of linear programming, mixed integer modeling, and a heavy dose of cutting plane theory. The chapter then moves to continuous relaxations for disjunctions of linear and nonlinear inequality systems, and from there to relaxations of several global constraints. There is a certain parallelism between Chapters 3 and 4, in that both take the reader through linear and integer inequalities, duality, propositional logic, and roughly the same set of global constraints. The main difference, of course, is that Chapter 3 presents inference methods for each of these constraint types, and Chapter 4 presents relaxation methods.

Chapter 5 is something of an appendix. It lists about forty metaconstraints, including about twenty global constraints from the CP community, as a starting point for a menu of constraints in an integrated solver. When possible, it says something about usage, inference methods, relaxation methods, and related constraints, providing pointers to the literature and to relevant sections of the book.

1.7 Bibliographic Notes

Integrated methods have developed over the last fifteen years or so in both the constraint programming (CP) and operations research (OR) communities. While a fuller history can be found elsewhere [187], a very brief synopsis might go as follows,

On the OR side, it is interesting that implicit enumeration [139], an early 1960s technique for integer programming, can be seen as anticipating the use of constraint propagation in an integer programming context. Constraint programming is explicitly mentioned in the OR literature as early as 1989 [65], but integrated methods were yet to develop.

CP began to investigate integrated methods in a serious way during the 1990s. They were initially conceived as double-modeling approaches, in which some constraints receive both CP and MILP formulations that exchange domain reduction and/or infeasibility information [225]. This mechanism was implemented in the constraint logic programming system ECL^iPS^e [288, 329]. The constraints community also began to recognize the parallel between constraint solvers and mixed integer solvers, as evidenced by [57].

In later work, such OR ideas as reduced-cost variable fixing, linear relaxations of global constraints, and convex hull relaxations of piecewise linear functions were brought into CP-based algorithms [127, 128, 129, 259, 279, 280]. ILOG's OPL Studio [334] provided a modeling language that invokes CP and MILP solvers.

While this research was underway in CP, the OR community introduced hybrid methods as generalizations of branch and cut or a logic-based form of Benders decomposition. Integer variables were replaced with logical disjunctions and their relaxations as early as 1990 [33]. A series of papers appearing the 1990s integrated CP and logic-based methods with branch and cut [161, 182, 185, 198]. The logic-based Benders approach was developed during the same period, initially for circuit verification [201] and later as a general method [185, 199]. A Benders method that joins MILP and CP was proposed [185] and successfully implemented [203]. CP-based branch and price, a very different approach, was also developed [209, 343].

The First International Joint Workshop on AI and OR was organized in 1995 to provide an early forum for discussion of integrated methods. The idea was revived in 1999 with the annual CP-AI-OR workshop (Integration of AI and OR Techniques in CP for Combinatorial Optimization), now a conference series with published proceedings.

Software that helps integrate CP and OR continues to evolve. Mosel [88, 89] is both a modeling and programming language that interfaces with various solvers, including MILP and CP solvers. SCIP [2] is a callable library that gives the user control of a solution process that can involve both CP and MILP solvers. In SIMPL [12], a high-level modeling language invokes integrated methods at the micro level, using the idea of constraint-based control described above. The global optimization system BARON [312] is to a large degree an integrated solver, since it combines relaxation with bounds reduction and constraint propagation.

Chapter 2

SEARCH

The main issue in search is deciding where to look next. There are many ways to do this, but exhaustive search methods have generally taken one of two forms—branching and constraint-directed search.

Branching is a divide-and-conquer strategy that is essentially guided by problem difficulty. If a problem is too hard to solve, it is broken into two or more subproblems that are more highly restricted and perhaps easier to solve. The process generates a search tree whose leaf nodes correspond to subproblems that can be solved or shown infeasible. The best solution found is optimal for the original problem.

Constraint-directed search is guided by past experience. Whenever a solution is examined, a nogood constraint is generated that excludes it and perhaps other solutions that can be no better. The next solution examined must satisfy the nogood constraints generated so far, to avoid covering the same ground. The search is over when there is no ground left to cover, and the best solution found is optimal.

This chapter explores the two basic search strategies and the roles played by inference and relaxation in each. But this is only part of its purpose. An equally important objective is to show how an integrated approach to optimization actually plays itself out in some concrete cases. To this end, it presents several small examples that cover a wide range of application areas. It each case it formulates a model that is appropriate for integrated solution, and in most cases it carries the solution procedure to completion.

The chapter examines three branching schemes: branch and infer, branch and relax, and branch and price. It then describes three types of constraint-directed search: constraint-directed branching, partial-order dynamic backtracking, and logic-based Benders decomposition. It concludes by showing that branching and constraint-directed search schemes can be seen at work in heuristic as well as exact methods. This allows one to understand exhaustive and local search as belonging to a single algorithmic framework.

The chapter begins with a general description of the search-infer-and-relax solution process that is developed in this book. This provides an opportunity to introduce some of the terminology and notation used thereafter.

2.1 The Solution Process

For the purposes of this book, an optimization problem can be written

$$\min (\text{or max}) \ f(x)$$
$$\mathcal{S} \qquad (2.1)$$
$$x \in D$$

where $f(x)$ is a real-valued function of variable x and D is the *domain* of x. The function $f(x)$ is to be minimized (or maximized) subject to a set $\mathcal{S}$ of constraints, each of which is either satisfied or violated by any given $x \in D$. Generally, x is a tuple $(x_1, \ldots, x_n)$ and D is a Cartesian product $D_1 \times \cdots \times D_n$, where each $x_j \in D_j$.

Any $x \in D$ is a *solution* of (2.1). A *feasible* solution is one that satisfies all the constraints in $\mathcal{S}$, and the *feasible set* of (2.1) is the set of feasible solutions. A feasible solution x^* is *optimal* if $f(x^*) \leq f(x)$ for all feasible x. An *infeasible* problem is one with no feasible solution. If (2.1) has no feasible solution, it is convenient to say that it has optimal value ∞ (or $-\infty$ in the case of a maximization problem). The problem is *unbounded* if there is no lower bound on $f(x)$ for feasible values of x, in which case the optimal value is $-\infty$ (or ∞ for maximization).

It is assumed throughout this book that (2.1) is either infeasible, unbounded, or has a finite optimal value. Thus such problems as minimizing x subject to $x > 0$ are not considered. An optimization problem is considered to be *solved* when an optimal solution is found, or when the problem is shown to be unbounded or infeasible. In incomplete search methods that do not guarantee an optimal solution, the problem is solved when a solution is found that is acceptable in some sense, or when the problem is shown to be unbounded or infeasible.

A constraint C can be *inferred* from $\mathcal{S}$ if it is *valid* for $\mathcal{S}$, meaning that all $x \in D$ satisfying $\mathcal{S}$ satisfy C. A *relaxation* of the minimization problem (2.1) is obtained, roughly speaking, by dropping constraints or replacing the objective function with a lower bound. A more precise

The Solution Process

definition allows for the possibility that a relaxed problem may introduce additional variables. Thus the problem

$$\begin{aligned} \min \ & f'(x,y) \\ & S' \\ & x \in D, \ y \in D' \end{aligned} \quad (2.2)$$

is a relaxation of (2.1), if for any x feasible in (2.1) there is a $y \in D'$ such that (x,y) is feasible in (2.2) and $f'(x,y) \leq f(x)$. The optimal value of a relaxation is always a lower bound on the optimal value of the original problem.

Search, inference, and relaxation interact to provide a general scheme for solving (2.1). The search procedure solves a series of restrictions or special cases of the problem. If the set of restrictions is exhaustive, the best optimal solution of a restriction is optimal in the original problem. Inference and relaxation provide opportunities to exploit problem structure, a key element of any successful approach to solving combinatorial problems. Inference and relaxation also obtain information that helps direct the search.

2.1.1 Search

Search is carried out by solving a series of problem *restrictions*, which may be denoted $P_1, P_2, \ldots, P_m$. Each P_k is obtained by adding constraints to the original problem P. The search is *exhaustive* if the restrictions are *exhaustive*; that is, the feasible set of P is equal to the union of the feasible sets of $P_1, \ldots, P_m$.

The rationale for search is that restrictions may be easier to solve than the original problem. Thus it may be practical to solve P by solving $P_1, \ldots, P_m$ to optimality and picking the best solution. Any optimal solution of a restriction, which might be called a *candidate solution*, is feasible in P (some restrictions may be infeasible and have no solution). If the search is complete, the best candidate solution is an optimal solution for P. In incomplete search, the restrictions may not be solved to optimality, as is often the case in local search schemes.

The most basic kind of search simply enumerates elements of the domain D and selects the best feasible solution. This can be viewed as a search over problem restrictions P_k, each of which is defined by fixing x to a particular value. It is generally more practical, however, to define restrictions by branching, constraint-directed search, or an incomplete version of these. Table 2.1 indicates briefly how they fit into the search-infer-and-relax framework. The concept of a selection function is discussed below in connection with constraint-directed search.

Table 2.1. How branching, constraint-directed search, and heuristic methods fit into the search-infer-and-relax framework.

Solution method	Restriction P_k	Relaxation R_k	Selection function $s(R_k)$	Inference
Branching	Corresponds to node of search tree	LP, Lagrangean, convex NLP, domain store	Any optimal (feasible) solution of R_k	Domain filtering, cutting planes
Constraint-directed search	Original problem plus processed nogoods	Processed nogoods generated so far	Solution that results in easy R_{k+1}, $R_{k+2}, \ldots$	Nogood generation and processing
Heuristic method	Neighborhood of current solution	Problem specific	Random solution, best solution, etc.	Tabu list

Note: LP and NLP refer to linear programming and nonlinear programming, respectively.

2.1.2 Inference

Search can often be accelerated by inference—that is, by inferring new constraints from the constraint set of each P_k. The new constraints are then added to the constraint set, which may make the restrictions easier to solve by describing the feasible set more explicitly. Common forms of inference are domain filtering in constraint programming methods and cutting plane generation in integer programming methods. Adding inferred constraints may also result in a stronger relaxation, as in the case of cutting planes.

Inference is most effective when it exploits problem structure. When a group of constraints has special characteristics, the model can indicate this by combining the constraints into a single *metaconstraint*, known as a *global constraint* in constraint programming. This allows inference algorithms to exploit the global structure of the group in order to derive strong implications. For instance, if variables $x_1, \ldots, x_n$ must take distinct values, one can write a single all-different constraint for them rather than writing $x_i \neq x_j$ for each pair. Inference algorithms developed specifically for this metaconstraint might deduce, for example, that x_j must take a certain value, thus obviating the necessity of branching on x_j. Similarly, a group of inequalities having common structure may allow the inference of specialized cutting planes (implied inequalities).

Developing special-purpose inference procedures has been a major theme of research in both constraint programming and integer programming.

Inference procedures applied to individual constraints or metaconstraints can miss implications of the constraint set as a whole. For instance, the inequalities $x_1 + x_2 \geq 1$ and $x_1 - x_2 \geq 0$ (where each $x_j \in \{0, 1\}$) jointly imply $x_1 = 1$, a fact that follows from neither inequality alone. This implication will be overlooked if the inequalities are processed individually. The constraint programming community partially addresses this problem by collecting at least some of the implications derived from individual constraints into a *constraint store* $\mathcal{D}$. When inferences are drawn from a constraint C, they are actually drawn from C and the current constraint store—that is, from the constraint set $\{C\} \cup \mathcal{D}$. Processing a constraint may enlarge D and thereby strengthen the implications that can be derived from the next constraint. This is a form of *constraint propagation*.

Propagation of this sort is possible only if the constraint store contains very elementary constraints that all of the inference algorithms can accommodate. In constraint programming practice, the constraint store typically consists of domain constraints, namely constraints of the form $x_i \in D_i$ that restrict the domain of x_i. The constraint store is therefore typically a *domain store*. Since any practical inference algorithm considers the variable domains, it already accommodates constraints in the domain store. As a result of this practice, inference algorithms developed by constraint programmers are generally *domain reduction* or *filtering* algorithms that remove values from variable domains. These domain filters are no less useful for integrated problem solving than for constraint programming in particular.

2.1.3 Relaxation

Relaxation is the third element of the solution scheme. It is often used when the subproblems P_k are themselves hard to solve. A relaxation R_k of each P_k is normally created by dropping some constraints in such a way as to make R_k easier than P_k. For instance, one might form a continuous relaxation by allowing integer-valued variables to take any real value.

If P is a minimization problem, the optimal value of the relaxation R_k is a lower bound on the optimal value of P_k. By solving R_k and making use of this bound, one may be able to accelerate the search by avoiding solution of the generally harder problem P_k. For instance, if the optimal value v of R_k is greater than or equal to the value of the best candidate solution found so far, then there is no need to solve P_k, since its optimal value can be no better than v. One can regard P_k as having been *enumerated*, even though it is not actually solved.

In many cases the solution of a relaxation also guides the search by helping to determine the next restriction P_{k+1} or the inferences that are derived from P_{k+1}.

Relaxation also provides a valuable opportunity to exploit special structure in individual constraints or groups of constraints, and this has been a perennial theme of the optimization literature. For example, strong implied inequalities (cutting planes) can be inferred from sets of inequalities that define certain types of polyhedra. The constraints that relax individual constraints or groups of constraints are pooled into a constraint set to obtain a relaxation of the original problem.

The relaxation, like the constraint store, must contain fairly simple constraints, but for a different reason: they must allow easy optimal solution of the relaxed problem. In traditional optimization methods, these are generally linear inequalities in continuous variables, or perhaps nonlinear inequalities that define a convex feasible set.

Both inference and relaxation provide the solver a somewhat global view over the entire problem even while exploiting local structure. The domain store allows propagation of constraints and therefore the deduction of some the global implications of the problem. The relaxation gathers individual constraint relaxations into a single optimization problem that reflects some of the problem's global characteristics. Although the constraint or domain store can be viewed as a relaxation of the problem, it plays a different role, because its constraints are an input to filtering, while the constraints in a relaxation are not.

2.1.4 Exercises

1 An optimization problem (2.1) can be viewed as defining a function $f : S \to \mathbb{R}$, where S is the feasible set of (2.1). The *epigraph* of f is the set $\{(z, x) \mid z \geq f(x),\ x \in S\}$. Show that (2.2) is a relaxation of (2.1) if and only if the projection of the epigraph of f' onto (z, x) contains the epigraph of f.[1]

2 A relaxation can have a different objective function than the problem relaxed. Can a restriction also be defined to allow this? How would exhaustive search be characterized? Hint: Use the concept of an epigraph.

---

[1] The projection of the epigraph E of f' onto (z, x) is $\{(z, x) \mid (z, x, y) \in E\}$.

2.2 Branching Search

Branching search uses a recursive divide-and-conquer strategy. If the original problem P is too hard to solve as given, the branching algorithm creates a series of restrictions $P_1, \ldots, P_m$ and tries to solve them. In other words, it *branches* on P. Each P_k is solved, if possible. If it is feasible, its solution becomes a candidate solution. If P_k is too hard to solve, the search procedure attacks P_k in a similar manner by branching on it, and so on, recursively.

A given restriction P_k is considered to be enumerated when it has been solved, or all of its restrictions have been enumerated. The search terminates when the original problem P has been enumerated. The optimal solution is the best candidate solution found.

The most popular branching mechanism is to branch on a variable x_j. The domain of x_j is divided into two or more subsets, and restrictions are created by successively restricting x_j to each of these subsets.

To ensure an exhaustive search, the restrictions $P_1, \ldots, P_m$ created at each branch should be exhaustive. Normally, their feasible sets also partition the feasible set of P (i.e., they are pairwise disjoint), which is unnecessary but more efficient, as it avoids covering the same ground more than once.

To ensure that the search terminates, the branching mechanism must be designed so that problems become easy enough to solve as they are increasingly restricted. For instance, if the variable domains are finite, then branching on variables will eventually reduce the domains to singletons, thus fixing the value of each x_j and making the restriction trivial to solve. Figure 2.1 displays a generic branching algorithm.

2.2.1 Branch and Infer

Inference may be combined with branching by inferring new constraints for each P_k before P_k is solved. When inference takes the form of do-

---

Let $S = \{P_0\}$ and $v_{\text{UB}} = \infty$.
While S is nonempty repeat:
   Select a problem restriction $P \in S$ and remove P from S.
   If P is too hard to solve then
     Define restrictions $P_1, \ldots, P_m$ of P and add them to S.
   Else
     Let v be the optimal value of P and let $v_{\text{UB}} = \min\{v, v_{\text{UB}}\}$.
The optimal value of P_0 is v_{UB}.

Figure 2.1. Generic branching algorithm for solving a minimization problem P_0. Set S contains the problem restrictions so far generated but not yet attempted, and v_{UB} is the best solution value obtained so far.

main filtering, for example, some of the variable domains are reduced in size. When one branches on variables, this tends to reduce the size of the branching tree because the domains are more rapidly reduced to singletons or the empty set.

2.2.2 Branch and Relax

Relaxation can also be combined with branching in a process that is known in the operations research community as *branch and bound* and in the constraint programming community as *branch and relax*. One solves the relaxation R_k of each restriction, rather than P_k itself. If the solution of R_k happens to be feasible in P_k, it is optimal for P_k and becomes a candidate solution. If R_k is infeasible, then so is P_k. Otherwise, the algorithm branches on P_k. To ensure termination, the branching mechanism must be designed so that solving R_k becomes sufficient to solve P_k if one descends deeply enough into the search tree.

The choice of how to branch depends on which constraint is violated by the solution of the relaxation. For example, if the domain of x_j is $\{1,2,3\}$, and the solution value x_j in the relaxation is 1.5, the domain constraint $x_j \in \{1,2,3\}$ is violated. This would trigger a branching scheme that the solver associates with the domain constraint, perhaps splitting the domain into $\{1\}$ and $\{2,3\}$. If two or more constraints are violated, the solver must decide which one will control the branching.

Branch and relax also uses the bounding mechanism described earlier. If the optimal value of R_k is greater than or equal to the value of the best candidate solution found so far, then there is no point in solving P_k and no need to branch on P_k. In this case, the search tree is said to be "pruned" at P_k. Thus, P_k is enumerated when the solution of R_k is feasible in P_k, R_k is infeasible, or the tree is pruned at P_k.

Inference and relaxation can work together effectively in branch-and-relax methods. The addition of inferred constraints (e.g., cutting planes) to P_k can result in a tighter bound when one solves its relaxation R_k. Conversely, the solution of R_k can provide guidance for generating further constraints, as for instance when separating cuts are generated to exclude a solution of R_k that is infeasible in P_k. A generic branching algorithm with inference and relaxation appears in Figure 2.2.

The most widely used solvers in constraint programming, integer programming, and continuous global optimization combine inference and relaxation with branching. Table 2.2 shows how they fit into the search-infer-and-relax framework.

The following sections present six examples of branching methods. A freight transfer problem shows how inference and relaxation techniques from both constraint and integer programming can work together. A

Branching Search

Let $S = \{P_0\}$ and $v_{\text{UB}} = \infty$.
While S is nonempty repeat:
  Select a problem restriction $P \in S$ and remove P from S.
  Repeat as desired:
    Add inferred constraints to P.
    Let v_R be the optimal value of a relaxation R of P.
  If $v_R < v_{\text{UB}}$ then
    If R's optimal solution is feasible for P then let $v_{\text{UB}} = \min\{v_R, v_{\text{UB}}\}$.
    Else define restrictions $P_1, \ldots, P_m$ of P and add them to S.
The optimal value of P_0 is v_{UB}.

Figure 2.2. Generic branching algorithm, with inference and relaxation, for solving a minimization problem P_0. The inner repeat loop is typically executed only once, but it may be executed several times, perhaps until no more constraints can be inferred or R becomes infeasible. The inference of constraints can be guided by the solution of previous relaxations.

Table 2.2. How some selected branching methods fit into the search-infer-and-relax framework.

Solution method	Restriction P_k	Relaxation R_k	Selection function $s(R_k)$	Inference
Constraint programming	Created by splitting domain, etc.	Domain store	Any feasible solution of R_k	Domain filtering, cutting planes
MILP branch and cut	Created by branching on fractional variables	LP relaxation + cutting planes	Optimal solution of R_k	Cutting planes, preprocessing
Continuous global optimization	Created by splitting intervals	LP or convex NLP relaxation	Optimal solution of R_k	Interval propagation, Lagrangean bounding

Note: NLP refers to nonlinear programming.

production planning problem introduces disjunctive constraints. An employee scheduling problem illustrates several modeling and filtering concepts developed in constraint programming. A small example of continuous global optimization provides a glimpse of how global solvers combine domain filtering and relaxation. A fifth example shows how an integrated method can solve a realistic product configuration problem more efficiently than constraint or integer programming alone. A final

2.2.3 Example: Freight Transfer

A simple freight transfer problem illustrates how inference and relaxation can interact with branching search. Forty-two tons of freight must be conveyed overland. The shipper has a choice of trucks from a fleet of twelve vehicles. The trucks come in four sizes, with three vehicles of each size (Table 2.3). No more than eight trucks may be used altogether. The problem is to select trucks to carry the freight at minimum cost.

Formulating the Problem

If x_i is the number of trucks of type i used, the requirement that 42 tons be transported can be written

$$7x_1 + 5x_2 + 4x_3 + 3x_4 \geq 42$$

This is a knapsack covering constraint, which in general has the form $ax \geq \alpha$, where each $a_i > 0$ and the variables x_i must take nonnegative integer values. Such a constraint can be interpreted as saying that one must select x_i knapsacks of each size a_i in such a way as to carry a total load of α. In the freight problem, the knapsacks are trucks.

One can also formulate a knapsack packing constraint of the form $bx \leq b_0$, where each $b_i > 0$. It arises when one wishes to pack a single knapsack with several types of items, and each type i consumes space b_i. The constraint says that the total space consumed must not exceed the capacity b_0 of the knapsack. In the freight problem, the limit of eight trucks can be conceived as a knapsack packing constraint

$$x_1 + x_2 + x_3 + x_4 \leq 8$$

in which the knapsack is a fleet of at most eight trucks.

Table 2.3. Data for a small instance of a freight transfer problem.

Truck type	Number available	Capacity (tons)	Cost per truck
1	3	7	90
2	3	5	60
3	3	4	50
4	3	3	40

The freight problem can now be formulated

$$\text{Integer linear:} \begin{cases} \min 90x_1 + 60x_2 + 50x_3 + 40x_4 \\ 7x_1 + 5x_2 + 4x_3 + 3x_4 \geq 42 & (a) \\ x_1 + x_2 + x_3 + x_4 \leq 8 & (b) \end{cases} \quad (2.3)$$

Domains: $x_i \in \{0, 1, 2, 3\}$, $i = 1, \ldots, 4$

In accord with the spirit of revealing problem structure, the constraints are grouped by type to provide the solver guidance as to how to process them. Since the objective function becomes an integer linear inequality whenever it is bounded, it is grouped with the integer linear constraints to form an integer linear metaconstraint. An optimal solution is $(x_1, \ldots, x_4) = (3, 2, 2, 1)$, with a minimum cost of 530.

The problem may be solved by a branching algorithm that uses inference and relaxation. The following sections first show how to carry out the inference and relaxation steps and then how to conduct the search.

Inference: Bounds Propagation

Domain filtering is a popular form of inference, in part because the reduced domains inferred from one constraint can be used as a starting point for domain reduction in another constraint. One can, in principle, cycle through the constraints in this fashion until no further filtering is possible. This allows one to draw some inferences that do not follow from any single constraint, even though the constraints are processed individually.

A type of domain filtering known as *bounds propagation* is useful for the freight transport problem. The current domain of each variable x_i is represented as a set $\{L_i, L_i + 1, \ldots, U_i\}$ of consecutive integers, since the domains are reduced solely by tightening lower and upper bounds on the variables.

Bounds propagation for a knapsack covering constraint $ax \geq a_0$ is based on the observation that, for any i, the constraint implies

$$x_i \geq \frac{a_0 - \sum_{j \neq i} a_j x_j}{a_i} \geq \frac{a_0 - \sum_{j \neq i} a_j U_j}{a_i}$$

The second inequality is due to $x_j \leq U_j$. This allows one to update each lower bound L_i as follows:

$$L'_i = \max \left\{ L_i, \left\lceil \frac{a_0 - \sum_{j \neq i} a_j U_j}{a_i} \right\rceil \right\}$$

where $\lceil \delta \rceil$ is δ rounded up. In other words, one *propagates* the upper bounds U_j to obtain new lower bounds. If there is a knapsack packing constraint $bx \leq b_0$ in the problem, one can use it to propagate the revised lower bounds L'_j in analogous fashion and obtain new upper bounds:

$$U'_i = \min\left\{U_i, \left\lfloor \frac{b_0 - \sum_{j \neq i} b_j L'_j}{b_i} \right\rfloor\right\}$$

where $\lfloor \delta \rfloor$ is δ rounded down.

At this point all of the upper and lower bounds have been updated. They can now be viewed as the original bounds L_i, U_i and propagated by cycling through $ax \geq a_0$ and $bx \leq b_0$ again. If desired, the process is repeated until a fixed point is obtained; that is, until the bounds are unchanged from the previous update.

Bounds propagation is easily applied to constraints (a) and (b) of the problem instance (2.3). Using the initial bounds $(L_i, U_i) = (0, 3)$, inequality (a) raises the lower bound on x_1 from 0 to

$$L'_1 = \max\left\{0, \left\lceil \frac{42 - 5U_2 - 4U_3 - 3U_4}{7} \right\rceil\right\} = 1 \qquad (2.4)$$

yielding a revised domain $\{1, 2, 3\}$ for x_1. No other domain reductions are possible.

Inference: Valid Inequalities

In addition to deducing smaller domains, one can deduce *valid inequalities* or *cutting planes* from the knapsack constraints. The inferred inequalities are normally added to the original constraint set in order to produce a stronger continuous relaxation.[2]

A simple class of valid inequalities for $ax \geq a_0$ are *knapsack cuts*, which are obtained as follows. Again, let $\{L_i, L_i + 1, \ldots, U_i\}$ be the current domain of x_i. Let a *packing* for $ax \geq a_0$ be a set I of indices for which the corresponding terms $a_i x_i$ cannot by themselves satisfy $ax \geq a_0$, even with each x_i at its upper bound. That is,

$$\sum_{i \in I} a_i U_i < a_0$$

---

[2] Reducing a domain to $\{L_i, \ldots, U_i\}$ can be viewed as a special case of deducing inequalities, because it, in effect, generates the valid inequalities $L_i \leq x_i \leq U_i$. It is nonetheless useful to treat this special case separately, since reduced domains are normally maintained in a different data structure than valid inequalities.

Thus, the remaining terms $a_i x_i$ for $i \notin I$ must cover the gap:

$$\sum_{i \notin I} a_i x_i \geq a_0 - \sum_{i \in I} a_i U_i$$

This yields the valid knapsack cut

$$\sum_{i \notin I} x_i \geq \left\lceil \frac{a_0 - \sum_{i \in I} a_i U_i}{\max_{i \notin I}\{a_i\}} \right\rceil$$

Since there can be a large number of packings, one may want to consider only cuts that correspond to maximal packings, which are packings of which no proper superset is a packing. One can also neglect any packing that contains all indices but one, say i, since the corresponding cut imposes a lower bound on x_i that is identical to that obtained by domain filtering.

There are analogous knapsack cuts for the knapsack packing inequality $bx \leq b_0$, but they add nothing to the freight problem because $b = (1, \ldots, 1)$.

In the above problem instance, the constraint (a) in (2.4) has three maximal packings that omit more than one index. They correspond to knapsack cuts as follows:

$$\begin{aligned} I = \{1,2\}: & \quad x_3 + x_4 \geq 2 \\ I = \{1,3\}: & \quad x_2 + x_4 \geq 2 \\ I = \{1,4\}: & \quad x_2 + x_3 \geq 3 \end{aligned} \quad (2.5)$$

Relaxation: Linear Programming

A continuous relaxation of the problem instance (2.4) can be obtained by allowing each variable x_i to take any value in the continuous interval $[L_i, U_i]$, where $\{L_i, L_i + 1, \ldots, U_i\}$ is its current domain. One can also add the knapsack cuts (2.5) before relaxing the problem. This yields the relaxation below:

$$\text{Linear:} \begin{cases} \min 90x_1 + 60x_2 + 50x_3 + 40x_4 & \\ 7x_1 + 5x_2 + 4x_3 + 3x_4 \geq 42 & (a) \\ x_1 + x_2 + x_3 + x_4 \leq 8 & (b) \\ x_3 + x_4 \geq 2 & (c) \\ x_2 + x_4 \geq 2 & (d) \\ x_2 + x_3 \geq 3 & (e) \\ L_i \leq x_i \leq U_i, \ i = 1, \ldots, 4 & \end{cases} \quad (2.6)$$

Domains: $x_i \in \Re, \ i = 1, \ldots, 4$

which can be solved by linear programming. The knapsack cuts (c)–(e) are those obtained when (L_i, U_i) are the original bounds $(0, 3)$. In general, stronger knapsack cuts can be obtained when the bounds are tightened during the branching search, but this is not done here.

Branching Search

A search tree for problem instance (2.3) appears in Figure 2.3. Each node of the tree below the root corresponds to a restriction of the orig-

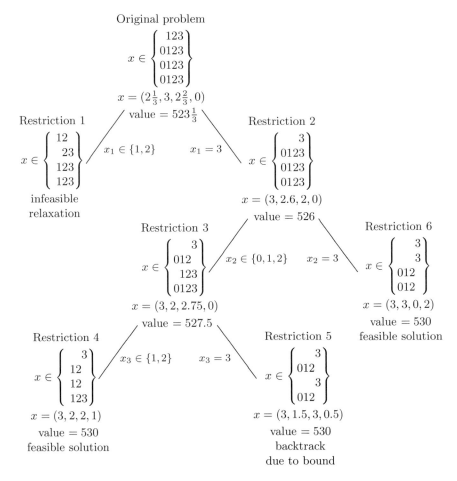

Figure 2.3. Branch-and-relax tree for the freight transfer problem. Each node of the tree shows, in braces, the filtered domains after domain reduction. The rows inside the braces correspond to $x_1, \ldots, x_4$. The solution of the continuous relaxation appears immediately below the domains.

inal problem. A restriction is processed by applying inference methods (bounds propagation and cut generation), then solving a continuous relaxation, and finally branching if necessary.

Bounds propagation applied to the original problem at the root node reduces the domain of x_1 to $\{1, 2, 3\}$, as described earlier. Figure 2.3 shows the resulting domains in braces. Next, three knapsack cuts (2.5) are generated, using the lower and upper bounds L_i, U_i that define the current domains. They are added to the constraint set in order to obtain the continuous relaxation (2.6). The optimal solution $x = (2\frac{1}{3}, 3, 2\frac{2}{3}, 0)$ of the relaxation has value $523\frac{1}{3}$.

This solution is infeasible in the original problem because x_1 and x_3 do not belong to their respective domains (they are nonintegral). It is therefore necessary to branch on one of the domain constraints $x_1 \in \{1, 2, 3\}$, $x_3 \in \{0, 1, 2, 3\}$. Branching on the first splits the domain into $\{1, 2\}$ and $\{3\}$, as the solution value $2\frac{1}{3}$ of x_1 lies between 2 and 3. This generates restrictions 1 and 2, which become the contents of set S in the branching algorithm of Figure 2.2.

The tree will be traversed in a depth-first manner. Moving first to restriction 1, where x_1's domain is $\{1, 2\}$, bounds propagation yields the smaller domains shown in Figure 2.3. These new bounds yield no further knapsack cuts. Due to the tighter bounds L_i, U_i, the continuous relaxation (2.6) of the current restriction is infeasible, and no branching is necessary. This leaves restriction 2 in S.

Moving now to restriction 2, no further domain filtering is possible, and solution of the relaxation (2.6) yields $x = (3, 2.6, 2, 0)$. Branching on x_2 creates restrictions 3 and 6.

Continuing in a depth-first manner, restriction 3 is processed next. Inference yields the domains shown, and branching on x_3 produces restrictions 4 and 5. The continuous relaxation of restriction 4 has the integral solution $x = (3, 2, 2, 1)$, which is feasible in the restriction. It becomes the *incumbent* solution (the best feasible solution so far), and no branching is necessary.

Restriction 5 is processed next. Here, the relaxation has a nonintegral solution, but its optimal value 530 is no better than the value of the incumbent solution. Since 530 is a lower bound on the optimal value of any further restriction of restriction 5, there is no need to branch. The tree is therefore "pruned" at restriction 5, and the search proceeds to restriction 6.

The continuous relaxation of this restriction has an integral solution, and there is no need to branch, thus completing the search. Since the solution is no better than the incumbent (in fact it is equally good), the incumbent solution $x = (3, 2, 2, 1)$ is optimal.

In some cases, one can prove the infeasibility of a restriction by inference alone, perhaps by reducing a variable domain to the empty set. When this occurs, there is no need to solve a relaxation of the restriction.

2.2.4 Example: Production Planning

A very simple production planning problem illustrates how logical and continuous variables can interact. A manufacturing plant has three operating modes, each of which imposes different constraints. The objective is to decide in which mode to run the plant, and how much of each of two products to make, so as to maximize net income.

Formulating the Problem

Let x_A, x_B be the production levels of products A and B, respectively. In mode 0 the plant is shut down, and $x_A = x_B = 0$. In mode 1, it incurs a fixed cost of 35, and the production levels must satisfy the constraint $2x_A + x_B \leq 10$. Mode 2 incurs a fixed cost of 45 and the constraint $x_A + 2x_B \leq 10$ is imposed. The company earns a net income of 5 for each unit of product A manufactured, and 3 for each unit of product B.

A natural modeling approach is to let a boolean variable δ_k be true when the plant runs in mode k. Then, the model is immediate:

$$\begin{aligned}
&\text{Linear: } \max 5x_A + 3x_B - f \\
&\text{Logic: } \delta_0 \vee \delta_1 \vee \delta_2 \\
&\text{Conditional: } \begin{cases} \delta_0 \Rightarrow (x_A = x_B = f = 0) \\ \delta_1 \Rightarrow (2x_A + x_B \leq 10, \ f = 35) \\ \delta_2 \Rightarrow (x_A + 2x_B \leq 10, \ f = 45) \end{cases} \\
&\text{Domains: } x_A, x_B \geq 0, \ \delta_k \in \{true, false\}, \ k = 0, 1, 2
\end{aligned} \quad (2.7)$$

where variable f represents the fixed cost. The logic constraint means that at least one of the three boolean variables must be true.

Relaxation

The model has an interesting relaxation because each production mode k enforces constraints that define a polyhedron in the space of the continuous variables x_A, x_B, f. So the projection of the feasible set onto this space is the union of the three polyhedra, illustrated in Figure 2.4. The best possible continuous relaxation of this set is its *convex hull*, which is itself a polyhedron and is also shown in the figure. The convex hull of a set is union of all line segments connecting any two points of the set.

There is a general procedure, given in Section 4.7.1, for writing a linear constraint set that describes the convex hull of a disjunction of

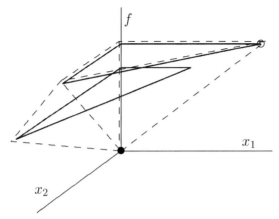

Figure 2.4. Feasible set (two triangular areas and black dot) of a production planning problem, projected onto the space of the continuous variables. The convex hull is the volume inside the dashed polyhedron. The open circle marks the optimal solution.

linear systems. In this case, the convex hull description is

$$
\begin{aligned}
& 2x_{A1} + x_{B1} \le 10y_1 \\
& x_{A2} + 2x_{B2} \le 10y_2 \\
& f \ge 35y_1 + 45y_2 \\
& x_A = x_{A1} + x_{A2}, \quad x_B = x_{B1} + x_{B2} \\
& y_0 + y_1 + y_2 = 1, \quad y_k \ge 0, \ k = 0, 1, 2 \\
& x_{Ak} \ge 0, \ x_{Bk} \ge 0, \ k = 1, 2
\end{aligned}
\quad (2.8)
$$

where the variables y_k correspond to the boolean variables δ_k, and fixing $\delta_k = \text{true}$ is equivalent to setting $y_k = 1$. Since there are new variables x_{Ak}, x_{Bk}, y_k in the relaxation, one should, strictly speaking, say that it describes a set whose projection onto x_A, x_B, f is the convex hull of (2.7)'s feasible set. A lower bound on the optimal value can now be obtained by minimizing $5x_A + 3x_b - f$ subject to (2.8).

To take advantage of this relaxation, the solver must somehow recognize that the feasible set in continuous space is a union of polyhedra. This can be done by collecting the constraints of (2.7) into a single *linear disjunction*, resulting in the model:

Linear: $\max 5x_A + 3x_B - f$

Linear disjunction:

$$
\begin{bmatrix} \delta_0 \\ x_A = x_B = 0 \\ f = 0 \end{bmatrix} \vee \begin{bmatrix} \delta_1 \\ 2x_A + x_B \le 10 \\ f = 35 \end{bmatrix} \vee \begin{bmatrix} \delta_2 \\ x_A + 2x_B \le 10 \\ f = 45 \end{bmatrix}
$$

Domains: $x_A, x_B \ge 0, \ \delta_k \in \{\text{true}, \text{false}\}, \ k = 0, 1, 2$

The linear disjunction has precisely the same meaning as the logic and conditional constraints of (2.7). Its presence tells the solver to formulate a relaxation for the union of polyhedra described by its three disjuncts. In general the user should be alert for metaconstraints that can exploit problem structure, although in this case an intelligent modeling system could automatically rewrite (2.7) as (2.8).

Branching Search

The problem can now be solved by branch and relax, where branching takes place on the boolean variables. In this case, the solution of the relaxation is $(y_1, y_2) = (0, 1)$, with $(x_A, x_B, f) = (10, 0, 35)$. Because this solution is feasible in (2.7) with $(\delta_0, \delta_1, \delta_2) = (\textit{false}, \textit{false}, \textit{true})$, there is no need to branch. One should operate the plant in mode 2 and make product A only.

In this problem, there is no branching because (2.8) is a convex hull relaxation not only for the corresponding linear disjunction but for the entire problem. In more complex problems, an auxiliary variable y_k may take a fractional value, in which case one can branch by setting $\delta_k = \textit{true}$ and $\delta_k = \textit{false}$.

Inference

Inference plays a role in such problems when the logical conditions become more complicated. Suppose, for example, that Plant 1 can operate in 2 modes (indicated by boolean variables δ_{1k} for $k = 0, 1$) and Plant 2 in 3 modes (indicated by variables δ_{2k}). There is also a rule that if plant 1 operates in mode 1, then plant 2 cannot operate in mode 2. The model might be as follows (where $\neg$ means *not*).

$$\text{Linear: max } cx$$

$$\text{Logic: } \begin{cases} \delta_{10} \vee \delta_{11} \\ \delta_{20} \vee \delta_{21} \vee \delta_{22} \\ \delta_{11} \rightarrow \neg \delta_{22} \end{cases}$$

$$\text{Conditional: } \delta_{ik} \Rightarrow A^{ik} x \geq b^{ik}, \text{ all } i, k$$

$$\text{Domains: } x \geq 0, \ \delta_{ik} \in \{\textit{true}, \textit{false}\}$$

To extract as many linear disjunctions as possible, one can use the *resolution* algorithm (discussed in Section 3.5.2) to compute the *prime implications* of the logical formulas in the above model:

$$\begin{array}{ll} \delta_{10} \vee \delta_{11} & \delta_{10} \vee \neg \delta_{22} \\ \delta_{20} \vee \delta_{21} \vee \delta_{22} & \neg \delta_{11} \vee \delta_{20} \vee \delta_{21} \\ \neg \delta_{11} \vee \neg \delta_{22} & \delta_{10} \vee \delta_{20} \vee \delta_{21} \end{array}$$

These are the undominated disjunctions implied by the logical formulas. Three of the prime implications contain no negative terms, and they provide the basis for linear disjunctions. The model becomes:

Linear: max cx

Logic: $\begin{cases} \neg\delta_{11} \vee \neg\delta_{22} \\ \delta_{10} \vee \neg\delta_{22} \\ \neg\delta_{11} \vee \delta_{20} \vee \delta_{21} \end{cases}$

Linear disjunction:

$$\bigvee_{k \in \{0,1\}} \begin{bmatrix} \delta_{1k} \\ A^{1k}x \geq b^{1k} \end{bmatrix}$$

$$\bigvee_{k \in \{0,1,2\}} \begin{bmatrix} \delta_{2k} \\ A^{2k}x \geq b^{2k} \end{bmatrix}$$

$$\begin{bmatrix} \delta_{10} \\ A^{10}x \geq b^{10} \end{bmatrix} \vee \begin{bmatrix} \delta_{20} \\ A^{20}x \geq b^{20} \end{bmatrix} \vee \begin{bmatrix} \delta_{21} \\ A^{21}x \geq b^{21} \end{bmatrix}$$

Domains: $x \geq 0$, $\delta_{ik} \in \{true, false\}$

Again, an intelligent modeling system can carry out this process automatically.

Inference plays a further role as branching proceeds. When branching fixes a boolean variable to true or false, it may be possible to deduce that some other variables must be true or false, thus reducing their domains to a singleton. For example, if δ_{10} is fixed to false, then δ_{11} must be true and δ_{22} false. The resolution method draws all such inferences.

2.2.5 Example: Employee Scheduling

An employee scheduling problem is useful for introducing several of the modeling and inference techniques that have been developed in the constraint programming field. Since the objective is to find a feasible rather than an optimal solution, bounds do not play a role, and relaxation is relatively unimportant. Nonetheless, the integration of technologies is a key ingredient of the solution method, since the filtering algorithms rely on such optimization techniques as maximum flow algorithms and dynamic programming.

A certain hospital ward requires that a head nurse be on duty seven days a week, twenty-four hours a day. There are three eight-hour shifts, and on a given day each shift must be staffed by a different nurse. The schedule must be the same every week. Four nurses (denoted A, B, C and D) are available, all of whom must work at least five days a week.

Since there are 21 eight-hour periods a week, this implies that three nurses will work five days and one will work six. For continuity, no shift should be staffed by more than two different nurses during the week.

Two additional rules reduce the burden of adjusting to shift changes. No employee is asked to work different shifts on two consecutive days; there must be at least one day off in between. Also, an employee who works Shift 2 or 3 must do so at least two days in a row. Shift 1 is the daytime shift and requires less adjustment.

Formulating the Problem

There are two natural ways to think about an employee schedule. One, shown in Table 2.4, is to view it as assigning a nurse to each shift on each day. Another is to see it as assigning a shift (or day off) to each nurse on each day; this is illustrated by Table 2.5, in which Shift 0 represents a day off. Either viewpoint gives rise to a different problem formulation, neither of which is convenient for expressing all the constraints of the problem. Fortunately, there is no need to choose between them. One can use both formulations in the same model and connect them with *channeling constraints*. This not only accommodates all the constraints but makes propagation more effective.

To proceed with the first formulation, let variable w_{sd} be the nurse that is assigned to shift s on day d. Three of the scheduling requirements are readily expressed in this formulation, using metaconstraints

Table 2.4. Employee scheduling viewed as assigning workers to shifts.

	Sun	Mon	Tue	Wed	Thu	Fri	Sat
Shift 1	A	B	A	A	A	A	A
Shift 2	C	C	C	B	B	B	B
Shift 3	D	D	D	D	C	C	D

Table 2.5. Employee scheduling viewed as assigning shifts to workers.

	Sun	Mon	Tue	Wed	Thu	Fri	Sat
Worker A	1	0	1	1	1	1	1
Worker B	0	1	0	2	2	2	2
Worker C	2	2	2	0	3	3	0
Worker D	3	3	3	3	0	0	3

that are well known to constraint programmers. One is the *all-different* constraint, alldiff(x), where x denotes a list $x_1, \ldots, x_n$ of variables. It simply requires that $x_1, \ldots, x_n$ take different values. It can express the requirement that three different nurses be scheduled each day:

$$\text{alldiff}(w_{\cdot d}), \text{ all } d$$

The notation $w_{\cdot d}$ refers to (w_{1d}, w_{2d}, w_{3d}).

The *cardinality* constraint can be used to require that every nurse be assigned at least five days of work. The constraint in general is written

$$\text{cardinality}(x \mid v, \ell, u)$$

where $x = (x_1, \ldots, x_n)$ and v is an m-tuple $(v_1, \ldots, v_m)$ of values. $\ell = (\ell_1, \ldots, \ell_m)$ and $u = (u_1, \ldots, u_m)$ contain lower and upper bounds respectively. The vertical bar in the argument list indicates that everything before the bar is a variable and everything after is a parameter. The constraint requires, for each $i = 1, \ldots, m$, that at least ℓ_i and at most u_i of the variables in x take the value v_i. To require that each nurse work at least five and at most six days, one can write

$$\text{cardinality}(w_{\cdot\cdot} \mid (A, B, C, D), (5, 5, 5, 5), (6, 6, 6, 6))$$

where $w_{\cdot\cdot}$ refers to a tuple of all the variables w_{sd}.

The *nvalues* constraint is written

$$\text{nvalues}(x \mid \ell, u)$$

and requires that the variables $x = (x_1, \ldots, x_n)$ take at least ℓ and at most u different values. To require that at most two nurses work any given shift, one can write

$$\text{nvalues}(w_{s\cdot} \mid 1, 2), \text{ all } s$$

Both nvalues and cardinality generalize the alldiff constraint, but one should use alldiff when possible, since it invokes a more efficient filtering algorithm.

The remaining constraints are not easily expressed in the notation developed so far, because they relate to the pattern of shifts worked by a given nurse. For this reason, it is useful to move to the formulation suggested by Table 2.5. Let y_{id} be the shift assigned to nurse i on day d, where shift 0 denotes a day off. It is first necessary to ensure that all three shifts be assigned for each day. The alldiff constraint serves the purpose:

$$\text{alldiff}(y_{\cdot d}), \text{ all } d$$

This condition is implicit in the constraints already written, but it is generally good practice to write redundant constraints when one is aware of them, in order to strengthen propagation.

The *stretch* constraint was expressly developed to impose conditions on stretches or contiguous sequences of shift assignments in employee scheduling problems. Given a tuple $x = (x_1, \ldots, x_n)$ of variables, a stretch is a maximal sequence of consecutive variables that take the same value. Thus, if $x = (x_1, x_2, x_3) = (a, a, b)$, x contains a stretch of value a having length 2 and a stretch of b having length 1, but it does not contain a stretch of a having length 1. The stretch constraint is written

$$\text{stretch}(x \,|\, v, \ell, u, P)$$

where x, v, ℓ, and u are defined as in the cardinality constraint. P is a set of *patterns*, each of which is a pair (v, v') of distinct values. The stretch constraint requires, for each v_i in v, that every stretch of value v_i have length at least ℓ_i and at most u_i. It also requires that whenever a stretch of value v comes immediately before a stretch of value v', the pair (v, v') must occur in the pattern set P. The requirements concerning consecutive nursing shifts can now be written

$$\text{stretch-cycle}(y_i. \,|\, (2,3), (2,2), (6,6), P), \text{ all } i$$

where P contains all patterns that include a day off:

$$P = \{(s, 0), (0, s) \,|\, s = 1, 2, 3\}$$

A cyclic version of the constraint is necessary because every week must have the same schedule. The cyclic version treats the week as a cycle and allows a single stretch to extend across the weekend.

Finally, the two formulations must be forced to have the same solution. This is accomplished with *channeling constraints*, which in this case take the form

$$w_{y_{id}d} = i, \text{ all } i, d \tag{2.9}$$

and

$$y_{w_{sd}s} = s, \text{ all } s, d \tag{2.10}$$

Constraint (2.9) says that on any given day d, the nurse assigned to the shift to which nurse i is assigned must be nurse i, and similarly for

(2.10). The subscripts y_{id} in (2.9) and w_{sd} in (2.10) are *variable indices* or *variable subscripts*.

The model is now complete. It can be written with five metaconstraints, plus domains:

$$\text{Alldiff:} \left\{ \begin{array}{l} (w_{\cdot d}) \\ (y_{\cdot d}) \end{array} \right\}, \text{ all } d$$

Cardinality: $(w_{\cdot\cdot} \mid (A, B, C, D), (5, 5, 5, 5), (6, 6, 6, 6))$

Nvalues: $(w_{s\cdot} \mid 1, 2)$, all s

Stretch-cycle: $(y_{i\cdot} \mid (2, 3), (2, 2), (6, 6), P)$, all i \hfill (2.11)

$$\text{Linear:} \left\{ \begin{array}{l} w_{y_{id}d} = i, \text{ all } i \\ y_{w_{sd}d} = s, \text{ all } s \end{array} \right\}, \text{ all } d$$

$$\text{Domains:} \left\{ \begin{array}{l} w_{sd} \in \{A, B, C, D\}, \ s = 1, 2, 3 \\ y_{id} \in \{0, 1, 2, 3\}, \ i = A, B, C, D \end{array} \right\}, \text{ all } d$$

The linear constraints are so classified because they are linear aside from the variable indices, which will be eliminated as described below. One can appreciate the convenience of metaconstraints by attempting to formulate this problem with, say, a mixed integer linear programming model.

Inference: Domain Filtering

The alldiff constraint poses an interesting filtering problem that, fortunately, can be easily solved. Suppose, for example, that the current domains of assignment variables w_{s1} are: The alldiff constraint poses an interesting filtering problem that, fortunately, can be easily solved. Suppose, for example, that the current domains of assignment variables w_{s1} are:

$$w_{11} \in \{A, B\}, \ w_{21} \in \{A, B\}, \ w_{31} \in \{A, B, C, D\}$$

The days are numbered $1, \ldots, 7$ so that the subscript 1 in w_{s1} refers to Sunday. Thus, only nurse A or B can be assigned to Shift 1 or 2 on Sunday, while any nurse can be assigned to shift 3. Since

$$\text{alldiff}(w_{11}, w_{21}, w_{31})$$

must be enforced, one can immediately deduce that neither A nor B can be assigned to Shift 3. Thus, the domain of w_{31} can be reduced to $\{C, D\}$. This type of reasoning can be generalized by viewing the solution of alldiff as a maximum flow problem on a network, for which there are very fast algorithms. Optimality conditions for maximum flows allow one to identify values that can be removed from domains. These ideas are presented in Section 3.9. Straightforward generalizations of this flow model can provide filtering for the cardinality and nvalues constraints.

Filtering for the stretch constraint is more complicated, but nonetheless tractable. Suppose, for example, that the domains of y_{Ad} contain the values listed beneath each variable:

y_{A1}	y_{A2}	y_{A3}	y_{A4}	y_{A5}	y_{A6}	y_{A7}
0	0		0		0	0
1				1	1	1
2	2	2		2		2
3	3	3	3		3	3

(2.12)

This means, for instance, that nurse A must work either Shift 2 or Shift 3 on Tuesday. After some thought, one can deduce from the stretch constraint in (2.11) that several values can be removed, resulting in the following domains:

y_{A1}	y_{A2}	y_{A3}	y_{A4}	y_{A5}	y_{A6}	y_{A7}
0			0		0	0
				1	1	1
2	2	2				2
3	3	3				3

(2.13)

Note that two variables are fixed. A polynomial-time dynamic programming algorithm can remove all infeasible values for any given stretch constraint. It is described in Section 3.12.

Inference for Variable Indices

The variably indexed expression $w_{y_{id}d}$ in (2.11) can be processed with the help of an *element* constraint. Domain reduction algorithms for this constraint are well known in the constraint programming community, and elementary polyhedral theory provides a continuous relaxation for the constraint.

The type of element constraint required here has the form

$$\text{element}(y, x, z) \qquad (2.14)$$

where y is an integer-valued variable, x is a tuple $(x_1, \ldots, x_m)$ of variables, and z is a variable. The constraint requires that z be equal to the yth variable in the list $x_1, \ldots, x_m$. One can therefore deal with a variably indexed expression like x_y by replacing it with z and adding

Branching Search

the element constraint (2.14). In particular, the constraint $w_{y_{id}d} = i$ is parsed by replacing it with $z_{id} = i$ and generating the constraint

$$\text{element}\,(y_{id}, (w_{0d}, \ldots, w_{3d}), z_{id})$$

Similarly, $y_{w_{sd}d} = s$ is replaced with $\bar{z}_{sd} = s$ and

$$\text{element}\,(w_{sd}, (y_{Ad}, \ldots, y_{Dd}), \bar{z}_{sd})$$

Filtering for element is a simple matter. An example will illustrate this and show how propagation of channeling constraints between two models can reduce domains. Focusing on Sunday (Day 1), suppose the domains of w_{s1} and y_{i1} contain the elements listed beneath each variable:

w_{01}	w_{11}	w_{21}	w_{31}	y_{A1}	y_{B1}	y_{C1}	y_{D1}
A		A	A	0	0		0
B	B		B	1	1	1	
C	C	C		2	2	2	2
	D	D	D	3		3	3

Suppose that no further propagation is possible among the constraints involving only the variables w_{sd}, and similarly for the constraints involving only the variables y_{id}. Nonetheless, the channeling constraints can yield further domain reduction. For instance, since y_{A1} has domain $\{0,1\}$, the constraint

$$\text{element}\,(y_{A1}, (w_{01}, \ldots, w_{31}), z_{A1})$$

implies that y_{A1} must select either w_{01} or w_{11} to be equated with z_{A1}. But $z_{A1} = A$, and only w_{01}'s domain contains A. Thus y_{A1} must select w_{01}, and y_{A1}'s domain can be reduced to $\{0\}$. Similarly, the constraints $\bar{z}_{01} = 0$ and

$$\text{element}\,(w_{01}, (y_{A1}, \ldots, y_{D1}), \bar{z}_{01})$$

remove C from the domain $\{A, B, C\}$ of w_{01}. Deductions of this kind reduce the domains to:

w_{01}	w_{11}	w_{21}	w_{31}	y_{A1}	y_{B1}	y_{C1}	y_{D1}
A			A	0	0		
B	B				1	1	
	C	C				2	2
		D	D	3			3

Search, Relaxation, and Symmetry

The nurse scheduling problem is solved through a combination of branching and propagation. Branching proceeds by splitting domains with more than one element. The search process is too long to reproduce here, but a feasible solution is exhibited in Tables 2.4 and 2.5.

The choice of which domain to split, and how to split it, is more an art than a science, but in feasibility problems a common rule is to use *first fail* branching: to branch on a variable that is more likely to lead quickly to infeasibility in the branch that is explored first. The rationale for this strategy is that it helps to avoid exploring large subtrees that contain no solution. The likelihood of causing infeasibility is gauged in various ways, perhaps by the size of the domain. Branching on a smaller domain may result in more propagation and therefore detection of infeasibility before fixing too many variables.

As remarked earlier, the domain store normally plays a different role than a relaxation because its constraints are an input to filtering. Yet, the domain store is nonetheless a relaxation and can play that role as well. One can "solve" the domain store, and if its solution happens to be feasible, the search for a feasible solution is over. The domain store can be solved simply by selecting a value from each domain, but in practice, one can use a heuristic method to construct and examine several promising solutions. This is sometimes called *probing* because it, in effect, probes deeper into the tree for feasible solutions. Such an incomplete search within a larger complete search sometimes proves very effective.

Symmetry is a important feature of the employee scheduling problem. The four employees, for example, are indistinguishable, and their roles can be interchanged in any solution to obtain another feasible solution. Similarly, Shifts 2 and 3 are subject to the same rules and can be interchanged in any solution. The days of the week can be rotated with no effect on feasibility. The presence of symmetry can prolong the search, because one may waste time enumerating several subtrees that are identical up to an interchange of variables or values. There are several means of *breaking* symmetry, the simplest of which is to fix some variables without loss of generality. For instance, employees A, B, and C can be assigned at the outset to work Shifts 1, 2 and 3, respectively, on Sunday.

Symmetry is not discussed further in this book, but it is an active research area that may pay dividends for accelerating search. One should bear in mind, however, that symmetry tends to occur more often in textbook problems than real-world problems, and more often in feasibility problems than optimization problems. In a real employee scheduling problem, for example, weekends are likely to have different rules than weekdays, and employees are likely to have different privileges. If the objective is to minimize cost or maximize the satisfaction of preferences, employees will probably have different shift-dependent wage rates as well as different preferences.

2.2.6 Example: Continuous Global Optimization

Optimization problems need not contain discrete variables to be combinatorial in nature. A continuous optimization problem may have a large number of locally optimal solutions, which are solutions that are optimal in a neighborhood about them. Nonlinear programming solvers, highly developed as they are, are geared to finding only a local optimum. To identify a global optimum, there is often no alternative but to enumerate the local optima, explicitly or implicitly, in search of the best one.

The most popular and effective global solvers use a branch-and-relax approach that is closely analogous to the one presented for the freight transfer problem. The main difference is that the variable domains are continuous intervals rather than finite sets, the algorithm branches on a variable by splitting an interval into two or more intervals. This sort of branching divides continuous space into increasingly smaller boxes until a global solution can be isolated in a very small box. Constraint propagation and relaxation are also somewhat different than in discrete problems, because they employ techniques that are specialized to the nonlinear functions that typically occur in continuous problems.

Global optimization is best illustrated with a very simple example that is chosen to highlight some of the simpler techniques rather than to represent a practical application. The problem is:

$$\begin{aligned}&\text{Linear: }\begin{cases}\max x_1 + x_2\\ 2x_1 + x_2 \leq 2\end{cases}\\ &\text{Bilinear: } 4x_1x_2 = 1\\ &\text{Domains: } x_1 \in [0,1],\ x_2 \in [0,2]\end{aligned} \quad (2.15)$$

The feasible set is illustrated in Figure 2.5. There are two locally optimal solutions,

$$(x_1, x_2) = (\tfrac{1}{2} + \tfrac{1}{4}\sqrt{2}, 1 - \tfrac{1}{2}\sqrt{2}) \approx (0.853553, 0.292893)$$
$$(x_1, x_2) = (\tfrac{1}{2} - \tfrac{1}{4}\sqrt{2}, 1 + \tfrac{1}{2}\sqrt{2}) \approx (0.146447, 1.707107)$$

the latter of which is globally optimal.

Inference: Bounds Propagation

Bounds propagation can be applied to nonlinear constraints as well as linear inequalities. This is easiest when a constraint can be "solved" for

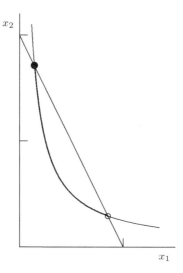

Figure 2.5. Illustration of a continuous global optimization problem. The heavy curve represents the feasible set. The dot marks the optimal solution, and the circle marks a second locally optimal solution.

a bound on each variable in terms of a monotone function of the other variables, as is possible for linear inequalities. To obtain a new bound for a given variable, one need only substitute the smallest or largest possible values for the other variables.

In the example, $4x_1x_2 = 1$ can be solved for $x_1 = 1/4x_2$. By substituting the upper bound of 2 for x_2, one can update the lower bound on x_1 to $\frac{1}{8}$, resulting in the domain $[0.125, 1]$. Similarly, the domain for x_2 can be reduced to $[0.25, 2]$.

More generally, suppose constraint C can be solved for a lower bound $g_1(x_2, \ldots, x_n)$ on x_1, where g_1 is monotone nonincreasing in each argument (as is commonly the case). That is, any solution $x = (x_1, \ldots, x_n)$ that is feasible for C satisfies $x_1 \geq g_1(x_2, \ldots, x_n)$. If each x_j has domain $[L_j, U_j]$, the lower bound L_1 can be updated to $g_1(U_2, \ldots, U_n)$, and analogously for an upper bound.

Each constraint of the problem can be used to reduce the domains obtained from the previous constraint. In the example, the linear constraint $2x_1 + x_2 \leq 2$ allows further reduction of the domains $[0.125, 1]$ and $[0.25, 2]$ to $[0.125, 0.875]$ and $[0.25, 1.75]$, respectively. These domains can be still further reduced by cycling back through the two constraints repeatedly, asymptotically reaching a fixed point that is approximately $[0.146, 0.854]$ for x_1 and $[0.293, 1.707]$ for x_2. Global solvers typically truncate this process quite early, however, because the marginal gains of

Relaxation: Factored Functions

Linear relaxations can often be created for nonlinear constraints by *factoring* the functions involved into more elementary functions for which linear relaxations are known. For instance, the constraint $x_1 x_2 / x_3 \leq 1$ can be written $y_1 \leq 1$ by setting $y_1 = y_2/x_3$ and $y_2 = x_1 x_2$. This factors the function $x_1 x_2 / x_3$ into the elementary operations of multiplication and division, for which tight linear relaxations have been derived.

The constraint $4x_1 x_2 = 1$ in the example (2.15) can be written $4y = 1$ by setting $y = x_1 x_2$. The product $y = x_1 x_2$ has the well-known relaxation

$$L_2 x_1 + L_1 x_2 - L_1 L_2 \leq y \leq L_2 x_1 + U_1 x_2 - U_1 L_2 \\ U_2 x_1 + U_1 x_2 - U_1 U_2 \leq y \leq U_2 x_1 + L_1 x_2 - L_1 U_2 \tag{2.16}$$

where again $[L_i, U_i]$ is the current interval domain of x_i.

The example (2.15) therefore has the relaxation

$$\text{Linear:} \begin{cases} \max\ x_1 + x_2 \\ 4y = 1 \\ 2x_1 + x_2 \leq 2 \\ L_2 x_1 + L_1 x_2 - L_1 L_2 \leq y \leq L_2 x_1 + U_1 x_2 - U_1 L_2 \\ U_2 x_1 + U_1 x_2 - U_1 U_2 \leq y \leq U_2 x_1 + L_1 x_2 - L_1 U_2 \\ L_i \leq x_i \leq U_i,\ \ i = 1, 2 \end{cases} \tag{2.17}$$

The constraint $4y = 1$ can be eliminated by substituting $1/4$ for y in the other constraints. Initially, $[L_1, U_1] = [0.125, 0.875]$ and $[L_2, U_2] = [0.25, 1.75]$, since these domains result from the bounds propagation described in the previous section.

Branching Search

At the root node of a branch-and-relax tree (Figure 2.6), the initial linear relaxation (2.17) is solved to obtain the solution $(x_1, x_2) = (\frac{1}{7}, \frac{41}{24}) \approx (0.143, 1.708)$. This solution is infeasible in the original problem (2.15), since $4x_1 x_2 = 1$ is not satisfied. It is therefore necessary to branch by splitting the domain of a variable.

The choice of branching heuristic can be crucial to fast solution, but for purposes of illustration one can simply split the domain $[0.25, 1.75]$ of x_2 into two equal parts. Restriction 1 in Figure 2.6 corresponds to the lower branch, $x_2 \in [0.25, 1]$. The domains reduce as shown. The

44   Search

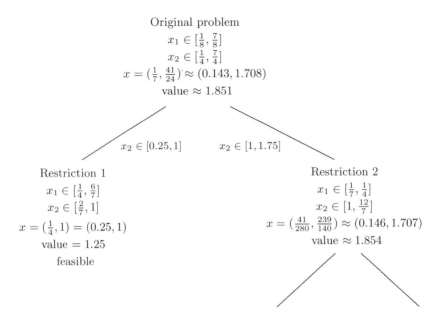

Figure 2.6. *A portion of a branch-and-relax tree for a global optimization problem. The reduced domains and solution of the relaxation are shown at each node.*

solution $(x_1, x_2) = (0.25, 1)$ of the relaxation is feasible and becomes the incumbent solution. The objective function value is 1.25, which provides a lower bound on the optimal value of the problem. No further branching on the left side is necessary. This, in effect, rules out the locally optimal point $(0.854, 0.293)$ as a global optimum, and it remains only to find the other locally optimal point to a desired degree of accuracy.

Moving to Restriction 2, the solution of the relaxation is $(x_1, x_2) = (\frac{41}{280}, \frac{239}{140}) \approx (0.146429, 1.707143)$. This is slightly infeasible, as $4x_1 x_2$ is 0.999898 rather than 1. One could declare this solution to be feasible within tolerances and terminate the search. Alternatively, one could continue branching to find a strictly feasible solution that is closer to the optimum than the incumbent.

Before branching, however, it may be possible to reduce domains by using Lagrange multipliers and the lower bound 1.25 on the optimal value. The constraints $2x_1 + x_2 \leq 2$ and $1/4 \leq U_2 x_1 + L_1 x_2 - L_1 U_2$, respectively, have Lagrange multipliers 1.1 and 0.7 in the solution of the relaxation (2.17) (the remaining constraints have vanishing multipliers since they are not satisfied as equalities). The multiplier 1.1 indicates that any reduction Δ in the right-hand side of $2x_1 + x_2 \leq 2$ reduces the optimal value of the relaxation, currently 1.854, by at least 1.1Δ. Thus any reduction Δ in the left-hand side of the constraint, currently equal to 2, has the same effect. Since the optimal value should not be reduced below 1.25 (the current lower bound on the optimal value of the original problem), we should have $1.1\Delta \leq 1.854 - 1.25$, or

$$1.1[2 - (2x_1 + x_2)] \leq 1.854 - 1.25$$

This implies $2x_1 + x_2 \geq 1.451$, an inequality that must be satisfied by any optimal solution found in the subtree rooted at the current node. Propagation of this inequality reduces the current domain $[1, 1.714]$ of x_2 to $[1.166, 1.714]$. Similar treatment of the inequality $1/4 \leq U_2 x_1 + L_1 x_2 - L_1 U_2$, which currently is $(12/7)x_1 + (1/7)x_2 \geq 97/196$, has no effect on the domains.

In general, if a constraint $ax \leq a_0$ has Lagrange multiplier λ in the optimal solution of the current relaxation, v is the optimal value of the relaxation, and L is a lower bound on the optimal value of the original problem, then the inequality

$$ax \geq a_0 - \frac{v - L}{\lambda}$$

can be propagated at the current node. This can be particularly useful when the constraint $ax \leq a_0$ is a bound on a single variable, because it provides a direct reduction of that variable's domain.

Unless proper care is exercised, global optimization methods can produce "feasible" solutions that are not actually feasible, due to truncation and other computational errors. The design of *safe* methods that are guaranteed to produce strictly feasible solutions has been the focus of much research in recent years.

2.2.7 Example: Product Configuration

A product configuration example shows how inference and relaxation combine in a slightly more complex problem. It also introduces indexed linear constraints, which occur when the coefficients of a linear constraint are indexed by variables.

The problem is to decide what type of power supply, disk drive, and memory chip to install in a laptop computer, and how many of each. The objective is to minimize total weight while meeting the computer's requirements—upper and lower bounds on memory, disk space, net power supplied, and weight. Only one type of each component may be installed. One power supply, at most three disk drives, and at most three memory chips may be used. Data for a small instance of the problem appear in Table 2.6.

Table 2.6. Data for a small instance of the product configuration problem.

Component i	Type k	Net power generation A_{i1k}	Disk space A_{i2k}	Memory capacity A_{i3k}	Weight A_{i4k}	Max number used
1. Power supply	A	70	0	0	200	1
	B	100	0	0	250	
	C	150	0	0	350	
2. Disk drive	A	-30	500	0	140	3
	B	-50	800	0	300	
3. Memory chip	A	-20	0	250	20	3
	B	-25	0	300	25	
	C	-30	0	400	30	
Lower bound L_j		0	700	850	0	
Upper bound U_j		∞	∞	∞	∞	
Unit cost c_j		0	0	0	1	

Formulating the Problem

To formulate the problem as an optimization model, let variable t_i be the type of component i that is chosen, and q_i the quantity. It is convenient to let variable r_j denote the total amount of attribute j that is produced, where $j = 1, 2, 3, 4$ correspond respectively to net power, disk space, memory, and weight. The computer requires a minimum L_j and maximum U_j of each attribute j.

In general, the objective in such a problem is to minimize the cost of supplying the attributes, where c_j is the unit cost of attribute j. In this case, only the weight is of concern, so that $(c_1, \ldots, c_4) = (0, 0, 0, 1)$.

The interesting aspect of the problem is modeling the connection between the amount r_j of attribute j supplied and the configuration chosen. The component characteristics can be arranged in a three-dimensional array A, in which A_{ijk} is the net amount of attribute j produced by type k of component i. A natural way to write r_j is to sum the contributions of each component

$$r_j = \sum_i A_{ijt_i} q_i$$

where A_{ijt_i} is the amount contributed by the selected type of component i. This leads immediately to the optimization model

$$\text{Linear:} \begin{cases} \min \sum_j c_j r_j & (a) \\ r_j = \sum_i A_{ijt_i} q_i, \text{ all } j & (b) \end{cases} \quad (2.18)$$

$$\text{Domains:} \begin{cases} t_1 \in \{\text{A,B,C}\}, t_2 \in \{\text{A,B}\}, t_3 \in \{\text{A,B,C}\} \\ r_j \in [L_j, U_j], \text{ all } j \\ q_1 \in \{1\}, \; q_2, q_3 \in \{1, 2, 3\} \end{cases}$$

The optimal solution of this model is $t = (t_1, t_2, t_3) = (\text{C,A,B})$ and $q = (1, 2, 3)$, which results in weight 705. This calls for installation of the largest power supply (type C), two of the smaller disk drives (type A), and three of the medium-sized memory chips (type B).

Constraints (a) of (2.18) are linear equations with some variable indices t_i. They are processed using a specially-structured element constraint. In general, a constraint like (a) contains inequalities of the form

$$\sum_{i \in I} A_{iy_i} x_i + \sum_{i \in I'} A_i x_i \geq b \quad (2.19)$$

where each A_{ik}, each A_i, and b are tuples of equal length. Two such inequalities are used to enforce an equation. (2.19) can be compiled by replacing it with the metaconstraint

$$\text{indexed linear} \begin{cases} \sum_{i \in I} z_i + \sum_{i \in I'} A_i x_i \geq b \\ \text{element}\,(y_i, x_i, z_i \,|\, (A_{i1}, \ldots, A_{im})), \text{ all } i \in I \end{cases} \quad (2.20)$$

Each element constraint sets z_i equal to the y_i^{th} tuple in the list $A_{i1}x_i, \ldots, A_{im}x_i$. Filtering and relaxation techniques can be devised to exploit the special structure of this constraint, which can be called an *indexed linear element* constraint.

Following this pattern, model (2.18) becomes:

$$
\begin{aligned}
&\text{Linear: } \min \sum_j c_j r_j \\
&\text{Indexed linear: } \begin{cases} r_j = \sum_i z_{ij}, \text{ all } j \\ \text{element}(t_i, q_i, z_{ij} \mid (A_{ij1}, \ldots, A_{ijm})), \text{ all } i,j \end{cases} \\
&\text{Domains: } \begin{cases} t_1 \in \{\text{A,B,C}\}, t_2 \in \{\text{A,B}\}, t_3 \in \{\text{A,B,C}\} \\ r_j \in [L_j, U_j], \text{ all } j \\ q_1 \in \{1\}, \; q_2, q_3 \in \{1,2,3\} \end{cases}
\end{aligned}
\tag{2.21}
$$

In practice, the solution software should generate this reformulation auto-matically from (2.18). The modeler would not have to be aware of the reformulation.

Inference: Propagation of the Element Constraint

Domain reduction for the linear constraints in (2.21) can be achieved by bounds propagation, as discussed earlier. It remains to propagate the indexed linear metaconstraint.

Domain filtering for the element constraint in an indexed linear metaconstraint can take advantage of its special form. The procedure is quite straightforward if tedious, and the details are presented in Sections 3.8.1–3.8.2. Additional filtering may be derived from the fact that the indexed linear constraint implies integer knapsack inequalities. If $[L_{rj}, U_{rj}]$ is the current domain of r_j, then

$$L_{rj} \leq \sum_i A_{ijt_i} q_i \leq U_{rj}$$

This in turn yields the integer knapsack inequalities

$$L_{rj} \leq \sum_i \max_{k \in D_{t_i}} \{A_{ijk}\} q_i$$

$$\sum_i \min_{k \in D_{t_i}} \{A_{ijk}\} q_i \leq U_{rj}$$

which can be used to reduce the domains of the variables q_i.

Three of the attributes in the example problem (power, disk space, and memory), respectively, yield the following knapsack inequalities:

$$0 \leq \max\{70, 100, 150\} q_1 + \max\{-30, -50\} q_2 + \max\{-20, -25, -30\} q_3$$
$$700 \leq \max\{0, 0, 0\} q_1 + \max\{500, 800\} q_2 + \max\{0, 0, 0\} q_3$$
$$850 \leq \max\{0, 0, 0\} q_1 + \max\{0, 0\} q_2 + \max\{250, 300, 400\} q_3$$

which simplify to

$$0 \leq 150 q_1 - 30 q_2 - 20 q_3, \quad 700 \leq 800 q_2, \quad 850 \leq 400 q_3 \qquad (2.22)$$

Weight yields no useful inequality as its lower bound is zero. Propagation of these inequalities reduces the domain of q_3 to $\{3\}$ but does not affect the other domains. When all the constraints of (2.21) are propagated, however, the domains are reduced to the following:

$$\begin{aligned}
& q_1 \in \{1\} && q_2 \in \{1, 2\} && q_3 \in \{3\} \\
& t_1 \in \{C\} && t_2 \in \{A, B\} && t_3 \in \{B, C\} \\
& z_{11} \in [150, 150] && z_{21} \in [-75, -30] && z_{31} \in [-90, -75] \\
& z_{12} \in [0, 0] && z_{22} \in [700, 1600] && z_{32} \in [0, 0] \\
& z_{13} \in [0, 0] && z_{23} \in [0, 0] && z_{33} \in [900, 1200] \\
& z_{14} \in [350, 350] && z_{24} \in [140, 600] && z_{34} \in [75, 90] \\
& r_1 \in [0, 45] && r_2 \in [700, 1600] && r_3 \in [900, 1200] \\
& r_4 \in [565, 1040]
\end{aligned} \qquad (2.23)$$

Relaxation: Convex Hull of the Element Constraint

Problem (2.21) is linear except for the indexed linear constraints and the integrality restriction on q_i. It can therefore be relaxed by dropping the integrality condition and finding a linear relaxation for the indexed linear constraints.

The key to relaxing an element constraint of the form

$$\text{element}(y, x, z \mid (A_1, \ldots, A_m))) \qquad (2.24)$$

where $x \geq 0$ is to note that for any given domain of y, the constraint implies a disjunction of linear equations:

$$\bigvee_{k \in D_y} (z = A_k x) \qquad (2.25)$$

This says that at least one of the equations $z = A_k x$ must hold. Relaxations for disjunctions of linear systems in general are well studied, and

(2.25) in particular has the relaxation
$$z = \sum_{k \in D_y} A_k x_k, \quad x = \sum_{k \in D_y} x_k$$
where $x_k \geq 0$ are new variables. This is, in fact, a convex hull relaxation, which is the tightest possible linear relaxation. A solution of the relaxation is feasible for (2.24) if exactly one of the x_ks (say x_{k^*}) is positive and therefore equal to x. In this case, z is equated with $A_{k^*} x$.

Based on this idea, the relaxation of (2.21) becomes:

$$\text{Linear:} \begin{cases} \min \sum_j c_j r_j & (a) \\ r_j = \sum_i \sum_{k \in D_{t_i}} A_{ijk} q_{ik}, \text{ all } j & (b) \\ q_i = \sum_{k \in D_{t_i}} q_{ik}, \text{ all } i & (c) \\ L_{r_j} \leq r_j \leq U_{r_j}, \text{ all } j & (d) \\ L_{q_i} \leq q_i \leq U_{q_i}, \text{ all } i & (e) \\ L_{r_j} \leq \sum_i \max_{k \in D_{t_i}} \{A_{ijk} q_i\}, \text{ all } j & (f) \\ \sum_i \min_{k \in D_{t_i}} \{A_{ijk} q_i\} \leq U_{r_j}, \text{ all } j & (g) \\ q_{ik} \geq 0, \text{ all } i, k & (h) \end{cases} \quad (2.26)$$

Domains: $q_i, q_{ik}, r_j \in \Re$, all i, j, k

The relaxation can be strengthened with cuts derived from the knapsack inequalities (f) and (g). For example, the first inequality in (2.22) yields the cuts $q_1 + (1 - q_3) \geq 1$ and $q_1 + (1 - q_2) \geq 1$, while the second and third, respectively, yield $q_2 \geq 1$ and $q_3 \geq 3$. In this particular example, knapsack cuts do not strengthen the relaxation.

The solution of (2.26), given the reduced domains (2.23), is $q_1, q_{13} = 1$, $q_2, q_{2A} = 2$, and $q_3, q_{3B} = 3$, with the other q_{ik}s equal to zero. Since the q_is are integer, and exactly one $q_{ij} > 0$ for each i, the solution is feasible and no branching is necessary. The nonzero q_{ij}s indicate that $(t_1, t_2, t_3) = (C, A, B)$. Since $(r_1, \ldots, r_4) = (15, 1000, 900, 705)$, the minimum weight is $r_4 = 705$.

Branching Search

A solution of the relaxation (2.26) is feasible for the original problem (2.21) unless it violates the integrality constraint for some q_i or fails to satisfy the indexed linear constraint. In the former case, branching follows the usual pattern. If $\hat{q}_i$ is a noninteger value in the solution of the relaxation, the domain $\{L_{q_i}, \ldots, U_{q_i}\}$ of q_i is split into $\{L_{q_i}, \ldots, \lfloor \hat{q}_i \rfloor\}$ and $\{\lceil \hat{q}_i \rceil, \ldots, U_{q_i}\}$. At either branch, the smaller domains are propagated and the resulting relaxation solved.

The solution of the relaxation violates the indexed linear constraint for some i when at least two q_{ik}s are positive, say q_{ik_1} and q_{ik_2}. In this case, the search can branch by splitting the domain of t_i into two sets, one excluding k_1 and the other excluding k_2.

2.2.8 Branch and Price

Branch and price is a special case of branch and relax designed for problems with an integer programming formulation—that is, a formulation consisting of linear constraints and integer-valued variables. Despite this restriction, the method is of interest here because it has become one of the more popular settings for integrated problem solving. It has proved particularly successful for airline crew scheduling and other transport-related applications.

Branch and price can be attractive when the integer programming model has a huge number of variables. In such cases the variables are added to the model only as needed to solve the continuous relaxation. A variable that does not occur in the current model can be *priced* to determine whether its addition would result in a better solution. Variables are added until no further improvement is possible, at which point the continuous relaxation has been solved. Typically, only a small fraction of the total variable set is needed to reach an optimal solution. Once the continuous relaxation is solved, one can branch as in ordinary branch-and-relax methods and solve the continuous relaxations at the branches by adding further variables.

The pricing problem can be solved by any convenient method, and this is where other technologies may enter the picture. Constraint programming, in particular, can be useful for this purpose, because it can deal with the often complex constraints that must be observed when the variables are generated. Airline crew scheduling, for example, is normally constrained by a host of complicated work rules.

An integer programming problem has the form

$$\min cx \\ Ax \geq b, \; x \geq 0 \text{ and integral} \tag{2.27}$$

where A is an $m \times n$ matrix. Its continuous relaxation is

$$\min cx \\ Ax \geq b, \; x \geq 0 \tag{2.28}$$

Since the expression Ax can be written $\sum_j A_j x_j$, where A_j is the jth column of A, adding a variable x_j to the problem can be viewed as adding a column to the coefficient matrix. It is for this reason that branch-and-price methods are often called *column generation* methods.

The problem restriction at each node of the branch-and-relax tree includes some constraints due to branching and likewise has the form (2.27). The task is to solve its continuous relaxation (2.28). When A has a huge number of columns, (2.28) is initially formulated with only a small subset of the columns:

$$\min\ cx$$
$$\sum_{j \in J} A_j x_j \geq b \tag{2.29}$$
$$x_j \geq 0,\ \text{all } j \in J$$

This is the *restricted master problem*. It is a restriction of (2.27) because any feasible solution of it is a feasible solution of (2.27) with the remaining variables set to zero.

When (2.29) is solved by linear programming software, one typically obtains not only optimal values for each x_j but an optimal *dual multiplier* u_i for each inequality constraint $\sum_j A_{ij} x_j \geq b_j$. Dual multipliers are a special case of Lagrange multipliers and will be discussed in Section 3.3. For present purposes, they can be viewed as the marginal rate of increase in the optimal cost cx for each unit increase in the right-hand side b_i. It will be seen that adding a new variable x_j ($j \notin J$) to the problem and re-solving can improve the optimal solution only if

$$c_j - \sum_i u_i A_{ij} < 0 \tag{2.30}$$

or more succinctly, $c_j - u A_j < 0$, where u is the row vector $[u_1 \cdots u_m]$. Intuitively, this is because increasing the value of x_j (now zero since x_j is not in the problem) by one unit increases cost by c_j units but has the same effect as forcing each right-hand side b_i down by A_{ij} units, which reduces the cost by $\sum_i u_i A_{ij}$ units. The net effect on cost is the left-hand side of (2.30), which is known as the *reduced cost* of x_j. If the reduced cost is negative, then the cost is brought down and the solution improved by introducing x_j to the problem and giving it some positive value. The process of evaluating the reduced cost of x_j is known as *pricing* the variable.

The problem of finding a variable with negative reduced cost can be solved in any number of ways. For instance, one can use a heuristic algorithm that generates columns of A until it finds one with a negative reduced cost. If no such column is found, then one must switch to an exact algorithm to make sure there is no improving column. If there is none, then the current solution of the restricted master problem (2.27) is optimal for (2.28). Another approach, explored in the next section, is to solve the pricing problem with a combination of optimization and constraint programming techniques.

2.2.9 Example: Airline Crew Scheduling

A simplified airline crew scheduling problem shows how constraint programming and mathematical programming ideas can be combined in a branch-and-price framework. It also illustrates set-valued variables, which have become important in constraint programming solvers.

The goal is to assign flight crew members to flights so as to minimize cost while covering all the flights and observing a number of work rules. Each flight j starts at time s_j, finishes at time f_j, and requires n_j crew members. A small example with six flights appears in Table 2.7. Whenever a crew member staffs two consecutive flights, the rest period between the flights must be at least $\Delta_{\min}$ and at most $\Delta_{\max}$. The total flight time assigned to a crew member must be between $T_{\min}$ and $T_{\max}$. There may be other restrictions on scheduling as well.

Integer Programming Formulation

The set of flights assigned to a crew member is known as a *roster*. The problem can in principle be formulated by generating all possible rosters and assigning one roster to each crew member in such a way as to cover every flight. The cost c_{ik} of assigning crew member i to roster k depends on a number of factors, such as seniority, the timing of flights and rest periods, and so forth.

Let δ_{jk} be 1 when flight j is part of roster k, and 0 otherwise. Also let 0-1 variable $x_{ik} = 1$ when crew member i is assigned roster k. Then, the

Table 2.7. Flight data for a small crew scheduling problem.

j	s_j	f_j
1	0	3
2	1	3
3	5	8
4	6	9
5	10	12
6	14	16

problem can be formulated with an integer programming model in which dual multipliers u_i, v_{ik} are associated with the constraints as shown:

$$\text{0-1 linear:} \begin{cases} \min \sum_{ik} c_{ik} x_{ik} \\ \sum_k x_{ik} = 1, \text{ all } i \quad (u_i) \\ \sum_i \sum_k \delta_{jk} x_{ik} \geq n_j, \text{ all } j \quad (v_j) \end{cases} \quad (2.31)$$

Domains: $x_{ik} \in \{0, 1\}$, all i, k

Suppose for the example of Table 2.7 that the minimum and maximum gap between flights are $(\Delta_{\min}, \Delta_{\max}) = (2, 3)$, and the minimum and maximum flight times are $(T_{\min}, T_{\max}) = (6, 10)$. Thus, Flight 6 cannot immediately follow Flight 1, since the gap is too large (11), and Flight 5 cannot immediately follow Flight 4, since the gap is too small (1). The constraints permit four possible rosters: $\{1, 3, 5\}$, $\{1, 4, 6\}$, $\{2, 3, 5\}$, and $\{2, 4, 6\}$. If there are two crew members, and each flight requires one crew member, the continuous relaxation of problem (2.31) becomes:

$$\min z$$

$$\begin{bmatrix} 10 & 12 & 7 & 13 & 9 & 11 & 6 & 12 \\ 1 & 1 & 1 & 1 & 0 & 0 & 0 & 0 \\ 0 & 0 & 0 & 0 & 1 & 1 & 1 & 1 \\ 1 & 1 & 0 & 0 & 1 & 1 & 0 & 0 \\ 0 & 0 & 1 & 1 & 0 & 0 & 1 & 1 \\ 1 & 0 & 1 & 0 & 1 & 0 & 1 & 0 \\ 0 & 1 & 0 & 1 & 0 & 1 & 0 & 1 \\ 1 & 0 & 1 & 0 & 1 & 0 & 1 & 0 \\ 0 & 1 & 0 & 1 & 0 & 1 & 0 & 1 \end{bmatrix} \begin{bmatrix} x_{11} \\ x_{12} \\ x_{13} \\ x_{14} \\ x_{21} \\ x_{22} \\ x_{23} \\ x_{24} \end{bmatrix} \begin{matrix} = \\ = \\ = \\ \geq \\ \geq \\ \geq \\ \geq \\ \geq \\ \geq \end{matrix} \begin{bmatrix} z \\ 1 \\ 1 \\ 1 \\ 1 \\ 1 \\ 1 \\ 1 \\ 1 \end{bmatrix} \quad (2.32)$$

$$x_{ik} \geq 0, \text{ all } i, k$$

The problem is written in matrix form to show the eight columns, which correspond to the four possible rosters for each of the two crew members. The top row of the matrix contains the costs c_{ik}.

Branching Search

Rather than solve the complete problem (2.32), the problem is first solved with a subset of the columns, perhaps the following:

$$\min z$$

$$\begin{bmatrix} 10 & 13 & 9 & 12 \\ 1 & 1 & 0 & 0 \\ 0 & 0 & 1 & 1 \\ 1 & 0 & 1 & 0 \\ 0 & 1 & 0 & 1 \\ 1 & 0 & 1 & 0 \\ 0 & 1 & 0 & 1 \\ 1 & 0 & 1 & 0 \\ 0 & 1 & 0 & 1 \end{bmatrix} \begin{bmatrix} x_{11} \\ x_{14} \\ x_{21} \\ x_{24} \end{bmatrix} \begin{matrix} = \\ = \\ = \\ \geq \\ \geq \\ \geq \\ \geq \\ \geq \\ \geq \end{matrix} \begin{bmatrix} z \\ 1 \\ 1 \\ 1 \\ 1 \\ 1 \\ 1 \\ 1 \\ 1 \end{bmatrix} \begin{matrix} (10) \\ (9) \\ (0) \\ (0) \\ (0) \\ (0) \\ (0) \\ (3) \end{matrix} \quad (2.33)$$

$$x_{ik} \geq 0, \text{ all } i, k$$

The optimal solution is $(x_{11}, x_{14}, x_{21}, x_{24}) = (0, 1, 1, 0)$. The corresponding dual multipliers $u = (10, 9)$ and $v = (0, 0, 0, 0, 0, 3)$ are shown on the right.

The Pricing Problem

The pricing problem is to identify a variable of (2.31) with a negative reduced cost. A column associated with x_{ik} contains a 1 in the row corresponding to u_i and the rows corresponding to v_j for each flight j in the roster. The reduced cost of x_{ik} is therefore

$$c_{ik} - u_i - \sum_{j \text{ in roster } k} v_j \qquad (2.34)$$

One way to formulate the pricing problem is to model it with a directed graph. The graph contains a node j for each flight, plus a source node s and sink node t. The graph contains a directed arc (j, j') when flight j can immediately precede j' in a roster; that is, $\Delta_{\min} \leq s_{j'} - f_j \leq \Delta_{\max}$. There are also arcs (s, j) and (j, t) for every flight j. Every possible roster corresponds to a path from s to t, although not every path corresponds to a roster, because the total fight time may not be in the range $[T_{\min}, T_{\max}]$.

The graph corresponding to the example appears in Figure 2.7. Only some of the arcs incident to s and t are shown, since the others can be removed by elementary preprocessing of the graph. For example, arc $(s, 3)$ is removed because every path from 3 to t results in a total flight time less than the minimum of $T_{\min} = 6$. This can be ascertained by computing the longest path from 3 to t when the length of each arc (j, j') is set to the duration of flight j.

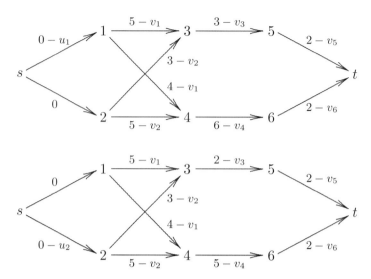

Figure 2.7. Path model for crew rostering. The top graph pertains to crew member 1, and the bottom graph to crew member 2. The arc lengths reflect costs offset by dual multipliers u_i, v_j.

The graph-based model assumes that the cost c_{ik} associated with x_{ik} can be equated with the length of the corresponding path, if the arc lengths are suitably defined. This is possible if the cost depends only on the cost $\alpha_{jj'}$ of staffing each flight j with crew member i and transferring him or her to the next flight j'. The length of each arc (j, j') is set equal to $\alpha_{jj'}$, the length of each (s, j) to the cost α_{sj} of transferring from home to flight j, and the length of (j, t) to the cost α_{jt} of operating flight j and returning home. The arc costs associated with each crew member in the example problem are shown in Figure 2.7 (ignore the terms u_i and v_j for the moment).

The reduced cost (2.34) of a variable x_{ik} can also be equated with the length of the corresponding path, if the arc lengths are offset by the dual multipliers. The cost term c_{ik} in (2.34) is the path length using the arc lengths just defined. The path length becomes the reduced cost if the length of each (j, j') is set to $\alpha_{jj'} - v_j$, the length of each (s, j) to $\alpha_{sj} - u_i$, and the length of each (j, t) to $\alpha_{jt} - v_j$. These adjustments are shown in Figure 2.7.

The pricing problem is now the problem of finding a path in the directed graph G with negative length, for which the total flight time lies in the interval $[T_{\min}, T_{\max}]$. Let X_i be the set of flights assigned to crew member i. Since each path is specified by some X_i, the problem can be written as follows:

$$\text{Path: } (X_i, z_i \mid G, c, s, t), \text{ all } i$$

$$\text{Set sum: } T_{\min} \leq \sum_{j \in X_i} p_j \leq T_{\max}, \text{ all } i \tag{2.35}$$

$$\text{Domains: } X_i \subset \{\text{flights}\}, \ z_i < 0, \text{ all } i$$

where $p_j = f_j - s_j$ is the duration of flight j. The path metaconstraint requires that the nodes in X_i define a path from s to t in G with length z_i, where c contains the edge lengths c_{ij}. Since z_i's domain elements are negative, the constraint enforces a negative reduced cost. The set sum constraint simply requires that the total flight duration be within the prescribed bounds.

In the example of Figure 2.7, the arc lengths are defined by the dual multipliers $u = (10, 9)$ and $v = (0, 0, 0, 0, 0, 3)$. The shortest path in the graph for Crew Member 1 is s-1-4-6-t, which has length -1. This defines the column corresponding to variable x_{12} in (2.32). The shortest path for Crew Member 2 is s-2-3-5-t with length -2, which defines the column corresponding to x_{23}. So variables x_{12} and x_{23} have negative reduced costs, and their columns are added to the restricted master problem (2.33). The new solution is $(x_{11}, x_{12}, x_{14}, x_{21}, x_{23}, x_{24}) = (0, 1, 0, 0, 1, 0)$ with dual multipliers $u = (10, 5)$ and $v = (0, 1, 0, 0, 0, 2)$. When the arc lengths are updated accordingly, the shortest path for both crew members has length zero. This means there is no improving column, and the solution of the restricted master is optimal. Since this solution is integral, there is no need to branch. The optimal rostering assigns Flights 1, 4 and 6 to Crew Member 1, and the remaining flights to Crew Member 2.

Bounds Propagation for Set-Valued Variables

The pricing problem (2.35) can, in general, be solved by a combination of bounds propagation and branching. Bounds propagation must be reinterpreted for the variables X_i since they are set valued, but the idea is straightforward. The domain of each X_i is stored in the form of bounds $[L_{X_i}, U_{X_i}]$, where L_{X_i} is a set that X_i must contain, and U_{X_i} is a set that must contain X_i. Initially L_{X_i} is empty and U_{X_i} is the set of all flights, but branching and bounds propagation may tighten these bounds. For instance, if branching fixes $j \in X_j$, then $L_{X_i} = \{j\}$.

The path constraint in (2.35) tightens bounds for both z_i and X_i. It updates the lower bound L_{z_i} on z_i by finding a shortest path from s to t whose node set lies in the range $[L_{X_i}, U_{X_i}]$. If ℓ is the length of the shortest path, then L_{z_i} is updated to $\max\{L_{z_i}, \ell\}$. If $\ell \geq 0$, the domain of z_i becomes empty because U_{z_i} is initially negative, and there is no variable with negative reduced cost (the constraint also updates U_{z_i} by finding a longest path, but this is not relevant here). Since G is acyclic, there are very fast algorithms for finding a shortest path, despite the possibility of negative arc lengths. These algorithms can be adapted to ensure that the node set lies in the range $[L_{X_i}, U_{X_i}]$ by temporarily modifying G. To make sure the path nodes belong to U_{X_i}, simply delete all nodes in G that are not in U_{X_i}. To make sure all nodes in L_{X_i} belong to the path, avoid routing around nodes in L_{X_i}. That is, whenever G contains an arc (j_1, j_3) and a path from j_1 to j_3 through a third node $j_2 \in L_{X_i}$, delete the arc (j_1, j_3) from G.

The modified graph G' can be used to tighten the bounds for X_i as well. If every path from s to t in G' contains node j, then add j to L_{X_i}. If no path contains j, then remove j from U_{X_i}. This can be accomplished by supposing that the arcs of G' carry flow in the direction of their orientation. It is assumed that a flow of 1 unit volume enters G' at node s and exits at node t. If the minimum flow through node j subject to these conditions is 1, then j contains all paths from s to t, and if the maximum is 0, it contains no paths. Minimum and maximum flow problems of this sort can be solved very quickly. Thus, if $L_{X_i} = \{3\}$ in the example, the minimum flow through node 5 in G' is 1 and the maximum through 4 is 0, which indicates that node 5 may be added to L_{X_i} and node 4 removed from U_{X_i}.

The set sum constraint can also be used for propagation in the obvious way. Let $[L_{z_i}, U_{z_i}]$ be the current domain for z_i. One can update L_{z_i} to

$$\max \left\{ L_{z_i}, \sum_{j \in L_{X_i}} p_j + \sum_{j \in U_{X_i} \setminus L_{X_i}} \min\{0, p_j\} \right\} \quad (2.36)$$

and analogously for U_{z_i}. Also, if

$$p_k + \sum_{j \in L_{X_i}} p_j > U_{z_i} \quad \text{or} \quad p_k + \sum_{j \in L_{X_i}} p_j < L_{z_i}$$

for $k \notin L_{X_i}$, then k can be removed from U_{X_i}. One can also write a sufficient condition for adding k to L_{X_i}.

Branching as well as propagation may be required to solve the pricing problem. A standard way to branch on a set-valued variable X_i is to consider the options $j \in X_i$ and $j \notin X_i$, where $j \in U_{X_i} \setminus L_{X_i}$. The former branch is implemented by adding j to L_{X_i}, and the latter by removing j from U_{X_i}.

The pricing problem discussed here contains only two metaconstraints. Realistic problems must deal with many complicated work rules. Spe-

Branching Search 59

cialized constraints and associated filtering methods have been developed for some of these.

2.2.10   Exercises

1. Write two integer linear inequalities (with initial domains specified) for which bounds propagation reduces domains more than minimizing and maximizing each variable subject to a continuous relaxation of the constraint set. Now write two integer linear inequalities for which the reverse is true.

2. Identify all the packings of $6x_1 + 4x_2 + 3x_3 + x_4 \geq 29$, with $x_i \in \{0, 1, 2, 3, 4\}$ for $i = 1, \ldots, 4$. Write the corresponding knapsack cuts.

3. Define a *cover* for an inequality $ax \leq a_0$ that is analogous to a packing for $ax \geq a_0$. What is the knapsack cut corresponding to a given cover?

4. Solve the problem of minimizing $3x_1 + 4x_2$ subject to $2x_1 + 3x_2 \geq 10$ and $x_1, x_2 \in \{0, 1, 2, 3\}$ using branch and relax without bounds propagation. Now, solve it using branch and infer without relaxation. Finally, solve it using branch and relax with propagation. Which results in the smallest search tree? (When using branch and infer, follow the common constraint programming practice of branching on the variable with the smallest domain.)

5. Write the knapsack cuts that correspond to the nonmaximal packings $I = \{1\}, \{2\}, \{3\}, \{4\}$ for constraint (2.3a). If a linear programming solver is available, verify that when these cuts are added to the relaxation (2.6), the resulting branch-and-relax tree has five rather than seven nodes.

6. Formulate the following problem with the help of conditional constraints. A lumber operation wishes to build temporary roads from Forests 1 and 2 to Sawmills A and B. Due to topography and environmental regulations, roads can be built only in certain combinations: (a) from 1 to A, 1 to B, and 2 to B; (b) from 1 to A, 1 to B, and 2 to A; (c) from 1 to A, 2 to A, and 2 to B. Each sawmill j requires d_j units of timber, the unit cost of shipping timber over a road from forest i to sawmill j is c_{ij}, and the fixed cost of a road from i to j is f_{ij}. Choose which roads to build, and how much timber to transport over each road, so as to minimize cost while meeting the sawmill demands. Now, formulate the problem without conditional constraints and with a disjunctive linear constraint.

7 Consider the problem

$$\text{Linear: min } cx$$
$$\text{Logic: } \begin{cases} y_1 \vee y_2 \\ y_1 \rightarrow y_3 \end{cases}$$
$$\text{Conditional: } y_k \Rightarrow \left(A^k x \geq b^k\right), \ k = 1, 2, 3$$

where each y_k is a boolean variable. Write the problem without conditional constraints and using as many linear disjunctions as possible. Hint: the formula $y_1 \rightarrow y_3$ is equivalent to $\neg y_1 \vee y_3$.

8 Formulate the following problem using the appropriate global constraints. There are six security guards and four stations. Each station must be staffed by exactly one guard each night of the week. A guard must be on duty four or five nights a week. No station should be staffed by more than three different guards during the week. A guard must never staff the same station two or more nights in a row and must staff at least three different stations during the week. A guard must not staff Stations 1 and 2 on consecutive nights (in either order), and similarly for Stations 3 and 4. Every week will have the same schedule.

9 Write all feasible solutions of the stretch-cycle constraint in (2.11) to confirm that the domains in (2.12) can be reduced to (2.13) and cannot be reduced further.

10 Formulate this lot sizing problem using a combination of variable indices, linear constraints, and stretch constraints. In each period t, there is a demand d_{it} for product i, and this demand must be met from the stock of product i at the end of period t. At most one product can be manufactured in each period t, represented by variable y_t. If nothing is manufactured, $y_t = 0$, where product 0 is a dummy product. The quantity of product i manufactured in any period must be either q_i or zero. When product i is manufactured, its manufacture must continue no fewer than ℓ_i and no more than u_i periods in a row. The manufacture of product i in any period must be followed in the next period by the manufacture of one of the products in S_i (one may assume $0, i \in S_i$). The unit holding cost per period for product i is h_i, and the unit manufacturing cost is g_i. The setup cost of making a transition from product i in one period to product j in the next is c_{ij} (where possibly i and/or j is 0). Minimize total manufacturing, holding, and setup costs over an n-period horizon while

meeting demand. After formulating the problem, indicate how to replace the variable indices with element constraints. Hint: let variable x_{ij} represent the quantity of product i manufactured in period t, and s_{it} the stock at the end of the period. The element(y, x, z) constraint also has a form element$(y, z \mid a)$. where a is a tuple of constants. It sets z equal to a_y.

11 A difficulty with the model of the previous exercise is that the setup cost after an idle period is always the same, regardless of which (non-dummy) product was manufactured last. How can the model be modified to allow setup cost to depend on the last non-dummy product? Hint: define several dummy products.

12 Formulate the following problem using linear disjunctions, variable indices, and stretch-cycle constraints. The week is divided into n periods, and in each period t, d_t megawatts of electricity must be generated. Each power plant i generates at most q_i megawatts while operating. Once plant i is started, it must run for at least ℓ_i periods, and once shut down, it must remain idle for at least ℓ'_i periods. The cost of producing one megawatt of power for one period at plant i is g_i, and the cost of starting up the plant is c_i. Determine which plants to operate in each period to minimize cost while meeting demand. The schedule must follow a weekly cycle. Indicate how to replace the variable indices with element constraints. Hint: let $y_{it} = 1$ if plant i operates in period t, and let the setup cost incurred by plant i in period t be $c_{i y_{i,t-1} y_{it}}$, where $c_{i01} = c_i$ and $c_{i00} = c_{i01} = c_{i11} = 0$.

13 Derive a linear relaxation for $y = x^2$ from (2.16), and derive a linear relaxation for $y = a^x$ when $a > 0$. Now, write a factored relaxation for the problem

$$\max \ a^{x_1} + b^{x_2}$$
$$(x_1 + 2x_2)^2 \leq 1, \quad L_j \leq x_j \leq U_j, \ j = 1, 2$$

when $a, b > 0$.

14 Consider the problem:

$$\min \ x_1^2 + x_2^2$$
$$3x_1 + x_2 \leq 15$$
$$x_1, x_2 \geq 0 \text{ and integral}$$

Suppose that the best known feasible solution for the problem is $(x_1, x_2) = (4, 3)$. The continuous relaxation has the optimal solution

$(x_1, x_2) = (4.5, 1.5)$ with a Lagrange multiplier of 3 corresponding to the first constraint. Derive an inequality that must be satisfied by any optimal solution. (Round down the right-hand side, since the coefficients and variables are integral).

15 What knapsack covering inequality can be inferred from the meta-constraint (2.20)?

16 A farmer wishes to apply fertilizer to each of several plots. The additional crop yield from plot i per unit of type k fertilizer applied is a_{ik}, and the runoff into streams is c_{ik}. There are storage facilities on the farm for at most k different kinds of fertilizer. Formulate the problem of identifying which fertilizer to apply to each plot, and how much, to maximize total additional yield subject to an upper limit U on the amount of runoff. Now, reformulate the problem using an indexed linear constraint. Hint: the nvalues constraint is useful here.

17 Recall the crew rostering problem illustrated by Figure 2.7. Suppose the current domain of the set-valued variable X_1 is the interval $[L_{X_1}, U_{X_1}]$ where $L_{X_1} = \{6\}$ and $U_{X_1} = \{1, 2, 3, 4, 6\}$. Use the min and max flow test to filter the domain.

18 If the third column is omitted from (2.33), the resulting dual multipliers are $(u_1, u_2) = (10, 9)$ and $(v_1, \ldots, v_6) = (0, 3, 0, 0, 0, 0)$. Use the graphs of Figure 2.7 to identify one or more columns with negative reduced cost.

19 Section 2.2.9 describes how to modify an acyclic graph G to obtain a graph G' such that finding a shortest path in G' finds a shortest path in G that includes a given subset of nodes. Show that this method breaks down when the graph is not acyclic.

20 Suppose the graph G in the previous exercise is acyclic, and its nodes are numbered so that $i < j$ whenever (i, j) is an arc. Design an algorithm based on this ordering to find the modified graph G'.

21 Write an expression analogous to (2.36) for updating U_{z_i}.

22 Consider the set sum constraint $\ell \leq \sum_{j \in X} p_j \leq u$, where X is a set-valued variable with domain $[L_X, U_X]$. Write two conditions that are jointly necessary and sufficient for the set sum constraint to be feasible. If the constraint is feasible and $k \notin L_X$, write two conditions, either of which is sufficient to add k to L_X.

23 In a two-dimensional cutting stock problem, customers have placed orders for q_i rectangular pieces of glass having size i (which has dimensions $a_i \times b_i$) for $i = 1, \ldots, m$. These orders must be filled by

cutting them from standard $a \times b$ sheets. There are many patterns according to which a standard sheet can be cut into one or more of the pieces on order. Let y_j be the number of standard sheets that are cut according to pattern j, and let A_{ij} be the number of pieces of size i the pattern yields from one sheet. Write an integer programming problem that minimizes the number of standard sheets cut while meeting customer demand. Indicate how to solve the problem by column generation by formulating a subproblem to find a pattern with a negative reduced cost. The subproblem should use the diffn global constraint (see Chapter 5). Hint: To allow for multiple pieces of the same size, define several distinct 0-1 variables δ_k for each size i, and let $\delta_k = 1$ if a piece of size i is cut. Then for each i, and k corresponding to size i, there is a term $u_i \delta_k$ in the objective function of the subproblem, where u is the tuple of dual multipliers, and a linear disjunction

$$\left[\begin{array}{c} \delta_k = 1 \\ \Delta x_k = (a_i, b_i) \end{array} \right] \vee \left[\begin{array}{c} \delta_k = 0 \\ \Delta x_k = (0, 0) \end{array} \right]$$

in the constraint set.

2.3 Constraint-Directed Search

An ever-present issue when searching over problem restrictions is the choice of which restrictions to consider and in what order. Branching search addresses this issue in a general way by letting problem difficulty guide the search. If a given restriction is too hard to solve, it is split into problems that are more highly restricted, and otherwise one moves on to the next restriction, thus determining the sequence of restrictions in a recursive fashion.

Another general approach is to create the next restriction on the basis of lessons learned from solving past restrictions. At the very least, one would like to avoid solving restrictions that are no better than past ones, in the sense that they cannot produce solutions any better those already found. This can be accomplished as follows. Whenever a restriction is solved, one can generate a constraint that excludes that restriction, and perhaps others that are no better. Such a constraint is a *nogood*. Then, when the next restriction is selected, it must satisfy the nogoods generated so far. The process continues until no restriction satisfies the nogoods, indicating an exhaustive search. At this point, the best candidate solution found in the process of solving restrictions is optimal in the original problem.

To put this more precisely, the search proceeds by creating a sequence of restrictions $P_1, P_2, \ldots$ of the original problem P. The optimal value v_k of each restriction P_k is computed and a nogood N_k derived. The nogood is designed to exclude P_k and perhaps other restrictions with optimal values no better than v_k. The next restriction P_{k+1} is selected so that it satisfies the set $\mathcal{N}_k$ of nogoods $N_1, \ldots, N_k$ generated so far (the initial nogood set $\mathcal{N}_0$ is empty). The algorithm continues until no restriction satisfies $\mathcal{N}_k$, at which point the optimal value of P is the minimum of $v_1, \ldots, v_k$. If the optimal value is infinite, P is infeasible.

The search is exhaustive because it excludes a restriction only when the restriction is infeasible, or some restriction already solved has a solution that is at least as good. It necessarily terminates if there are a finite number of possible restrictions, because each iteration excludes at least one restriction. If there are infinitely many restrictions, some care must be taken in designing the nogoods to ensure that the search is finite.

This formal process does not indicate in what sense a restriction may be viewed as satisfying or violating a constraint N_k. In practice the restrictions are parameterized by the variables $x = (x_1, \ldots, x_n)$, so that each $x \in D$ determines a restriction $P(x)$ with optimal value $v(x)$. The nogoods are written as constraints on the problem variables x rather than constraints on restrictions. The nogoods exclude restrictions by excluding values of x that give rise to those restrictions. Thus, in iteration k, a solution x^k of the current nogood set $\mathcal{N}_k$ is found and the next restriction is $P(x^k)$.

The most obvious way to define a restriction $P(x^k)$ is simply to fix x to x^k. In this case, a nogood is generated to exclude x^k (and perhaps other solutions that are no better than x^k). The next solution examined must satisfy all the nogoods so far generated, which means that it must differ from all the solutions so far enumerated. The search terminates when the nogoods rule out all possible solutions.

Typically, however, x^k defines a restriction by fixing only some of the variables in x to their values in x^k. That is, $P(x^k)$ is defined by adding to P the constraint $(x_{j_1}, \ldots, x_{j_p}) = (x_{j_1}^k, \ldots, x_{j_p}^k)$, where $p \,(< n)$ and the index set $\{j_1, \ldots, j_p\}$ may vary from one iteration to the next. This occurs, for example, in constraint-directed branching and the various forms of Benders decomposition. The method by which nogoods are constructed is problem-dependent and ideally exploits the problem structure to obtain strong nogoods. A generic constraint-directed search algorithm appears in Figure 2.8.

There is normally a good deal of freedom in how to select a feasible solution x^k of $\mathcal{N}_k$, and a constraint-directed search is partly characterized

Constraint-Directed Search

Let $v_{\text{UB}} = \infty$ and $\mathcal{N} = \emptyset$.
Associate a restriction $P(x)$ of P with each $x \in D$, and let $v(x)$
   be the optimal value of $P(x)$.

While $\mathcal{N}$ is feasible repeat:
   Select a feasible solution $x = s(\mathcal{N})$ of $\mathcal{N}$.
   Compute $v(x)$ by solving $P(x)$ and let $v_{UB} = \min\{v(x), v_{UB}\}$.
   Define a nogood N that excludes x and possibly other solutions x'
       with $v(x') \geq v(x)$.
   Add N to $\mathcal{N}$ and process $\mathcal{N}$.
The optimal value of P is v_{UB}.

Figure 2.8. Generic constraint-directed search algorithm for solving a minimization problem P with variable domain D, where s is the selection function. $\mathcal{N}$ contains the nogoods generated so far.

by its *selection function*; that is, by the way it selects a feasible solution for a given $\mathcal{N}_k$. Certain selection functions can make subsequent $\mathcal{N}_k$s easier to solve—a theme that is illustrated below. In some constraint-directed methods, such as dynamic backtracking methods, it may be necessary to *process* the nogood set $\mathcal{N}_k$ to make it easier to solve. In such cases, the selection function is designed to make processing easier. In partial-order dynamic backtracking, for example, a solution of $\mathcal{N}_k$ is selected to *conform* to previous solutions, so that $\mathcal{N}_k$ can be processed by a fast version of resolution (parallel resolution).

It may not be evident at this point how constraint-directed search reflects the search-infer-and-relax paradigm. There is obviously an enumeration of problem restrictions, but no mention has been made of relaxation. Section 2.3.5 will show, however, that each nogood set $\mathcal{N}_k$ is naturally viewed as a relaxation R_k of the problem. The search process, therefore, solves a series of relaxations whose solutions guide the choice of the next restriction. Unlike branching methods, constraint-directed search requires the solution of every restriction, regardless of the outcome of solving the previous relaxation.

Subsequent sections describe three instances of constraint-directed search: constraint-directed branching, partial-order dynamic backtracking, and logic-based Benders decomposition. Table 2.8 indicates briefly how these methods exemplify the search-infer-and-relax scheme.

2.3.1 Constraint-Directed Branching

Constraint-directed branching stems from the observation that branching on variables is a special case of constraint-directed search. The leaf nodes of the branching tree correspond to problem restrictions. The no-

Table 2.8. How some selected constraint-directed search methods fit into the search-infer-and-relax framework.

Solution method	Restriction P_k	Relaxation R_k	Selection function $s(R_k)$	Inference
DPL for SAT	Created by adding conflict clauses	Processed conflict clauses	Unit clause rule + greedy solution of R_k	Nogood generation + parallel resolution
Partial-order dynamic backtracking	Created by adding nogoods	Processed nogoods	Greedy, but consistent with partial order	Nogood generation + parallel resolution
Logic-based Benders decomposition	Subproblem defined by solution of master	Master problem (Benders cuts)	Optimal solution of master	Derivation of Benders cuts (nogoods)

goods help guide future branching and contain information about why the search backtracked at previous leaf nodes. Well-chosen nogoods can permit the search to prune large portions of the enumeration tree.

Search algorithms of this sort are widely used in artificial intelligence and constraint programming. Nogoods in the form of *conflict clauses* have played a particularly important role in fast algorithms for the propositional satisfiability problem—a key problem for combinatorial optimization.

Branching can be understood as constraint-directed search in the following way. To simplify discussion, suppose that a feasible solution, rather than an optimal solution, is sought. The initial nogood set $\mathcal{N}_0$ is empty. The branching process reaches the first leaf node by fixing some of the variables $(x_1, \ldots, x_p)$ to certain values $(v_1, \ldots, v_p)$. These assignments partially define a feasible solution x^0 for $\mathcal{N}_0$. The remaining variables $x_{p+1}, \ldots, x_n$ can be assigned values from their current domains, as desired. Thus, the branching mechanism helps to define the selection function due to its role in choosing a feasible solution for $\mathcal{N}_0$.

The restriction P_1 corresponding to the first leaf node fixes $(x_1, \ldots, x_p)$ to $(v_1, \ldots, v_p)$. Thus, P_1 contains the original constraints of P plus the branching constraints $x_j = v_j$ that bring the search to the leaf node.

Constraint-Directed Search 67

If the solution $(x_1, \ldots, x_n) = (v_1, \ldots, v_n)$ is feasible in P, the search terminates at the first leaf node. If the search backtracks, it is because fixing $(x_1, \ldots, x_p)$ to $(v_1, \ldots, v_p)$ violates some constraint. In other words, P_1 is infeasible. Typically, only some of the fixed variables $x_1, \ldots, x_p$ are actually responsible for the violations, perhaps $x_{j_1}, \ldots, x_{j_r}$. A nogood

$$(x_{j_1}, \ldots, x_{j_r}) \neq (v_{j_1}, \ldots, v_{j_r})$$

is constructed to prevent assigning the same values to these critical variables again. From here out the search avoids taking branches that are inconsistent with the nogood.

Each subsequent leaf node corresponds to a problem restriction P_k and nogood set $\mathcal{N}_k$, where P_k contains the constraints of P and the branching constraints that define the leaf node. Since the search avoids branches that violate previous nogoods, the fixed variables at the leaf node partially define a feasible solution for $\mathcal{N}_k$. A nogood is generated as before. If the search tree is complete at this point, the accumulated nogoods are jointly unsatisfiable. This means that the next nogood set $\mathcal{N}_{k+1}$ will be infeasible and the search will terminate. Otherwise the search continues to the next leaf node.

As remarked earlier, an appropriate selection function can result in nogoods that make subsequent nogood sets easy to solve. In constraint-directed branching, the branching mechanism defines the selection function in such a way that future branching can easily avoid violating the nogoods. This idea is best explained by example, presented in the next section.

2.3.2 Example: Propositional Satisfiability

Propositional satisfiability is one of the fundamental problems of combinatorial optimization, partly because a wide range of combinatorial problems can be formulated in the language of propositional logic. The currently fastest algorithms for propositional satisfiability use a form of the Davis-Putnam-Loveland (DPL) algorithm with clause learning. This algorithm can be interpreted as constraint-directed branching, and a small example will illustrate the basic idea. The example also prepares the ground for the discussion of partial-order dynamic backtracking in the next section.

The example is artificial but is contrived to show how nogoods in the form of conflict clauses help solve a satisfiability problem. Let's suppose I must hire workers to complete a job and have workers 1, ..., 6 to choose from. Workers 3 and 4 are temporary workers. I am constrained by the following conditions:

(a) I must hire at least one of the workers 1, 5 and 6.
(b) I cannot hire 6 unless I hire 1 or 5.
(c) I cannot hire 5 unless I hire 2 or 6.
(d) If I hire 5 and 6, I definitely must hire 2.
(e) If I hire 1 or 2, then I must hire at least one temporary worker.
(f) I can hire neither 1 nor 2 if I hire any temporary workers.

I wish to know whether it is possible to satisfy these conditions simultaneously.

Formulating the Problem

Let x_j have the value T (for *true*), if I hire worker j, and F (for *false*) otherwise. Thus, x_j can be viewed as a proposition that asserts that I am hiring worker j. Conditions (a)–(f) can be written in logical form by using some standard notation ($\vee$ for *or*, $\wedge$ for *and*, $\neg$ for *not*, and $\rightarrow$ for *implies*).

$$\begin{array}{ll} x_1 \vee x_5 \vee x_6 & (a) \\ x_6 \rightarrow (x_1 \vee x_5) & (b) \\ x_5 \rightarrow (x_2 \vee x_6) & (c) \\ (x_5 \wedge x_6) \rightarrow x_2 & (d) \\ (x_1 \vee x_2) \rightarrow (x_3 \vee x_4) & (e) \\ (x_3 \vee x_4) \rightarrow (\neg x_1 \wedge \neg x_2) & (f) \end{array} \quad (2.37)$$

The conjunction of these formulas is a proposition Q that must be true if I am to meet the conditions.

A logical proposition is *satisfiable* if some assignment of truth values to its variables makes it true. To check the satisfiability of Q, it is convenient to write Q in *conjunctive normal form* (CNF), which is to say as a conjunction of logical clauses. A *clause* is a disjunction of variables or their negations, such as $\neg x_1 \vee x_2$. The disjuncts $\neg x_1$, x_2 are known as *literals*. An implication $x_1 \rightarrow x_2$ can be written $\neg x_1 \vee x_2$ because it is interpreted as a material conditional—that is, it states that either x_2 is true or x_1 is false.

Thus, formula (b) in (2.37) can be written $x_1 \vee x_5 \vee \neg x_6$, and similarly for (c). Formula (d) is equivalent to $\neg(x_5 \wedge x_6) \vee x_2$, which can be written $\neg x_5 \vee \neg x_6 \vee x_2$. Formula (e) is equivalent to the conjunction of two conditionals, $x_1 \rightarrow (x_3 \vee x_4)$ and $x_2 \rightarrow (x_3 \vee x_4)$. Formula (f) is

Constraint-Directed Search

equivalent to four conditionals:

$$x_3 \rightarrow \neg x_1$$
$$x_3 \rightarrow \neg x_2$$
$$x_4 \rightarrow \neg x_1$$
$$x_4 \rightarrow \neg x_2$$

Proposition Q can be put in CNF by writing it as the conjunction of the clauses obtained from the propositions in (2.37).

The satisfiability problem for Q now can be stated

$$\text{Logic:} \begin{cases} x_1 & \vee\ x_5 \vee\ x_6 & (a) \\ x_1 & \vee\ x_5 \vee \neg x_6 & (b) \\ x_2 & \vee \neg x_5 \vee\ x_6 & (c) \\ x_2 & \vee \neg x_5 \vee \neg x_6 & (d) \\ \neg x_1 \vee\ x_3 \vee\ x_4 & & (e1) \\ \neg x_2 \vee\ x_3 \vee\ x_4 & & (e2) \\ \neg x_1 & \vee \neg x_3 & (f1) \\ \neg x_1 & \vee \neg x_4 & (f2) \\ \neg x_2 \vee \neg x_3 & & (f3) \\ \neg x_2 & \vee \neg x_4 & (f4) \end{cases} \quad (2.38)$$

Domains: $x_j \in \{T, F\}$, $j = 1, \ldots, 6$

where *logic* indicates that the constraint is a conjunction of the clauses listed.

Inference: Unit Clause Rule

A simple form of propagation for logical clauses is the *unit clause rule*. The rule says that when all but one of the literals in a clause have been determined to be false, the remaining literal must be true. In other words, the clause has been reduced to a *unit clause* (a clause containing one literal) by fixing all other literals to false. For instance, if $(x_1, x_2) = (T, F)$, then the third literal in $\neg x_1 \vee x_2 \vee \neg x_3$ must be true, which fixes $x_3 = F$. Thus, the domain of x_3 is reduced to $\{F\}$.

The unit clause rule is applied repeatedly until the domains can be reduced no further. This process need not detect all variables that can be fixed, as in the case of clauses $x_1 \vee x_2$, $x_1 \vee \neg x_2$. Variable x_1 must be true, but the unit clause rule fixes no variables.

Search

Figure 2.9 depicts a branching tree for problem (2.38). The tree will first be described as a branching search and subsequently reinterpreted as a constraint-directed search.

The branching order is $x_1, \ldots, x_6$, and the left branch (F) is taken first. The unit clause rule is applied at each node. The resulting search scheme is essentially the DPL method with conflict clauses.

At Node 5, the unit clause rule uses clauses (a) and (b) to remove both T and F from x_6's domain, resulting in an empty domain and infeasibility. Since the branching assignments $x_1 = F$ and $x_5 = F$ are enough to result in this empty domain, either x_1 or x_5 must be true in any feasible solution. So, a nogood $(x_1, x_5) \neq (F, F)$ is created. It can be written as the *conflict clause* $x_1 \vee x_5$.

Due to the empty domain at Node 5, the search backtracks to Node 4 and takes the branch to Node 6, where conflict clause $x_2 \vee \neg x_5$ is generated. At this point, the search backtracks again to Node 4 and notes that the entire subtree of Node 4 is infeasible. Since setting $(x_1, x_2) = (F, F)$ is enough to create this infeasibility, a second conflict clause $x_1 \vee x_2$ is associated with Node 6. The search must now backtrack all the way to

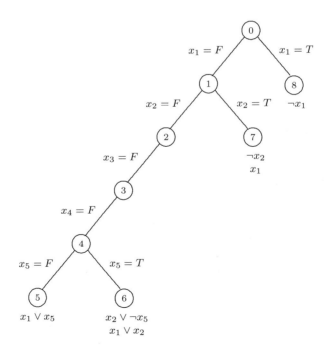

Figure 2.9. Branching tree for a propositional satisfiability problem. Conflict clauses are shown below the nodes with which they are associated.

Constraint-Directed Search 71

Node 1 in order to find a right branch that is consistent with this clause. As a result, a large portion of the search tree is bypassed. This sort of move is sometimes called *backjumping*.

At Node 7, the unit clause rule uses (f3), (f4) and (e2) to derive a contradiction. The conflict clause $\neg x_2$ is created, and the search backtracks to Node 1, where a second conflict clause x_1 is created. Finally, at Node 8, the unit clause rules uses (f1), (f2) and (e1) to derive a contradiction. This creates the conflict clause $\neg x_1$, and the search terminates without finding a feasible solution. It must be concluded that conditions (a)-(f) cannot be simultaneously satisfied.

When the search is interpreted as a constraint-directed search, the leaf nodes correspond to nogood sets $\mathcal{N}_k$ as shown in Table 2.9. $\mathcal{N}_k$ is solved by assigning variables the values they receive in the branching tree along the path from the root to the leaf node. (Assignments to any remaining variables can be made arbitrarily.) Thus, the branching pattern defines the selection function. The problem restriction P_k consists of the original clauses and the branching constraints at the leaf node. The nogoods are the conflict clauses generated at each leaf node in the branching search.

The initial nogood set $\mathcal{N}_0$ is empty. It is solved by setting $x_1, \ldots, x_5$ to F, as in the branching tree, and x_6 to either value. P_1 contains the original clauses and the constraint $(x_1, \ldots, x_5) = (F, F, F, F, F)$. P_1 is infeasible because these assignments already falsify a clause. The conflict clause $x_1 \vee x_5$ mentioned earlier is generated as a nogood. A solution for $\mathcal{N}_1$ is similarly obtained, and the nogood $x_2 \vee \neg x_5$ generated.

The third nogood $x_1 \vee x_2$ is generated in the branching scheme by backtracking to Node 4 and noting that the subtree at that node is infeasible. In constraint-directed search, however, $x_1 \vee x_2$ is generated

Table 2.9. Interpretation of the Davis-Putnam-Loveland procedure with conflict clauses as constraint-directed branching. Each problem restriction P_k consists of the original problem and all previously generated nogoods. The symbol $*$ indicates a value that can be either T or F.

k	Node	Relaxation R_k	Solution x^k of R_k	Nogoods generated
0	5		$(F, F, F, F, F, *)$	$x_1 \vee x_5$
1	6	$x_1 \vee x_5$	$(F, F, F, F, T, *)$	$x_2 \vee \neg x_5$
2	7	$x_1 \vee x_2$	$(F, T, *, *, *, *)$	$\neg x_2$
3	8	x_1	$(T, *, *, *, *, *)$	$\neg x_1$
4	—	$\emptyset$	R_4 is infeasible	

by *processing* the nogoods; that is, by inferring it from the nogoods already generated. It is not hard to see that $x_1 \vee x_5$ and $x_2 \vee \neg x_5$ imply $x_1 \vee x_2$. Formally, this can be obtained by *resolving* $x_1 \vee x_5$ and $x_2 \vee \neg x_5$. In general, two clauses can be resolved when exactly one variable (in this case, x_5) switches sign between them. The resolvent is obtained by removing the variable that switches sign and retaining all the other literals. Once the nogood $x_1 \vee x_2$ is obtained by resolution, the two nogoods $x_1 \vee x_5$ and $x_2 \vee \neg x_5$ become redundant, in a sense that will be clarified in the next section, and are dropped.

In similar fashion, the nogood set $\mathcal{N}_3$ is obtained by resolving $x_1 \vee x_2$ and $\neg x_2$ to obtain x_1, and dropping $x_1 \vee x_2$. Nogood set $\mathcal{N}_4$ consists of the resolvent of x_1 and $\neg x_1$, which is the necessarily false empty clause $\emptyset$. Since $\mathcal{N}_4$ is infeasible, the search terminates at this point without having found a solution.

A key property of constraint-directed branching is that the selection function and nogood processing work together to make the nogood sets easy to solve. The selection function simply mimics the branching process. It assigns each variable its current value in the branching tree. When the nogoods are processed as described above, this assignment always solves $\mathcal{N}_k$ (unless it is infeasible). This idea is generalized in partial-order dynamic backtracking, to be discussed next.

2.3.3 Partial-Order Dynamic Backtracking

A slight change in the selection function and nogood processing converts constraint-directed branching to a more general search procedure: partial-order dynamic backtracking.

In the chronological backtracking algorithm of the previous section, the order in which variables are instantiated and uninstantiated is fixed by the branching order. In partial-order dynamic backtracking, the order of instantiation can change dynamically in the course of the algorithm. In fact, the instantiation order is determined only in part by a constantly changing partial-order on the variables, and in part by choices made dynamically by the user. The algorithm, therefore, allows more freedom in the search than branching does, without sacrificing completeness.

When a nogood (conflict clause) is generated, the user selects one of the variables in the clause to be *last*. The remaining variables in the clause are *penultimate*. The last variables in the current nogood set define a partial order in which each penultimate variable in a nogood precedes the last variable in that nogood. The last variable in a new nogood must be selected so as to be consistent with the partial order defined by the existing nogoods. In constraint-directed branching, the last variable is always the one on which the search last branched.

Constraint-Directed Search

When conflict clauses are resolved, the eliminated variable must be the last variable in both clauses (*parallel* resolution). Note that the resolvents generated in constraint-directed branching are always parallel resolvents. A conflict clause is redundant and can be eliminated when one of its penultimate literals is last in another clause.

Since there is no fully defined branching order, the selection function cannot mimic the branching order as in constraint-directed branching. Rather, the solution of the current relaxation must *conform to* the nogoods in the current nogood set. This means that whenever a variable occurs penultimately in a nogood, it must be assigned a value opposite to the sign in which it occurs. Thus, a penultimate variable is set to false if it occurs positively, and to true if it is negated. This criterion is well defined, because it can be shown that a variable will have the same sign in all its penultimate occurrences. If a variable does not occur penultimately in any nogood, it can be set to any value that violates none of the nogoods.

The nogood set can always be solved in this fashion without backtracking, unless of course the nogoods are infeasible, in which case the search is over. Note that the selection function used in constraint-directed branching observes this same rule of conformity.

A partial-order dynamic backtracking algorithm for feasibility problems is stated in Figure 2.10. In this algorithm, $\leq_\mathcal{N}$ denotes the partial order that is defined by the penultimate and last variables in each clause.

Let $\mathcal{N} = \emptyset$.
For a given nogood N let x_N be the last variable in N.
For a given nogood set $\mathcal{N}$ let the partial order $\leq_\mathcal{N}$ be the transitive closure of the relations $x_j \leq_\mathcal{N} x_N$ for all penultimate variables x_j in N and every $N \in \mathcal{N}$.

While $\mathcal{N}$ is feasible repeat:
    Select a feasible solution x of $\mathcal{N}$ that conforms to the nogoods in $\mathcal{N}$.
    If x is feasible in P then stop.
    Else
        Define a nogood N that excludes x and possibly other solutions that are infeasible in P.
        Select a variable x_N in N to be last in N, so that $x_N \not\leq_\mathcal{N} x_j$ for all x_j that are penultimate in N.
        Add N to $\mathcal{N}$ and process $\mathcal{N}$ with parallel resolution.
P is infeasible.

Figure 2.10. *Partial-order dynamic backtracking algorithm for a feasibility problem P with boolean variables.*

2.3.4 Example: Propositional Satisfiability

The satisfiability instance (3.47) solved earlier is convenient for illustrating partial-order dynamic backtracking. Table 2.10 summarizes the procedure. Initially, the conformity principle imposes no restriction since the nogood set is empty. For purposes of this illustration, the nogood sets are solved by setting variables unaffected by the conformity principle to false, if possible, or to true, if necessary, to avoid violating a nogood. The unit clause rule is applied after each setting.

For $k = 0$ in Table 2.10, the variables are assigned values in the order in which they are indexed, but any order would be acceptable. When x_5 is reached, a clause in the original constraint set is violated, and the nogood $x_1 \vee x_5$ is generated. The variable x_1 is arbitrarily selected as last, as is indicated by writing the nogood as $x_5 \vee x_1$.

At this point, x_5 occurs positively as a penultimate variable, and it must therefore be set to false in the solution of $\mathcal{N}_1$. Variable x_1 is arbitrarily assigned next, and it must be assigned true to avoid violating the nogood $x_5 \vee x_1$. At this point a clause in the original constraint set is already violated. The restriction P_2, which contains the original clauses and $(x_1, x_5) = (T, F)$, is infeasible, and the nogood $x_5 \vee \neg x_1$ is generated (one could generate the stronger nogood $\neg x_1$, since it alone creates an infeasibility). Variable x_1 must be selected as last, since it occurs after x_5 in the partial order defined by the one existing nogood. Now the two nogoods $x_5 \vee x_1$ and $x_5 \vee \neg x_1$ can be parallel-resolved, resulting in the new nogood x_5, whose only variable is necessarily chosen as last. The other two nogoods are now redundant and are dropped, since the last literal of the clause x_5 occurs penultimately in both.

Table 2.10. Partial-order dynamic backtracking solution of a propositional satisfiability problem. The symbol ∗ indicates a value that can be either T or F. The "last" variable in each nogood is written last.

k	Nogood set $\mathcal{N}_k$	Solution x^k of $\mathcal{N}_k$	Nogoods generated
0		$(F, F, F, F, F, *)$	$x_5 \vee x_1$
1	$x_5 \vee x_1$	$(T, *, *, *, F, *)$	$x_5 \vee \neg x_1$
2	x_5	$(*, F, *, *, T, *)$	$\neg x_5 \vee x_2$
3	$\begin{cases} x_5 \\ \neg x_5 \vee x_2 \end{cases}$	$(*, T, *, *, T, *)$	$\neg x_2$
4	$\emptyset$	$\mathcal{N}_4$ is infeasible	

Nogood set $\mathcal{N}_2$ can be solved without regard to conformity, since no variables occur penultimately in it. Variable x_5 must be set to true, and variable x_2 is arbitrarily set to false next. This already violates a constraint and yields the nogood $\neg x_5 \vee x_2$, in which x_2 is arbitrarily chosen to be last. The current nogoods x_5 and $\neg x_5 \vee x_2$ have the resolvent x_2, but they do not have a parallel resolvent since x_5 does not occur last in both clauses. Both clauses are therefore retained in $\mathcal{N}_3$. When nogood x_2 is generated by the solution of $\mathcal{N}_3$, two steps of parallel resolution yield the empty clause, and the search terminates without finding a feasible solution.

2.3.5 Relaxation in Constraint-Directed Search

Constraint-directed search has so far been characterized as an enumeration of problem restrictions, but relaxations play a central role as well. In fact the nogood set $\mathcal{N}_k$ can be reinterpreted as a problem relaxation, and nogoods R_k if the nogoods are written to impose bounds on the optimal value, rather than excluding unwanted solutions. The bounds, in effect, state that the excluded solutions cannot yield an optimal value better than the current solution. So, to achieve a better value, one must choose a solution that is allowed by the nogoods.

Recall that in step k of the search, a feasible solution x^k of the nogood set $\mathcal{N}_k$ is found, and the next restriction $P_{k+1} = P(x^k)$ is solved for its optimal value $v(x^k)$. Suppose that the resulting nogood N_{k+1} excludes the solutions in set T, so that $v(x) \geq v(x^k)$ for all $x \in T$. If v represents the optimal value of the original problem, the nogood can be written as a *nogood bound*

$$v \geq B_{k+1}(x)$$

where

$$B_{k+1}(x) = \begin{cases} v(x^k) & \text{if } x \in T \\ -\infty & \text{otherwise} \end{cases}$$

The nogood bound says that any $x \in T$ must result in a restriction $P(x)$ with optimal value at least $v(x^k)$. Thus, in order to do better than $v(x^k)$, one must avoid solutions in T.

A pair (v, x) satisfies the nogood bounds accumulated so far if and only if it is feasible in the following optimization problem R_k:

$$\begin{aligned} \min \ & v \\ & v \geq B_i(x), \quad i = 1, \ldots, k \\ & x \in D \end{aligned} \quad (2.39)$$

It is straightforward to show that (2.39) is a relaxation of P (and therefore each P_i) in the sense defined earlier. In fact, $B_i(x)$ need not go to

$-\infty$ for $x \notin T$. It is enough that $B_i(x)$ provide a valid lower bound on the optimal value of $P(x)$.

THEOREM 2.1 *Let $v(x)$ be the optimal value of restriction $P(x)$ for all $x \in D$. Then (2.39) is a relaxation of (2.1) if $B_i(x) \leq v(x)$ for all $x \in D$ and $i = 1, \ldots, k$.*

In some constraint-directed methods, such as Benders decomposition, the bound $B_i(x)$ remains fairly tight when x is in the vicinity of solutions in T.

Rather than find a solution x^k that satisfies the nogood set $\mathcal{N}_k$, the search algorithm now finds a solution (v_k, x^k) of R_k for which v_k is less than the value of the best candidate solution found so far. That is,

$$v_k < \min \left\{ v(x^1), \ldots, v(x^{k-1}) \right\}$$

It is therefore not necessary to solve R_k to optimality—only to find a sufficiently good solution. If no such solution exists, the search terminates. A generic algorithm appears in Figure 2.11.

Solving R_k to optimality can be useful, however, because its optimal value provides a lower bound on the optimal value of the original problem. The best candidate solution provides an upper bound, which means that one can halt the algorithm at any point and bracket the optimal value between two bounds. The search terminates when the two bounds converge.

2.3.6 Logic-Based Benders Decomposition

Benders decomposition is a constraint-directed search in which all the problem restrictions are defined by fixing the same subset of variables.

Let $v_{\text{UB}} = \infty$. Initially the relaxation R minimizes v subject to $x \in D$.
Associate a restriction $P(x)$ of P with each $x \in D$, and let $v(x)$ be the
   optimal value of $P(x)$.

While R has a feasible solution $(v, \bar{x})$ with $v < v_{\text{UB}}$ repeat:
   Compute $v(\bar{x})$ by solving $P(\bar{x})$.
   Let $v_{UB} = \min\{v(\bar{x}), v_{UB}\}$.
   Add to R a nogood bound $v \geq B(x)$ for which $B(\bar{x}) = v(\bar{x})$
      and $B(x) \leq v(x)$ for all $x \in D$.
The optimal value of P is v_{UB}.

Figure 2.11. Generic constraint-directed search algorithm for solving a minimization problem P with variable domain D.

That is, when a solution x^k of the current relaxation is found, the next restriction is defined by fixing the variables $(x_1, \ldots, x_p)$ to the solution values $(x_1^k, \ldots, x_p^k)$ in each iteration, where the same variables $x_1, \ldots, x_p$ are fixed in each iteration. They might be called the *search variables*, because the search procedure in effect enumerates values of these variables. In a Benders context, the nogood bounds are known as *Benders cuts*, the relaxation R_k as the *master problem*, and the restriction $P(x^k)$ as the *subproblem* or *slave problem*.

The primary rationale for Benders decomposition is that the problem may have special structure that allows it to simplify considerably when certain variables are fixed to any value. The choice of search variables is therefore crucial to the success of the method, as is the ability to formulate strong Benders cuts.

Classical Benders decomposition is defined for the case in which the subproblem is a continuous linear or nonlinear programming problem. The Benders cuts are obtained from Lagrange multipliers associated with the constraints of the subproblem. The root idea of the Benders method, however, can be generalized to great advantage. The Benders cuts can be obtained from a logical analysis of the subproblem, resulting in a *logic-based* Benders method. This, in principle, allows the subproblem to take any form, but a separate analysis must be conducted for each class of subproblems.

To prepare a problem for solution by a Benders method, the variables are decomposed into a vector x of search variables and a vector y of subproblem variables. The problem P can now be written

$$\min \ f(x,y)$$
$$S(x,y)$$
$$x \in D_x, \ y \in D_y$$

where $S(x,y)$ is a constraint set that contains variables x, y.

In step k of the algorithm, an optimal solution x^k is computed for relaxation R_k. The problem restriction (subproblem) $P(x^k)$ is obtained by fixing x to x^k in P. $P(x^k)$ is therefore

$$\min \ f(x^k, y)$$
$$S(x^k, y)$$
$$y \in D_y$$

where $S(x^k, y)$ is the constraint set that remains when x is fixed to x^k in $S(x,y)$. The resulting Benders cut $v \geq B_{k+1}(x)$ involves only the variables x_j, since only they are relevant to defining $P(x^k)$. The cut is added to R_k to obtain the next master problem.

Thus the master problem R_k at step k is

$$\min v$$
$$v \geq B_i(x), \quad i = 1, \ldots, k$$
$$x \in D_x$$

The algorithm appears in Figure 2.12. In practice, the first relaxation R_0 may be augmented with precomputed nogoods for a "warm start." R_0 may also contain constraints from problem P that involve only the search variables, as well as other constraints that involve only x and are valid for P.

2.3.7 Example: Machine Scheduling

A simple machine scheduling problem illustrates logic-based Benders decomposition. It is a small representative of an important class of planning and scheduling problems that frequently occur in manufacturing and supply chain management. The goal in these problems is to assign tasks to facilities and then schedule the tasks. The facilities might be factories, machines in a factory, transport modes, delivery vehicles, or computer processors.

In the problem instance at hand, five jobs are to be allocated to two machines, named A and B, and scheduled on them. Each job j has a release time r_j and a deadline d_j. The time required to process job j on machine i is p_{ij}. The specific problem data appear in Table 2.11. Note that machine A is faster than machine B. The objective is to minimize makespan; that is, to minimize the finish time of the last job to finish.

Let $v_{\text{LB}} = -\infty$ and $v_{\text{UB}} = \infty$.
Initially the relaxation R minimizes v subject to $x \in D_x$
   and possibly other valid constraints involving x.
Let $\bar{x}$ be a feasible solution of R.

While $v_{\text{LB}} < v_{\text{UB}}$ repeat:
   Let $v(\bar{x})$ be the minimum value of $f(\bar{x}, y)$ subject to $S(\bar{x}, y)$ and $y \in D_y$.
   Let $v_{UB} = \min\{v(\bar{x}), v_{UB}\}$.
   Add to R a Benders cut $v \geq B(x)$.
   Compute the optimal value v_{LB} of R.
The optimal value of P is v_{UB}.

Figure 2.12. Generic logic-based Benders algorithm for minimizing $f(x, y)$ subject to $S(x, y)$ and $(x, y) \in D_x \times D_y$.

Table 2.11. Data for a machine scheduling problem.

Job j	Release time r_j	Dead-line d_j	Processing time	
			p_{Aj}	p_{Bj}
1	0	10	1	5
2	0	10	3	6
3	2	7	3	7
4	2	10	4	6
5	4	7	2	5

Formulating the Problem

It is convenient to use a metaconstraint *disjunctive* to represent the scheduling portion of the problem. The constraint may, in general, be written

$$\text{disjunctive}(s \,|\, p)$$

where $s = (s_1, \ldots, s_n)$ are the start times of the jobs to be scheduled, and $p = (p_1, \ldots, p_n)$ are the processing times. The constraint is satisfied when the jobs do not overlap. That is,

$$s_j + p_j \leq s_k \text{ or } s_k + p_k \leq s_j, \text{ all jobs } j, k \text{ with } j \neq k$$

The name *disjunctive* derives from the fact that the task of scheduling jobs sequentially is commonly known as disjunctive scheduling, as opposed to cumulative scheduling, in which several jobs can run simultaneously subject to resource constraints.

Only two types of decision variables are needed to formulate the problem—the start time s_j already mentioned, and the machine x_j to which job j is assigned.

If there are n jobs and m machines, the formulation is

Linear: $\begin{cases} \min M \\ M \geq s_j + p_{x_j j}, \text{ all } j \\ s_j + p_{x_j j} \leq d_j, \text{ all } j \end{cases}$

Disjunctive: $((s_j \,|\, x_j = i) \,|\, (p_{ij} \,|\, x_j = i)),$ all i

Domains: $s_j \in [r_j, \infty),\ x_j \in \{1, \ldots, m\},$ all j

In the objective function, M represents the makespan. The linear constraints define the makespan and enforce the deadlines. The release times are observed in the domain constraints. A disjunctive scheduling constraint is imposed for each machine. In the disjunctive constraints, the notation $(s_j \mid x_j = i)$ denotes the tuple of start times s_j for jobs assigned to machine i, and similarly for the processing times.

A natural decomposition for this problem distinguishes the assignment portion from the scheduling portion. One can search over various assignments of jobs to machines and, for each, try to find a feasible schedule for the jobs assigned to each machine. The assignment variables x_j are therefore the search variables, and each subproblem P_k is a scheduling problem that decouples into separate scheduling problems for the individual machines.

Relaxation: The Master Problem

The master problem (relaxation) R_k minimizes makespan v subject to the Benders cuts generated so far. It can be solved by whatever method is most suitable for its structure. One option is to solve it as an MILP problem, since it is naturally expressed in this form. For this purpose, the variables x_i can be replaced with 0-1 variables x_{ij}, where $x_{ij} = 1$ when job i is assigned to machine j.

The master problem can be strengthened by adding valid constraints. One can observe, for example, that the jobs assigned to a machine must fit within the earliest release time and latest deadline of those jobs. In fact, this is true of any subset of the jobs assigned to a given machine. To formulate this condition, let $J(t_1, t_2)$ be the set of jobs whose time windows lie in the interval $[t_1, t_2]$. So $J(t_1, t_2) = \{j \mid [r_j, d_j] \subset [t_1, t_2]\}$. The total processing times of the jobs in $J_i(t_1, t_2)$ that are assigned to a given machine i must not exceed $t_2 - t_1$:

$$\sum_{j \in J(t_1,t_2)} p_{ij} x_{ij} \leq t_2 - t_1 \tag{2.40}$$

It suffices to consider release times for t_1 and deadlines for t_2. In the problem instance at hand,

$$J(0,7) = \{3,5\} \qquad J(2,10) = \{3,4,5\}$$
$$J(0,10) = \{1,2,3,4,5\} \quad J(4,7) = \{5\}$$
$$J(2,7) = \{3,5\} \qquad J(4,10) = \{5\}$$

Some of these sets give rise to vacuous or redundant inequalities (2.40). For instance, the inequality for $J(0,7)$ is $p_{i3}x_{i3} + p_{i5}x_{i5} \leq 7$, which is $3x_{A3} + 2x_{A5} \leq 7$ for $i = A$ and $7x_{B3} + 5x_{B5} \leq 7$ for $i = B$. The former is obviously redundant since $x_{ij} \in \{0,1\}$. The latter is dominated by another inequality for Machine B (namely, $7x_{B3} + 5x_{B5} \leq 5$). The nonredundant inequalities for Machine A are

$$J(0,10): \sum_{j\in\{1,2,3,4,5\}} p_{Aj} x_{Aj} \leq 10$$

$$J(2,10): \sum_{j\in\{3,4,5\}} p_{Aj} x_{Aj} \leq 8 \qquad (2.41)$$

and those for Machine B are

$$J(0,10): \sum_{j\in\{1,2,3,4,5\}} p_{Bj} x_{Bj} \leq 10$$

$$J(2,7): \sum_{j\in\{3,5\}} p_{Bj} x_{Bj} \leq 5$$

$$J(2,10): \sum_{j\in\{3,4,5\}} p_{Bj} x_{Bj} \leq 8 \qquad (2.42)$$

$$J(4,7) \quad \sum_{j\in\{5\}} p_{Bj} x_{Bj} \leq 3$$

Section 4.16.3 shows how to identify nonredundant inequalities of this sort in a systematic way.

Further inequalities can be added to the master problem to constrain the makespan v:

$$v \geq \sum_{j\in\{1,2,3,4,5\}} p_{ij} x_{ij}, \quad i = A, B$$

$$v \geq 2 + \sum_{j\in\{3,4,5\}} p_{ij} x_{ij}, \quad i = A, B \qquad (2.43)$$

$$v \geq 4 + \sum_{j\in\{5\}} p_{ij} x_{ij}, \quad i = A, B$$

The three sets of inequalities correspond to the release times 0, 2, and 4. The first includes the jobs in $J(0,\infty)$, the second the jobs in $J(2,\infty)$, and the third the jobs in $J(4,\infty)$.

The master problem is now

$$\min v$$
$$\text{inequalities (2.41)–(2.43)}$$
$$\text{Benders cuts}$$
$$x_{ij} \in \{0,1\}, \text{ all } i,j$$

An optimal solution of the initial master problem (without Benders cuts) sets $x_{A1} = x_{A2} = x_{A3} = x_{A5} = x_{B4} = 1$, with all other $x_{ij} = 0$. This assigns Jobs 1, 2, 3 and 5 to the faster Machine A and Job 4 to Machine B. The objective function value is 9. It will be seen shortly, however, that 9 is not a feasible makespan.

Inference: Benders Cuts

The inference stage consists of inferring Benders cuts from the subproblem that results when the master problem variables are fixed to their current values.

The subproblem separates into an independent scheduling problem on each machine. Thus, if $\bar{x}$ is the solution of the previous master problem, the subproblem on each machine i is

$$\min M_i$$
$$s_j + p_{ij} \leq d_j, \text{ all } j \text{ with } \bar{x}_{ij} = 1$$
$$\text{disjunctive}((s_j \mid \bar{x}_{ij} = 1) \mid (p_{ij} \mid \bar{x}_{ij} = 1))$$
$$s_j \in [r_j, \infty), \text{ all } j$$

If M_i^* is the optimal makespan on machine i for each i, then the optimal makespan overall is $\max_i\{M_i^*\}$.

The subproblem does not separate in this way if there are precedence constraints between jobs, because the time at which a job can be scheduled may depend on the times at which jobs are scheduled on other machines. Yet even when there are precedence constraints, separability can be preserved if they involve only jobs that must be scheduled on the same machine. Thus, if jobs j and k must be scheduled on the same machine, and j must precede k, one can add constraint $x_j = x_k$ to the master problem and the constraint $s_j + d_{ij} \leq s_k$ to the scheduling problem on every machine.

As noted above, the initial solution of the master problem assigns Job 4 to Machine B and the other jobs to Machine A. The minimum makespan schedule on Machine B simply starts Job 4 at its release time, resulting in a makespan of 8. Thus, whenever Job 4 is assigned to Machine B (i.e., whenever $x_{4B} = 1$), the makespan will be at least 8. This gives rise to the Benders cut

$$v \geq 8x_{B4}$$

Any solution that achieves a makespan better than 8 must avoid assigning Job 4 to Machine B.

The minimum makespan schedule for Jobs 1, 2, 3 and 5 on Machine A appears in Figure 2.13. The resulting makespan is 10, which produces the Benders cut

$$v \geq 10(x_{A1} + x_{A2} + x_{A3} + x_{A5} - 3) \tag{2.44}$$

Constraint-Directed Search

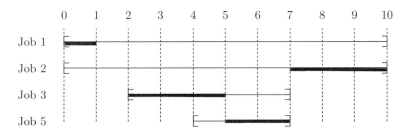

Figure 2.13. A minimum makespan schedule on Machine A for Jobs 1, 2, 3 and 5 of Table 2.11. The horizontal lines represent time windows.

Thus, any solution that obtains a makespan better than 10 must avoid assigning at least one of these jobs to machine A.

The master problem is now

$$\min v$$
$$\text{inequalities } (2.41)–(2.43)$$
$$v \geq 10$$
$$v \geq 10(x_{A1} + x_{A2} + x_{A3} + x_{A5} - 3)$$
$$x_{ij} \in \{0, 1\}, \text{ all } i, j$$

Solution of this problem results in the same machine assignments as before, but the optimal value is now 10. Since this is a lower bound on the minimum makespan, and a feasible solution with makespan 10 was found by solving the subproblem, the algorithm terminates. The schedule found by solving the subproblem is optimal.

In practice, the success of a Benders method often rests on finding strong Benders cuts that rule out as many infeasible solutions as possible. One way to deduce stronger cuts is to examine more closely the reasoning process that proves infeasibility in the subproblem. In the present case, the initial infeasibility on Machine A can be proved using a well-known *edge-finding* technique, and it provides the basis for a stronger cut.

Edge finding is a procedure for identifying jobs that must precede, or that must follow, a set J of other jobs. The basic idea is that if there is too little time to complete both job k and the jobs in J when k starts no earlier than the jobs in J start, then k must finish before the jobs in J start. To state this more precisely, let E_j and L_j be the earliest and latest possible start times for job j, which is to say that the current domain of start time t_j is the interval $[E_j, L_j - p_{ij}]$. If

$$\max_{j \in J \cup \{k\}} \{L_j\} - \min_{j \in J} \{E_j\} < \sum_{j \in J \cup \{k\}} p_{ij} \qquad (2.45)$$

then job k must finish before any job in J starts. A similar rule can be used to establish that job k cannot start until all the jobs in J have finished. This process is called *edge finding* because, by establishing precedence relations between the jobs, it adds edges to a directed graph that represents these relations.

Edge finding may permit one to reduce some of the domains. Initially, the domain of t_j is the interval $[r_j, d_j - p_{ij}]$, but as edge finding proceeds the current domain $[E_j, L_j - p_{ij}]$ may become a smaller interval. If a domain is eventually reduced to the empty set, then there is no feasible schedule.

In particular, if edge finding determines that job k must precede the jobs in J, then for any subset J' of J, k must finish by $L'_k = \max_{j \in J'}\{L_j\} - \sum_{j \in J'} p_{ij}$. This means the latest finish time L_k can be tightened to L'_k if this number is smaller than E_j. Similar reasoning can be used when job k must follow the jobs in J.

The key to forming a strong Benders cut is to keep track of which jobs are actually involved in the domain reduction operations that lead to an empty domain. These jobs alone are sufficient to create infeasibility. A Benders cut that involves fewer jobs is a stronger cut.

To keep track of the relevant jobs, one can associate with each latest finish time L_j the set J_j^L of jobs that helped to tighten that bound, and a similar set J_j^E for an earliest start time. Initially, $J_j^L = J_j^E = \{j\}$. If edge finding determines that job k precedes the jobs in J and tightens L_k as a result, then the jobs in J_j^L are added to J_k^L. If E_k is tightened, the jobs in J_j^E are added to J_k^E. When the edge finding process is finished, and the domain of some t_j is found to be empty, one can conclude that the jobs in $J_j^L \cup J_j^E$ are sufficient to create the infeasibility.

In the example, edge finding deduces that Jobs 2, 3, and 5 are sufficient to create infeasibility on Machine A, whether or not Job 1 is assigned to the machine. The reasoning can be traced as follows. Since a feasible schedule with makespan 10 exists (Figure 2.13), one can show that 10 is the optimal makespan by proving the infeasibility of a makespan of 9. Imposing a maximum makespan of 9 reduces the deadlines of Jobs 1 and 2 to 9. The initial time windows for jobs 2, 3, and 5 are therefore

$$[E_2, L_2] = [0, 9] \qquad [E_3, L_3] = [2, 7] \qquad [E_5, L_5] = [4, 7]$$

and the domains of t_2, t_3 and t_5 are

$$[E_2, L_2 - p_{A2}] = [0, 6] \qquad [E_3, L_3 - p_{A3}] = [2, 4] \qquad [E_5, L_5 - p_{A5}] = [4, 5]$$

Edge finding deduces that Job 2 must precede Jobs 3 and 5, as can be seen from (2.45):

$$\max\{L_2, L_3, L_5\} - \min\{E_3, E_5\} < p_{A2} + p_{A3} + p_{A5}$$

Figure 2.14 illustrates the situation. At this point the latest finish time L_2 for job 2 can be tightened to $\max\{L_3, L_5\} - (p_{13} + p_{15}) = 2$. Since Jobs 3 and 5 brought this about, they are added to J_2^L, which becomes $\{2, 3, 5\}$. But now the domain of t_2 is the interval $[0, -1]$, which is the empty set. This proves infeasibility, and the jobs involved in the proof are those in $J_2^L \cup J_2^E = \{2, 3, 5\}$. Because Job 1 was not involved in the proof, one can deduce a stronger Benders cut than (2.44):

$$v \geq 10(x_{A2} + x_{A3} + x_{A5} - 2) \qquad (2.46)$$

The stronger Benders cut would not have accelerated the solution of this particular problem, but strong cuts can be very helpful in general.

In many cases, edge finding must be combined with other procedures, such as branching, to prove infeasibility. But the above ideas can be extended to such cases. For instance, if edge finding is combined with branching, there is a proof of infeasibility at every leaf node ℓ of the search tree. Along the path from node ℓ to the root there is a series of domain reductions that contribute to this proof, and one can note the set J_ℓ of jobs that contribute to these reductions. Only the jobs in $\bigcup_\ell J_\ell$ need be included in the Benders cut, where the union is taken over all leaf nodes.

One difficulty with a cut of the form (2.46) is that it imposes no bound on v when a proper subset of Jobs 2, 3, and 5 is assigned to Machine A. One should be able to say something about the resulting minimum makespan if only Jobs 2 and 3 are assigned to this machine, for example. In fact, one can often derive a cut that remains useful in such cases. Section 3.13.3 shows how to do this when all the release times are equal.

Actually it is not necessary to re-solve the master problem each time a Benders cut is generated. If the master is solved by branching, one can suspend the branching as soon as a feasible solution is discovered

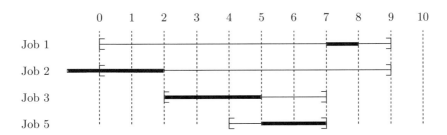

Figure 2.14. Edge-finding proof of infeasibility of scheduling Jobs 1, 2, 3 and 5 on Machine A to achieve a makespan of 9.

and use that solution to define a subproblem. The resulting Benders cut is added to the master problem, and the search continues until another feasible solution is discovered. The search terminates when the search tree is exhaustive and its optimal solution has the same value as the best subproblem solution. This process has been called *branch and check*.

2.3.8 Exercises

1 A group of medications are commonly used to treat a form of cancer, but they can be taken only in certain combinations. A patient who takes Medications 1 and 2 must take Medication 5 as well. Medication 1 can be taken if and only 5 is not taken. At least one of Medications 3, 4, and 5 must be taken. If 5 is taken, then 3 or 4 must be taken. If 4 is taken, then 3 or 5 must be taken. Medication 3 must be taken if both 4 and 5 are taken. Medication 3 cannot be taken without 4, and 5 cannot be taken without 1. Let x_j be true when medication j is taken, and write these conditions in propositional form. Convert them to CNF without adding variables.

2 Find a feasible solution of the CNF expression in Exercise 1 using a DPL algorithm with clause learning. Branch on variables in the order $x_1, \ldots, x_5$, and take the false branch first.

3 Interpret the branching search of Exercise 2 as constraint-directed search by writing a table similar to Table 2.9.

4 Find a feasible solution of the problem in Exercises 1, 2, and 3 by partial-order dynamic backtracking. Experiment with various choices of the last literal in a nogood, and with various heuristics for solving the problem restriction.

5 Find an optimal solution of Exercise 4 using partial-order dynamic backtracking, where the objective is to minimize the number of medications taken. Solve the current nogood set by setting a variable to false whenever possible. When a feasible solution is found, generate a nogood that rules it out, and continue the search. Thus, if the solution $x = (T, F, T, T, F)$ is found, generate the nogood $\neg x_1 \vee x_2 \vee \neg x_3 \vee \neg x_4 \vee x_5$. Continue until the search is exhaustive, and the optimal solution is the best feasible solution found.

6 Prove Theorem 2.1.

7 Interpret the solution of the nogood set in Exercise 5 as the solution of a relaxation. For each step k, write the resulting nogood in the form of an inequality $v \geq B_k(x)$. In the simplest scheme, $B_k(x)$ is

either 0 or ∞ when P_k is infeasible, depending on x. If a feasible solution is found for P_k, $B_k(x)$ is either zero or a finite value (i.e., the number of true variables in the feasible solution).

8 Write (2.40) for each release time t_1 and each deadline t_2 ($t_1 < t_2$) in the problem of Table 2.11. Verify that (2.41) are the nonredundant inequalities.

9 Write the symmetric form of the edge-finding rule (2.45) that is sufficient to establish that job k must start after all the jobs in J finish. Indicate how E_k can be updated if k must follow J.

10 Suppose that in the problem of Table 2.11, Jobs 2, 3, and 4 are assigned to Machine A. Use the edge-finding rules to find jobs that must precede, or follow, subsets of jobs, and update the bounds accordingly. Note that when bounds have been updated, it may be possible to find additional edges. For example, initially one cannot deduce that Job 3 must follow 2 (so that E_3 is not updated), even though one can deduce that Job 2 must precede $\{3, 4\}$ (which updates L_1). However, after L_1 is updated, one can deduce that 3 follows 2 and update E_3. In this case, edge finding identifies all possible bound updates, but this is not true in general.

11 What is the minimum makespan on Machine A in the Exercise 10? What jobs play a role in deriving the minimum? Trace the algorithm that computes J_j^L and J_j^E to verify this.

12 Write a Benders cut that corresponds to the minimum makespan solution of Exercise 11.

13 Exhibit a disjunctive scheduling problem in which edge finding fails to discover all precedence relations.

14 A number of projects must be carried out in a shop, and each project j must start and finish within a time window $[r_j, d_j]$. Once started, the project must run p_j days without interruption. Only one project can be underway at any one time. Every month, the shop is shut down briefly to clean and maintain the equipment, and no project can be in process during this period. The goal is to find a feasible schedule. Formulate this problem and indicate how to solve it with logic-based Benders decomposition. Hint: let each month's schedule be a subproblem. Note that logic-based Benders can provide a scheme for optimizing a master schedule (here, assignment of jobs to months) and daily schedules simultaneously. The monthly resource constraints

take the form of a relaxation of the subproblem within the master problem.

15 A vehicle routing problem requires that a fleet of vehicles make deliveries to customers within specified time windows. Each customer j must take delivery between e_j and d_j, and vehicle i requires time p_{ijk} to travel from customer j to customer k. The objective is to minimize the number of vehicles. Describe a Benders-based solution method in which the master problem assigns customers to vehicles and the subproblem routes each vehicle. The subproblem for each vehicle is a traveling salesman problem with time windows. The subproblem can be written with variable indices and the circuit constraint (see Chapter 5), although it would be desirable to design a metaconstraint specifically for traveling salesman problems with time windows.

2.4 Local Search

Local search methods solve a problem by solving it repeatedly over small subsets of the solution space, each of which is a *neighborhood* of the previous solution. The neighborhood consists of solutions obtained by making small changes in the previous solution, perhaps by changing the value of one variable or swapping the values of two variables.

The motivation for local search is that a neighborhood is more easily searched than the entire solution space. By moving from neighborhood to neighborhood, the search may happen upon a good solution. Well-designed local search methods can, in fact, deliver remarkably good solutions within a reasonable time, although tuning them to work efficiently is more an art than a science. Local search has become indispensable for attacking many practical problems that are too large to solve by exact methods.

In general, the neighborhoods examined during the search cover only a small portion of the solution space. Even the neighborhoods themselves may not be examined exhaustively. Local search therefore provides no guarantee that the solution is optimal or even lies within any given distance from the optimum.

Local search fits naturally into the solution scheme presented here. Because each neighborhood is the feasible set of a problem restriction, local search in effect solves a sequence of problem restrictions. Inference and relaxation can also play a role. In fact, many local search strategies can be viewed as analogs of branching or constraint-directed search, and these analogies suggest how techniques from exhaustive search can

Local Search 89

be transferred to heuristic methods. The role of relaxation in branching, for example, can be mirrored in such branching-related local search methods as greedy randomized adaptive search procedures (GRASPs). Inference is already a part of local search methods related to constraint-directed search, such as tabu search, where the tabu list can be viewed as consisting of nogoods. The analogy can be exploited further, because ideas from such techniques as partial-order dynamic backtracking can be imported into tabu search, resulting in a more sophisticated heuristic method.

2.4.1 Some Popular Metaheuristics

Such popular local search schemes or *metaheuristics* as simulated annealing, tabu search, and GRASP algorithms are easily seen to be searches over problem restrictions (Table 2.12). Simulated annealing randomly chooses a solution x' in the neighborhood of the current solution x. If x' is better than x, then x' is *accepted* and becomes the current solution, whereupon the process repeats. If x' is no better than x, x' is nonetheless accepted with a certain probability p. If x' is not accepted, another solution x' is chosen randomly from the neighborhood of x, and the process repeats. The algorithm mimics a cooling process in which molecules seek a minimum energy configuration. The probability p decreases with the *temperature* as the process continues. The search may be terminated at will, and it may be rerun with several different starting points. Clearly the neighborhoods are not examined exhaustively in this

Table 2.12. How some selected heuristic methods fit into the search-infer-and-relax framework.

Solution method	Restriction P_k	Relaxation R_k	Selection function $s(R_k)$	Inference
Simulated annealing	Neighborhood of current solution	P_k	Random solution in neighborhood	None
Tabu search	Neighborhood minus tabu list	P_k	Best solution in neighborhood	Addition of nogoods to tabu list
GRASP	Neighborhood of partial solution	Problem specific	Random or greedy selection of solution in neighborhood	None

method. Each restriction is "solved" simply by selecting a solution, or at most a few solutions, randomly from the current neighborhood.

Tabu search differs in that it exhaustively searches each neighborhood. The best solution x' in the neighborhood of the current solution x becomes the current solution. To reduce the probability of cycling repeatedly through the same solutions, a *tabu list* of the last few solutions is maintained. Solutions on the tabu list are excluded from the neighborhood of x (the tabu list can also contain the types of alterations or *moves* performed on the last few solutions to obtain the next solution, rather than the solutions themselves). The items on the tabu list can be viewed as nogoods that rule out solutions or moves that have recently been examined. Tabu search is therefore an inexhaustive form of constraint-directed search.

Each iteration of a GRASP has two phases, the first of which constructs a solution in a greedy fashion, and the second of which uses this solution as a starting point for a local search. The greedy algorithm of the first phase assigns values to one variable at a time until all variables are fixed. The possible values that might be assigned to each variable x_k are ranked according to an easily computable criterion. The algorithm is adaptive in the sense that this ranking depends on what values were assigned to $x_1, \ldots, x_{k-1}$. One of the highly ranked values is then randomly selected as the value of x_k. This random component allows different iterations of the GRASP to construct different starting solutions.

The local search phase can be seen as a search over problem restrictions for reasons already discussed. The greedy phase is likewise a search over problem restrictions in a sense that is reminiscent of a branching search. Recall that a branching search typically branches on a problem P by assigning some variable its possible values. This creates a series of restrictions $P_1, \ldots, P_m$ whose feasible sets partition the feasible set of P. The search may then create restrictions of each P_i by branching on a second variable, and so on recursively.

The greedy algorithm is analogous, except that it generates only one restriction of P rather than an exhaustive list of restrictions $P_1, \ldots, P_m$. Specifically, it creates a restriction P_1 by setting x_1 to a value that is highly ranked. It then restricts P_1 by setting x_2 to a highly ranked value (given the value of x_1), and so forth, until all variables are assigned values.

2.4.2 Local Search Conceived as Branching

Simulated annealing and GRASPs can be seen as special cases of a generic local search procedure that is analogous to branching but does

Local Search

not explore all possible branches. This interpretation of local search also incorporates relaxation in a natural way.

The generic local search algorithm of Figure 2.15 keeps "branching" until it arrives at a problem that is easy enough to solve, at which point it solves the problem (by searching a neighborhood) and backtracks. When branching on a given problem restriction P, however, the algorithm creates only one branch. The search may backtrack to P later and generate additional branches. The branches eventually created at P differ in two ways, however, from those in a normal branching search: (a) they need not be exhaustive, which is to say the union of their feasible sets need not be the feasible set F of P, and (b) their feasible sets need not partition F.

Local search and GRASPs are special cases of this generic algorithm in which each restriction P is specified by setting one or more variables. If all the variables $x = (x_1, \ldots, x_n)$ are set to values $v = (v_1, \ldots, v_n)$, P's feasible set is a neighborhood of v. P is easily solved by searching the neighborhood. If only some of the variables $(x_1, \ldots, x_k)$ are set to $(v_1, \ldots, v_k)$, P is regarded as too hard to solve.

A pure local search algorithm, such as simulated annealing, branches on the original problem P_0 by setting all the variables at once to $v = (v_1, \ldots, v_n)$. The resulting restriction P is solved by searching a neighborhood of v. Supposing P's solution is v', the search backtracks to P_0 and branches again by setting $x = v'$. Thus, in pure local search, the search tree is never more than one level deep. The algorithm stops generating branches whenever the user terminates the search, generally long before the search is exhaustive.

Let $v_{\text{UB}} = \infty$ and $S = \{P_0\}$.
While S is nonempty repeat:
   Select a restriction $P \in S$ and remove P from S.
   If P is too hard to solve then
     Add a restriction of P to S.
   Else
     Let v be the value of P's solution and let $v_{\text{UB}} = \min\{v, v_{\text{UB}}\}$.
     Remove P from S.
The best solution found for P_0 has value v_{UB}.

Figure 2.15. Generic algorithm for local search conceived as branching. The algorithm solves a minimization problem P_0. Set S contains the problem restrictions generated so far. v_{UB} is the value of the incumbent solution. Note that the algorithm is almost identical to the generic branching algorithm of Figure 2.1.

In simulated annealing, P is "solved" by randomly selecting one or more elements of the neighborhood until one of them, say v', is accepted. The search backtracks to P_0 and branches by setting $x = v'$.

In a GRASP-like algorithm, the branching choices differ in the constructive and local search phases. In the constructive phase, the search branches by setting variables one at a time. At the original problem P_0, it branches by setting one variable, say x_1, to a value v_1 chosen in a randomized greedy fashion. It then branches again by setting x_2, and so forth. The resulting restrictions P are regarded as too hard to solve until all the variables x are set to some value v. When this occurs, a solution v' of P is found by searching a neighborhood of v, and the algorithm moves into the local search phase. It backtracks directly to P_0 and branches by setting $x = v'$ in one step. Local search continues as long as desired, whereupon the search returns to the constructive phase.

It was noted earlier that branching need not create a partition, and this is true in particular of a GRASP scheme. Figure 2.16, for instance, illustrates a small GRASP search in which the initial constructive phase assigns variables x_1, x_2, and x_3 the values A, B, and C, respectively, thus arriving at Restriction 3. At this point, the algorithm moves into the local search phase. It searches a neighborhood of $x = (A, B, C)$ by considering all interchanges of two components of x and selects $x = (B, A, C)$. It backtracks to the root and immediately generates a branch (Restric-

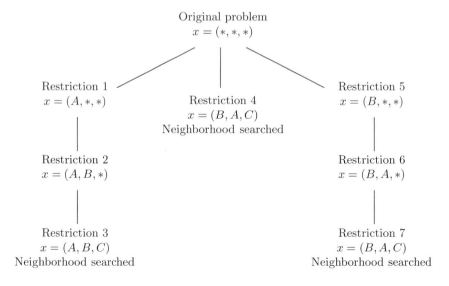

Figure 2.16. Branching tree for a GRASP search. $x = (A, *, *)$ indicates that x_1 is set to A, but x_2 and x_3 are not set.

Local Search 93

tion 4), at which the feasible set is a neighborhood of $x = (B, A, C)$. After searching this neighborhood, the local search is terminated and a new constructive phase assigns B, A, and C, respectively, to x_1, x_2, and x_3, thus arriving at Restriction 7. The neighborhood here is the same as for Restriction 4. Thus the branches at the root node do not create a partition: $x = (B, A, C)$ is consistent with two of the branches.

2.4.3 Relaxation

Conceiving local search as part of a quasi-branching scheme has the advantage of revealing an analogy with branch-and-relax algorithms, and thereby suggesting how relaxation can be used to accelerate the search.

The idea can be illustrated in the example of Figure 2.16. Suppose that an objective function $f(x)$ is to be minimized. Thus the solution of restriction 3 has value $f(B, A, C)$. Suppose that $x = (B, A, C)$ is still the incumbent solution when restriction 5 is encountered. If a *relaxation* of restriction 5 is solved and its value is no less than $f(B, A, C)$, there is no need to branch further at restriction 5. Restrictions 6 and 7 are pruned from the tree. Figure 2.17 contains a generic local search algorithm with relaxation.

In ordinary branch-and-relax algorithms, pruning the tree at some node ensures that no problem below the node will be solved. This is not true of local search. For example, Restriction 7 in Figure 2.16 is solved despite the pruning because it is identical to Restriction 4. In general, a restriction might be reached via several paths in the tree, and pruning one path may leave other access routes open. Nonetheless, pruning by relaxation reduces the size of the search tree that would otherwise be traversed. This is illustrated by the example in Section 2.4.4.

Let $v_{\text{UB}} = \infty$ and $S = \{P_0\}$.
While S is nonempty repeat:
   Select a restriction $P \in S$ and remove P from S.
   If P is too hard to solve then
      Let v_{R} be the optimal value of a relaxation of P.
      If $v_{\text{R}} < v_{\text{UB}}$ then
         Add a restriction of P to S.
   Else
      Let v be the value of P's solution and $v_{\text{UB}} = \min\{v, v_{\text{UB}}\}$.
The best solution found for P_0 has value v_{UB}.

Figure 2.17. Generic local-search-and-relax algorithm for solving a minimization problem P_0. The notation is the same as in Figure 2.15.

2.4.4 Constraint-Directed Local Search

Nothing in the generic local search algorithm of Figure 2.15 or 2.17 prevents enumeration of the same solution several times. Repetition can be reduced or eliminated by maintaining a list of nogoods (i.e., a tabu list) and rejecting any solution or partial solution that violates one of the nogoods. The list could be of finite length, as in tabu search, or it could remember all nogoods generated. In the latter case, the search would eventually become complete.

In exhaustive constraint-directed search, the algorithm terminates when the nogood set becomes infeasible. An inexhaustive version can be obtained by generating less than a complete set of nogoods. For instance, older nogoods can be dropped as in tabu search. This means that the nogood set may never become infeasible, and some other termination condition must be used. A generic constraint-directed local search algorithm appears in Figure 2.18.

In particular, partial-order dynamic backtracking can be converted to an inexhaustive search method by dropping older nogoods from the relaxation. In fact, since the nogood set remains small, it may be practical to process the nogoods more intensely than with the parallel resolution method described in Section 2.3.3. This allows more freedom in the solution of the current nogood set, because the solution may be allowed to *conform* to previous solutions in a weaker sense. By carrying over such ideas from constraint-directed search to heuristic methods, one can obtain an entire family of generalizations and extensions of tabu search.

Let $v_{\text{UB}} = \infty$, and let R be a relaxation of P.
While R is feasible repeat as desired:
   Select a feasible solution $x = s(R)$ of R.
   If x is feasible in P then
     Let $v_{UB} = \min\{v_{UB}, f(x)\}$.
     Define a nogood N that excludes x and possibly other solutions x'
       with $f(x') \geq f(x)$.
   Else
     Define a nogood N that excludes x and possibly other solutions
       that are infeasible in P.
   Remove nogoods from R as desired.
   Add N to R and process R.
The optimal value of P is v_{UB}.

Figure 2.18. Generic constraint-directed local search algorithm for solving a minimization problem P with objective function $f(x)$, where s is the selection function. R is the relaxation of the current problem restriction.

Local Search

A Benders method can also be converted to a heuristic method by dropping older Benders cuts from the master problem, or perhaps generating cuts that are too weak to ensure termination. Such techniques may be used by practitioners when a Benders algorithm bogs down.

2.4.5 Example: Single-Vehicle Routing

The idea of a local-search-and-relax algorithm can be illustrated with a single-vehicle routing problem with time windows, also known as a traveling salesman problem with time windows. A vehicle must deliver packages to several customers and then return to its home base. Each package must be delivered within a certain time window. The truck may arrive early, but it must wait until the beginning of the time window before it can drop off the package and proceed to the next stop. The problem is to decide in what order to visit the customers so as to return home as soon as possible, while observing the time windows.

The data for a small problem appear in Table 2.13. The home base is at location A, and the four customers are located at B, C, D and E. The travel times are symmetric, and so the time from A to B and from B to A is 5, for instance. The time windows indicate the earliest and latest time at which the package may be dropped off. The vehicle leaves home base (location A) at time zero and returns when all packages have been delivered.

Exhaustive enumeration of the twenty-four possible routings would reveal six feasible ones: ACBDEA, ACDBEA, ACDEBA, ACEDBA, ADCBEA, and ADCEBA. The last one is optimal and requires thirty-four time units to complete.

Local Search

A simple heuristic algorithm adds one customer at a time to the route in a greedy fashion, by adding the customer that can be served the earliest. The search creates a branch whenever a customer is added. When all customers have been served, or when it is no longer possible to observe

Table 2.13. Data for a small single-vehicle routing problem with time windows.

Origin	Travel time to:				Customer	Time window
	B	C	D	E		
A	5	6	3	7	B	[20,35]
B		8	5	4	C	[15,25]
C			7	6	D	[10,30]
D				5	E	[25,35]

time windows, the search jumps to a random node N in the current search tree. It deletes from the tree all successors of N to keep memory requirements under control. It creates a branch at N by adding a random customer. At subsequent branches, customers are added according to the greedy criterion. The process can start over repeatedly as desired by returning to the root node.

This algorithm can be viewed as a generalized GRASP. It is a GRASP in the sense that it alternates between a greedy phase and a local search phase. The greedy phase constructs a solution as in an ordinary GRASP. The local search phase, however, does not necessarily select the next solution from a neighborhood of the current solution, as in a conventional GRASP. Rather, it randomly jumps to a previously enumerated partial solution and randomly instantiates one more variable. If the random jump is restricted to a jump to the immediate successor of the current leaf node, then the random instantiation is equivalent to randomly selecting a solution in a neighborhood of the current solution, where the neighborhood consists of solutions that differ in one variable. Thus, when the random jump is restricted in this way, a generalized GRASP becomes a conventional GRASP.

Figure 2.19 illustrates a possible search. Starting from the home base (Node 0), the earliest possible delivery is to Customer D at time 10. The travel time to D is only 3, but D's time window starts at 10. The search therefore branches to Node 1. Departing Customer D at time 10, the earliest possible delivery is to Customer C at time 17, and so forth. The greedy procedure is fortunate enough to obtain a feasible solution at Node 3 without backtracking. The search jumps randomly to Node 1, whereupon Nodes 2 and 3 are deleted. A randomly chosen customer, E, is added to the route, and the greedy criterion adds Customer B at node 5. This violates the time windows, and the search randomly jumps to Node 0, where it randomly adds Customer B. Nodes 1–5 are deleted, and the greedy process obtains another infeasible routing at Node 8. The search is arbitrarily terminated at this point.

This can be viewed as a local search algorithm in the sense that the greedy procedure searches a neighborhood in the space of problem restrictiontions. The neighborhood consists of all restrictions that can be formed from the current restriction by adding a customer to the end of the route. A deleted node can reappear due to subsequent branching.

Local Search with Relaxation

A relaxation mechanism can help the search avoid unproductive areas of the search tree. One way to relax the problem is to replace the travel times for unscheduled trip segments with lower bounds on the travel

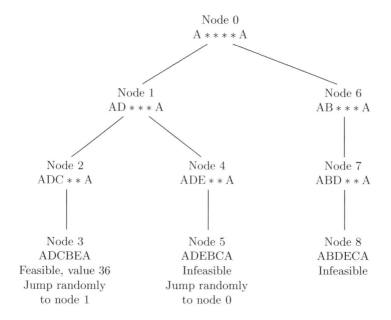

Figure 2.19. Local-search tree for a single-vehicle routing problem with time windows. The notation AD ∗ ∗ ∗ A indicates a partial routing that runs from A to D, through 3 unspecified stops, and back to A.

times. A segment is the portion of the trip between two customers, i and j, that are adjacent on the route. If customer j has not been scheduled, then the preceding customer i and the segment travel time are unknown. Yet a simple lower bound on this time is the travel time to j from the nearest customer that could precede j.

To make this more precise, let t_{ij} be the travel time between customers i and j, and let variable x_i be the ith customer visited (where x_0 is fixed to be the home base, Customer 0). Suppose a partial route consisting of the first k customers has been formed, so that $x_0, \ldots, x_k$ have been assigned distinct values. For $j \notin \{x_0, \ldots, x_k\}$, the travel time to customer j from the customer that precedes it in the route will be at least

$$L_j = \min_{i \notin \{j, x_0, \ldots, x_{k-1}\}} \{t_{ij}\}$$

and the travel time from the last customer served to the base will be at least

$$L_0 = \min_{j \notin \{x_0, \ldots, x_k\}} \{t_{j0}\}$$

Then, if T is the earliest time the vehicle can depart customer k,

$$T + L_0 + \sum_{j \notin \{x_0,...,x_k\}} L_j$$

is a lower bound on the duration of any completion of the partial route. If this value is greater than or equal to the value of the incumbent solution, there is no need to branch further.

The search algorithm can be amended so that whenever the tree can be pruned at some node by bounding, that node is deleted from the tree. The search then proceeds exactly as it does when it constructs a feasible route or encounters infeasibility: it jumps to a randomly chosen node that remains in the tree and branches by adding a random customer to the end of the route. A node deleted by bounding may reappear due to subsequent branching, whereupon it will again be deleted. Unlike a conventional GRASP, this particular algorithm will never find an alternate route to solutions below a node that is pruned by bounding.

It is illustrative to rerun the search of Figure 2.19 with bounding, and the result appears in Figure 2.20. In the partial route ADE at Node 4, the vehicle cannot depart E before time 25. Since B and C are unscheduled, the lower bound on the duration of the completed route is

$$25 + \min\{t_{CB}, t_{EB}\} + \min\{t_{BC}, t_{EC}\} + \min\{t_{BA}, t_{CA}\} = 40$$

Since this is larger than the incumbent value of 36, Node 4 is deleted. The search randomly jumps to Node 0 and randomly adds customer B at Node 5. Here the relaxation value is 38, which again allows the node to be pruned.

Constraint-Directed Search

The vehicle routing problem can solved in a manner similar to partial-order dynamic backtracking, as illustrated in Table 2.14. However, since the size of the nogood set will be limited, it is practical to process the nogood set more thoroughly to avoid backtracking while solving it. This, in turn, allows one to solve the relaxation without any restrictions other than the nogoods themselves.

Initially there are no nogoods, and a greedy algorithm selects the first solution ADCBEA by moving from each customer to the next customer that can be served most quickly. The greedy solution is feasible, and the nogood ADCB is generated to rule out this particular nogood. The meaning of the nogood ADCB is that no solution beginning ADCB can be considered. In Iteration 1, the greedy algorithm is constrained by the nogood ADCB and selects ADCEBA, which generates nogood ADCE.

Local Search   99

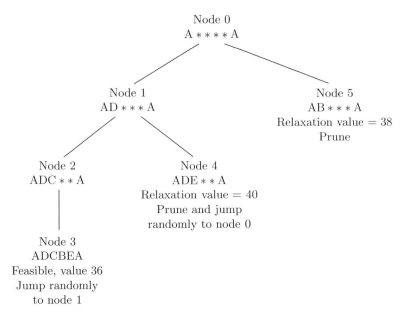

Figure 2.20. Local-search-and-relax tree for a single-vehicle routing problem with time windows. The notation is the same as in Figure 2.19.

Table 2.14. Solution of a single-vehicle problem with time windows by incomplete constraint-directed search.

k	Relaxation R_k	Solution of R_k	Value	New nogoods
0		ADCBEA	36	ADCB
1	ADCB	ADCEBA	34	ADCE
2	ADC	ADBECA	infeas.	EC
3	ABDE, ABEC ADBE, ADC ADEC, AEC	ADBCEA	infeas.	BC
4	ABC, ABEC AD, AEC, AEDB	ACDBEA	38	ACDB
5	ABC, ACDB AD, AEDB	ACDEBA	36	ACDE
⋮				

These two nogoods obtained so far exclude all solutions beginning ADC, and so the nogood ADC comprises the relaxation in Iteration 2. In effect ADCB and ADCE are resolved to yield ADC.

The greedy solution subject to ADC is the infeasible solution ADBECA. Some analysis reveals that the cause of the infeasibility is the subsequence EC, which is therefore generated as a nogood. To avoid backtracking in the solution of the next relaxation R_3, all excluded subsequences beginning with A must be spelled out: AEC, ABEC, ADEC, ABDE, ADBE. These are added to R_3, which has solution ADBCEA, which is again infeasible. Because subsequence BC is the cause of the infeasibility, all excluded subsequences beginning with A are added to the nogood set, and all possible resolutions performed to obtain the relaxation shown for Iteration 4. The resulting feasible solution generates nogood ACDB.

At this point, some of the older nogoods are dropped before adding ACDB to keep the nogood list short. Since the nogoods in R_1 and R_2 are no longer present in R_4, the nogoods in R_3 that are still present are dropped, leaving ABC, AD, and AEDB. Now, the new nogood ACDB is added to obtain R_5. The process continues in this fashion until one wishes to terminate it. As it happens, the algorithm discovers the optimal solution ADCEBA in Iteration 1.

2.4.6 Exercises

1 Consider a knapsack packing problem in which the objective is to maximize cx subject to $ax \leq 26$ and each $x_j \in \{0, 1\}$, where the data appear in Table 2.15. Generate part of a local-search-and-relax-tree similar to that of Figure 2.20. At each step of the greedy phase, fix to 1 the variable x_i that has not already been fixed and that has the largest ratio c_i/a_i. A leaf node is reached when no more variables can be fixed to 1. Thus every leaf node will correspond to a feasible solution. After evaluating a leaf node, backtrack to a random node in the current tree, and randomly select the next variable to instantiate before resuming the greedy approach. As a relaxation, maximize cx subject to $ax \leq 26$, $x_j \in [0, 1]$, and the currently fixed values (this is trivial to solve). For instance, after finding the first feasible solution (which has value 53), one might randomly backtrack to the

Table 2.15. Data for a small knapsack packing problem.

i	1	2	3	4	5
c_i	24	14	15	9	14
a_i	11	7	8	5	9
c_i/a_i	2.182	2.000	1.875	1.800	1.556

Local Search

root node, which causes the other nodes created so far to be deleted. If one randomly selects $x_5 = 1$, the optimal value of the relaxation is 50. This is worse than the incumbent value 53, thus allowing the tree to be pruned. The search backtracks to the root node (the only other node in the tree) and randomly selects another variable to instantiate to 1.

2 Solve the problem of Exercise 1 with constraint-directed search. The heuristic for solving the current problem restriction is to fix the variables in the order $x_1, \ldots, x_5$, fixing each x_i to 1 if this, in combination with the variables already fixed, does not violate a nogood. If the solution is infeasible, the nogood is

$$x_1^{1-v_1} \vee \cdots \vee x_j^{1-v_j} \qquad (2.47)$$

where v_i is the value to which x_i is fixed, $x_i^1 = x_i$, $x_i^0 = \neg x_i$, and j is the smallest index such that $\sum_{i=1}^{j} a_i v_i > 26$. If the solution is feasible, the nogood is (2.47) with $j = 5$. Since the order of instantiation is constant, parallel resolution of the nogoods is adequate. Solve the problem by a complete search, which implements a depth-first search with conflict clauses. The first few steps of the search appear in Table 2.16. If the older nogoods are dropped as the search proceeds, the result is an incomplete constraint-directed search. Then, instantiate the variables in any order, but apply full resolution to the nogoods, dropping the older nogoods as the search proceeds. One could also use an incomplete form of partial-order dynamic backtracking. All of these incomplete searches might be described as sophisticated forms of tabu search.

Table 2.16. First few iterations of constraint-directed search for a knapsack packing problem.

k	Relaxation R_k	Solution x of R_k	Value	New nogoods
0		(1, 1, 1, 1, 1)	infeas.	$\neg x_1 \vee \neg x_2 \vee \neg x_3 \vee \neg x_4$
1	$\neg x_1 \vee \neg x_2 \vee \neg x_3 \vee \neg x_4$	(1, 1, 1, 0, 1)	infeas.	$\neg x_1 \vee \neg x_2 \vee \neg x_3 \vee \neg x_5$
2	$\begin{cases} \neg x_1 \vee \neg x_2 \vee \neg x_3 \vee \neg x_4 \\ \neg x_1 \vee \neg x_2 \vee \neg x_3 \vee \neg x_5 \end{cases}$	(1, 1, 1, 0, 0)	53	$\neg x_1 \vee \neg x_2 \vee \neg x_3 \vee x_4 \vee x_5$
3	$\neg x_1 \vee \neg x_2 \vee \neg x_3$	(1, 1, 0, 1, 1)	infeas.	$\neg x_1 \vee \neg x_2 \vee x_3 \vee \neg x_4 \vee \neg x_5$
⋮				

3 A *genetic algorithm* mimics evolution by natural selection. It begins with a set of solutions (i.e., a population) and allows some pairs of solutions, perhaps the best ones, to mate. A crossover operation produces an offspring that inherits some characteristics of the parent solutions. At this point, the less desirable solutions are eliminated from the population so that only the fittest survive. The process repeats for several generations, and the best solution in the resulting population is selected. Indicate how this algorithm can be viewed as examining a sequence of problem restrictions. In what way does generation of offspring produce a relaxation of the current restriction? If the "solution" of the relaxation is the selection of some good solutions, how does solution of the relaxation guide the selection of the next restriction? Why is relaxation bounding, however, unhelpful in this algorithm? Hint: relaxation bounding is helpful when it obviates the necessity of solving the current restriction. Think about how the current relaxation is obtained.

4 *Ant colony optimization* can be applied to the traveling salesman problem on n cities as follows. Initially all the ants of the colony are in City 1. In each iteration, each ant crawls from its current location i to city j with probability proportional to u_{ij}/d_{ij}, where u_{ij} is the density of accumulated pheromone deposit on the trail from i to j, and d_{ij} is the distance from i to j. Each ant deposits pheromone at a constant rate while crawling, and a certain fraction of the pheromone evaporates between each iteration and the next. Each ant remembers where it has been and does not visit the same city twice until all cities have been visited. After returning to City 1, the ants forget everything and start over again. When the process terminates, the shortest tour found by an ant is selected. Show how this algorithm can be understood as enumerating problem restrictions. How can relaxation bounding be introduced into the algorithm?

5 *Particle swarm optimization* can be applied to global optimization, as follows. The goal is to search a space of many dimensions for the best solution. A swarm of particles are initially distributed randomly through the space. Certain particles have two-way communication with certain others. In each iteration, each particle moves randomly to another position, but with higher probability of moving closer to a communicating particle that occupies a good solution. After many iterations, the best solution found is selected. How can this process be viewed as enumerating a sequence of problem restrictions? Why is there no role for relaxation bounding here?

2.5 Bibliographic Notes

Section 2.1. Various elements of the search-infer-and-relax framework presented here were proposed in [12, 57, 182, 184, 185, 188, 198, 200]. An extension to dynamic backtracking and heuristic methods is given in [192, 193].

Section 2.2. Cuts for general integer knapsack constraints are discussed in [75, 234, 268, 269]. A comprehensive treatment can be found in [14], which strengthens the inequalities discussed here. Nearly all development of knapsack cuts, however, has been concerned with 0-1 knapsack constraints, beginning with [15, 166, 262, 335].

Conditional modeling is advocated in [198]. The idea of disjunctive modeling goes back at least to [146, 159] and is developed in [16, 17, 33, 60, 161, 198, 222, 277, 322] and elsewhere.

The employee timetable model is similar to one presented in [285]. An overview of employee timetabling appears in [238].

Continuous global optimization is extensively surveyed in [252], and the integrated approach taken here is similar to that described in [312] and implemented in the solver BARON. Factorization of functions for purposes of relaxation was introduced by [236].

The product configuration model is based on [315]. The generic element constraint was introduced by [174]. The form of the constraint used here appears in [315].

Column generation methods have been used for decades. A unifying treatment of branch and price for mixed integer programming can be found in [27]. Branch-and-price with CP-based column generation originated with [209, 343], and the area is surveyed in [115]. The airline crew rostering example described here is based on [120]. CP-based branch-and-price methods are surveyed in [115, 290].

Section 2.3. Constraint-directed search is discussed in connection with dynamic backtracking in [144, 145, 235], which also point out the connection between nogood-based search and branching. The ideas are further developed in [185].

The Davis-Putnam-Loveland (DPL) method for the propositional satisfiability problem was originally a resolution method proposed by Davis and Putnam [102]. Loveland [227] replaced the resolution step with branching. The fastest satisfiability algorithms, such as CHAFF [242], combine DPL with clause learning [32].

Partial-order dynamic backtracking was introduced in [235] and generalized in [56]. It is unified with other forms of dynamic backtracking and further generalized in [185], which also proves the completeness and polynomial complexity of parallel resolution for partial-order dynamic backtracking.

Classical Benders decomposition is due to [43] and was generalized to nonlinear programming in [142]. Logic-based Benders decomposition was introduced in [201] and developed in [185, 199]. Its application to planning and scheduling, with a CP subproblem, was proposed in [185], first implemented in [203], and extended in [168]. The machine scheduling example described here is adapted from [189]. Edge finding originates in [69, 70]. Branch and check is proposed in [185] and successfully implemented in [314].

Section 2.4. The integrated approach to heuristic methods presented here follows [193]. Tabu search is due to [150, 167]. GRASP originated with [307]. The idea of using relaxations in local search appears in [271].

Chapter 3

INFERENCE

Inference brings hidden information to light. When applied to a constraint set, it deduces valid constraints that were only implicit. These constraints can reveal that certain regions of the search space contain no solutions, or at least no optimal solutions, and one wastes less time in unproductive search.

Inference is not only useful, but is conceptually fundamental to optimization, because optimization can be regarded as an inference problem. The minimum value of an objective function subject to a constraint set is the maximum lower bound on the function that can be inferred from the constraint set. So, the minimum value can be found in either of two ways: by searching through feasible solutions, or by generating proofs.

This duality of search and inference is a unifying theme that runs through optimization methods. The two concepts interweave in both mathematical programming and constraint programming, both of which typically enumerate solutions while deducing valid constraints in the form of cutting planes or reduced domains. Their dual relationship can be formalized as *inference duality*, whose special cases have contributed much to optimization, particularly the linear programming dual in its various forms, the Lagrangean dual, and the superadditive dual. Inference duality underlies the constraint-directed search methods surveyed in Chapter 2, which range from Benders decomposition in mathematical programming to nogood-based search in artificial intelligence. Inference duality also provides a basis for sensitivity analysis.

Because it is hard to draw useful inferences from a general constraint set, inference is most useful when applied to specially structured subsets of constraints, or metaconstraints. Most of this chapter is therefore organized around inference methods that are tailored to specific types of constraints.

The chapter begins by defining the fundamental concept of completeness for inference methods and constraint sets. The concept has played a major role in constraint programming, due to the fact that generalized arc consistency and bounds consistency are forms of completeness. It then defines inference duality and shows how it relates to sensitivity analysis and constraint-directed search.

At this point, the chapter takes up inference methods for specific constraint classes, starting with linear inequalities. The Farkas lemma, which is the fundamental theorem of linear optimization, is actually a completeness theorem for a particular inference method. This result gives rise to linear programming duality and sensitivity analysis, as well as classical Benders decomposition and domain reduction methods. General inequality constraints provide the setting for two well-known inference duals—the surrogate and Lagrangean duals. Next comes propositional logic, for which inference is, of course, well developed. Integer and 0-1 inequalities can be regarded as logical propositions, and a theory of inference is developed for them.

Following this, inference methods in the form of domain filters are developed for several global constraints, including the element, all-different, cardinality, nvalues, circuit, and stretch constraints. The chapter concludes with domain reduction methods for disjunctive and cumulative scheduling, which are among the most successful application areas for constraint programming.

The inference of nogoods and Benders cuts, for purposes of constraint-directed search, is discussed when it has been studied—for linear programming, integer linear inequalities, and disjunctive and cumulative scheduling. The use of nogoods for solving propositional satisfiability problems is described in Chapter 2, as are constraint-directed methods for local search. One should bear in mind, however, that whenever an inference dual can be defined, a constraint-directed search method can be developed.

3.1 Completeness

If the purpose of inference is to make explicit what is implicit, one measure of an inference method is whether it derives all implied constraints of a given form—that is to say, whether it is *complete* with respect to constraints of a given form. The concept of completeness is particularly

Completeness

important in constraint programming, where it goes by the name *consistency*, although not all types of consistency are completeness properties.

Following some basic definitions, several forms of completeness are reviewed below. Domain completeness and bounds completeness are the simplest and perhaps the most important for constraint solvers. Domain completeness can be generalized as k-domain completeness, which affects the amount of backtracking necessary to solve a problem. A slightly weaker property, strong k-consistency, is not a form of completeness but achieves the same effect on backtracking. Strong k-consistency can actually eliminate backtracking if the problem's dependency graph has width less than k—a result that is presented here mainly for its theoretical interest.

3.1.1 Basic Definitions

It is necessary to begin by clarifying what a constraint is. Suppose there is a stock of variables $x_1, \ldots, x_n$, and each variable x_j takes values in its *domain* D_{x_j}. It is convenient to write $x \in D$, where $x = (x_1, \ldots, x_n)$ and $D = D_{x_1} \times \cdots \times D_{x_n}$. A *constraint* C is associated with a function $C(x_C)$ whose value is *true* or *false*. The variables $x_C = (x_{j_1}, \ldots, x_{j_d})$ are the variables that appear in C. An assignment $x_C = v_C$ satisfies C if $C(v_c) = true$, and it *violates* C otherwise.

It will be useful to say that an assignment $(x_1, \ldots, x_k) = (v_1, \ldots, v_k)$ to a subset of variables satisfies C if it assigns values to all the variables in C and makes C true. That is, $\{x_{j_1}, \ldots, x_{j_d}\} \subset \{x_1, \ldots, x_k\}$ and $C(v_{j_1}, \ldots, v_{j_d}) = true$. The assignment violates C if it assigns values to all the variables in x_C and makes C false.

Two constraints associated with the same function are *equivalent* but are not necessarily the same constraint. For instance, the constraints $x_1^2 = -1$ and $x_1^2 = -2$ are equivalent if x_1 is a real-valued variable, because they define a function that is always false, and yet they are regarded here as distinct constraints. This departs from the convention in the CP community, where a constraint is defined extensionally as a relation among its variables, and two constraints associated with the same relation are identical. This convention is not followed here because it seems too distant from the common-sense notion of a constraint.

A fundamental concept of inference is *implication*. Constraint C_1 implies constraint C_2 (with respect to domain D) if any $x \in D$ that satisfies C_1 also satisfies C_2. A constraint set $\mathcal{S}$ implies constraint C if any $x \in D$ that is feasible for $\mathcal{S}$ (i.e., satisfies all the constraints in $\mathcal{S}$) also satisfies C. Two constraints are equivalent if they imply each other, and similarly for two constraint sets.

Completeness can be defined for both constraint sets and inference methods. A constraint set S is *complete* with respect to a family $\mathcal{F}$ of constraints if S contains all of its implications in $\mathcal{F}$. More precisely, every constraint in $\mathcal{F}$ that is implied by S is implied by a constraint in $\mathcal{F} \cap S$. Thus, if constraints in $\mathcal{F}$ are viewed as those having a certain form, a complete constraint set S is one whose every implication of this form is captured by an individual constraint in S of the same form.

An *inference method* is a procedure that derives implied constraints, and a *complete* inference method makes a constraint set complete. To make this more precise, suppose that an inference method is *applied* to constraint set S when one adds to S all constraints in some constraint family $\mathcal{F}$ that the method can derive from S, and that are not already implied by some constraint in $\mathcal{F} \cap S$. The operation repeats until no further constraints can be added to S in this fashion.[1] An inference method is complete for S with respect to $\mathcal{F}$ if applying the method to S makes S complete with respect to $\mathcal{F}$. So, a complete inference method in some sense brings out all the relevant information, where relevance is understood as expressibility by a constraint in $\mathcal{F}$.

3.1.2 Domain Completeness

In a search procedure, it is often useful to know whether an individual variable assignment $x_j = v$ is feasible; that is, whether x_j takes value v in at least one feasible solution. If not, then no time should be wasted enumerating solutions in which $x_j = v$.

The desired property is *domain completeness*. A constraint set S is domain complete if it is complete with respect to *domain constraints*, which are constraints of the form $x_j \in D$. For historical reasons, domain completeness is known as *hyperarc consistency* or *generalized arc consistency* in the CP community.

Domain completeness is generally achieved by removing all infeasible values from the domain D_{x_j} of each variable x_j and including the domain constraints $x_j \in D_{x_j}$ in S. This ensures that any domain constraint $x_j \in D$ implied by S is implied by a domain constraint in S, namely $x_j \in D_{x_j}$. This makes S complete with respect to domain constraints.

Domain completeness is closely related to projection. If S is a set of tuples $(x_1, \ldots, x_n)$, the *projection* of S onto variables $x_1, \ldots, x_k$ is the set of all $(x_1, \ldots, x_k)$ that can be extended to a tuple in S; that is, the set of all $(x_1, \ldots, x_k)$ such that $(x_1, \ldots, x_k, x_{k+1}, \ldots, x_n) \in S$ for some

---

[1] If the procedure does not terminate, then S is set equal to the infinite union of all constraint sets obtained by the procedure. This can occur, for instance, when applying interval propagation to linear inequalities.

$(x_{k+1}, \ldots, x_n)$. Then S is domain complete if each domain D_{x_j} is equal to the projection of S's feasible set onto x_j.

The process of removing infeasible values from domains is known as *domain reduction* or *domain filtering*. Domain filtering, combined with its weaker counterpart bounds propagation is the workhorse of CP solvers and can play a key role in integrated solvers.

Domain completeness can be illustrated with a small instance of the traveling salesman problem. A salesman must decide in which order to visit four cities so that the distance traveled is at most 30 km, including the distance from the last city back to the city of origin. Let x_j denote the city visited immediately after city j, and c_{ij} the distance between cities i and j in either direction (Table 3.1). The problem can be written

$$\text{Linear: } \sum_{j=1}^{4} c_{jx_j} \leq 30 \tag{3.1}$$
$$\text{Circuit: } (x_1, x_2, x_3, x_4)$$
$$\text{Domains: } x_j \in \{1, 2, 3, 4\} \setminus \{j\}, \; j = 1, \ldots, 4$$

The circuit constraint requires that the sequence of cities visited form a single circuit that covers all cities. Only six solutions satisfy the circuit constraint: $(x_1, \ldots, x_4) = (2,3,4,1), (2,4,1,3), (3,1,4,2), (3,4,2,1), (4,1,2,3), (4,3,1,2)$. Of these, only two satisfy the distance constraint:

$$(x_1, \ldots, x_4) = (2,3,4,1), (4,1,2,3) \tag{3.2}$$

These two solutions form the feasible set.

The constraint set (3.1) is not domain complete. For instance, $x_1 = 3$ is infeasible but $3 \in D_{x_1}$. Domain completeness can be achieved by removing the one infeasible value in each domain, resulting in the domain constraints

$$x_1 \in \{2, 4\}, \quad x_2 \in \{1, 3\}, \quad x_3 \in \{2, 4\}, \quad x_4 \in \{1, 3\} \tag{3.3}$$

Each D_{x_j} is now the projection of the feasible set onto x_j.

Table 3.1. Distances $c_{ij} = c_{ji}$ between cities i and j in a small instance of the traveling salesman problem.

		\multicolumn{3}{c}{j}		
		2	3	4
	1	5	8	7
i	2		6	9
	3			9

3.1.3 Bounds Completeness

Bounds completeness is useful when domain elements have a natural ordering, as in the case of real numbers or integers. Bounds completeness is a relaxation of domain completeness and easier to achieve, but it can nonetheless accelerate solution.

Let I_{x_j} be the interval that spans the domain D_{x_j} of variable x_j. So, $I_{x_j} = \{L_j, \ldots, U_j\}$ if x_j is discrete, and $I_{x_j} = [L_j, U_j]$ if x_j is continuous, where $L_j = \min D_{x_j}$ and $U_j = \max D_{x_j}$. A constraint set is *bounds complete* if it is complete with respect to constraints of the form $L \leq x_j \leq U$ when the domains D_{x_j} are taken to be the intervals I_{x_j}. Bounds completeness is known as *bounds consistency* in the CP community.

Like domain completeness, bounds completeness can be maintained by reducing domains. Let $\mathcal{S}'$ be the result of replacing the variable domains D_{x_j} by the interval domains I_{x_j} in the original constraint set $\mathcal{S}$. $\mathcal{S}$ is bounds complete when, for every j, L_j is the smallest value x_j can take in a feasible solution of $\mathcal{S}'$, and U_j is the largest. In other words, L_j is the smallest value in the projection of $\mathcal{S}'$ onto x_j, and U_j is the largest.

In practice, it is common for solvers not to maintain the precise domain of an integer- or real-valued variable, but only lower and upper bounds. In this case bounds completeness is achieved by making the bounds as tight as possible. Domain completeness is a stronger property because it requires the solver to keep track of *holes* in the domain, or intermediate values that the variable cannot take in a feasible solution.

Sections 2.2.3 and 2.2.6 provide examples of bounds filtering for arithmetical inequalities and equations.

3.1.4 k-Completeness

A constraint set is domain complete when every infeasible assignment to some variable violates some constraint in the set. A natural extension of the concept ensures that every infeasible assignment to k variables violates some constraint. This allows one to avoid unpromising assignments to subsets of variables. In particular, it can reduce backtracking in a branching algorithm.

Define a constraint set $\mathcal{S}$ to be *k-complete* if every infeasible assignment $(x_{j_1}, \ldots, x_{j_k}) = (v_1, \ldots, v_k)$ to k variables (where each $v_i \in D_{x_{j_i}}$) violates some constraint in $\mathcal{S}$. Thus, a domain complete constraint set is 1-complete.

Like domain completeness, k-completeness is closely related to projection. A constraint set $\mathcal{S}$ can be made k-complete by including within it a description of the projection of $\mathcal{S}$ onto every set of k variables. That is, for any set J of k variables, $\mathcal{S}$ contains a subset of constraints with variables in J whose feasible set is precisely the projection of $\mathcal{S}$ onto J. This achieves k-completeness because any infeasible assignment to a set J of k variables lies outside the projection of $\mathcal{S}$ onto J and therefore violates a constraint in $\mathcal{S}$.

The traveling salesman problem of Section 3.1.2 is domain complete after the addition of domain constraints (3.3). Yet it is not 2-complete. For example, the assignment $(x_1, x_2) = (2, 1)$ violates no constraint but is not part of a feasible solution. The problem becomes 2-complete if constraints are added to require each pair of variables to belong to the projection of the feasible set onto that pair:

$$\begin{array}{ll} (x_1, x_2) \in \{(2, 3), (4, 1)\} & (x_2, x_3) \in \{(1, 2), (3, 4)\} \\ (x_1, x_3) \in \{(2, 4), (4, 2)\} & (x_2, x_4) \in \{(1, 3), (3, 1)\} \\ (x_1, x_4) \in \{(2, 1), (4, 3)\} & (x_3, x_4) \in \{(4, 1), (2, 3)\} \end{array} \quad (3.4)$$

An attractive feature of k-completeness is that it can reduce backtracking. Suppose that at level $k-1$ of the branching tree, the branching process has assigned values to $k-1$ variables An attractive feature of k-completeness is that it can reduce backtracking. Suppose that at level $k-1$ of the branching tree, the branching process has assigned values to $k-1$ variables

$$(x_{j_1}, \ldots, x_{j_{k-1}}) = (v_1, \ldots, v_{k-1}) \quad (3.5)$$

without violating any constraints in $\mathcal{S}$ so far. But, suppose no value can be assigned to the next variable x_{j_k} without violating a constraint. That is, there is no $v_k \in D_{x_{j_k}}$ such that $(x_{j_1}, \ldots, x_{j_{k-1}}, x_{j_k}) = (v_1, \ldots, v_{k-1}, v_k)$ violates no constraint. It is then necessary to backtrack.

This backtrack could have been avoided if some constraint in $\mathcal{S}$ explicitly ruled out the assignment (3.5), so that this point in the branching tree was never reached. $\mathcal{S}$ contains such a constraint if it is $(k-1)$-complete.

Let $\mathcal{S}$ be *strongly k-complete* if it is i-complete for $i \leq k$. If $\mathcal{S}$ is strongly $(k-1)$-complete, one can branch at least down to level k without backtracking. The procedure for doing so might be called *zero-step lookahead*. At the root node (Level 1) of the search tree, assign x_1 any value that violates no constraint in $\mathcal{S}$; such a value exists unless $\mathcal{S}$ is infeasible. At Level 2, 1-completeness implies that $x_1 = v_1$ can be extended to a feasible solution of $\mathcal{S}$. So, in particular, there is some value $v_2 \in D_{x_2}$ such that $(x_1, x_2) = (v_1, v_2)$ violates no constraint. The procedure continues until variables $x_{j_1}, \ldots, x_{j_k}$ are all assigned values.

The traveling salesman problem (3.1) becomes strongly 3-complete if it includes the domain constraints (3.3) and (3.4). Thus, one can find a feasible solution without backtracking. Arbitrarily choose the branching order to be $x_1, \ldots, x_4$. At the root node of a search tree x_1 can be set to 2 or 4. Arbitrarily choose 2. The first constraint in (3.4) forces $x_2 = 3$, so that $(x_1, x_2) = (2, 3)$. Continuing in similar fashion yields a feasible solution $(x_1, \ldots, x_4) = (2, 3, 4, 1)$ without backtracking.

3.1.5 k-Consistency

A property that slightly relaxes strong $(k-1)$-completeness suffices to avoid backtracking down to level k. It is known as strong k-consistency, and one can see why it suffices by reexamining zero-step lookahead.

To avoid backtracking at level $k-1$, it is enough that some constraint in $\mathcal{S}$ rule out assignment (3.5) when it cannot be extended to a kth variable; it is not necessary to rule out all assignments that cannot be extended to a feasible solution. The constraint set $\mathcal{S}$ has the required property if it is k-*consistent*. Thus, $\mathcal{S}$ is k-consistent if for every assignment $(x_{j_1}, \ldots, x_{j_{k-1}}) = (v_1, \ldots, v_{k-1})$ that violates no constraints in $\mathcal{S}$, and for every variable $x_{j_k} \notin \{x_{j_1}, \ldots, x_{j_{k-1}}\}$, there is a value $v_k \in D_{x_{j_k}}$ for which setting $(x_{j_1}, \ldots, x_{j_{k-1}}, x_{j_k}) = (v_1, \ldots, v_{k-1}, v_k)$ violates no constraints in $\mathcal{S}$.

$\mathcal{S}$ is *strongly k-consistent* if it is i-consistent for $i = 1, \ldots, k$. Thus, if $\mathcal{S}$ is feasible and strongly k-consistent, zero-step lookahead can reach level k of the search tree without backtracking.

$(k-1)$-completeness obviously implies k-consistency, but the traveling salesman example shows that the reverse is not true. The original constraint set (3.1) is 3-consistent but not 2-domain complete. It is 3-consistent simply because no constraint contains fewer than four variables. Because no assignment to three variables violates any constraints, any assignment to two variables can be extended to another variable by assigning it any value one wants. However, (3.1) is not 2-complete because solution $(x_1, x_2) = (2, 4)$ violates no constraint but is not part of a feasible solution.

3.1.6 Backtracking and Width

If the variables are loosely coupled, strong k-consistency can eliminate backtracking altogether, even when k is relatively small. To achieve this, however, it is necessary to branch on variables in a certain order.

Two variables are *coupled* in a constraint set when they occur in a common constraint. The pattern of variable coupling is indicated by its *dependency graph*, sometimes called its *primal graph*. The dependency graph contains a vertex for each variable and an edge connecting two variables when they occur in a common constraint.

Completeness

For instance, the following constraint set has the dependency graph in Figure 3.1, where each vertex j corresponds to variable x_j.

$$
\begin{aligned}
-3x_1 + 2x_2 - x_3 &\geq -8 \\
2x_1 - 3x_2 \phantom{{}+{}} + x_4 &\geq -4 \\
x_3 \phantom{{}+{}} + x_5 &\geq 4 \\
2x_4 - x_5 + x_6 &\geq 2 \\
x_j \in \{1, 2, 3\}, \text{ all } j
\end{aligned}
\tag{3.6}
$$

The amount of backtracking depends on the branching order, as well as on the nature of the variable coupling. To account for this, it is useful to define the *width* of the dependency graph with respect to a given ordering of the vertices. Let an edge connecting vertices i and j be directed from i to j when i occurs before j in the ordering. The *in-degree* of a vertex j is the number of edges incident to j that are directed toward j. The width of the graph, with respect to the ordering, is the maximum in-degree of its vertices. Thus, the graph in Figure 3.1 has width 2 with respect to ordering $1, \ldots, 6$ and width 3 with respect to ordering $6, 5, \ldots, 1$.

A strongly k-consistent problem can be solved without backtracking if its dependency graph has width less than k with respect to some ordering.

THEOREM 3.1 *If a feasible constraint set S is strongly k-consistent, and its dependency graph has width less than k with respect to some ordering of the variables, then zero-step lookahead with respect to that ordering obtains a feasible solution for S.*

Proof. Suppose the variables are ordered $x_1, \ldots, x_n$. Zero-step lookahead assigns x_1 any value that violates no constraint; such a value exists because S is feasible. Arguing by induction, suppose that the first $i-1$ variables have been assigned values $(x_1, \ldots, x_{i-1}) = (v_1, \ldots, v_{i-1})$ that violate no constraint. It suffices to show that x_i can be assigned a value

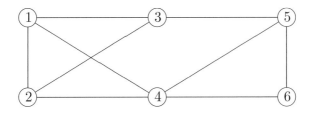

Figure 3.1. Dependency graph for an integer knapsack problem.

that, together with the previous assignments, violate no constraint. It is enough to check the constraints that contain x_i, because by hypothesis none of the other constraints are violated so far. But since the dependency graph of $\mathcal{S}$ has width less than k with respect to the ordering $1,\ldots,n$, variable x_i occurs in a common constraint with fewer than k other variables $x_{j_1},\ldots,x_{j_d}$ in the set $\{x_1,\ldots,x_{i-1}\}$. Because $\mathcal{S}$ is strongly k-consistent, the assignment $(x_{j_1},\ldots,x_{j_d}) = (v_{j_1},\ldots,v_{j_d})$ can be extended to an assignment $(x_{j_1},\ldots,x_{j_d},x_i) = (v_{j_1},\ldots,v_{j_d},v_i)$ without violating any constraint. Thus, $(x_1,\ldots,x_{i-1}) = (v_1,\ldots,v_{i-1})$ can be extended to $(x_1,\ldots,x_{i-1},x_i) = (v_1,\ldots,v_{i-1},v_i)$ without violating any constraint. □

An immediate corollary is that a strongly $(k-1)$-complete problem can be solved without backtracking if its dependency graph has width less than k for some ordering.

It is easily checked that the integer knapsack instance (3.6) is strongly 3-consistent (it is 2-complete as well). Thus since its dependency graph has width 2 with respect to the ordering $1,\ldots,6$, zero-step lookahead always finds a feasible solution. Suppose, for instance, that each x_i is assigned the smallest value that violates no constraint. This produces the feasible solution $(x_1,\ldots,x_6) = (1,1,1,1,3,3)$ without backtracking. However, the same heuristic applied to the variables in reverse order results in $(x_6,\ldots,x_1) = (1,1,1,3,1,?)$, with no feasible value for x_1, thus requiring a backtrack.

3.1.7 Exercises

1 Consider the constraint set $\mathcal{C}$ consisting of the equation $x_1+x_2 = 2x_3$ and domains $x_j \in \{0,1,2,3,4,5\}$ for $j = 1,2,3$. Reduce the domains so as to achieve bounds completeness. Reduce them further to achieve domain completeness. Add to $\mathcal{C}$ one or more 2-variable inequalities that, together with the reduced domains, make the set 2-complete.

2 Show by counterexamples that a k-complete constraint set is not necessarily $(k-1)$-complete, and not necessarily $(k+1)$-complete.

3 Show that a $(k-1)$-complete constraint set is k-consistent.

4 Show by counterexample (other than the example in the text) that a k-consistent constraint set is not necessarily $(k-1)$-complete.

5 Show by counterexamples that a k-consistent constraint set is not necessarily $(k-1)$-consistent and not necessarily $(k+1)$-consistent.

6 Consider the constraint set $\mathcal{C}$, consisting of

$$x_1 + x_2 + x_4 \geq 1$$
$$x_1 + (1 - x_2) + x_3 \geq 1$$
$$x_1 + (1 - x_4) \geq 1$$

with domains $x_j \in \{0, 1\}$ for $j = 1, 2, 3, 4$. Draw the dependency graph and note that it has width 2 with respect to the ordering 1,2,3,4. Show that the constraint set is not 3-consistent, and show that a branching algorithm that follows the ordering 1,2,3,4 may be required to backtrack. Recall that a constraint is not violated until all of its variables are fixed. Add the constraints $x_1 + x_2 \geq 1$ and $x_1 + x_3 \geq 1$ to $\mathcal{C}$ and verify that $\mathcal{C}$ is now strongly 3-consistent. Check that the sequence of branches that led to the backtrack is no longer possible.

3.2 Inference Duality

A general optimization problem with constraint set $\mathcal{S}$

$$\min \ f(x)$$
$$\mathcal{S} \tag{3.7}$$
$$x \in D$$

can be viewed as posing a *search* task—finding a feasible x that minimizes $f(x)$. The *inference dual* of (3.7) poses a complementary task: maximizing the lower bound on $f(x)$ that can be *deduced* from the constraints.

The dual problem can be written

$$\max \ v$$
$$\mathcal{S} \overset{Q}{\vdash} (f(x) \geq v) \tag{3.8}$$
$$v \in \mathbb{R}, \ Q \in \mathcal{Q}$$

where $\mathcal{S} \overset{Q}{\vdash} (f(x) \geq v)$ indicates that proof Q deduces $f(x) \geq v$ from $\mathcal{S}$. The domain of variable Q is a family $\mathcal{Q}$ of proofs, and the dual solution is a pair (v, Q). When $\mathcal{S}$ is infeasible and (3.8) therefore unbounded, the dual solution can be understood to contain, for any given x, a proof schema Q that derives $f(x) \geq v$ from $\mathcal{S}$ for any given v.

The inference dual is defined only when the constraint language is rich enough to include such inequalities as $f(x) \geq v$, as it is in linear or integer programming. Section 3.3.4 will show, for example, that when (3.7) is a linear programming problem, the family $\mathcal{Q}$ of proofs can be taken to be nonnegative linear combinations of inequalities in the constraint set that derive bounds $f(x) \geq v$. In this case, the dual solution can be identified with the vector of multipliers in the linear combination that derives the tightest bound.

3.2.1 Strong Duality and Completeness

The optimal value $\bar{v}$ of the dual (3.8) may, in general, be less than the optimal value v^* of the *primal* problem (3.7). Even though $\mathcal{S}$ implies $f(x) \geq v^*$, there may not be a proof in $\mathcal{Q}$ that deduces $f(x) \geq v^*$ from $\mathcal{S}$. The difference between v^* and $\bar{v}$ is the *duality gap*.

There is no duality gap if the family $\mathcal{Q}$ of proofs is complete with respect to constraints of the form $f(x) \geq v$, where $f(x)$ is the objective function in the primal and v is any real number. In linear programming, for example, nonnegative linear combination (plus domination of one inequality by another) is a complete inference method for linear inequalities, and there is no duality gap. The absence of a duality gap is known as *strong duality* and holds almost by definition when the inference method is complete.

To see this, suppose that the primal problem has finite minimum value v^*. Then $\mathcal{S}$ implies $f(x) \geq v$ for $v = v^*$ but not for any $v > v^*$. Thus by completeness, some $Q \in \mathcal{Q}$ deduces $f(x) \geq v$ from $\mathcal{S}$ for $v = v^*$ but not for any $v > v^*$, which means that the optimal dual value is v^*, as well. If the primal problem is infeasible, then $\mathcal{S}$ implies $f(x) \geq v$ for any v, because an infeasible constraint set implies everything. Thus by completeness, the dual is unbounded, and by convention both the primal and dual have optimal value ∞. If the primal is unbounded, then no v is a valid lower bound on $f(x)$, and both the primal and dual have optimal value $-\infty$. This is summed up in the following theorem.

THEOREM 3.2 *Suppose the family $\mathcal{Q}$ of proofs is complete with respect to constraints of the form $f(x) \geq v$. Then, the primal problem (3.7) and dual problem (3.8) have the same optimal value. Furthermore, a feasible solution x^* of (3.7) is optimal if and only if the inference dual (3.8) has a feasible solution (v, Q) with $v = f(x^*)$.*

Typically, an optimization problem and its inference dual are solved simultaneously. Although search may yield an optimal solution of the primal problem, a proof of optimality—that is, a solution of the inference dual—is generally desired as well. The simplex method for linear programming, for instance, yields *dual multipliers* that specify a linear combination proving optimality. In integer programming, the branch-

Inference Duality 117

and-cut tree that identifies an optimal solution also provides a proof of optimality that combines cutting planes (a form of inference) with argument by cases. CP solvers similarly combine domain filtering (a form of inference) with enumeration by cases. This interleaving of search and inference is characteristic of many successful optimization methods and provides part of the motivation for the search-infer-and-relax strategy explored in this book.

3.2.2 Certificates and Problem Complexity

A solution of the inference dual typically has a very different character than a solution of the primal problem. It takes the form of a proof rather than an assignment of values to variables, and it tends to be much longer, perhaps growing exponentially with the problem size. This asymmetry reflects a fundamental property of combinatorial problems. They typically belong to NP, but not to co-NP. This idea may be briefly developed as follows.

A *certificate* of feasibility for a problem is a piece of information that allows one to verify that a given feasible instance of the problem is, in fact, feasible. For example, a certificate might be a set of variable values that satisfy the constraints. A certificate of infeasibility is a parallel concept: it is a piece of information, normally a proof of some kind, that allows one to verify that a problem instance has no feasible solution.

If there is a certificate that allows one to verify *feasibility* in polynomial time, the problem *belongs to NP*. Polynomial time means that the time required to verify feasibility is bounded by a polynomial function of the size of the problem instance (i.e., the number of binary digits required to encode the instance). In particular, the certificate itself must have polynomial length. If there is a certificate that allows one to verify *infeasibility* in polynomial time, the problem *belongs to co-NP* (NP abbreviates *nondeterministic polynomial*, a concept from complexity theory that need not be developed here).

An optimization problem (3.7) can be associated with a feasibility problem that contains an additional constraint of the form $f(x) \leq v_0$. One can say that the optimization problem belongs to NP when the associated feasibility problem belongs to NP. The inference dual (3.8) can similarly be associated with a feasibility problem with the additional constraint $v \geq v_1$. The dual problem belongs to NP when the associated feasibility problem belongs to NP.

A certificate of feasibility for the dual problem can be regarded as a certificate of infeasibility for the primal problem, because it proves that the primal is infeasible for any given $v_0 < v_1$. Conversely, a certificate of infeasibility for the primal provides a certificate of feasibility for the dual when $v_1 > v_0$. This connects inference duality with complexity theory.

THEOREM 3.3 *An optimization problem (3.7) belongs to co-NP if and only if its inference dual (3.8) belongs to NP for some $\mathcal{Q}$.*

It is an interesting fact that most of the better known combinatorial problems belong to NP but not to co-NP. In other words, a combinatorial problem typically belongs to NP, and its inference dual typically does not. The primal problem might be expected to belong to NP, because a set of feasible variable values is a polynomial-size certificate. It is less predictable that a proof of infeasibility is typically much longer.

There are some notable exceptions, such as linear programming, which belongs to both NP and co-NP. Such problems are sometimes said to have *good characterizations*, in the sense that feasible solutions and proofs of infeasibility are easily encoded. In fact, one can plausibly conjecture that any problem belonging to both NP and co-NP is easy to solve, perhaps in the sense that it can be solved in polynomial time, as can a linear programming problem. No such conjecture has been proved, however.

3.2.3 Sensitivity Analysis

Optimization models used in practice often require more data than can be accurately determined. Fortunately, most of the numbers in a model typically have little bearing on the solution. It is enough to focus on a small subset of the parameters and make sure they are correct. The data that really matter, however, generally cannot be identified in advance. In such cases, *sensitivity analysis* can be performed after the problem is solved to find the data to which the solution is sensitive. The key numbers are then adjusted or corrected and the problem re-solved. The cycle can be repeated until modelers are confident of the model's realism.

The inference dual plays a central role in sensitivity analysis because it can bound the effect of changes in the problem data. Assume $\mathcal{Q}$ is complete, and let $\bar{x}$ be an optimal solution of (3.7). By Theorem 3.2, the inference dual (3.8) has a solution $(\bar{v}, \bar{Q})$ in which $\bar{v}$ is the optimal value $f(\bar{x})$ of the primal. The proof $\bar{Q}$ derives $f(x) \geq \bar{v}$ using the constraint set $\mathcal{S}$ as premises. Now if the problem data are changed, one can investigate whether *this same proof* $\bar{Q}$ still derives the bound $\bar{v}$. Perhaps the premises that are essential to the proof are still intact. More generally, one can investigate what kind of alterations in the data can be made without invalidating $\bar{Q}$ as a proof of bound $\bar{v}$. Even if $\bar{v}$ can no longer be deduced, proof $\bar{Q}$ may nonetheless derive some other useful bound when it is applied to the altered premises.

Inference Duality   119

3.2.4 Duality and Constraint-Directed Search

Inference duality is closely related to constraint-directed search. Nogoods can always be obtained by solving a strong inference dual of each problem restriction formulated during the search. The idea is very similar to that underlying sensitivity analysis.

Recall that constraint-directed search solves a series of restrictions $P_1, P_2, \ldots$ of a given problem P. In each iteration, a solution x^k of the current nogood set $\mathcal{N}_k$ determines the next restriction $P_{k+1} = P(x^k)$. A nogood N_{k+1} is somehow generated that excludes x^k and perhaps other solutions that are no better than x^k. The nogood is added to $\mathcal{N}_k$ to obtain $\mathcal{N}_{k+1}$.

One way to generate the nogood N_{k+1} is to solve an inference dual of the restricted problem $P(x^k)$. The proof family $\mathcal{Q}$ associated with the dual is assumed to be complete, which means that the dual is a strong dual and there is no duality gap. The dual solution therefore has the form $(v_{k+1}, \bar{Q})$, where v_{k+1} is the optimal value of $P_{k+1} = P(x^k)$. The constraint $x \neq x^k$ is trivially a nogood, because there is no point in examining x^k again. This nogood can be strengthened by observing which characteristics of $P(x^k)$ are actually used in the proof $\bar{Q}$ that $f(x) \geq v_{k+1}$. Perhaps there is a set T_{k+1} of values of x such that $\bar{Q}$ remains valid when $P(x^k)$ is changed to $P(x)$ for any $x \in T_{k+1}$. Then the constraint $x \notin T_{k+1}$ can serve as the next nogood N_{k+1}.

All of the examples of constraint-directed search in the previous chapter illustrate this process. The Davis-Putnam-Loveland (DPL) algorithm of Section 2.3.2, for instance, forms each restriction $P(x^k)$ by fixing the variables on which the search branched to reach the current leaf node of the search tree. When $P(x^k)$ is infeasible, a nogood or conflict clause is formed to exclude the variable assignments that are actually responsible for the unsatisfiability. Thus, at Node 5 in Figure 2.9, $P(x^0)$ is formed by fixing $(x_1, \ldots, x_5)$ to (F, F, F, F, F). The conflict clause $x_1 \lor x_5$ is generated because the assignments $(x_1, x_5) = (F, F)$ are enough to create the infeasibility.

Formally, the restriction $P(x^k)$ used in the DPL algorithm is a feasibility problem that minimizes 0 subject to the original clause set and the fixed variables. The inference dual is the problem of proving that $P(x^k)$ is infeasible, or more precisely, proving a lower bound of ∞ on the objective function ($v \geq \infty$). Because DPL uses the unit clause rule to check for feasibility, the dual solution is $(\infty, \bar{Q})$, where $\bar{Q}$ is a proof of infeasibility using the unit clause rule. This proof is examined to determine which fixed variables are actually used as premises, perhaps $(x_{j_1}, \ldots, x_{j_p}) = (x^k_{j_1}, \ldots, x^k_{j_p})$. Then there is no need to consider solu-

tions in
$$T_{k+1} = \left\{ x \in \{T,F\}^n \mid (x_{j_1}, \ldots, x_{j_p}) = (x_{j_1}^k, \ldots, x_{j_p}^k) \right\}$$
and $x \notin T_{k+1}$ is the desired nogood.

At Node 5, the unit clause rule proves infeasibility by deriving the unit clause x_5 from the constraint $x_1 \vee x_5$ and $x_1 = F$, and then deriving the empty clause from x_5 and $x_5 = F$. Since only the fixed variables $(x_1, x_5) = (F, F)$ are used,
$$T_1 = \{(x_1, \ldots, x_5) \mid (x_1, x_5) = (F, F)\}$$
and the nogood is $x \notin T_1$, or equivalently $x_1 \vee x_5$.

The same principle can be used to derive nogood bounds when the nogood set is solved in the form of a relaxation R_k. Recall that in this case a nogood bound $v \geq B_{k+1}(x)$ is designed to exclude x^k and perhaps other solutions x that produce restrictions $P(x)$ with optimal values that are no better than x^k. So, $B_{k+1}(x)$ must have the property that $B_{k+1}(x^k) = v_{k+1}$ and $B_{k+1}(x) \leq v(x)$ for all $x \in D$, where $v(x)$ is the optimal value of $P(x)$.

Suppose again that $(v_{k+1}, \bar{U})$ solves the inference dual of $P(x^k)$. Perhaps the proof $\bar{U}$ can be used to derive a useful bound $B_{k+1}(x)$ on $v(x)$ for values of x other than x^k.

The machine scheduling problem of Section 2.3.7 provides an example. This is an instance of Benders decomposition, because the subproblem $P(x^k)$ is always obtained by fixing the same subset of variables, namely the variables that assign jobs to machines. It fixes these variables to the values obtained in the master problem R_k.

In iteration $k = 0$ of the machine scheduling example, the solution x^0 of the initial master problem R_0 assigns Jobs 1, 2, 3, and 5 to Machine A and Job 4 to Machine B. The resulting subproblem $P(x^0)$ is to schedule these jobs to minimize makespan. The optimal makespan is 10, which is shown to be optimal by an edge-finding proof of infeasibility when the makespan is constrained to be at most 9. This edge-finding proof is the solution $\bar{U}$ of the inference dual of $P(x^0)$. By tracing the steps of the proof, it is determined that Jobs 2, 3, and 5 are responsible for the infeasibility, which means that the makespan remains 10 when only these jobs are assigned to Machine A. This yields the Benders cut
$$v \geq B_1(x) = 10(x_{A2} + x_{A3} + x_{A5} - 2)$$
This cut remains useful for machine assignments x other than x^0, so long as x assigns Jobs 2, 3 and 5 to Machine A. Section 3.13.3 presents easy way to obtain stronger cuts when all release times are the same.

As remarked in Section 2.3.7, edge finding alone may fail to detect infeasibility, because it is an incomplete inference method. To ensure that the inference dual of $P(x^k)$ is a strong dual, it is necessary to combine edge finding with branching. Section 2.3.7 briefly indicates how Benders cuts can be obtained from an infeasibility proof that involves branching as well as edge finding.

3.2.5 Exercises

1 Consider the optimization problem

$$\begin{aligned}
\min\ & 2x_1 + x_2 \\
& x_1 + x_2 \geq 1 \\
& x_1 - x_2 \geq 0 \\
& x_1, x_2 \geq 0
\end{aligned} \quad (3.9)$$

where each x_j is a real number. Suppose that an inequality can be inferred from a constraint set if and only if it is a sum of one or more constraints in the set. Solve the inference dual of (3.9) using this family of proofs. Exhibit the proof that solves the dual. What is the duality gap?

2 Recall the constraint-directed branching search of Figure 2.9. At Node 6, what is the restricted problem P_{k+1} (here $k = 1$)? Exhibit a proof (a series of steps using the unit clause rule) that solves the inference dual of P_{k+1} by proving infeasibility. What is T_{k+1}, and what is the resulting nogood? Do the same analysis at Node 7, where $k = 2$.

3 Identify the inference dual of each P_{k+1}, its solution, and T_{k+1} for each step of the partial-order dynamic backtracking algorithm of Table 2.10. In what way does the nogood for $k = 1$ fail to reflect a full analysis of the dual solution?

4 In the machine scheduling problem of Section 2.3.7, Job 4 is assigned to Machine B in iteration $k = 0$. What is the resulting subproblem $P(x^k)$ on this machine, and what is its inference dual? Exhibit a (trivial) solution of the inference dual. What is the Benders cut $v \geq B_{k+1}(x)$ for this machine?

3.3 Linear Inequalities

The success and versatility of linear programming has proved the usefulness of linear inequalities as a modeling device for problems with continuous variables. Furthermore, the theory of inference for linear inequalities is simple, and linear programming solvers are extremely effective. Linear programming duality is an elegant and useful theory that provides a basis for sensitivity analysis and economic analysis, as well as Benders decomposition. Bounds propagation is also a simple matter, although it may converge only asymptotically. The dual solution can provide additional domain filtering, as in the case of reduced-cost variable fixing.

Without loss of generality, a linear inequality set can be written

$$Ax \geq b, \quad x \geq 0 \tag{3.10}$$

If one wishes a variable x_j to take negative values, it can be replaced wherever it occurs with the difference of two nonnegative variables.

When dealing with linear inequalities, it is convenient to regard $x \geq 0$ as defining the domain of x. Thus, an inequality $ax \geq a_0$ implies $cx \geq c_0$ when any $x \geq 0$ that satisfies the former also satisfies the latter. Similarly, a linear system $Ax \geq b$ implies $cx \geq c_0$ when any $x \geq 0$ that satisfies $Ax \geq b$ also satisfies $cx \geq c_0$.

A linear inequality $ax \geq a_0$ can be regarded as feasible when some $x \geq 0$ satisfies it. Clearly, $ax \geq a_0$ is infeasible if and only if $a \leq 0$ and $a_0 > 0$. An inequality $ax \geq a_0$ *dominates* $cx \geq c_0$ when $a \leq c$ and $a_0 \geq c_0$, or $ax \geq a_0$ is infeasible. Thus, a linear inequality $ax \geq a_0$ implies $cx \geq c_0$ if and only if some nonnegative multiple of the first dominates the second (an infeasible inequality implies all inequalities). Implication of $cx \geq c_0$ by a system $Ax \geq b$ is characterized in the next section.

3.3.1 A Complete Inference Method

The classical Farkas Lemma provides a complete inference method for linear inequalities. The lemma says, in effect, that any inequality implied by a system of linear inequalities is implied by some nonnegative linear combination of the system. More precisely, any inequality implied by a feasible system $Ax \geq b$ is dominated by a *surrogate* of that system, which is a linear combination $uAx \geq ub$ with $u \geq 0$.

The Farkas Lemma, as classically stated, characterizes feasible systems of linear equations, but the result can be extended to implication by systems of linear inequalities.

THEOREM 3.4 (FARKAS LEMMA) *The linear system $Ax = b$, $x \geq 0$ is insoluble if and only if $uA \leq 0$ and $ub > 0$ for some real vector u.*

Linear Inequalities

The system $Ax = b$ is clearly insoluble if such a u exists, because in this case it implies a surrogate $uAx \geq ub$ that is nonpositive on the left-hand side and positive on the right-hand side.

The converse can be understood geometrically. If (3.10) is insoluble, b lies outside the polyhedral cone $C = \{Ax \mid x \geq 0\}$ spanned by the columns of A. The classical Weierstrass theorem implies that C contains a point $b-u$ that is the closest point in C to b (Figure 3.2). Furthermore, the hyperplane $\{y \mid uy = 0\}$ *separates* b *from* C, in the sense that $ub > 0$ and $uy \leq 0$ for all $y \in C$. But since any point of the form Ax for $x \geq 0$ lies in C, $uAx \leq 0$ for all $x \geq 0$. Because this could not be the case if any component of uA were positive, $uA \leq 0$ as claimed.

The Farkas Lemma can be applied to inequalities as follows. The system (3.10) is insoluble if and only if

$$\begin{bmatrix} A & -I \end{bmatrix} \begin{bmatrix} x \\ s \end{bmatrix} = b, \quad x, s \geq 0$$

is insoluble, where s is a vector of *surplus variables* and I is an identity matrix. Applying the Farkas Lemma to this yields

COROLLARY 3.5 *(3.10) is insoluble if and only if $Ax \geq b$ has an infeasible surrogate; that is, there is a real vector $u \geq 0$ for which $ub > 0$ and $uA \leq 0$.*

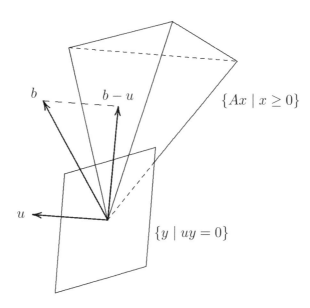

Figure 3.2. *Illustration of the proof of the Farkas Lemma.*

This implies that nonnegative linear combination is a complete inference method for linear inequalities.

COROLLARY 3.6 *A linear system $Ax \geq b$ implies all and only inequalities dominated by its surrogates.*

Proof. Clearly $Ax \geq b$ implies any surrogate and therefore any inequality dominated by a surrogate. For the converse it suffices to consider any inequality $cx \geq c_0$ that is implied by $Ax \geq b$, and show that some surrogate of $Ax \geq b$ dominates $cx \geq c_0$. Suppose first that (3.10) is feasible. Then (3.10) together with $cx \leq c_0 - \epsilon$ is insoluble for any $\epsilon > 0$. That is, the following is insoluble for any $\epsilon > 0$.

$$\begin{bmatrix} A \\ -c \end{bmatrix} x \geq \begin{bmatrix} b \\ \epsilon - c_0 \end{bmatrix}, \quad x \geq 0$$

Applying Corollary 3.5, there is a vector $[\bar{u} \ \ u_0] \geq [0 \ \ 0]$ such that $\bar{u}A - u_0 c \leq 0$ and $\bar{u}b + u_0\epsilon - u_0 c_0 > 0$ for any $\epsilon > 0$. One may suppose $u_0 > 0$, because otherwise (3.10) is infeasible by Corollary 3.5. Dividing by u_0 yields $uA \leq uc$ and $ub \geq c_0$, where $u = \bar{u}/u_0 \geq 0$. Thus, $cx \geq c_0$ is dominated by the surrogate $uAx \geq ub$ of (3.10).

If (3.10) is infeasible, then by Corollary 3.5, $Ax \geq b$ has an infeasible surrogate, which dominates all inequalities, including $cx \geq c_0$. □

3.3.2 Domain and Bounds Completeness

Domain and bounds completeness are identical for linear inequalities, because the projection of a linear system (3.10) onto a variable x_j is always an interval $[L_j, U_j]$ of real numbers. The bound L_j can be computed by minimizing x_j subject to the constraints of (3.10) and U_j by maximizing x_j. Since both are linear programming problems, domain completeness can be achieved by solving $2n$ linear programming problems over the same constraint set.

Bounds filtering is often faster than solving $2n$ linear programming problems, although in general it does not achieve completeness. Each round of filtering propagates the inequalities of (3.10) one at a time. Prior to propagating inequality i, let the current domain of each x_j be $[L_j, U_j]$. Inequality i is used to compute an updated lower bound for each x_j in the obvious way:

$$L'_j = \begin{cases} \max\left\{L_j, \dfrac{1}{A_{ij}}\left(b_i - \sum_{k \in J_i^+} A_{ik} U_k - \sum_{k \in J_i^-} A_{ik} L_k\right)\right\} & \text{if } A_{ij} > 0 \\ L_j & \text{otherwise} \end{cases}$$

Linear Inequalities

and similarly for the upper bound:

$$U'_j = \begin{cases} \min\left\{U_j, \dfrac{1}{A_{ij}}\left(b_i - \sum_{k \in J_i^+} A_{ik}L_k - \sum_{k \in J_i^-} A_{ik}U_k\right)\right\} & \text{if } A_{ij} < 0 \\ U_j & \text{otherwise} \end{cases}$$

Here, $J_i^+ = \{k \neq j \mid A_{ik} > 0\}$, and analogously for J_i^-. An illustration is presented in Section 2.2.3.

An example in which bounds filtering does not achieve completeness is the system $x_1 + x_2 \geq 1$, $x_1 - x_2 \geq 0$. If the initial domain is $[0,1]$ for each variable, bounds filtering has no effect, but the projection onto x_1 is $[\frac{1}{2}, 1]$.

Bounds filtering need not converge to a fixed point in finitely many iterations. Consider, for example, the system

$$\begin{aligned} \alpha x_1 - x_2 &\geq 0 \\ -x_1 + x_2 &\geq 0 \end{aligned} \quad (3.11)$$

with $0 < \alpha < 1$ and with initial domain $[0,1]$ for each variable. In each round, the first inequality yields $U'_2 = \alpha U_1$, and the second inequality yields $U'_1 = U'_2 = \alpha U_1$. So the upper bound U_1 converges asymptotically to zero.

3.3.3 k-Completeness

One can achieve k-completeness for a linear system (3.10) by projecting its feasible set onto each subset of k variables. This is the polyhedral projection problem, which is traditionally approached in two ways—Fourier-Motzkin elimination and generation of surrogates. Both tend to be computationally difficult.

The general polyhedral projection problem is to project a linear system of the form

$$\begin{aligned} Ax + By &\geq b \\ x \in \mathbb{R}^p, \; y &\in \mathbb{R}^q \end{aligned} \quad (3.12)$$

onto x. The projection is itself a polyhedron and is therefore described by a finite set of linear inequalities.

Fourier-Motzkin elimination computes the projection by removing one variable at a time. It first projects (3.12) onto $(x, y_1, \ldots, y_{q-1})$. Then, it projects the system thereby obtained onto $(x, y_1, \ldots, y_{q-2})$, and so forth until $y_1, \ldots, y_q$ are eliminated.

The idea may be illustrated by projecting the following system onto $x = (x_1, x_2)$:

$$\begin{aligned} x_1 + 2x_2 + y_1 + 2y_2 &\geq 10 \\ 2x_1 - x_2 - 2y_1 - y_2 &\geq 4 \\ 3y_1 + 2y_2 &\geq 2 \\ 3x_1 - x_2 + y_1 &\geq 6 \end{aligned} \quad (3.13)$$

First, y_2 is eliminated by "solving" for y_2 the inequalities that contain y_2:

$$\begin{aligned} & y_2 \geq -\tfrac{1}{2}x_1 - x_2 - \tfrac{1}{2}y_1 + 5 \\ 2x_1 - x_2 - 2y_1 - 4 \geq y_2 & \\ & y_2 \geq -\tfrac{3}{2}y_1 + 1 \end{aligned}$$

The expression(s) on the left are paired with those on the right:

$$\begin{aligned} 2x_1 - x_2 - 2y_1 - 4 &\geq -\tfrac{1}{2}x_1 - x_2 - \tfrac{1}{2}y_1 + 5 \\ 2x_1 - x_2 - 2y_1 - 4 &\geq -\tfrac{3}{2}y_1 + 1 \end{aligned}$$

The resulting inequalities are simplified and combined with the inequalities in (3.13) that do not contain y_2:

$$\begin{aligned} 5x_1 - 3y_1 &\geq 18 \\ 4x_1 - 2x_2 - y_1 &\geq 10 \\ 3x_1 - x_2 + y_1 &\geq 6 \end{aligned} \quad (3.14)$$

Next, y_1 is eliminated in the same manner, leaving the system

$$\begin{aligned} 7x_1 - 3x_2 &\geq 16 \\ 14x_1 - 3x_2 &\geq 36 \end{aligned} \quad (3.15)$$

This is the projection of (3.12) onto x.

A second method of projection eliminates all of the y_j's at once. It takes advantage of the completeness of nonnegative linear combination as an inference method. The projection of (3.12) onto x is described by all the linear inequalities containing only x that can be inferred from (3.12). By Corollary 3.6, these are all dominated by the surrogates that contain only x. These surrogates, in turn, are all nonnegative linear combinations $uAx \geq ub$ for which $uB = 0$. Thus, the projection of (3.12) onto x is the set

$$P(x) = \{x \mid uAx \geq ub, \text{ all } u \in C\}$$

where C is the polyhedral cone $C = \{u \geq 0 \mid uB = 0\}$.

This defines $P(x)$ in terms of infinitely many surrogates, but fortunately only finitely many are necessary. The *extreme rays* of C are vectors that define the edges of the cone; all vectors in C are nonnegative

Linear Inequalities   127

linear combinations of a finite set of extreme rays. Thus, the projection can be finitely computed

$$P(x) = \{x \mid uAx \geq ub, \text{all } u \in \text{extr}(C)\}$$

where $\text{extr}(C)$ is the set of extreme rays of the cone C.

In the example above, the cone is

$$C = \left\{ [u_1\ u_2\ u_3\ u_4] \geq [0\ 0\ 0\ 0] \,\middle|\, [u_1\ u_2\ u_3\ u_4] \begin{bmatrix} 1 & 2 \\ -2 & -1 \\ 3 & 2 \\ 1 & 0 \end{bmatrix} = \begin{bmatrix} 0 \\ 0 \end{bmatrix} \right\}$$

C has two extreme rays, $u = [0\ 2\ 1\ 1]$ and $u = [1\ 2\ 0\ 3]$. The resulting surrogates $uA \geq ub$ are the two inequalities (3.15) already found to define the projection $P(x)$.

A step of Fourier-Motzkin elimination is actually a special case of this approach that eliminates only one variable. The first inequality in (3.14), for example, was obtained from the first two inequalities of (3.13). It is a linear combination of these inequalities with multipliers 1 and 2, respectively, which causes y_2 to vanish.

3.3.4   Linear Programming Duality

The classical linear programming dual can be viewed as an inference dual or as a relaxation dual. It is developed here as an inference dual and applied to sensitivity analysis. Section 4.2.2 will reinterpret it as a relaxation dual.

A *linear programming problem* minimizes (or maximizes) a linear function subject to linear inequalities. The problem can be written

$$\begin{aligned} \min\ & cx \\ & Ax \geq b,\ x \geq 0 \end{aligned} \tag{3.16}$$

where A is an $m \times n$ matrix. If $x \geq 0$ is viewed as defining the domain of x, the inference dual of (3.16) is

$$\begin{aligned} \max\ & v \\ & (Ax \geq b) \overset{Q}{\vdash} (cx \geq v) \\ & v \in \mathbb{R},\ Q \in \mathcal{Q} \end{aligned} \tag{3.17}$$

Since nonnegative linear combination and domination is an inference method for linear inequalities, the proof variable Q can range over nonnegative linear combinations of the inequalities in $Ax \geq b$. Each proof can be encoded as a vector $u \geq 0$ of multipliers that define the linear combination. Thus, the inference dual seeks the strongest possible inequality $cx \geq v$ that is dominated by a surrogate.

The completeness of nonnegative linear combination and domination as an inference method for linear inequalities (Corollary 3.6) implies strong duality for linear programming. The primal (3.16) and the inference dual (3.17) always have the same optimal value.

Since the proof u that solves the dual has polynomial size and can be checked in polynomial time, the inference dual (3.17) belongs to NP. This and strong duality imply, by Theorem 3.3, that linear programming belongs to both NP and co-NP.

As an example, consider the linear programming problem

$$\begin{aligned} \min\ & 4x_1 + 7x_2 \\ & 2x_1 + 3x_2 \geq 6 \quad (a) \\ & 2x_1 + x_2 \geq 4 \quad (b) \\ & x_1, x_2 \geq 0 \end{aligned} \qquad (3.18)$$

which is illustrated in Figure 3.3. The optimal solution is $(x_1, x_2) = (3, 0)$ with value 12. The inference dual seeks the largest v for which $4x_1 + 7x_2 \geq v$ can be inferred from the constraints. The dual solution is $(u_1, u_2) = (2, 0)$ with $v = 12$. One can verify that a linear combination of constraints (a) and (b) using multipliers 2 and 0, respectively, yields an inequality $4x_1 + 6x_2 \geq 12$ that dominates $4x_1 + 7x_2 \geq 12$.

The classical linear programming dual of (3.16) is itself a linear programming problem:

$$\begin{aligned} \max\ & ub \\ & uA \leq c, \ u \geq 0 \end{aligned} \qquad (3.19)$$

It is equivalent to the inference dual when the primal is feasible.

THEOREM 3.7 *The inference dual (3.17) and the classical dual (3.19) of a feasible linear programming problem (3.16) have the same optimal value, and any finite optimal solution of one is optimal for the other.*

Proof. A linear combination u derives $cx \geq v$ from $Ax \geq b$ precisely when $uAx \geq ub$ dominates $cx \geq v$; that is, when $uA \leq c$ and $ub \geq v$. Since the inference dual is the problem of finding the strongest inequality $cx \geq v$ dominated by a surrogate of $Ax \geq b$, it is the problem of finding a $u \geq 0$ that maximizes v (and therefore maximizes ub) subject to $uA \leq c$. This latter problem is the classical dual problem (3.19). □

Thus, the inference dual of a linear programming problem not only belongs to NP but can be solved by linear programming (when the pri-

Linear Inequalities

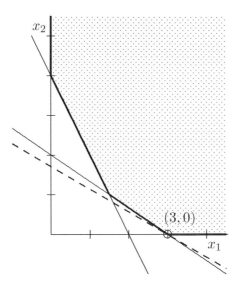

Figure 3.3. Feasible set (shaded area) of a linear programming problem. The small circle marks the solution $(3,0)$ of the primal problem. The dashed line represents the strongest inequality of the form $4x_1 + 7x_3 \geq v$ that can be inferred from the constraint set (i.e., $v = 12$). The dual solution is a proof that derives $4x_1 + 7x_3 \geq 12$.

mal is feasible).[2] For example, the classical dual of (3.18) is the linear programming problem

$$\max \ 6u_1 + 4u_2$$
$$2u_1 + 2u_2 \leq 4$$
$$3u_1 + u_2 \leq 7$$
$$u_1, u_2 \geq 0$$

This problem can be solved to obtain the multipliers $(u_1, u_2) = (2, 0)$.

Although strong duality holds without exception for the inference dual of a linear programming problem, the classical strong duality theorem is subject to a constraint qualification.

THEOREM 3.8 *A linear programming problem (3.16) and its classical dual (3.19) have the same optimal value, unless both problems are infeasible.*

Proof. If the primal (3.16) is feasible, then the classical dual and inference dual are equivalent by Theorem 3.7, so that (3.16) and (3.19)

---

[2]The inference dual can be solved by linear programming even when the primal is infeasible, provided the classical dual is feasible. In this case, the classical dual is unbounded and is "solved" by a ray u for which $uA \leq 0$ and $ub > 0$. Thus, the surrogate $uAx \geq ub$ is infeasible and therefore dominates $cx \geq v$ for any v, however large.

have the same optimal value. If the classical dual (3.19) is feasible, then one can take advantage of the fact that, for classical duality, the dual of the dual is the primal. Thus, (3.19) can be regarded as the feasible primal and (3.16) the dual. These have the same optimal value, again by Theorem 3.7. □

An important corollary of strong duality is the *complementary slackness principle*. It states that a primal constraint is slack only if the corresponding dual variable is zero. That is, if $\bar{x}$ is optimal in the primal and $\bar{u}$ is optimal in the dual, then $\bar{u}_i(A^i\bar{x} - b_i) = 0$ for all i, where A^i is row i of A. This is implied by the following.

COROLLARY 3.9 *If $\bar{x}$ is optimal in the primal problem (3.16), and $\bar{u}$ is optimal in the dual problem (3.19), then $\bar{u}(A\bar{x} - b) = 0$.*

3.3.5 Sensitivity Analysis

Due to the simplicity of inference duality for linear programming, sensitivity analysis is straightforward. Suppose that the linear programming problem (3.16) is feasible and is altered as follows:

$$\min (c + \Delta c)x$$
$$(A + \Delta A)x \geq b + \Delta b, \quad x \geq 0$$

Let u^* be an optimal dual solution for the original problem. It defines a valid surrogate inequality for the altered problem:

$$u^*(A + \Delta A)x \geq u^*(b + \Delta b)$$

This inequality dominates and therefore proves the bound

$$(c + \Delta c)x \geq u^*b + u^*\Delta b \qquad (3.20)$$

if $u^*(A + \Delta A) \leq c + \Delta c$. Since $u^*A \leq c$ by dual feasibility, it suffices that $u^*\Delta A \leq \Delta c$. Noting that u^*b in (3.20) is the optimal value of the original problem, one can state the following.

THEOREM 3.10 *Let v^* be the optimal value of a linear programming problem (3.16) and u^* an optimal solution of its dual. The optimal value of the altered problem (3.3.5) is bounded below by $v^* + u^*\Delta b$, provided $u^*\Delta A \leq \Delta c$.*

The components of u^* are variously known as Lagrange multipliers or shadow prices, because they indicate the marginal cost of increasing the right-hand side of $Ax \geq b$. If b_i represents an output requirement, $u_i^*\Delta b_i$ is the cost of raising that requirement by Δb_i. In linear programming, $u_i^*\Delta b_i$ is exactly the cost, not merely a bound on the cost, if the perturbation Δb_i is not too large. One can determine what "too large" means if additional information about the primal solution is available, as discussed in the next section.

Returning to the example (3.19), the optimal value is 12 and the optimal dual solution is $(u_1^*, u_2^*) = (2, 0)$. Applying Theorem 3.10, the perturbed problem (3.3.5) has an optimal value of at least $12 + 2\Delta b_1$, provided $2\Delta A_{11} \leq \Delta c_1$ and $2\Delta A_{12} \leq \Delta c_2$. For instance, if the right-hand side of the first inequality is reduced from 6 to 3, and the problem is otherwise unchanged, the minimum cost is bounded below by $12 + 2(-3) = 6$ (it is actually 8). Thus the solution is rather sensitive to the first constraint's right-hand side. Perturbations of the second constraint have no effect on the bound, because $u_2^* = 0$.

3.3.6 Basic Solutions

A sharper sensitivity analysis requires a closer look at the algebra of linear programming, particularly the role of basic solutions. An algebraic discussion will prove useful in several other contexts as well, due to the popularity of linear programming as a relaxation.

Without loss of generality, a linear optimization problem can be written with equality constraints.

$$\begin{aligned} \min \; & cx \\ & Ax = b, \quad x \geq 0 \end{aligned} \qquad (3.21)$$

An inequality constraint $ax \geq a_0$ can always be converted to an equality constraint by introducing a surplus variable $s_0 \geq 0$ and writing $ax - s_0 = a_0$.

The constraint set of (3.21) defines a polyhedron. If (3.21) has an optimal solution, then some vertex of this polyhedron is an optimal solution. It therefore suffices to restrict one's attention to vertex solutions, also known as basic feasible solutions, to find an optimal solution.

If A has rank m, a *basic* solution of (3.21) is one in which at most m variables are positive. Each basic solution corresponds to a choice of m linearly independent columns of A. After removing dependent rows from A, let A be partitioned into a square submatrix B consisting of m independent columns and a submatrix N that contains the remaining columns. Then, (3.21) may be written

$$\begin{aligned} \min \; & c_B x_B + c_N x_N \\ & B x_B + N x_N = b, \quad x_B, x_B \geq 0 \end{aligned} \qquad (3.22)$$

where x and c are similarly partitioned. The variables in x_B are the basic variables, and those in x_N are the nonbasic variables. Since B is a nonsingular $m \times m$ matrix, the equations may be solved for x_B in terms

of x_N:
$$x_B = B^{-1}b - B^{-1}Nx_N \tag{3.23}$$

Any feasible solution of (3.21) may be obtained by setting x_N to some nonnegative value. If x_N is set to zero, the resulting solution $(x_B, x_N) = (B^{-1}b, 0)$ is a basic solution, because obviously at most m variables are positive. It is a basic feasible solution if $B^{-1}b \geq 0$.

If B^{-1} is known, it is easy to check whether a basic feasible solution is optimal. Using (3.23), the objective function $c_B x_B + c_N x_N$ may be written in terms of the nonbasic variables only. After collecting terms, this yields

$$c_B B^{-1} b + (c_N - c_B B^{-1} N) x_N = c_B B^{-1} b + r x_N$$

The quantity $c_B B^{-1} b$ is the objective function value of the basic solution $(x_B, x_N) = (B^{-1}b, 0)$, and $r = c_N - c_B B^{-1} N$ is the vector of *reduced costs*. If $r \geq 0$, no feasible solution can yield a smaller objective function value (since $x_N \geq 0$ in any feasible solution). Thus, a basic feasible solution is optimal if the corresponding reduced costs are nonnegative.

As an example, consider the problem (3.18) in equality form

$$\min \begin{bmatrix} 4 & 7 & 0 & 0 \end{bmatrix} x$$
$$\begin{bmatrix} 2 & 3 & -1 & 0 \\ 2 & 1 & 0 & -1 \end{bmatrix} x = \begin{bmatrix} 6 \\ 4 \end{bmatrix}, \quad x \geq 0 \tag{3.24}$$

where $x = (x_1, x_2, x_3, x_4)$ and x_3, x_4 are surplus variables. The basic solutions are illustrated in Figure 3.4. The optimal solution is obtained when x_1, x_4 are basic and

$$B = \begin{bmatrix} 2 & 0 \\ 2 & -1 \end{bmatrix}$$

The optimal solution is

$$x_B = \begin{bmatrix} x_1 \\ x_4 \end{bmatrix} = B^{-1} b = \begin{bmatrix} \frac{1}{2} & 0 \\ 1 & -1 \end{bmatrix} \begin{bmatrix} 6 \\ 4 \end{bmatrix} = \begin{bmatrix} 3 \\ 2 \end{bmatrix}$$
$$x_N = \begin{bmatrix} x_2 \\ x_3 \end{bmatrix} = \begin{bmatrix} 0 \\ 0 \end{bmatrix}$$

The reduced cost vector is

$$r = c_N - c_B B^{-1} N = \begin{bmatrix} 7 & 0 \end{bmatrix} - \begin{bmatrix} 4 & 0 \end{bmatrix} \begin{bmatrix} \frac{1}{2} & 0 \\ 1 & -1 \end{bmatrix} \begin{bmatrix} 3 & -1 \\ 1 & 0 \end{bmatrix} = \begin{bmatrix} 1 & 2 \end{bmatrix}$$

Since $r \geq 0$, the solution is indeed optimal.

Linear Inequalities

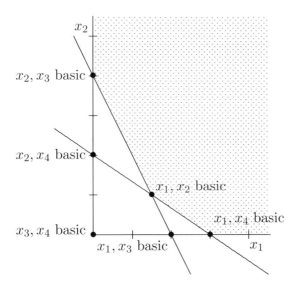

Figure 3.4. Basic solutions (black dots) of a linear programming problem, showing the basic variables in each case. Three of the basic solutions are feasible.

Linear programming solvers typically deliver an optimal basic solution when an optimal solution exists. They also provide such information as r, B^{-1}, and $c_B B^{-1}$. It will be seen in the next section that the last is the vector of dual variables.

3.3.7 More Sensitivity Analysis

The development of Section 3.3.6 permits a more precise sensitivity analysis than presented earlier. First, the classical dual of a linear programming problem in equality form (3.21) is

$$\max\ ub$$
$$uA \leq c$$

The dual variables u are not restricted to be nonnegative because the constraints are equalities. To see why, write the equality constraints $Ax = b$ as two systems of inequalities $Ax \geq b$, $-Ax \geq -b$ and associate with them dual variables u^+, u^-, respectively. The dual constraints then become $u^+A - u^-A \leq c$, which can be written $uA \leq c$ where $u = u^+ - u^-$ is an unrestricted variable.

The classical dual of the partitioned problem (3.22) is

$$\max\ ub$$
$$uB \leq c_B,\ \ uN \leq c_N \tag{3.25}$$

The following is easily shown.

THEOREM 3.11 *If $(x_B, x_N) = (B^{-1}b, 0)$ is an optimal basic solution of the primal problem (3.22), then $u = c_B B^{-1}$ is optimal in the dual (3.25).*

Thus $c_B B^{-1}$ is the vector of shadow prices. For further sensitivity analysis, consider the perturbed problem

$$\min (c_B + \Delta c_B) x_B + (c_N + \Delta c_N) x_N \\ B x_B + (N + \Delta N) x_N = b + \Delta b, \quad x_B, x_N \geq 0 \quad (3.26)$$

Note that the basic columns B are unchanged, since otherwise the analysis is quite difficult. Let $(x_B^*, 0)$ be an optimal solution of the original problem (3.22) with optimal value v^* and optimal dual solution $u^* = c_B B^{-1}$. If B remains the optimal basis after the perturbation, the new optimal solution is $B^{-1}(b + \Delta b)$, with objective function value $(c_B + \Delta c_B) B^{-1}(b + \Delta b)$. B remains the optimal basis so long as the current basic solution remains feasible (nonnegative) and the reduced costs remain nonnegative. The latter occurs when

$$(c_N + \Delta c_N) - (c_B + \Delta c_B) B^{-1}(N + \Delta N) \geq 0$$

The following is easily verified using the above and the definitions of x_B^*, v^*, u^* and r^*.

THEOREM 3.12 *Suppose that $x_B^* = B^{-1}b$ is an optimal solution of the linear programming problem (3.22), v^* the optimal value, u^* an optimal dual solution, and $r^* = c_N - u^* N$ the vector of reduced costs. If the problem is altered as in (3.26), the new optimal solution is $x_B^* + B^{-1} \Delta b$, the new objective function value is $v^* + \Delta c_B x_B^* + u^* \Delta b + \Delta c_B B^{-1} \Delta b$, and the optimal dual solution remains u^*, provided*

$$B^{-1} \Delta b \geq -x_B^* \\ u^* \Delta N - \Delta c_N + \Delta c_B B^{-1}(N + \Delta N) \leq r^*$$

For example, suppose the right-hand sides 6 and 4 of problem (3.24) are perturbed by Δb_1 and Δb_2, and the original costs 4 and 7 are perturbed by Δc_1 and Δc_2. Then,

$$\Delta b = (\Delta b_1, \Delta b_2), \quad \Delta c_B = [\Delta c_1 \; 0], \quad \Delta c_N = [\Delta c_2 \; 0]$$

The new solution is $(x_1, x_2) = (3 + \frac{1}{2} \Delta b_1, 0)$, and the new optimal cost is $12 + 3\Delta c_1 + 2\Delta b_1 + \frac{1}{2} \Delta b_1 \Delta c_1$, provided

$$\Delta b_1 \geq -6 \qquad 3\Delta c_1 - 2\Delta c_2 \leq 2 \\ \Delta b_1 - \Delta b_2 \geq -2 \qquad \Delta c_1 \geq -4$$

3.3.8 Domain Reduction with Dual Multipliers

Both integer programming and constraint programming frequently use reduced costs to reduce variable domains. This is generally done when the variables are required to be integral, and at least one feasible solution (with value U) is known—perhaps because it was found earlier in the search process.

First, a linear programming relaxation of the problem is solved. If the reduced cost of a nonbasic variable x_j indicates that increasing its value from zero to v would raise the optimal value of the relaxation above U, then only integral values of x_j less than v need be considered. If this is true for $v = 1$ in particular, then x_j can be fixed to zero, an operation known as *reduced-cost variable fixing*.

This kind of domain reduction is actually a special case of a general method for deriving valid inequalities that is based on the dual solution. It is convenient to state the general method first and then specialize it to domain reduction.

Suppose that (3.16) is a linear programming relaxation of the original problem with optimal solution x^*, optimal value v^*, and optimal dual solution u^*. Suppose further than $u_i^* > 0$, which means the ith constraint $A^i x \geq b_i$ is tight (i.e., $A^i x^* = b_i$), due to complementary slackness. If the solution x were to change to some value other than x^*, the left-hand side of $A^i x \geq b_i$ of (3.16) could change, say by an amount Δb_i. This would increase the optimal value of (3.16) as much as changing the constraint $A^i x \geq b_i$ to $A^i x \geq b_i + \Delta b_i$, which is to say it would increase the optimal value by at least $u_i^* \Delta b_i$. Since the optimal value should not rise to a value greater than U, one must have that $u_i^* \Delta b_i \leq U - v^*$, or $\Delta b_i \leq (U - v^*)/u_i^*$. Now, because $\Delta b_i = A^i x - A^i x^* = A^i x - b_i$, this yields the inequality

$$A^i x \leq b_i + \frac{U - v^*}{u_i^*} \qquad (3.27)$$

for each tight constraint i of (3.16). The inequality (3.27) can be propagated, which is particularly useful if some of the variables x_j have integer domains in the original problem.

One can reduce the domain of a particular nonbasic variable x_j by considering the nonnegativity constraint $x_j \geq 0$. Since the reduced cost r_j of x_j measures the effect on cost of increasing x_j, r_j can play the role of the dual multiplier in (3.27). So, (3.27) becomes $x_j \leq (U - v^*)/r_j$. If x_j has an integer domain in the original problem, one can say $x_j \leq \lfloor (U - v^*)/r_j \rfloor$. In particular, if $(U - v^*)/r_j < 1$, one can fix x_j to zero.

Suppose, for example, that (3.18) is a linear programming relaxation of a problem with integer-valued variables. The optimal dual solution of (3.18) is

$$u^* = c_B B^{-1} = [4 \ 0] \begin{bmatrix} \frac{1}{2} & 0 \\ 1 & -1 \end{bmatrix} = [2 \ 0]$$

Because the optimal value of (3.18) is $v^* = 12$ and the dual multiplier for the first constraint $2x_1 + 3x_2 \geq 6$ is $u_1^* = 2$, the inequality (3.27) becomes $2x_1 + 3x_2 \leq \frac{1}{2}U$. It was also found earlier that the nonbasic variable x_2 has reduced cost $r_2 = 2$, which yields the bound $x_2 \leq \lfloor \frac{1}{2}(U-12) \rfloor$. Thus, if $U < 14$, x_2 can be fixed to zero.

3.3.9 Classical Benders Cuts

Analysis of the inference dual can provide nogoods for constraint-directed search methods, including Benders decomposition (discussed in Section 3.2.4). If the problem becomes linear when certain variables are fixed, then the inference dual of the resulting subproblem is the classical linear programming dual. Its dual solution yields the linear cuts that were originally developed for Benders decomposition.

Classical Benders decomposition applies to problems of the form

$$\begin{aligned} \min \ & f(x) + cy \\ & g(x) + Ay \geq b \\ & x \in D_x, \ y \geq 0 \end{aligned}$$

which become linear when x is fixed to the solution x^k of the previous master problem R_k. The subproblem $P(x^k)$ is

$$\begin{aligned} \min \ & f(x^k) + cy \\ & Ay \geq b - g(x^k), \ y \geq 0 \end{aligned} \tag{3.28}$$

Its classical dual, ignoring the constant $f(x^k)$, is

$$\begin{aligned} \max \ & u\left(b - g(x^k)\right) \\ & uA \leq c, \ u \geq 0 \end{aligned} \tag{3.29}$$

Let v_{k+1} be the optimal value of (3.28), and suppose first that the dual (3.29) has a finite optimal solution u^{k+1}. Since the linear programming dual is a strong dual, the primal and dual have equal optimal values:

$$v_{k+1} - f(x^k) = u^{k+1}(b - g(x^k)) \tag{3.30}$$

But the value of any dual feasible solution is a lower bound on the optimal value of the primal (weak duality). This along with (3.30) implies that

$$B_{k+1}(x) = f(x) + u^{k+1}(b - g(x)) \tag{3.31}$$

is the tightest lower bound on cost when $x = x^k$. That is, u^{k+1} encodes a proof of the lower bound v_{k+1} by defining a linear combination
$$u^{k+1}Ay \geq u^{k+1}(b - g(x^k))$$
that dominates $cy \geq v_{k+1} - f(x^k)$.

The key to generating the classical Benders cut is the fact that u^{k+1} remains feasible in the dual problem (3.29) when x^k is replaced by any x. Thus, by weak duality, (3.31) remains a lower bound on cost for any x, or to put it differently, u^{k+1} encodes a proof of the lower bound $B_{k+1}(x)$ for any x. This yields the Benders cut
$$v \geq f(x) + u^{k+1}(b - g(x)) \tag{3.32}$$

If the dual (3.29) is unbounded, there is a direction or ray u^{k+1} along which its solution value can increase indefinitely. In this case, the Benders cut is
$$u^{k+1}(b - g(x)) \leq 0 \tag{3.33}$$
rather than (3.32).

Consider, for example, the problem
$$\min x_1 + 2x_2 + 3y_1 + 4y_2$$
$$4x_1 + x_2 - y_1 - 2y_2 \geq -2$$
$$-x_1 - x_2 - y_1 + y_2 \geq 2$$
$$x_j, y_j \geq 0, \ x_j \text{ integral}, \ j = 1, 2$$

Since the problem becomes linear when $x = (x_1, x_2)$ is fixed, the problem naturally decomposes into a master problem containing x and a linear subproblem containing y. If x^k is the solution of the master problem, the resulting subproblem $P(x^k)$ is:
$$\min x_1^k + x_2^k + 3y_1 + 4y_2$$
$$-y_1 - 2y_2 \geq -2 - 4x_1^k - x_2^k$$
$$-y_1 + y_2 \geq 2 + x_1^k + x_2^k$$
$$y_j \geq 0, \ j = 1, 2$$

The initial master problem R_0 minimizes v subject to no constraints and is solved by setting $v = -\infty$ and x to, say, $x^0 = (0, 0)$. The subproblem $P(x^0)$ is infeasible, and its dual is unbounded with an extreme ray solution $u^1 = (1, 2)$. The resulting Benders cut (3.33) is added to the master problem to obtain R_1:
$$\min v$$
$$2x_1 - x_2 \geq 2$$
$$x_j \geq 0 \text{ and integer}$$

An optimal solution is $x^1 = (1, 0)$ with $v = -\infty$. The next subproblem $P(x^1)$ has optimal solution $y = (0, 3)$ with dual solution $u^2 = (0, 4)$ and value $v_2 = 13$. The resulting Benders cut (3.32) is added to the master problem:

$$\min v$$
$$2x_1 - x_2 \geq 2$$
$$v \geq 9 + 4x_1 + 4x_2$$
$$x_j \geq 0 \text{ and integer}$$

The optimal solution of this problem is $x^2 = (1, 0)$ with value 13. Since this is equal to the value of a previous subproblem, the algorithm terminates with optimal solution $x = (1, 0)$ and $y = (0, 3)$.

3.3.10 Exercises

1. Suppose the domain of $x \in \mathbb{R}^n$ is given by $\ell \leq x \leq u$. Under what conditions does $ax \geq a_0$ imply $cx \geq c_0$?

2. Consider the linear system $Ax = b, x \geq 0$ given by

$$-x_1 + 2x_2 = -1$$
$$2x_1 - x_2 = 0$$
$$x_1, x_2, \geq 0$$

   Draw the polyhedral cone $C = \{Ax \mid x \geq 0\}$. What is the point $u - b$ in C closest to b? What is u? Prove that the linear system is infeasible.

3. Use the Farkas Lemma to prove Corollary 3.5.

4. Use the Farkas Lemma to prove that a feasible system $Ax = b, x \geq 0$ implies all and only inequalities dominated by linear combinations of the system.

5. Consider the inequalities $x_1 + x_2 \geq 1$, $x_1 - x_2 \geq 0$ with each $x_j \in [0, \alpha]$. For what values of $\alpha \geq 0$ does bounds propagation achieve bounds completeness?

6. Use Fourier-Motzkin elimination to project the feasible set of (3.11) and $x_1, x_2 \in [0, 1]$ onto x_1, assuming $0 < \alpha < 1$.

7. Suppose that every component of A is strictly positive. Use Fourier-Motzkin elimination to verify that bounds propagation is enough to achieve bounds completeness for $Ax \geq b, x \geq 0$.

Linear Inequalities

8 Suppose that the system $Ax + By \geq b, x \geq 0$ is given by

$$x + 2y \geq 3$$
$$x - y \geq 1$$
$$x - 2y \geq 0$$
$$x, y \geq 0$$

and let $P(x)$ be the projection of $P = \{x \mid Ax + Bx \geq b, x \geq 0\}$ onto x. Draw the polyhedral cone $C = \{u \geq 0 \mid uB = 0\}$, in u_1, u_2-space. (Eliminate u_3 in the expression for C by solving $uB = 0$ for u_3 in terms of u_1, u_2.) What are the two extreme rays of C in u-space? What are the two resulting projection cuts $uAx \geq ub$? What is $P(x)$?

9 Let $P(x) = \{x \mid uAx \geq ub$, all $u \in C\}$ be the projection of $P = \{x \mid Ax + By \geq b, x \geq 0\}$ onto x, where $C = \{u \geq 0 \mid uB = 0\}$. Show that

$$P(x) = \{x \mid uAx \geq ub, \text{ all } u \in \text{extr}(C)\}$$

by showing that any inequality $uAx \geq ub$ for $u \in C$ is implied by inequalities $uAx \geq ub$ for $u \in \text{extr}(C)$.

10 Consider the linear programming problem

$$\min 4x_1 + 4x_2 + 3x_3$$
$$x_1 + 2x_2 + x_3 \geq 2$$
$$2x_1 + x_2 + x_3 \geq 3$$
$$x_1, x_2, x_3 \geq 0$$

Solve the classical dual by hand and use the solution to obtain the surrogate that provides the tightest bound on the optimal value of the primal. What is the optimal value? (There is no need to solve the primal directly.) Now use complementary slackness to find an optimal solution of the primal by solving two simultaneous equations.

11 Exhibit a linear programming problem with two variables for which both the primal and the dual are infeasible.

12 Show that the classical dual of $\min\{cx \mid Ax = b, x \geq 0\}$ is the problem $\max\{ub \mid uA \leq c\}$, where u is not restricted to be nonnegative.

13 Prove Corollary 3.9 by first showing $\bar{u}b \leq \bar{u}A\bar{x} \leq c\bar{x}$.

14 Compute the reduced costs for (3.24) when x_1 and x_2 are basic.

15 Prove Theorem 3.11.

16 Prove Theorem 3.12.

17 Suppose that the problem in Exercise 10 is the continuous relaxation of an integer programming problem (i.e., the same problem but with the restriction that the variables take integral values). Suppose further that the best known integral solution has value 8. Derive two inequalities from the dual solution that can be propagated to reduce domains. Also, derive an upper bound on the nonbasic variable x_3 from its reduced cost. (You can deduce the reduced cost from the slack in the corresponding dual constraint. Why?)

18 Solve the following problem by Benders decomposition:
$$\min\ 6x_1 + 5x_2 + 4y_1 + 3y_2$$
$$7x_1 + 5x_2 + 3y_1 + 2y_2 \geq 16$$
$$x_1, x_2 \geq 0 \text{ and integral},\ y_1, y_2 \geq 0$$

19 Show that (3.33) is a valid Benders cut when the dual of the subproblem is unbounded.

20 Show that the classical Benders algorithm converges if D_x is finite. Assume the dual of the subproblem is never infeasible.

3.4 General Inequality Constraints

General inequality constraints have too little structure for the development of specialized inference methods. They provide the right level of abstraction, however, for defining the well-known surrogate dual and Lagrangean dual, both of which are generalizations of the linear programming dual. The Lagrangean dual, in particular, is widely used for reducing domains and obtaining tight bounds.

These two duals appear quite different on the surface, but their close relationship becomes evident when they are viewed as inference duals. In both duals, the inference method is nonnegative linear combination plus domination, but the Lagrangean dual uses a slightly stronger form of domination. It follows immediately that the surrogate dual, in general, provides a tighter bound. The Lagrangean dual, however, has some convenient properties that make it far more popular.

Like the linear programming dual, the Lagrangean and surrogate duals can be conceived as relaxation duals as well as inference duals. They are derived here as inference duals, and in Section 4.6 as relaxation duals. Domain reduction is discussed here because it is an inference technique. The bounding function will be taken up when the duals are derived as relaxation duals.

General Inequality Constraints 141

3.4.1 The Surrogate Dual

An inequality-constrained problem may be written

$$\begin{aligned} & \min \ f(x) \\ & g(x) \geq 0 \\ & x \in S \end{aligned} \quad (3.34)$$

where $g(x)$ is a tuple of functions $g_1(x), \ldots, g_m(x)$, which may be linear or nonlinear. The domain constraint $x \in S$ represents an arbitrary constraint set that does not necessarily consist of inequalities. The variables themselves can be discrete or continuous. Linear programming is a special case of (3.34) in which $f(x) = cx$, $g(x) = Ax - b$, and $S = \{x \in \mathbb{R}^n \mid x \geq 0\}$.

A single inequality $g_0(x) \geq 0$ is regarded as infeasible when no $x \in S$ satisfies it. An inequality $g_0(x) \geq 0$ *dominates* inequality $h_0(x) \geq 0$ when $g_0(x) \leq h_0(x)$ for all $x \in S$, or when $g_0(x) \geq 0$ is infeasible. It is important to note that, unlike the situation with linear inequalities, a feasible inequality $g_0(x) \geq 0$ may imply $h_0(x) \geq 0$ even when no nonnegative multiple of the former dominates the latter. Thus, domination by an inequality is stronger than implication by a nonnegative multiple of that inequality.

The surrogate dual results when one formulates an inference dual for (3.34) in which the inference method is nonnegative linear combination and implication (not domination):

$$\begin{aligned} & \max v \\ & (g(x) \geq 0) \overset{Q}{\vdash} (f(x) \geq v) \\ & Q \in \mathcal{Q}, \ v \in \Re \end{aligned} \quad (3.35)$$

Each proof $Q \in \mathcal{Q}$ corresponds to a vector $u \geq 0$ of multipliers. Q deduces $f(x) \geq v$ from $g(x) \geq 0$ when the inequality $ug(x) \geq 0$ implies the inequality $f(x) \geq v$ for $x \in S$, which is to say that the minimum of $f(x)$ subject to $ug(x) \geq 0$ and $x \in S$ is at least v. So, (3.35) is equivalent to

$$\max_{u \geq 0} \left\{ v \ \middle| \ \min_{x \in S} \{ f(x) \mid ug(x) \geq 0 \} \geq v \right\}$$

This can be written

$$\max_{u \geq 0} \{\sigma(u)\}, \quad \text{where } \sigma(u) = \min_{x \in S} \{ f(x) \mid ug(x) \geq 0 \} \quad (3.36)$$

The surrogate dual therefore seeks a surrogate of $g(x) \geq 0$ for which $\min f(x)$ subject to the surrogate is as large as possible.

THEOREM 3.13 *The surrogate dual, conceived as in inference dual in which inference is nonnegative linear combination and implication, is equivalent to (3.36).*

As an example, consider the integer programming problem

$$\begin{aligned} & \min\ 3x_1 + 4x_2 \\ & -x_1 + 3x_2 \geq 0, \\ & 2x_1 + x_2 - 5 \geq 0 \\ & x_1, x_2 \in \{0, 1, 2, 3\} \end{aligned} \quad (3.37)$$

The optimal solution of the problem is $(x_1, x_2) = (2, 1)$, with optimal value 10. The surrogate dual maximizes $\sigma(u_1, u_2)$ subject to $u_1, u_2 \geq 0$, where $\sigma(u_1, u_2)$ is

$$\min_{x_j \in \{0,1,2,3\}} \{3x_1 + 4x_2 \mid (-u_1 + 2u_2)x_1 + (3u_1 + u_2)x_2 - 5u_2 \geq 0\}$$

For example, $\sigma(2, 5) = 10$ because the minimum value of $3x_1 + 4x_2$ subject to the surrogate $8x_1 + 11x_2 \geq 25$ and $x_j \in \{0, 1, 2, 3\}$ is 10 (Figure 3.5). Since the optimal value of the dual can be no greater than the optimal value 10 of the primal, $(u_1, u_2) = (2, 5)$ must be an optimal solution of the dual. In this case, there is no duality gap, although in general there is a gap.

Surrogate duality is identical to linear programming duality when the problem is linear.

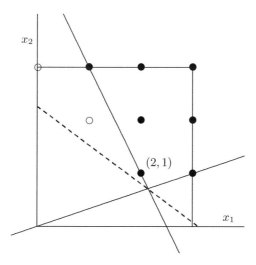

Figure 3.5. Feasible set (black dots) of two integer linear inequalities, and feasible set (black dots and open circles) of the surrogate relaxation $8x_1 + 11x_2 \geq 25$ (dashed line). The point $(2, 1)$ is optimal for both the original problem and the relaxation.

3.4.2 The Lagrangean Dual

The Lagrangean dual is an inference dual of (3.34) with a slightly different family $\mathcal{Q}$ of proofs than in the case of the surrogate dual. In the surrogate dual, each proof $Q \in \mathcal{Q}$ corresponds to a vector $u \geq 0$ for which

$$ug(x) \geq 0 \text{ implies } f(x) \geq v \text{ for } x \in S$$

In the Lagrangean dual, implication is replaced by domination. Thus in the Lagrangean dual, each proof $Q \in \mathcal{Q}$ corresponds to a vector $u \geq 0$ for which

$$ug(x) \geq 0 \text{ dominates } f(x) \geq v \quad (3.38)$$

This results in a stronger form of inference, since one inequality can imply another even when no nonnegative multiple of the first dominates the second.

The surrogate $ug(x) \geq 0$ dominates $f(x) \geq v$ when $ug(x) \leq f(x) - v$ for all $x \in S$, or when $ug(x) \geq 0$ is an infeasible inequality. Suppose first that $ug(x) \geq 0$ is a feasible inequality for all $u \geq 0$. Then, due to the definition of domination, the Lagrangean dual maximizes v subject to $u \geq 0$, $ug(x) \leq f(x) - v$ and $x \in S$. Since the constraint $ug(x) \leq f(x) - v$ can be written $v \leq f(x) - ug(x)$, one can let

$$\theta(u, x) = f(x) - ug(x)$$

and write the Lagrangean dual as

$$\max_{u \geq 0}\{\theta(u)\}, \quad \text{where } \theta(u) = \min_{x \in S}\{\theta(u, x)\} \quad (3.39)$$

The problem $\min_{x \in S}\{\theta(u, x)\}$ can be referred to as the *inner problem*. The constraints $g(x) \geq 0$ are said to be *dualized* because they are moved to the objective function. The components of u are *Lagrange multipliers*.

If $ug(x) \geq 0$ is an infeasible inequality for some $u \geq 0$, then the Lagrangean dual is unbounded. But in this case, (3.39) is likewise unbounded. The following theorem has been shown.

THEOREM 3.14 *The Lagrangean dual, conceived as an inference dual in which inference is nonnegative linear combination and domination, is equivalent to (3.39).*

Since domination by a nonnegative multiple is stronger than implication, the family $\mathcal{Q}$ of proofs in the Lagrangean dual is a subset of the corresponding family in the surrogate dual. This implies the following theorem.

THEOREM 3.15 *The optimal value of the surrogate dual of a problem (3.34) is at least as large as the optimal value of the Lagrangean dual.*

In other words, the surrogate dual in general provides a tighter bound than the Lagrangean dual. The Lagrangean dual, however, enjoys two useful properties that the surrogate dual lacks, and these are established in the next section.

The Lagrangean dual for the problem instance (3.37) finds the maximum of $\theta(u_1, u_2)$ subject to $u_1, u_2 \geq 0$, where

$$\theta(u_1, u_2) = \min_{x_j \in \{0,1,2,3\}} \{(3 + u_1 - 2u_2)x_1 + (4 - 3u_1 - u_2)x_2 + 5u_2\} \quad (3.40)$$

Note that the value $\theta(u_1, u_2)$ of the inner problem is easy to compute, because values of x_1, x_2 that achieve the minimum are

$$x_1 = \begin{cases} 0 & \text{if } 3 + u_1 - 2u_2 \geq 0 \\ 3 & \text{otherwise} \end{cases} \quad x_2 = \begin{cases} 0 & \text{if } 4 - 3u_1 - u_2 \geq 0 \\ 3 & \text{otherwise} \end{cases} \quad (3.41)$$

as illustrated in Figure 3.6. It will be seen in the next section that the optimal solution of the dual is $(u_1, u_2) = (\frac{5}{7}, \frac{13}{7})$, with optimal value $9\frac{2}{7}$. This establishes a lower bound of $9\frac{2}{7}$ on the optimal value of (3.37). Since the optimal value of (3.37) is 10, there is a duality gap, and the Lagrangean dual provides a weaker bound than obtained from the surrogate dual in the previous section.

In this example, the Lagrangean dual provides the same bound $9\frac{2}{7}$ as the continuous relaxation of (3.37), obtained by replacing $x_1, x_2 \in \{0, 1, 2, 3\}$ with $x_1, x_2 \in [0, 3]$. This is because the inner problem has the same solution as its continuous relaxation and can therefore be solved as a linear programming problem. In general, the Lagrangean dual of an integer programming problem gives the same bound as the continuous relaxation when the inner problem can be solved as a linear programming problem. The Lagrangean dual is therefore useful for integer programming only when the inner problem does not reduce to linear programming. In practice, the choice of which constraints to dualize is made so that the inner problem is harder than linear programming but is nonetheless reasonably easy to solve, perhaps because it separates into smaller problems or has special structure.

3.4.3 Properties of the Lagrangean Dual

A simple but important property of the Lagrangean dual is weak duality.

THEOREM 3.16 *If $\bar{x}$ is feasible in (3.34) and $\bar{u} \geq 0$, then $\theta(\bar{u}) \leq f(\bar{x})$.*

This is because

$$\theta(\bar{u}) = \min_{x \in S}\{\theta(\bar{u}, x)\} \leq \theta(\bar{u}, \bar{x}) = f(\bar{x}) - \bar{u}g(\bar{x}) \leq f(\bar{x})$$

General Inequality Constraints

where the second inequality is due to $\bar{u} \geq 0$ and $g(\bar{x}) \geq 0$.

Perhaps the most useful property of the Lagrangean dual is the following corollary.

COROLLARY 3.17 *The function $\theta(u)$ is concave.*

This allows one to maximize $\theta(u)$ by a steepest ascent search (Section 4.6.5). The concavity of $\theta(u)$ derives from the fact that $\theta(u,x)$, for fixed x, is an affine function of u. This makes $\theta(u)$ a minimum over affine functions and therefore a concave function. If S is finite, $\theta(u)$ is a minimum of finitely many affine functions and is therefore concave and piecewise linear.

In the example (3.37), $\theta(u_1, u_2)$ is linear except where $3 + u_1 - 2u_2 = 0$ or $4 - 3u_1 - u_2 = 0$. So, in this small example, the maximum can be found by examining points in the nonnegative quadrant at which the lines described by these two equations intersect with each other or with the axes (Figure 3.6). The maximum occurs at their intersection with each other, namely at $(u_1, u_2) = (\frac{5}{7}, \frac{13}{7})$.

A second property of Lagrangean duality is that a complementary slackness property holds when there is no duality gap.

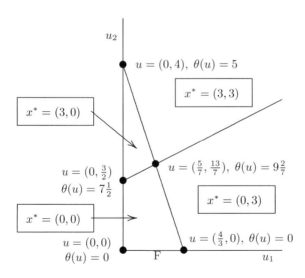

Figure 3.6. *Some values of the Lagrangean function $\theta(u)$ for the integer programming problem of Figure 3.5. Regions in which x^* takes four possible values are shown, where $x = x^*$ maximizes $\theta(u, x)$ and thus $\theta(u) = \theta(u, x^*)$. The function $\theta(u)$ is linear within each of the four regions and is concave overall.*

COROLLARY 3.18 *Let x^* be an optimal solution of (3.34) and u^* an optimal solution of the Lagrangean dual of (3.34). If there is no duality gap, then $u^*g(x^*) = 0$.*

This follows from
$$f(x^*) = \theta(u^*) = \min_{x \geq 0}\{\theta(u^*, x)\} \leq \theta(u^*, x^*) = f(x^*) - u^*g(x^*) \leq f(x^*)$$
where the first equation is due to the lack of a duality gap, and the last inequality due to $u^* \geq 0$ and $g(x^*) \geq 0$.

Lagrangean duality, like surrogate duality, is identical to linear programming duality when the problem is linear. This is because, when the problem is linear, the form of domination (3.38) used in the Lagrangean dual reduces to the form of domination used to define the linear programming dual.

3.4.4 Domain Reduction with Lagrange Multipliers

Section 3.3.8 used dual multipliers to filter domains. Lagrange multipliers can be used in a similar fashion.

Let u^* be an optimal solution of the Lagrangean dual of (3.34), and let $v^* = \theta(u^*)$ be the optimal value of the dual. If x^* solves the inner optimization problem $\min_{x \in S}\{\theta(u^*, x)\}$, then $\theta(u^*) = \theta(u^*, x^*)$. Suppose further that constraint i is tight, so that $g_i(x^*) = 0$. If x^* were changed to some other value, the function value $g_i(x)$ could increase, say by an amount Δ_i. This would increase the optimal value of (3.34) as much as changing the constraint $g_i(x) \geq 0$ to $g_i(x) - \Delta_i \geq 0$. The function $\theta(u)$ for this altered problem is
$$\theta'(u) = \min_{x \in S}\{\theta'(u, x)\}, \quad \text{where } \theta'(u, x) = f(x) - ug(x) + u_i\Delta_i$$
Since $\theta'(u^*, x)$ differs from $\theta(u^*, x)$ only by a constant, any x that minimizes $\theta(u^*, x)$ also minimizes $\theta'(u^*, x)$. So
$$\theta'(u^*) = \theta'(u^*, x^*) = \theta(u^*) + u_i^*\Delta_i = v^* + u_i^*\Delta_i$$
is a lower bound on the optimal value of the altered problem, due to weak duality (Theorem 3.16).

Now suppose some known feasible solution of (3.34) has value U, so that U is an upper bound on the optimal value of (3.34). Then any optimal solution x must satisfy $v^* + u_i^*\Delta_i \leq U$, or
$$\Delta_i \leq \frac{U - v^*}{u_i^*}$$

General Inequality Constraints

Since $\Delta_i = g_i(x) - g_i(x^*) = g_i(x)$, this becomes

$$g_i(x) \leq \frac{U - v^*}{u_i^*} \tag{3.42}$$

This inequality can now be propagated.

There is also an analog of reduced cost variable fixing. If constraint i imposes a lower bound L_j on a variable x_j (i.e., $x_j - L_j \geq 0$), and the constraint is tight, then (3.42) becomes

$$x_j \leq L_j + \frac{U - v^*}{u_i^*} \tag{3.43}$$

If constraint i imposes an upper bound U_j on x_j and is tight, then

$$x_j \geq U_j - \frac{U - v^*}{u_i^*} \tag{3.44}$$

Both (3.43) and (3.44) can be used to reduce the domain of x_j.

In the example (3.37), suppose that the best known feasible solution is $x = (3, 1)$, which means the upper bound is $U = 13$. The optimal dual solution is $u^* = (u_1^*, u_2^*) = (\frac{5}{7}, \frac{13}{7})$ with $\theta(u^*) = 9\frac{2}{7}$. The solution $x^* = (0, 0)$ of the inner problem given by (3.41) makes the first constraint tight and therefore allows one to infer (3.45) for $i = 1$:

$$-x_1 + 3x_2 \leq 5\tfrac{1}{5} \tag{3.45}$$

Bounds propagation using the original constraints yields the domains $x_1, x_2 \in \{1, 2, 3\}$. Propagation using both the original constraints and (3.45) yields the smaller domains $x_1 \in \{2, 3\}$ and $x_2 \in \{1, 2\}$.

3.4.5 Exercises

1 Show that surrogate duality reduces to linear programming duality when the problem is linear.

2 State the Lagrangean dual of the problem

$$\min\ 15x_1 + 15x_2$$
$$3x_1 + 5x_2 \geq 9,\ x_1, x_2 \in \{0, 1, 2\}\ \text{and integral}$$

in which $3x_1 + 5x_2 \geq 9$ is dualized. Solve it by plotting $\theta(u)$ against u. What is the duality gap?

3 Let the system $g(x) \geq 0$ of inequalities imply $f(x) \geq v$ for $x \in S$ when there is a $u \geq 0$ such that $ug(x) \leq f(x) - v - \alpha \min\{0, ug(x)\}$

for all $x \in S$, where $\alpha \geq 0$. Write the corresponding inference dual of (3.34). Show that its solution provides a bound that is no better than that of the surrogate dual and no worse than that of the Lagrangean dual.

4 Show that Lagrangean duality reduces to linear programming duality when $g(x) = Ax - b$ and $S = \{x \mid x \geq 0\}$. It suffices to show that the form of domination (3.38) used in the Lagrangean dual reduces to the form of domination used to define the linear programming dual.

5 In an integer programming problem $\min_{x \geq 0}\{cx \mid Ax \geq b, x \text{ integral}\}$, the Lagrangean relaxation is

$$\theta(u) = \min_{x \geq 0}\{(c - uA)x + ub \mid x \text{ integral}\}$$

Show that when the Lagrangean relaxation can be solved as a linear programming problem, the Lagrangean dual gives the same bound as the linear programming relaxation. That is, if

$$\theta(u) = \min_{x \geq 0}\{(c - uA)x + ub\}$$

then $\max_{u \geq 0}\{\theta(u)\} = \min_{x \geq 0}\{cx \mid Ax \geq b\}$. Also show that the Lagrangean bound is greater than or equal to the linear programming bound.

6 Consider the problem of minimizing cx subject to $\sum_j a_j x_j^2 \geq b$, where each $a_j > 0$ and $b > 0$, and x_j is any real number. Solve the Lagrangean dual. Show that there is no duality gap and that the constraint is tight at the solution of the inner problem. Hint: $g(x)$ is convex. Now, suppose that this problem is the continuous relaxation of a nonlinear integer programming problem, and the best known solution has value U. Write an inequality that can be used to reduce domains.

3.5 Propositional Logic

The logic of propositions provides a convenient way to express logical relations among variables. Both inference methods and relaxations are well developed for propositional logic. The inference methods can achieve k-completeness and k-consistency, not only for logical propositions, but for general constraint sets with two-valued variables.

Constraints expressed in propositional logic appear as logical formulas in which variables take boolean values (such as 1 and 0) corresponding

Propositional Logic

to true and false. Formulas are constructed by prefixing ¬ (*not*), or by joining subformulas with ∧ (*and*) and ∨ (inclusive *or*). Formulas can be defined recursively as consisting of (a) the empty formula, which is false by definition; (b) a single variable x_j, which is an unanalyzed or *atomic* proposition; or (c) expressions of the form ¬F, $F \wedge G$, and $F \vee G$, where F and G are formulas. The *material conditional* $F \to G$ is frequently introduced by defining it to mean ¬$F \vee G$. The *equivalence* $F \equiv G$ means $(F \to G) \wedge (G \to F)$.

The truth functions corresponding to the formulas are recursively defined in the obvious way. ¬F is true if and only if F is false, $F \wedge G$ is true if and only if F and G are true, and $F \vee G$ is true if and only if at least one of the formulas F, G is true.

3.5.1 Logical Clauses

It is useful for purposes of both inference and relaxation to convert propositional formulas to *conjunctive normal form* (CNF) or *clausal form*.

A *clause* is a disjunction of zero or more *literals*, each of which is a variable or its negation. Thus, $x_1 \vee \neg x_2 \vee x_3$ is a clause of three literals. A formula in CNF or clausal form is a conjunction of zero or more logical clauses. An attractive property of clauses is that it is easy to tell when one implies another. Clause C implies clause D when C *absorbs* D, meaning that all the literals of C occur in D.

Any propositional formula can be converted to an equivalent formula in clausal form. One way to do so is to bring negations inward by using De Morgan's laws

$$\neg(F \wedge G) \equiv (\neg F \vee \neg G)$$
$$\neg(F \vee G) \equiv (\neg F \wedge \neg G)$$

and to distribute disjunctions using the equivalence

$$(F \vee (G \wedge H)) \equiv ((F \vee G) \wedge (F \vee H))$$

For instance, the formula

$$\neg(x_1 \vee \neg x_2) \vee (x_1 \wedge \neg x_3)$$

can be put in CNF by first applying De Morgan's law

$$(\neg x_1 \wedge x_2) \vee (x_1 \wedge \neg x_3)$$

and then distributing the disjunction :

$$(\neg x_1 \vee x_1) \wedge (\neg x_1 \vee \neg x_3) \wedge (x_2 \vee x_1) \wedge (x_2 \vee \neg x_3)$$

and finally deleting the first clause since it is a *tautology* (necessarily true).

Distribution can lead to an exponential explosion, however. For example, the formula

$$(x_1 \wedge y_1) \vee \cdots \vee (x_n \wedge y_n) \tag{3.46}$$

converts to the conjunction of 2^n clauses of the form $F_1 \vee \cdots \vee F_n$, where each F_j is x_j or y_j. The explosion can be avoided by adding variables as follows. Rather than distribute a disjunction $F \vee G$ (when neither F nor G is a literal), replace it with the conjunction

$$(z_1 \vee z_2) \wedge (\neg z_1 \vee F) \wedge (\neg z_2 \vee G)$$

where z_1, z_2 are new variables, and the clauses $\neg z_1 \vee F$ and $\neg z_2 \vee G$ encode the implications $z_1 \rightarrow F$ and $z_2 \rightarrow G$, respectively. The conversion requires only linear time and space. Formula (3.46), for example, yields the conjunction

$$(z_1 \vee \cdots \vee z_n) \wedge \bigwedge_{j=1}^{n} (\neg z_j \vee x_j) \wedge (\neg z_j \vee y_j)$$

3.5.2 A Complete Inference Method

A simple inference method, *resolution*, is complete with respect to logical clauses. Given any two clauses for which exactly one variable x_j occurs positively in one clause and negatively in the other, one can infer the *resolvent* of the clauses, which consists of all the literals in the clauses except x_j and $\neg x_j$. For instance, the two clauses

$$\begin{array}{c} x_1 \vee x_2 \vee x_3 \\ \neg x_1 \vee x_2 \vee \neg x_4 \end{array} \tag{3.47}$$

imply their resolvent

$$x_2 \vee x_3 \vee \neg x_4 \tag{3.48}$$

The resolvent is obtained by resolving "on x_1."

Resolution is a valid inference because it reasons by cases. In the example, x_1 is either false or true. If it is false, then the first clause of (3.47) implies $x_2 \vee x_3$. If it is true, the second clause implies $x_2 \vee \neg x_4$. In either case, (3.48) follows by absorbsion.

Section 3.1.1 defines what it means to apply an inference method to a constraint set. Thus, to apply the resolution method to a clause set $\mathcal{S}$ is to add to $\mathcal{S}$ all resolvents from $\mathcal{S}$ that are not already absorbed by a clause in $\mathcal{S}$, and to repeat the procedure until no further resolvents

Propositional Logic

can be added. S is unsatisfiable if and only if resolution eventually generates the *empty clause*, (the clause with zero literals) by resolving two *unit clauses* x_j and $\neg x_j$.

THEOREM 3.19 *Resolution is a complete inference method with respect to logical clauses.*

Proof. Let S' be the result of applying the resolution method to a clause set S containing variables $x_1, \ldots, x_n$. It suffices to show that any clause implied by S is absorbed by a clause in S'. Suppose to the contrary. Let C be the longest clause with variables in $\{x_1, \ldots, x_n\}$ that is implied by S but absorbed by no clause in S', where length is measured by the number of literals in the clause. One can suppose without loss of generality that no variables in C are negated, because any negated variable x_j can be replaced in every clause of S with $\neg x_j$ without changing the problem. Note first that C cannot contain all the variables $x_1, \ldots, x_n$. If it did, setting all variables to false would violate C and must therefore violate some clause C' of S', because S' implies C. This means that C' contains only positive literals and therefore absorbs C, contrary to the definition of C.

Thus, C must lack some variable x_j. But in this case, the clauses $x_j \vee C$ and $\neg x_j \vee C$, which S implies (since it implies C), must be absorbed respectively by clauses in S', say D and $\bar{D}$ (because $x_j \vee C$ and $\neg x_j \vee C$ are longer than C). Furthermore, D must contain x_j, because otherwise it would absorb C, contrary to C's definition, and similarly $\bar{D}$ must contain $\neg x_j$. So, the resolvent of D and $\bar{D}$ absorbs C, which means that some clause in S' absorbs C, which is again inconsistent with C's definition. □

If all absorbed clauses are deleted at each step of the resolution algorithm, the clause set S' that remains contains the *prime implications* of the original clause set S. They are the undominated clauses implied by S; that is, the implications of S that are absorbed by no implication of S. An example appears in Section 2.2.4.

The resolution algorithm has exponential complexity in the worst case and can explode in practice if there are too many atomic propositions. There are cases in which resolution is well worth a substantial time investment, however, as when each node of the branching tree incurs a high computational cost, perhaps due to complex nonlinearities in the problem. The resolution algorithm may fix some of the logical variables and avoid expensive branches that would otherwise be explored.

3.5.3 Unit Resolution and Horn Clauses

When the full resolution algorithm is too slow, one can use weaker variants of resolution that run more rapidly and may nonetheless fix some of the logical variables. The simplest variant is *unit resolution*, which is ordinary resolution restricted to the case in which one of the clauses resolved is a unit clause.

The running time of unit resolution on a clause set S is proportional to the number of literals in S, because it is essentially a form of back substitution. An efficient implementation of unit resolution begins by counting the number n_i of literals in each clause $C_i \in S$, and placing in set Q the unit clauses of S. Each step of the algorithm removes a literal L from Q and removes from S every clause that contains L. It also removes $\neg L$ from every clause C_i that contains it and reduces n_i by 1; if $n_i = 1$, it adds C_i to Q. The procedure stops when Q is empty or some $n_i = 0$. S is unsatisfiable in the latter case.

Unit resolution can be used to check the satisfiability of an important class of clauses—renamable Horn clauses. A clause is *Horn* if it contains at most one positive literal. A Horn clause with exactly one positive literal, such as $x_1 \vee \neg x_2 \vee \neg x_3 \vee \neg x_4$, can be viewed as a conditional statement or *rule* whose consequent is a unit clause: $(x_2 \wedge x_3 \wedge x_4) \rightarrow x_1$. Such propositions are called *definite clauses* and are the norm in practical rule bases. A clause set is *renamable Horn* if all its clauses become Horn after some set of zero or more variables x_j are replaced by their negations $\neg x_j$.

Unit resolution is not a complete inference method even for Horn clauses. For example, it fails to derive $x_2 \vee \neg x_3$ from $\neg x_1 \vee x_2$ and $x_1 \vee \neg x_3$. However, the following is easily shown.

THEOREM 3.20 *A renamable Horn clause set S is unsatisfiable if and only if the unit resolution algorithm derives the empty clause from S.*

3.5.4 Domain Completeness and k-Completeness

A clause set S is k-complete if its projection onto any set of k variables is described by some set of clauses in S containing only those variables. One can therefore achieve k-completeness by computing the projection of S onto every subset of k variables.

The projection of S onto a given subset $\{x_1, \ldots, x_k\}$ of variables can be computed by applying the resolution method to S and selecting the resulting clauses that contain only variables in $\{x_1, \ldots, x_k\}$. In fact, a slight restriction of the resolution method accomplishes the same task.

Propositional Logic 153

THEOREM 3.21 *A restricted resolution method that resolves only on variables in $\{x_{k+1}, \ldots, x_n\}$ derives all clauses that describe the projection of a clause set onto variables $x_1, \ldots, x_k$.*

Proof. It is convenient to begin by proving a lemma. Suppose that clause set S implies clause C, and let S_C be the result of removing from S all clauses that contain a variable with sign opposite to its sign in C. Then S_C also implies C. To show this, suppose to the contrary. Then some truth assignment $v = (v_1, \ldots, v_n)$ satisfies every clause in S_C, violates C, and violates some clause $C' \in S$. Since $C' \notin S_C$, C' contains some variable that occurs in C with the opposite sign. But in this case it is impossible that v violate both C and C'.

Now, take any clause C implied by S that contains only variables in $\{x_1, \ldots, x_k\}$. It suffices to show that the restricted form of resolution derives from S a clause that absorbs C. But S_C implies C, due to the lemma. Since full resolution is complete, it derives from S_C a clause that absorbs C. This derivation never resolves on a variable in $\{x_1, \ldots, x_k\}$, because these variables occur in S_C always with the same sign. It therefore derives C from S using only restricted resolution. □

The theorem can be illustrated by projecting the following clause set onto x_1, x_2:

$$\begin{array}{l} x_1 \vee x_2 \\ \neg x_1 \vee x_3 \\ \neg x_1 \vee \neg x_2 \vee \neg x_3 \\ x_1 \vee x_3 \vee x_4 \\ \phantom{x_1 \vee{}} x_2 \vee x_3 \vee \neg x_4 \end{array} \qquad (3.49)$$

Only one resolvent on x_3 or x_4 that is not already absorbed by a clause can be generated, namely $\neg x_1 \vee \neg x_2$. The projection is therefore described by the clauses $x_1 \vee x_2$ and $\neg x_1 \vee \neg x_2$.

Unfortunately, full resolution is needed to achieve k-completeness. Since the projection must be computed for every subset of k variables, it is necessary to resolve on every variable, which achieves k-completeness for every k. Even simple domain completeness requires full resolution. Strong k-consistency is easier to achieve, however, as seen in the next section.

Resolution can achieve k-completeness not only for logical clauses but for any constraint set S that contains only boolean variables. If the two values in each variable's domain are arbitrarily identified with *true* and *false*, then one can, at least in principle, identify the logical clauses implied by a constraint in S. Let S' consist of all clauses implied by some constraint in $\mathcal{S}$. Then $\mathcal{S}$ and $\mathcal{S}'$ are equivalent, and one can achieve k-completeness for all k by applying the resolution algorithm to $\mathcal{S}'$.

3.5.5 Strong k-Consistency

Let *k-resolution* be the resolution method modified so that only resolvents with fewer than k literals are generated. When full resolution is too slow, but unit resolution is ineffective, k-resolution for small k may be a useful alternative. It can reduce and even eliminate backtracking, due to the following theorem:

THEOREM 3.22 *The k-resolution method achieves strong k-consistency.*

Proof. It suffices to show, for any $i \leq k$, that i-resolution applied to a clause set $\mathcal{S}$ achieves i-consistency. Let $\mathcal{S}'$ be the set that results from applying i-resolution to $\mathcal{S}$, which means that no further i-resolvents can be generated for $\mathcal{S}'$. Suppose, contrary to the claim, that $\mathcal{S}'$ is not i-consistent. Then there is an assignment

$$(x_{j_1}, \ldots, x_{j_{i-1}}) = (v_1, \ldots, v_{j_{i-1}}) \tag{3.50}$$

that violates no clause in $\mathcal{S}'$, but for which the extended assignments

$$(x_{j_1}, \ldots, x_{j_i}) = (v_1, \ldots, v_{j_{i-1}}, 0)$$
$$(x_{j_1}, \ldots, x_{j_i}) = (v_1, \ldots, v_{j_{i-1}}, 1)$$

respectively violate two clauses in $\mathcal{S}'$. These clauses must contain only variables in $\{x_{j_1}, \ldots, x_{j_i}\}$ and must therefore have a resolvent R (on x_{j_i}) that contains fewer than i variables. No clause in $\mathcal{S}'$ absorbs R because (3.50) violates no clause in $\mathcal{S}'$. But in this case, i-resolution generates R, which contradicts the assumption that no further resolvents can be generated. □

Strong 3-consistency can be achieved for the clause set (3.49) by generating all resolvents with fewer than three literals. The first round of 3-resolution yields the new resolvents

$$\neg x_1 \vee \neg x_2$$
$$x_2 \vee x_3$$
$$x_3 \vee x_4$$

after which no further 3-resolvents can be added to the clause set. Adding the above to (3.49) therefore achieves strong 3-consistency.

It was observed in the previous section that resolution can achieve k-completeness for any constraint set $\mathcal{S}$ with boolean variables. Resolution can also achieve strong k-consistency for $\mathcal{S}$. Let $\mathcal{S}'$ consist of all logical clauses implied by some constraint in $\mathcal{S}$. Apply k-resolution to $\mathcal{S}'$, and let $\mathcal{S}''$ be the result. Then, by Theorem 3.22, $\mathcal{S} \cup \mathcal{S}''$ is strongly k-consistent.

Propositional Logic

3.5.6 Completeness of Parallel Resolution

Section 2.3.4 uses parallel resolution, a restricted form of resolution, to implement a constraint-directed search algorithm—namely, partial-order dynamic backtracking. The algorithm is applied to the propositional satisfiability problem, but the same method can in principle be applied to any problem whose variables are boolean and for which the nogoods can be written as logical clauses.

A key property of parallel resolution is that, after it is used to *process* the nogood set, one can solve the resulting clause set (if a solution exists) without backtracking. This is because parallel resolution has a completeness property in the context of partial-order dynamic backgrounding. Specifically, it is a complete inference method with respect to conforming clauses.

Recall that parallel resolution is applied to a clause set $\mathcal{N}$ in which each clause has a *last* variable, and the other variables are *penultimate*. It is assumed that when the penultimate variables in each clause are viewed as preceding the last variable in the clause, the resulting precedence relations determine a well-defined partial order on the variables (i.e., there are no cycles). It is further assumed that any given variable always has the same sign whenever it occurs in a penultimate literal, since this is a property of the partial-order dynamic backtracking algorithm.

A clause D *parallel-absorbs* clause C if D is the empty clause, or $D = C$, or some penultimate literal of C is last in D. Each step of the parallel resolution algorithm identifies a pair of clauses in $\mathcal{N}$ that have a resolvent R on a variable that is last in both clauses, where R is parallel-absorbed by no clause of $\mathcal{N}$. Then all clauses in $\mathcal{N}$ that are parallel-absorbed by R are removed from $\mathcal{N}$, and R is added to $\mathcal{N}$. The process continues until no such pair of clauses exists.

A partial assignment *conforms* to $\mathcal{N}$ if each variable that occurs penultimately in $\mathcal{N}$ is assigned a value opposite to the sign in which it occurs. Thus a partial assignment

$$(x_{j_1}, \ldots, x_{j_r}) = (v_{j_1}, \ldots, v_{j_r}) \tag{3.51}$$

conforms to $\mathcal{N}$ if for each literal x_j that occurs penultimately in $\mathcal{N}$, x_j is one of the variables x_{j_i} and $v_{j_i} = F$, and for each literal $\neg x_j$ that occurs penultimately in $\mathcal{N}$, x_j is one of the variables x_{j_i} and $v_{j_i} = T$. Conformity is well defined because x_j has the same sign whenever it occurs penultimately.

The negation of a partial assignment (3.51) is the clause C given by

$$L_1 \vee \cdots \vee L_r$$

where $L_i = \neg x_{j_i}$ if $v_{j_i} = T$ and $L_i = x_{j_i}$ if $v_{j_i} = F$. One can say that clause C conforms to $\mathcal{N}$ when it is the negation of a partial assignment (3.51) that conforms to $\mathcal{N}$. Then C conforms to $\mathcal{N}$ when every penultimate literal in $\mathcal{N}$ occurs in C.

A basic property of parallel absorbsion is the following.

LEMMA 3.23 *Suppose that clause C conforms to $\{D\}$. Then if C is absorbed by some clause that D parallel-absorbs, C is absorbed by D.*

Proof. Suppose that C is absorbed by a clause D' that D parallel-absorbs. If $D = D'$ the lemma is trivial, so suppose $D \neq D'$. Then D' contains a penultimate literal that is last in D. Thus every penultimate literal of D occurs in C, since C conforms to D, and the last literal of D occurs in C, since D' absorbs C. It follows that D absorbs C. □

In partial-order dynamic backtracking, the nogood set $\mathcal{N}$ is solved in a greedy fashion. The variables $x_{j_1}, \ldots, x_{j_r}$ that occur penultimately in $\mathcal{N}$ are first assigned values (3.51) that conform to $\mathcal{N}$. Then each remaining variable x_j is assigned a value in such a way that it, along with the assignments already made, does not violate any clause in $\mathcal{N}$. For this procedure to find a solution without backtracking (assuming a solution exists), it suffices that any partial assignment that conforms to $\mathcal{N}$ and falsifies no clause of $\mathcal{N}$ can be extended to a solution of $\mathcal{N}$.

This is equivalent to a completeness property. A partial assigment falsifies a clause D if and only if its negation is absorbed by D. Thus the desired property is this: any clause conforming to $\mathcal{N}$ that is implied by $\mathcal{N}$ is also absorbed by some clause in $\mathcal{N}$. In other words, $\mathcal{N}$ is complete with respect to clauses that conform to it. It can be shown that $\mathcal{N}$ has this property after parallel resolution is applied.

THEOREM 3.24 *Parallel resolution applied to a clause set $\mathcal{N}$ is complete with respect to clauses that conform to $\mathcal{N}$.*

Proof. Let $\mathcal{N}'$ be the result of applying parallel resolution to $\mathcal{N}$. The claim is that any clause that conforms to $\mathcal{N}$ and is implied by $\mathcal{N}$ is absorbed by some clause of $\mathcal{N}'$. Suppose to the contrary, and let C be the longest clause that conforms to $\mathcal{N}$ and is implied by $\mathcal{N}$ but is absorbed by no clause of $\mathcal{N}'$. Also suppose without loss of generality that all the literals of C are positive. As in the proof of Theorem 3.19, C must omit at least one variable x_j. Then the clauses $C \vee x_j$ and $C \vee \neg x_j$, which $\mathcal{N}$ implies, must be absorbed respectively by clauses in $\mathcal{N}'$, say D and $\bar{D}$, where D contains x_j and $\bar{D}$ contains $\neg x_j$. Clauses D and $\bar{D}$ therefore have a resolvent R on x_j that absorbs C. Furthermore, R is a parallel resolvent because x_j must be the last variable in D and

Propositional Logic 157

$\bar{D}$. If x_j were not last in D, conformity would require C to contain x_j, which it does not, and the same is true if x_j were not last in $\bar{D}$. Now because R is a parallel resolvent, $\mathcal{N}'$ must contain some clause R' that parallel absorbs R, since $\mathcal{N}'$ was obtained by applying parallel resoluton to $\mathcal{N}$. This means that C conforms to R'. Clearly $R' \neq R$, since otherwise R absorbs C, contrary to assumption. Thus every literal of R' is either penultimate in R' or, if it is last in R', penultimate in R. So, Lemma 3.23 and the fact that R absorbs C imply that R' absorbs C, contrary to assumption. The theorem follows. □

It can be shown that parallel resolution has polynomial complexity when applied to the nogood set $\mathcal{N}$ in any step of partial-order dynamic backtracking. In fact, the running time is linear in the summed lengths of the clauses of $\mathcal{N}$.

3.5.7 Exercises

1. Write the formula $(a \wedge b) \rightarrow (c \wedge d)$ in CNF, first without adding variables, and then by adding two new variables.

2. Use the resolution algorithm to compute the prime implications of the formulas $\delta_{10} \vee \delta_{11}$, $\delta_{20} \vee \delta_{21} \vee \delta_{22}$, and $\delta_{11} \rightarrow \neg \delta_{22}$ in the example of Section 2.2.4.

3. It is impossible to put three pigeons in two pigeon holes with at most one pigeon per hole. Prove this by the resolution algorithm. Let x_{ij} be true when pigeon i is placed in hole j and formulate the problem in clausal form. The resolution proof for pigeon hole problems explodes exponentially with the number of pigeons.

4. Write the inequalities in Exercise 6 of Section 3.1 as clauses. Use 3-resolution to achieve strong 3-consistency.

5. Prove Theorem 3.20. Hints: prove the theorem first for Horn clauses. Then note that a unit resolution proof before renaming is a unit resolution proof after renaming.

6. A 2-satisfiability (2SAT) problem is a CNF problem with at most two literals in each clause. Show that 3-resolution, which has polynomial complexity, is complete for 2SAT. Describe a faster method to check for satisfiability by formulating the problem on a graph. Create a vertex for every variable and its negation and two directed edges for each clause.

7 Formulate the problem of checking whether a clause set is renamable Horn as a 2SAT problem. For each of the n variables x_j, let y_j be true when x_j is renamed. Since the number of clauses is quadratic in n, show how to add variables to make it linear in n.

8 Suppose that constraint set S contains only boolean variables, and suppose that every clause that is implied by a constraint in S belongs to clause set S'. Show that if S' contains no other clauses, S and S' are equivalent.

9 Two families are meeting for party. One family consists of Jane, Robert, and Suzie, and the other consists of Juan, Maria, and Javier. Since the house is small, three will congregate outdoors and three indoors, with both families represented in each location. Jane is shy and insists on being in the same location as another family member, and similarly for Juan. Suzie wants to be with Juan or Maria, and Juan wants to be with Robert or Suzie. Since the choice for each family member is boolean, the feasibility problem is naturally modeled in propositional logic. Draw the dependency graph and use k-resolution to achieve the degree of consistency necessary to solve the problem without backtracking.

10 Let a *multivalent clause* have the form $\bigvee_j (x_j \in S_j)$ where each x_j has a finite domain D_{x_j} and each $S_j \subset D_{x_j}$. Generalize the resolution algorithm to obtain a complete inference method for multivalent clauses, and prove completeness. Define an analog of unit resolution. Hints: a resolvent can be derived from several clauses. Resolving on x_j requires taking an intersection. Multivalent resolution can be used in dynamic backtracking methods.

11 Let be $\mathcal{N}$ be a clause set for which conformity is well defined. Show that the following are equivalent: (a) any partial assigment that conforms to $\mathcal{N}$ and falsifies no clause in $\mathcal{N}$ can be extended to a solution of $\mathcal{N}$; (b) any clause conforming to $\mathcal{N}$ that is implied by $\mathcal{N}$ is also absorbed by some clause in $\mathcal{N}$.

12 Show that at any stage of partial-order dynamic backtracking, any variable x_j that occurs in the nogood set has the same sign whenever it occurs penultimately.

3.6   0-1 Linear Inequalities

0-1 linear inequalities are useful for modeling because they combine elements of both linear inequalities and propositional logic. They have the

0-1 Linear Inequalities 159

numerical character of linear inequalities and can therefore deal with costs, counting, and the like. At the same time, they contain 0-1 variables that can express logical relations and, in fact, include logical clauses as a subset.

A second advantage of 0-1 linear inequalities is that they provide a ready-made continuous relaxation, formed simply by replacing the 0-1 variable domains with continuous unit intervals. This relaxation can often be strengthened by the addition of cutting planes to the constraint set, which are implied 0-1 linear inequalities that *cut off* noninteger solutions of the continuous relaxation.

Inference methods for 0-1 linear inequalities can in fact be studied with two purposes in mind—making the constraint set more nearly complete, and making the continuous relaxation tighter. The former goal leads to a theory of inference for 0-1 linear inequalities, which is developed in this section, and the latter inspires cutting plane theory, which is taken up in Section 4.4.

3.6.1 Implication between Inequalities

A 0-1 linear inequality has the form $ax \geq a_0$, where $x \in \{0,1\}^n$. An inequality $ax \geq a_0$ implies a second 0-1 inequality $cx \geq c_0$ when any $x \in \{0,1\}^n$ that satisfies the first also satisfies the second. Thus $ax \geq a_0$ implies $cx \geq c_0$ if and only if the following *0-1 knapsack problem*[3] has an optimal value of at least c_0:

$$\min cx \\ ax \geq a_0, \quad x \in \{0,1\}^n \tag{3.52}$$

This problem can be solved in polynomial time relative to the number of variables, provided the coefficients a_j are bounded in absolute value; in other words, the problem is pseudopolynomial. The problem of checking implication between 0-1 linear inequalities is therefore pseudopolynomial.

To study implication further, it is convenient to write a 0-1 inequality in the form $ax \geq a_0 + n(a)$, where $n(a)$ is the sum of the negative components of a. In this case, a_0 is the *degree* of the inequality. The degree is what the right-hand side would be if all variables with negative coefficients were replaced by their complements. Thus, $-2x_1 + 3x_2 \geq 1$ is written $-2x_1 + 3x_2 \geq 3 - 2$ and has degree 3.

---

[3]Strictly speaking, (3.52) is a 0-1 knapsack problem only when each $a_j, c_j \geq 0$. But it is easily converted to a 0-1 knapsack problem by replacing each variable x_j for which $a_j, c_j < 0$ with its complement $1 - \bar{x}_j$. If $a_j < 0$ and $c_j \geq 0$, then clearly one can set $x_j = 0$ in any optimal solution, and x_j can be dropped from the problem.

There are two sufficient conditions for implication that can be quickly checked: absorbsion and reduction. A 0-1 inequality *absorbs* any inequality obtained by strengthening the left-hand side and/or weakening the right-hand side. Thus, if $a, b \geq 0$, $ax \geq a_0$ absorbs $bx \geq b_0$ when $a \leq b$ and $a_0 \geq b_0$. For arbitrary a and b, $ax \geq a_0 + n(a)$ absorbs $bx \geq b_0 + n(b)$ if $|a| \leq |b|$, $a_0 \geq b_0$, and $a_j b_j \geq 0$ for all j, where $|a| = (|a_1|, \ldots, |a_n|)$. For example, $-2x_1 + 3x_2 \geq 3 - 2$ absorbs $-3x_1 + 3x_2 + x_3 \geq 2 - 3$.

THEOREM 3.25 *A 0-1 linear inequality implies any 0-1 linear inequality it absorbs.*

Proof. Suppose that $ax \geq a_0 + n(a)$ absorbs $bx \geq b_0 + n(b)$. They may respectively be written

$$\sum_{j \in J_1} a_j x_j - \sum_{j \in J_0} |a_j| x_j \geq a_0 - \sum_{j \in J_0} |a_j| \qquad (3.53)$$

and

$$\sum_{j \in J_1} b_j x_j - \sum_{j \in J_0} |b_j| x_j \geq b_0 - \sum_{j \in J_0} |b_j| \qquad (3.54)$$

where $a_j, b_j \geq 0$ for $j \in J_1$ and $a_j, b_j \leq 0$ for $j \in J_0$. (Since $a_j b_j \geq 0$, a_j and b_j do not differ in sign.) The following inequality

$$\sum_{j \in J_0} (|a_j| - |b_j|) x_j \geq \sum_{j \in J_0} (|a_j| - |b_j|) \qquad (3.55)$$

is valid because $|a_j| - |b_j| \leq 0$. Adding (3.55) to (3.53), one obtains

$$\sum_{j \in J_1} a_j x_j - \sum_{j \in J_0} |b_j| x_j \geq a_0 - \sum_{j \in J_1} |b_j|$$

which implies (3.54) because $a_0 \geq b_0$ and $a_j \leq b_j$ for $j \in J_1$. □

Absorbsion for 0-1 inequalities is a generalization of absorbsion for logical clauses. A clause such as $x_1 \lor \neg x_2 \lor \neg x_3$ can be written as the 0-1 inequality $x_1 + (1 - x_2) + (1 - x_3) \geq 1$ or $x_1 - x_2 - x_3 \geq 1 - 2$, where 0 and 1 correspond to false and true. A 0-1 inequality that represents a clause is a *clausal inequality* and always has degree 1. Clearly, one clause absorbs another if and only if the same is true of the corresponding clausal inequalities.

A second sufficient condition for implication between 0-1 inequalities is reduction. An inequality $ax \geq a_0$ *reduces to* any inequality obtained by reducing coefficients on the left-hand side and adjusting the right-hand

0-1 Linear Inequalities

side accordingly. Assuming $a, b \geq 0$, the inequality $ax \geq a_0$ reduces to $bx \geq a_0 - \sum_j (a_j - b_j)$ if $a \geq b$. Thus, $3x_1 + x_2 + x_3 \geq 3$ reduces to $2x_1 + x_2 \geq 1$. More generally, $ax \geq a_0 + n(a)$ reduces to $bx \geq a_0 + n(b) - \sum_j (|a_j| - |b_j|)$ if $|a| \geq |b|$ and $a_j b_j \geq 0$ for all j. For instance, $-3x_1 + 3x_2 + x_3 \geq 4 - 3$ reduces to $-2x_1 + 2x_2 \geq 1 - 2$.

THEOREM 3.26 *A 0-1 linear inequality implies any 0-1 linear inequality to which it reduces.*

Proof. Suppose $ax \geq a_0 + n(a)$ reduces to:
$$bx \geq a_0 + n(b) - \sum_j (|a_j| - |b_j|)$$

These two inequalities may respectively be written
$$\sum_{j \in J_1} a_j x_j - \sum_{j \in J_0} |a_j| x_j \geq a_0 - \sum_{j \in J_0} |a_j| \tag{3.56}$$

and

$$\sum_{j \in J_1} b_j x_j - \sum_{j \in J_0} |b_j| x_j \geq \\ a_0 - \sum_{j \in J_0} |b_j| - \left(\sum_{j \in J_1} a_j + \sum_{j \in J_0} |a_j| \right) + \left(\sum_{j \in J_1} b_j + \sum_{j \in J_0} |b_j| \right) \tag{3.57}$$

where $a_j, b_j \geq 0$ for $j \in J_1$ and $a_j, b_j \leq 0$ for $j \in J_0$. Note that two terms in (3.57) cancel. The following inequality is valid
$$\sum_{j \in J_1} (b_j - a_j) x_j + \sum_{j \in J_0} (|a_j| - |b_j|) x_j \geq -\sum_{j \in J_1} a_j + \sum_{j \in J_1} b_j, \tag{3.58}$$

because $b_j - a_j \leq 0$ for $j \in J_1$ and $|a_j| - |b_j| \geq 0$ for $j \in J_0$. Adding (3.58) to (3.56) yields (3.57).

3.6.2 Implication of Logical Clauses

It is easy to check whether a 0-1 linear inequality implies a clause or clausal inequality. The idea can be seen in an example. The inequality
$$x_1 - 2x_2 + 3x_3 - 4x_4 \geq 2 \tag{3.59}$$

implies the clause
$$x_1 \vee \neg x_2 \vee x_5$$

because the falsehood of the clause implies that (3.59) is violated. The clause is false only when $(x_1, x_2, x_3) = (0, 1, 0)$, and if the first two values are substituted into (3.59), the left-hand side is maximized when $(x_3, x_4) = (1, 0)$. This means the maximum value of the left-hand side is 1, and (3.59) is necessarily violated. In general,

THEOREM 3.27 *The 0-1 inequality $ax \geq a_0$ implies the logical clause*

$$\bigvee_{j \in J_1} x_j \vee \bigvee_{j \in J_0} \neg x_j$$

if and only if

$$\sum_{\substack{j \notin J_1 \\ a_j > 0}} a_j + \sum_{\substack{j \in J_0 \\ a_j < 0}} a_j < a_0$$

A simple recursive function (Figure 3.7) generates all nonredundant logical clauses implied by a 0-1 inequality $ax \geq a_0$. It assumes that $a_1 \geq \cdots \geq a_n \geq 0$. When the function is called with arguments (J, k), it determines recursively whether clause $C = \bigvee_{j \in J} x_j$ is implied by $ax \geq a_0$ or can be extended to a clause that is implied (where x_k is the last variable in C). If so, it returns the value *true*, and otherwise *false*. The first step is to check whether C itself is implied by checking whether $\sum_{j \notin J} a_j \leq a_0$. If so, the procedure generates C and returns *true*. If not, it tries adding the next variable x_{k+1} to C, then x_{k+2}, and so on, until the function returns false to indicate failure to find an implied extension of C. If k is already n, and C is not implied, the function immediately returns *false* to indicate failure.

Let *succeed* = **Dominate**($\emptyset$, 0).

Function **Dominate**(J, k).
    If $\sum_{j \notin J} a_j < a_0$ then generate the implied clause $\bigvee_{j \in J} x_j$ and return *true*.
    Else if $k = n$ then return *false*.
    Else
        Let *succeed* = *true*.
        For $j = k+1, \ldots, n$ while *succeed*:
            Let *succeed* = **Dominate**($J \cup \{j\}, j$).
        Return *succeed*.

Figure 3.7. Recursive function procedure for generating all nonredundant clauses implied by a 0-1 inequality $ax \geq a_0$ with $a_1 \geq \cdots a_n \geq 0$. It returns true *if at least one implied clause exists.*

0-1 Linear Inequalities

For example, the 0-1 inequality $9x_1+6x_2+5x_3+3x_4+x_5 \geq 10$ implies the clauses
$$x_1 \vee x_2$$
$$x_1 \vee x_3 \vee x_4$$
$$x_2 \vee x_3 \vee x_4 \vee x_5$$

3.6.3 Implication of Cardinality Clauses

A *cardinality clause* is a generalization of a logical clause. It states that at least a certain number of literals must be true. Thus a 0-1 inequality $ax \geq a_0 + n(a)$ is a cardinality clause when each $a_j \in \{0, 1, -1\}$ and a_0 is a nonnegative integer. For instance, $x_1 + x_2 - x_3 \geq 2 - 1$ is a cardinality clause and states that at least 2 of the literals $x_1, x_2, \neg x_3$ must be true.

Cardinality clauses deserve attention because they preserve the counting ability of 0-1 inequalities, and yet inference is much easier for them. In particular, it is easy to determine whether a 0-1 linear inequality implies a cardinality clause, and whether one cardinality clause implies another.

Suppose, without loss of generality, that $a \geq 0$ in a given 0-1 inequality $ax \geq a_0 + n(a)$ (if some $a_j < 0$, replace x_j with its complement). This inequality, which can be written $ax \geq a_0$, implies a cardinality clause $bx \geq b_0 + n(b)$ if and only if it implies the clause without its negative terms; that is, if and only if it implies

$$\sum_{j=1}^{k} x_j \geq b_0 \qquad (3.60)$$

where $b_1, \ldots, b_k$ are the positive coefficients in b. But $ax \geq a_0$ implies (3.60) if and only if the $\min\{b_0, k\}$ largest coefficients in $\{a_1, \ldots, a_k\}$ are required to accumulate a sum of at least a_0 after $x_{k+1}, \ldots, x_n$ are set equal to one.

THEOREM 3.28 *Let $b_j > 0$ for $j \leq k$ and $b_j \leq 0$ for $j > k$ in cardinality clause $bx \geq b_0+n(b)$. Then if $a \geq 0$ and $a_1 \geq \cdots \geq a_k$, the 0-1 inequality $ax \geq a_0$ implies $bx \geq b_0 + n(b)$ if and only if*

$$\sum_{j<\min\{b_0,k\}} a_j + \sum_{j>k} a_j < a_0$$

For example, the 0-1 inequality
$$4x_1 + 3x_2 + 2x_3 + x_4 + 5x_6 \geq 13$$
implies the cardinality clause
$$x_1 + x_2 + x_3 + x_4 + x_5 - x_7 \geq 3 - 1$$

because $a_1 + a_2 + a_6 + a_7 = 12$ is less than 13 (note that $a_5 = a_7 = 0$).

Implication between cardinality clauses is also easy to check.

THEOREM 3.29 *Consider the cardinality clauses $ax \geq a_0 + n(a)$ and $bx \geq b_0 + n(b)$. Let $\Delta = \sum_j |a_j| \max\{a_j b_j, 0\}$, which means Δ is the number of terms in the first clause that do not appear in the second clause. The following are equivalent:*

(i) $ax \geq a_0 + n(a)$ implies $bx \geq b_0 + n(b)$.

(ii) $\Delta \leq a_0 - b_0$.

(iii) $ax \geq a_0 + n(a)$ reduces to a cardinality clause that absorbs $bx \geq b_0 + n(b)$.

Before proving the theorem, consider the following example:

$$x_1 - x_2 + x_3 - x_4 + x_5 \geq 4 - 2$$
$$x_1 - x_2 \geq 1 - 1$$
$$x_1 - x_2 - x_3 + x_4 + x_6 \geq 1 - 2$$

The first cardinality clause implies the third. The second is a clause that is a reduction of the first and that absorbs the third, as predicted by the theorem. Also $\Delta = 3 \leq 4 - 1$, as required by the theorem.

Proof of Theorem 3.29. Part (iii) of the theorem implies (i) by virtue of Theorems 3.25 and 3.26.

To show that (i) implies (ii), assume that $a \geq 0$. The proof is easily generalized to arbitrary a by complementing variables. The first cardinality clause may therefore be written $ax \geq a_0$. It suffices to show that for any set J of a_0 indices j with $a_j = 1$, setting $x_j = 1$ for $j \in J$ is enough to satisfy $bx \geq b_0 + n(b)$. But this assignment does satisfy $bx \geq b_0 + n(b)$ if $b_j \neq 1$ for at most $a_0 - b_0$ of the indices in J. This means that $\Delta \leq a_0 - b_0$.

To show that (ii) implies (iii), construct a third cardinality clause

$$cx \geq a_0 - \Delta + n(c) \qquad (3.61)$$

by setting

$$c_j = \begin{cases} a_j & \text{if } a_j = b_j \\ 0 & \text{otherwise} \end{cases}$$

This is a cardinality clause because $\Delta \leq a_0 - b_0$ and $b_0 \geq 1$ imply $a_0 - \Delta \geq 1$. In addition, Δ is the number of terms removed from the left-hand side of $ax \geq a_0 + n(a)$ to obtain (3.61), which means that (3.61) is a reduction of $ax \geq a_0 + n(a)$. Finally, (3.61) absorbs $bx \geq b_0 + n(b)$ by construction of c, and the fact that $b_0 \leq a_0 - \Delta$. □

3.6.4 0-1 Resolution

The resolution method for logical clauses can be generalized to a complete inference method for 0-1 linear inequalities, which might be called *0-1 resolution*.

0-1 resolution is analogous to the well-known Chvátal-Gomory cutting plane procedure, to be presented in Section 4.4.1, in that both procedures generate all implied 0-1 inequalities. 0-1 resolution relies on the ability to recognize when one inequality implies another, but it achieves completeness by using only two special cases of the cutting planes generated by the Chvátal-Gomory procedure.

Neither 0-1 resolution nor the Chvátal-Gomory procedure is a practical method for solving general 0-1 problems. However, they help provide a theoretical foundation for the theory of 0-1 inference in one case and cutting plane theory in the other.

The 0-1 resolution procedure consists of two repeated operations. One is essentially clausal resolution. The other exploits a diagonal pattern in coefficients and might be called diagonal summation. It is assumed that the coefficients and right-hand sides are integers. If they are rational numbers, they can be converted to integers by multiplying each inequality by a appropriate number.

A set $\mathcal{S}$ of 0-1 linear inequalities has a *resolvent* R when R is the resolvent of two clausal inequalities C_1 and C_2 such that C_1 is implied by some inequality in $\mathcal{S}$, and similarly for C_2. For example, the inequalities

$$
\begin{array}{ll}
3x_1 - 2x_2 \phantom{{}+3x_3} + 3x_4 \phantom{{}+4x_5} \geq 3 - 2 \\
-x_1 - 3x_2 + 3x_3 \phantom{{}+3x_4} + 4x_5 \geq 5 - 4
\end{array}
\tag{3.62}
$$

respectively imply the clausal inequalities

$$
\begin{array}{ll}
x_1 - x_2 \phantom{{}+x_3} + x_4 \geq 1 - 1 \\
-x_1 - x_2 + x_3 \phantom{{}+x_4} \geq 1 - 2
\end{array}
$$

which have the resolvent $-x_2 + x_3 + x_4 \geq 1 - 1$. So, an inequality set containing the inequalities (3.62) has the resolvent $-x_2 + x_3 + x_4 \geq 1-1$.

An inequality $cx \geq \delta + n(c)$ is a *diagonal sum* of the system

$$
c^i x \geq \delta - 1 + n(c^i), \quad i \in J \tag{3.63}
$$

where $J = \{j \mid c_j \neq 0\}$ and

$$
c^i_j = \begin{cases} c_j - 1 & \text{if } c_j > 0 \\ c_j + 1 & \text{if } c_j < 0 \\ 0 & \text{if } c_j = 0 \end{cases}
$$

Furthermore, $cx \geq \delta + n(c)$ is a *diagonal sum* of a 0-1 inequality set S if it is a diagonal sum of a system (3.63) in which each inequality is implied by some inequality in S.

For example, inequality (d) below is a diagonal sum of the first three inequalities.
$$\begin{aligned} 2x_1 - 2x_2 + 4x_3 &\geq 2 - 2 \quad (a) \\ 3x_1 - x_2 + 4x_3 &\geq 2 - 1 \quad (b) \\ 3x_1 - 2x_2 + 3x_3 &\geq 2 - 2 \quad (c) \\ 3x_1 - 2x_2 + 4x_3 &\geq 3 - 2 \quad (d) \end{aligned} \qquad (3.64)$$

Note that the diagonal coefficients of (a)–(c) are reduced by one in absolute value, and that diagonal summation raises the degree by one. Inequality (d) is also a diagonal sum of the system

$$\begin{aligned} 4x_3 + 2x_4 &\geq 4 \\ 3x_1 - 3x_2 + 3x_3 + x_4 &\geq 4 - 3 \end{aligned} \qquad (3.65)$$

because (a) and (b) of (3.64) are implied by the first inequality of (3.65), and (c) is implied by the second.

Each iteration of the 0-1 resolution method consists of the following. If some resolvent R of S is implied by no inequality in S, R is added to S. Then, if some diagonal sum D of S is implied by no inequality in S, D is added to S. The iterations continue until no such resolvent and no such diagonal sum can be generated.

It is first shown below that 0-1 resolution is complete with respect to clausal inequalities. This will lead to a proof that it is complete with respect to all 0-1 linear inequalities.

THEOREM 3.30 *The 0-1 resolution method is complete with respect to clausal inequalities.*

Proof. Given a 0-1 system $Ax \geq b$, let S be the set of clausal inequalities implied by some inequality in $Ax \geq b$. Then S is equivalent to $Ax \geq b$. 0-1 resolution generates all clausal inequalities that are generated by applying classical resolution to S, or inequalities that imply them. Thus, by Theorem 3.19, 0-1 resolution is complete with respect to clausal inequalities. $\square$

THEOREM 3.31 *The 0-1 resolution method is complete with respect to 0-1 linear inequalities.*

Proof. Let S' be the result of applying 0-1 resolution to a set S of 0-1 inequalities. It suffices to show that any 0-1 inequality $cx \geq \delta + n(c)$

implied by $\mathcal{S}$ is implied by an inequality in $\mathcal{S}'$. The proof is by induction on the degree δ.

First, suppose that $\delta = 1$. Note that any inequality $cx \geq 1 + n(c)$ is equivalent to the clausal inequality $c'x \geq 1 + n(c')$, where $c'_j = 1$ if $c_j > 0$, -1 if $c_j < 0$, and 0 if $c_j = 0$. Now Theorem 3.30 implies that $c'x \geq 1 + n(c')$, and therefore $cx \geq 1 + n(c)$, is implied by an inequality in $\mathcal{S}'$.

Assuming now that the theorem is true for all inequalities of degree $\delta - 1$, it can be shown that it is true for all inequalities of degree δ. Suppose otherwise, and let $cx \geq \delta + n(c)$ be an inequality that is implied by $\mathcal{S}$ but by no inequality in $\mathcal{S}'$. But $cx \geq \delta + n(c)$ is a diagonal sum of the system (3.63). It suffices to show that each inequality in (3.63) is implied by an inequality in $\mathcal{S}'$, since in this case $\mathcal{S}'$ contains an inequality that implies $cx \geq \delta + n(c)$, contrary to hypothesis. To show this, note first that $c^i x \geq \delta - 1 + n(c^i)$ is a reduction of $cx \geq \delta + n(c)$ and is therefore implied by $\mathcal{S}$. Because $c^i x \geq \delta - 1 + n(c^i)$ has degree $\delta - 1$, the induction hypothesis ensures that it is implied by some inequality in $\mathcal{S}'$, as claimed. □

A slightly stronger statement can be made. The above theorem states that 0-1 resolution generates an inequality I that implies any given implication $cx \geq \delta + n(c)$ of $\mathcal{S}$. An extra step of diagonal summation generates $cx \geq \delta + n(c)$ itself. Since each inequality in (3.63) is implied by $cx \geq \delta + n(c)$, and therefore by I, $cx \geq \delta + n(c)$ is another diagonal sum that can be obtained.

3.6.5 k-Completeness

Section 3.5.4 pointed out that k-completeness can be achieved for any constraint set with boolean variables by applying resolution to the logical clauses implied by the constraints. This idea can be applied to 0-1 linear inequalities.

Given a 0-1 system $Ax \geq b$, the algorithm of Figure 3.7 can be used to generate all clausal inequalities implied by each inequality in $Ax \geq b$. If the implied clauses are collected in a set $\mathcal{S}$, $Ax \geq b$ is equivalent to $\mathcal{S}$. One can now achieve k-completeness for $\mathcal{S}$ by computing the projection of $\mathcal{S}$ onto every subset of k variables. Using Theorem 3.21, this is accomplished by applying to $\mathcal{S}$ a restricted resolution algorithm that resolves only on $x_{k+1}, \ldots, x_n$.

If one wishes to achieve full k-completeness rather than simply project $Ax \geq b$ onto a given set of k variables, it is necessary to resolve on all variables. This, of course, achieves k-completeness for all k.

Consider, for example, the 0-1 system

$$\begin{aligned} 2x_1 - x_2 + 3x_3 - 2x_4 &\geq 4 - 3 \quad (a) \\ -x_1 + 2x_2 - x_3 + 3x_4 &\geq 3 - 4 \quad (b) \\ 3x_1 + 2x_2 - 2x_3 + x_4 &\geq 4 - 2 \quad (c) \end{aligned} \quad (3.66)$$

The feasible solutions are (1,0,1,1), (1,1,0,0), and (1,1,1,1).

Each of the three inequalities (a), (b), and (c) implies the clauses listed below:

$$\begin{array}{lrr} (a): & x_1 - x_2 & -x_4 \geq 1 - 2 \\ & x_1 & +x_3 \geq 1 \\ & & x_3 - x_4 \geq 1 - 1 \\ (b): & -x_1 & -x_3 + x_4 \geq 1 - 2 \\ & x_2 & +x_4 \geq 1 \\ (c): & x_1 + x_2 & \geq 1 \\ & x_1 & -x_3 \geq 1 - 1 \\ & & x_2 - x_3 + x_4 \geq 1 - 1 \end{array} \quad (3.67)$$

To project (3.66) onto x_1, x_2, it is enough to resolve only on x_3, x_4, which yields the following after deleting clauses that are absorbed by others:

$$\begin{array}{lr} x_1 & \geq 1 \\ x_2 + x_3 & \geq 1 \\ x_2 + x_4 & \geq 1 \\ x_3 - x_4 & \geq 1 - 1 \\ -x_3 + x_4 & \geq 1 - 1 \end{array} \quad (3.68)$$

The only clause that contains variables in $\{x_1, x_2\}$ is the unit clause x_1, which describes the projection $\{(1,0), (1,1)\}$ onto x_1, x_2.

One can achieve k-completeness for all k (i.e., strong 3-completeness) by resolving on all variables. As it happens, this also yields the clause set (3.68), which is therefore strongly 3-complete. One can therefore solve (3.66) without backtracking by solving (3.68) with zero-step lookahead.

3.6.6 Strong k-Consistency

Strong k-consistency for a 0-1 system $Ax \geq b$ can be achieved in a manner analogous to k-completeness. First, use the algorithm of Figure 3.7 to create the set $\mathcal{S}$ of clausal inequalities that are implied by inequalities in $\mathcal{S}$. Then apply the k-resolution algorithm to $\mathcal{S}$ by generating resolvents with fewer than k literals. Using Theorem 3.22, $Ax \geq b$ becomes strongly k-consistent after augmenting it with the resulting clauses in $\mathcal{S}$.

0-1 Linear Inequalities

Note that it is necessary to include in $\mathcal{S}$ only clausal inequalities with k or fewer terms, which can accelerate the process considerably.

For example, the 0-1 inequality set (3.66) can be made 2-consistent by first generating implied clauses with 2 or fewer terms:

$$\begin{aligned} x_1 \quad\quad\quad + x_3 \quad\quad\quad &\geq 1 \\ x_3 - x_4 &\geq 1 - 1 \\ x_2 \quad\quad + x_4 &\geq 1 \\ x_1 + x_2 \quad\quad\quad\quad &\geq 1 \\ x_1 \quad\quad\quad - x_3 \quad\quad\quad &\geq 1 - 1 \end{aligned}$$

Application of 2-resolution yields the following after dropping redundant clauses:

$$\begin{aligned} x_1 \quad\quad\quad\quad\quad\quad &\geq 1 \\ x_2 + x_3 \quad\quad\quad &\geq 1 \\ x_2 \quad\quad + x_4 &\geq 1 \\ x_3 - x_4 &\geq 1 - 1 \end{aligned} \quad (3.69)$$

The original constraint set (3.66) therefore becomes strongly 2-consistent after augmenting it with the clauses in (3.69).

Obtaining strong k-consistency can be practical if k is small or if the individual inequalities in $Ax \geq b$ contain only a few variables and therefore do not imply a large number of nonredundant clauses.

3.6.7 Exercises

1. Show that one clause absorbs another if and only the same is true of the corresponding clausal inequalities.

2. Show by counterexamples that neither absorbsion nor reduction is a necessary condition for implication between 0-1 linear inequalities.

3. Let a *roof point* for a 0-1 inequality $ax \geq a_0$ (with $a \geq 0$) be a minimal satisfier of $ax \geq a_0$; that is, a point $\bar{x}$ that satisfies $ax \geq a_0$ but fails to do so when any one $\bar{x}_j$ is flipped from 1 to 0. For any roof point $\bar{x}$, $\{j \mid \bar{x}_j = 1\}$ is a *roof set*. A *satisfaction set* for $bx \geq b_0$ is an index set J such that $bx \geq b_0$ whenever $x_j = 1$ for all $j \in J$. Show that the 0-1 inequality $ax \geq a_0$ (with $a \geq 0$) implies another 0-1 inequality $bx \geq b_0$ if and only if all roof sets of the former are satisfaction sets of the latter. Use this fact to prove Theorems 3.25 and 3.26.

4. Use the algorithm of Figure 3.7 to derive the four nonredundant clauses implied by $10x_1 + 8x_2 + 4x_4 + 3x_5 + 2x_6 \geq 12$.

5. Show that the algorithm of Figure 3.7 generates all nonredundant clauses implied by a 0-1 inequality $ax \geq a_0$ with $a_1 \geq \cdots \geq a_n \geq 0$.

6. What is the smallest a_0 for which $5x_1 + 4x_2 + 3x_3 + 3x_4 + 3x_5 \geq a_0$ implies the cardinality clause $x_1 + x_2 + x_3 + x_4 - x_5 \geq 3 - 1$?

7. Prove Theorem 3.28 by considering two cases: $b_0 \leq k$ and $b_0 > k$.

8. Show that the cardinality clause $x_1 + x_2 - x_3 - x_4 + x_5 \geq 4 - 2$ implies $-x_1 + x_2 - x_3 + x_4 + x_5 \geq 2 - 2$ by exhibiting a cardinality clause to which the former reduces and that absorbs the latter. Without looking at the proof of Theorem 3.29, indicate how such an intermediate clause can be identified in general.

9. Show that the resolvent of two clausal inequalities C_1, C_2 is the result of taking a nonnegative linear combination of C_1, C_2 and bounds $0 \leq x_j \leq 1$ and rounding up any fractions that result. Thus, clausal resolution is a special case of a Chvátal-Gomory cut. Hint: assign multiplier $\frac{1}{2}$ to C_1 and C_2.

10. Show that the diagonal sum $cx \geq \delta + n(c)$ of a system (3.63) is the result of taking a nonnegative linear combination of the inequalities of (3.63) and rounding up any fractions that result. Thus diagonal summation is a special case of a Chvátal-Gomory cut. Hint: assign each inequality $c^i \geq \delta - 1 + n(c^i)$ the multiplier $|c_i|/(\sum_j |c_j| - 1)$.

11. The proof of Theorem 3.31 states, "but $cx \geq \delta + n(c)$ is a diagonal sum of the system (3.63)." For what set J in (3.63) is this true?

12. Verify by resolution that (3.68) is the set of prime implications of (3.67). Solve (3.68) by zero-step lookahead.

13. Show that 0-1 resolution always terminates after finitely many iterations.

3.7 Integer Linear Inequalities

Inference methods for general integer linear inequalities remain substantially unexplored, but inference duals for this class of problems have been studied in some depth. Perhaps the best known inference dual is the *subadditive dual* (or *superadditive dual*, in maximization problems), which can also be viewed as a relaxation dual. Thanks to the completeness of Chvátal-Gomory cuts for integer programming, the subadditive dual is a strong dual and has some additional properties that make it appropriate for sensitivity analysis.

The subadditive dual can, in principle, be solved by computing a series of Gomory cuts, but this can become computationally burdensome. In such cases, the branching tree that has already been generated to solve the primal problem can be treated as a solution of a second type of inference dual. For lack of a better name, it might be called the *branching dual*. inxxbranching dual It is also a strong dual that can be used for sensitivity analysis that is more comprehensive than that delivered by subadditive duality, since it deals with perturbations in all of the problem data, rather than just the right-hand sides. The branching dual can also yield Benders cuts for an integer programming subproblem.

3.7.1 The Subadditive Dual

An integer linear programming problem may be written

$$\begin{aligned} \min\ & cx \\ Ax \geq b,\ & x \geq 0 \text{ and integer} \end{aligned} \quad (3.70)$$

For technical reasons, it is assumed that $Ax \geq b$ is a *bounded* system, meaning that the maximum and minimum of x_j subject to the constraints of (3.70) are finite for every j. An inference dual of (3.70) has the form

$$\begin{aligned} \max\ & v \\ (Ax \geq b) & \overset{Q}{\vdash} (cx \geq v) \\ v \in \mathbb{R},\ & Q \in \mathcal{Q} \end{aligned} \quad (3.71)$$

In the *subadditive dual*, each proof $Q \in \mathcal{Q}$ corresponds to a subadditive, homogeneous, nondecreasing function $h : \mathbb{R}^m \to \mathbb{R}$. Function h is *subadditive* if

$$h(d^1 + d^2) \leq h(d^1) + h(d^2)$$

for all d^1, d^2, and it is *homogeneous* if $h(0) = 0$. The bound $cx \geq v$ is deduced from $Ax \geq b$ when $h(Ax) \geq h(b)$ dominates $cx \geq v$ for some h satisfying these properties. Domination is defined so that $h(Ax) \geq h(b)$ dominates $cx \geq v$ when $h(Ax) \leq cx$ for all $x \in D$, and $h(b) \geq v$. Note that this form of domination is stronger than mere implication.

The mere fact that h is nondecreasing makes this a valid inference method. It will be seen shortly that it is a complete inference method as well, which means that the subadditive dual is a strong dual. The *superadditive dual*, also a strong dual, is defined in an analogous way for maximization problems., A function h is superadditive if $h(d^1 + d^2) \geq h(d^1) + h(d^2)$ for all d^1, d^2.

Due to the definition of domination, the subadditive dual (3.71) can be written

$$\max_{h \in H}\ \{h(b) \mid h(Ax) \leq cx \text{ for all integral } x \geq 0\} \quad (3.72)$$

where H is the set of subadditive, homogeneous, nondecreasing functions. Subadditivity and homogeneity are useful properties because they allow one to simplify the dual, due to the following fact. Let $h(A)$ denote the row vector of function values $[h(A_1) \cdots h(A_n)]$, where A_j is column j of A.

LEMMA 3.32 *If h is nondecreasing, subadditive and homogeneous, then $h(Ax) \leq cx$ for all integral $x \geq 0$ if and only if $h(A) \leq c$.*

Proof. First, if $h(Ax) \leq cx$ for all integral $x \geq 0$, then in particular $h(Ae^j) \leq ce^j$ for each unit vector e^j. This says $h(A_j) \leq c_j$ for each j, or $h(A) \leq c$. For the converse, suppose $h(A_j) \leq c_j$ for each j. Then

$$h(Ax) = h\left(\sum_j A_j x_j\right) \leq \sum_j h(A_j) x_j \leq \sum_j c_j x_j = cx$$

If $x \neq 0$, the first inequality is due to subadditivity and the fact that each x_j is a nonnegative integer. If $x = 0$, it is due to homogeneity. The second inequality follows from the hypothesis $h(A_j) \leq c_j$. □

Due to Lemma 3.32, the subadditive dual (3.72) can be written

$$\max_{h \in H} \{h(b) \mid h(A) \leq c\} \tag{3.73}$$

This is closely analogous to the classical linear programming dual of (3.70) without the integrality constraint:

$$\max_{u \geq 0} \{ub \mid uA \leq c\}$$

The subadditive dual is a strong dual because the associated inference method is complete. It is complete because any inequality implied by a bounded integer system $Ax \geq b$ is dominated by $h(Ax) \geq h(b)$ for some submodular, homogeneous, nondecreasing function h. This, in turn, is true because it holds for a subclass of such functions known as *Chvátal functions*. A Chvátal function $h(x) : \mathbb{R}^m \to \mathbb{R}$ has the form

$$h(x) = \lceil u \lceil M_{k-1} \lceil M_{k-2} \cdots \lceil M_2 \lceil M_1 x \rceil \rceil \cdots \rceil \rceil$$

for some sequence of matrices $M_1, \ldots, M_{k-1}$ and some row vector u in which all components are nonnegative.[4] Thus a Chvátal function

---
[4] The components are generally defined to be rational as well, but this condition is not required here.

Integer Linear Inequalities

operates on a tuple x by taking several nonnegative linear combinations of components of x, rounding up each result to the nearest integer, and creating a tuple of the resulting numbers. The process repeats a finite number of times. A scalar value is obtained at the end by taking a single linear combination.

Chvátal functions are defined in this way because of a fundamental result of cutting plane theory to be proved in Chapter 4 (Theorem 4.7). Namely, any inequality implied by a bounded integer system $Ax \geq b$ can be obtained by a repeated process of taking linear combinations of the inequalities in $Ax \geq b$ and rounding the coefficients and right-hand sides that result. This leads to the following.

COROLLARY 3.33 *Any inequality implied by a bounded integer system $Ax \geq b$ is dominated by $h(Ax) \geq h(b)$ for some Chvátal function h.*

Proof. Suppose $Ax \geq b$ implies $cx \geq v$. According to Theorem 4.7, $cx \geq v$ can be written $h(A)x \geq h(b)$ for some Chvátal function h so that $h(A) = c$ and $h(b) = v$. Then, by Lemma 3.32, $h(Ax) \geq h(b)$ dominates $cx \geq v$. □

It can also be shown that

LEMMA 3.34 *Chvátal functions are submodular, homogeneous, and nondecreasing.*

Due to Corollary 3.33 and Lemma 3.34, any bound $cx \geq v$ implied by the integer system $Ax \geq b$ is dominated by $h(Ax) \geq h(b)$ for some submodular, homogeneous, nondecreasing function h. This means that the submodular dual is associated with a complete inference method and is therefore a strong dual.

Since there is no duality gap, $h(b)$ is the optimal value of the primal problem (3.70) when h is an optimal solution of the dual (3.73). The dual can therefore be solved by identifying a Chvátal function h for which $h(b)$ is the optimal value of the primal. Section 4.5.2 will show that this can be accomplished by solving (3.70) with the Gomory cutting plane algorithm.

Sensitivity analysis for the right-hand side b is now straightforward. Let h be a Chvátal function that solves the dual. If the right-hand side of (3.71) is perturbed to $b + \Delta b$, weak duality implies that $h(b + \Delta b)$ is a lower bound on the optimal value of the perturbed problem.

Consider, for example, the linear integer programming problem

$$\begin{array}{ll} \min\ 2x_1 + 3x_2 & \\ x_1 + 3x_2 \geq 3 & (a) \\ 4x_1 + 3x_2 \geq 6 & (b) \\ x_1, x_2 \geq 0 \text{ and integral} & \end{array} \qquad (3.74)$$

The optimal solution is $(x_1, x_2) = (1, 1)$ with value 5. The following sequence of nonnegative linear combinations and roundings yields the optimal bound $2x_1 + 3x_2 \geq 5$ (Section 4.5.2 will illustrate how the sequence is obtained using Gomory's cutting plane method). First, take a linear combination of constraints (a) and (b) using multipliers $\frac{5}{9}$ and $\frac{1}{9}$. This yields $x_1 + 2x_2 \geq \frac{21}{9}$, which after rounding is

$$x_1 + 2x_2 \geq 3 \quad (c)$$

Now, take a linear combination of (b) and (c) using multipliers $\frac{3}{5}$ and $\frac{3}{5}$ to obtain the inequality $3x_1 + 3x_2 \geq \frac{27}{5}$, which rounds to

$$3x_1 + 3x_2 \geq 6 \quad (d)$$

Finally, take a linear combination of (c) and (d) using multipliers 1 and $\frac{1}{3}$, which yields the desired bound $2x_1 + 3x_2 \geq 5$. The corresponding Chvátal function is

$$h(x) = u \lceil M_2 \lceil M_1 x \rceil \rceil$$

where

$$u = [0 \ 0 \ 1 \ \tfrac{1}{3}], \quad M_2 = \begin{bmatrix} 1 & 0 & 0 \\ 0 & 1 & 0 \\ 0 & 0 & 1 \\ 0 & \tfrac{3}{5} & \tfrac{3}{5} \end{bmatrix}, \quad M_1 = \begin{bmatrix} 1 & 0 \\ 0 & 1 \\ \tfrac{5}{9} & \tfrac{1}{9} \end{bmatrix}$$

One can check that $h(A)x \geq h(b)$ is the desired inequality $2x_1 + 3x_2 \geq 5$. In particular, $h(b) = h(3, 6) = 5$ is the optimal value. This Chvátal function therefore solves the subadditive dual.

For sensitivity analysis, suppose that right-hand side is perturbed to $(3 + \Delta_1, 6 + \Delta_2)$. The resulting optimal value is at least

$$h(3 + \Delta_1, 6 + \Delta_2) = \\ \lceil \tfrac{7}{3} + \tfrac{5}{9}\Delta_1 + \tfrac{1}{9}\Delta_2 \rceil + \tfrac{1}{3}\lceil \tfrac{3}{5}\lceil 6 + \Delta_2 \rceil + \tfrac{3}{5}\lceil \tfrac{7}{3} + \tfrac{5}{9}\Delta_1 + \tfrac{1}{9}\Delta_2 \rceil \rceil$$

Thus, for instance, if the right-hand side is changed to $(5, 8)$, the optimal value is at least $h(5, 8) = 6\tfrac{1}{3}$. It is actually 7.

A practical difficulty with the subadditive dual is that the only evident method for solving it is to solve the original problem with the Gomory cutting plane algorithm. Although Gomory cuts are very useful in a branch-and-cut algorithm, problems are rarely solved with Gomory's method alone. The next section presents a dual whose solution can be read directly from a branching tree that is created to solve the problem.

3.7.2 The Branching Dual

Integer programming problems are most often solved by some version of a branch-and-relax algorithm. The branching dual regards the branching tree that results as an optimal solution of the inference dual. This has the obvious advantage that solving the primal problem yields a dual solution with no extra effort. The branching dual is obviously a strong dual, because an optimal solution can always be found, or infeasibility proved, by a sufficiently large branching tree. The dual solution can therefore provide the basis for sensitivity analysis along the lines described in Section 3.2.3.

Formally, the inference dual of an integer programming problem (3.70) is the problem (3.71), where $Q \in \mathcal{Q}$ when Q is a branch-and-relax tree showing that $cx \geq v$. That is, every feasible solution found in the tree has a value of at least v. If cutting planes are used in the branch-and-relax tree, the analysis below must be modified.

Sensitivity analysis can be viewed as follows. Let v^* be the optimal value of (3.71), which means that v^* is the value of the best solution found in the branch-and-relax tree. Given a tolerance Δ, one can ask how much the problem can be changed without reducing the optimal value more than Δ.

Let a *feasible node* of the tree be a leaf node at which a feasible solution is found, an *infeasible node* be a leaf node at which the relaxation is infeasible, and a *fathomed* node be a leaf node at which the search backtracks because the value of the relaxation is no better than the value of the best feasible solution found so far. Then, the branch-and-relax tree remains a proof that the new optimal value is at least $v^* - \Delta$ if two conditions are satisfied: (a) The altered relaxation at every infeasible node is infeasible, and (b) the optimal value of the altered relaxation at every feasible node and every fathomed node is at least $v^* - \Delta$.

For ease of exposition, the analysis is developed here for 0-1 programming, but it can be extended to general integer programming. It is best to begin with an example.

$$\begin{aligned} \min \; & 5x_1 + 6x_2 + 7x_3 \\ & 4x_1 + 3x_2 - x_3 \geq 2 \quad (a) \\ & -x_1 + x_2 + 4x_3 \geq 3 \quad (b) \\ & x_1, x_2, x_3 \in \{0, 1\} \end{aligned} \qquad (3.75)$$

A branch-and-relax tree appears in Figure 3.8. The optimal solution is $(x_1, x_2, x_3) = (1, 0, 1)$ with value 12, and is found at Leaf Node 2.

Consider Leaf Node 1, at which the relaxation contains constraints (a) and (b), $x_1 = x_2 = 0$, and $0 \leq x_3 \leq 1$. The relaxation is infeasible,

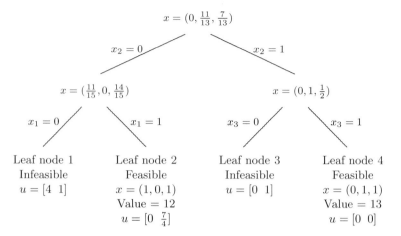

Figure 3.8. Branch-and-relax tree used for integer programming sensitivity analysis. The vector u contains the dual multipliers for the original constraints when the continuous relaxation is solved at each leaf node.

and the solution of its (unbounded) linear programming dual assigns multipliers $u = [u_1 \; u_2] = [4 \; 1]$ to constraints (a) and (b). This defines a surrogate $15x_1 + 13x_2 \geq 11$ that is violated by the branching constraints $(x_1, x_2) = (0, 0)$. Now suppose the problem data A, b, c are changed to $\tilde{A}, \tilde{b}, \tilde{c}$. This same dual solution remains a proof that the relaxation is infeasible if $(x_1, x_2) = (0, 0)$ violates the surrogate

$$(4\tilde{A}_{11} + \tilde{A}_{21})x_1 + (4\tilde{A}_{12} + \tilde{A}_{22})x_2 + (4\tilde{A}_{13} + \tilde{A}_{23})x_3 \geq 4\tilde{b}_1 + \tilde{b}_2 \quad (3.76)$$

It is easy to write a necessary and sufficient condition for when $(x_1, x_2) = (0, 0)$ violates (3.76). In general, a partial assignment $x_j = v_j$ for $j \in J$ violates a 0-1 inequality $dx \geq \delta$ if and only if

$$\sum_{j \in J} d_j v_j + \sum_{j \notin J} \max\{0, d_j\} < \delta$$

Thus, $(x_1, x_2) = (0, 0)$ violates (3.76) as long as the perturbed data satisfy

$$\max\{0, 4\tilde{A}_{13} + \tilde{A}_{23}\} < 4\tilde{b}_1 + \tilde{b}_2 \quad (3.77)$$

The dual multipliers $u = [0 \; 1]$ at Leaf Node 3 provide a proof of infeasibility as long as $(x_2, x_3) = (1, 0)$ violates the surrogate

$$\tilde{A}_{21}x_1 + \tilde{A}_{22}x_2 + \tilde{A}_{23}x_3 \geq \tilde{b}_2$$

Thus the node remains infeasible if

$$\max\{0, \tilde{A}_{21}\} + \tilde{A}_{22} < \tilde{b}_2 \quad (3.78)$$

The feasible leaf nodes require a slightly different treatment. The optimal value is at least $12 - \Delta$ at a feasible node if the relaxation is infeasible when $5x_1 + 5x_2 + 7x_3 < 12 - \Delta$ is added to its constraint set. This inequality can be written

$$-5x_1 - 6x_2 - 7x_3 > -12 + \Delta \tag{3.79}$$

By assigning to $Ax \geq b$ dual multipliers u from the solution of the relaxation at the current node, and assigning multiplier 1 to (3.79), one obtains a surrogate that is violated at the current node when $\Delta = 0$. At Leaf Node 2, for instance, the linear programming dual assigns $u = [0 \; \frac{7}{4}]$ to constraints (a) and (b), respectively. If multiplier 1 is assigned to (3.79), the resulting surrogate is

$$-\tfrac{27}{4}x_1 - \tfrac{17}{4}x_2 > \Delta - \tfrac{27}{4}$$

which is violated by the branching constraints $(x_1, x_2) = (1, 0)$. If the problem data are changed to $\tilde{A}, \tilde{b}, \tilde{c}$, the surrogate that results from this same dual solution is

$$(\tfrac{7}{4}\tilde{A}_{21} - \tilde{c}_1)x_1 + (\tfrac{7}{4}\tilde{A}_{22} - \tilde{c}_2)x_2 + (\tfrac{7}{4}\tilde{A}_{23} - \tilde{c}_3)x_3 > \tfrac{7}{4}\tilde{b}_2 - 12 + \Delta$$

The partial assignment $(x_1, x_2) = (1, 0)$ violates this surrogate when

$$\tfrac{7}{4}\tilde{A}_{21} - \tilde{c}_1 + \max\{0, \tfrac{7}{4}\tilde{A}_{23} - \tilde{c}_3\} \leq \tfrac{7}{4}\tilde{b}_2 - 12 + \Delta \tag{3.80}$$

At Leaf Node 4, the dual multipliers for both (a) and (b) vanish. The surrogate is therefore

$$-\tilde{c}_1 x_1 - \tilde{c}_2 x_2 - \tilde{c}_3 x_3 > -12 + \Delta$$

This is violated by $(x_2, x_3) = (1, 1)$ when

$$\min\{0, \tilde{c}_1\} + \tilde{c}_2 + \tilde{c}_3 \geq 12 - \Delta \tag{3.81}$$

The conclusion of this analysis is that for a given Δ, altered problem data $\tilde{A}, \tilde{b}, \tilde{c}$ will reduce the optimal value no more than Δ if the data satisfy (3.77), (3.78), (3.80) and (3.81). Also, for a given set of data one can find the best lower bound on the optimum that can be derived from this analysis. Simply minimize Δ subject to these four inequalities, which is a linear programming problem.

In general, sensitivity analysis proceeds as follows. At each infeasible node, let u be the vector of dual multipliers assigned to constraints $Ax \geq b$ when solution of the relaxation detects infeasibility. Let J be the set of indices j for which branching has fixed x_j to some value $v_j \in \{0,1\}$ at the current node. Then $x_J = v_J$ (i.e., $x_j = v_j$ for $j \in J$) violates the surrogate $uAx \geq ub$. For alternate data $\tilde{A}, \tilde{b}$, $x_J = v_J$ violates $u\tilde{A}x \geq u\tilde{b}$ if

$$\sum_{j \in J} u\tilde{A}_j v_j + \sum_{j \notin J} \max\{0, u\tilde{A}_j\} < u\tilde{b} \tag{3.82}$$

At each feasible node and fathoming node let u be the vector of dual multipliers assigned to $Ax \geq b$ in the optimal dual solution of the relaxation at that node, and let J be as before. Let v^* be the optimal value of the original problem (3.70) found in the tree. Then $x_J = v_J$ violates the surrogate $(uA - c)x > ub - v^*$. For any $\Delta \geq 0$ and alternate data $\tilde{A}, \tilde{b}, \tilde{c}$, $x_J = v_J$ violates the surrogate $(u\tilde{A} - \tilde{c})x > u\tilde{b} - v^* + \Delta$ if

$$\sum_{j \in J}(u\tilde{A}_j v_j - \tilde{c}_j) + \sum_{j \notin J} \max\{0, u\tilde{A}_j - \tilde{c}_j\} \leq u\tilde{b} - v^* + \Delta \tag{3.83}$$

Thus the new data $\tilde{A}, \tilde{b}, \tilde{c}$ reduce the optimal value no more than Δ if they satisfy (3.82) at every infeasible node and (3.83) at every feasible node and fathoming node.

If cutting planes are used in the tree search, then one must write conditions under which the cutting planes remain valid and impose these conditions as well on the problem data.

3.7.3 Benders Cuts

Since the subadditive and branching duals are inference duals, both can provide the basis for constraint-directed search. In particular, they provide a Benders method when applied to problems of the form

$$\begin{aligned}
\min\ & f(x) + cy \\
& g(x) + Ay \geq b \\
& x \in D_x,\ y \geq 0 \text{ and integral}
\end{aligned} \tag{3.84}$$

The problem becomes an integer programming subproblem when the variables x are fixed, and solution of its subadditive or branching dual can yield Benders cuts.

An important special case arises when the entire problem (3.84) is an integer programming problem but simplifies when x is fixed, perhaps by separating into a number of smaller problems. This occurs in stochastic integer programming, for instance, in which there are many scenarios or possible outcomes and a set of constraints associated with each. When certain variables are fixed, the problem separates into smaller problems corresponding to the scenarios. Benders decomposition is an attractive alternative for such problems.

Integer Linear Inequalities

Benders cuts based on the subadditive dual are straightforward. If x^k is the solution of the master problem in iteration k, the subproblem is

$$\begin{aligned} \min \ & f(x^k) + cy \\ & Ay \geq b - g(x^k) \\ & y \geq 0 \text{ and integral} \end{aligned} \qquad (3.85)$$

The subadditive dual of (3.85) is

$$\max_{h \in H} \left\{ h(b - g(x^k)) \mid h(A) \leq c \right\}$$

where H is the set of subadditive, homogeneous, nondecreasing functions. If a Chvátal function h_{k+1} solves the dual, then

$$f(x^k) + h_{k+1}(b - g(x^k))$$

is the optimal value of the subproblem. Since h_{k+1} remains a feasible dual solution for any x, by weak duality $f(x) + h_{k+1}(b - g(x))$ is a lower bound on the optimal value of (3.84) for any x. This yields the Benders cut

$$v \geq f(x) + h_{k+1}(b - g(x))$$

Such cuts have the practical shortcoming, however, that the dual is hard to solve and the Chvátal function that solves it is typically very complicated. It could be difficult to solve a master problem that contains them.

A more promising Benders approach is based on the branching dual. In this case, the inference dual of the subproblem (3.85) is solved by the branch-and-relax tree that solves (3.85) itself. The issue for generating a Benders cut is how one can bound the optimal value of (3.84) when x^k is replaced by some other value of x. Fortunately, this change only affects the right-hand sides, which simplifies the analysis.

Suppose, as in the previous section, that (3.85) is a 0-1 programming problem, and let v_{k+1} be its optimal value. Using the results of that section, the branch-and-relax tree that solves (3.85) proves that the optimal value remains at least $v_{k+1} - \Delta$ if the following conditions are met. First, using (3.82), at every infeasible leaf node ℓ the variable x must satisfy

$$u^\ell(b - g(x)) > B_\ell \qquad (3.86)$$

where

$$B_\ell = \sum_{j \in J_\ell} u^\ell A_j v_{\ell j} + \sum_{j \notin J_\ell} \max\{0, u^\ell A_j\}$$

Here, u^ℓ is the vector of dual multipliers for $Ax \geq b$ in the solution of the relaxation at node ℓ, and J_ℓ is the set of indices j for which x_j is fixed by branching to some value $v_{\ell j}$ at node ℓ. Second, using (3.83), at every feasible node and every fathoming node ℓ the variable x must satisfy

$$u^\ell(b - g(x)) - v_{k+1} + f(x^k) + \Delta \geq B_\ell \tag{3.87}$$

where

$$B_\ell = \sum_{j \in J_\ell}(u^\ell A_j v_{\ell j} - c_j) + \sum_{j \notin J_\ell} \max\{0, u^\ell A_j - c_j\}$$

The optimal value of (3.84) must be at least $v_{k+1} - \Delta$, or else one of the conditions (3.86) or (3.87) must be violated. Thus a Benders cut may be written in disjunctive form:

$$(v \geq v_{k+1} - \Delta) \vee \bigvee_{\ell \in L_1}\left(u^\ell(b - g(x)) \leq B_\ell\right) \vee \\ \bigvee_{\ell \in L_2}\left(u^\ell(b - g(x)) < v_{k+1} - f(x^k) - \Delta + B_\ell\right) \tag{3.88}$$

where L_1 is the set of infeasible leaf nodes and L_2 the set of the remaining leaf nodes. To ensure convergence, the Benders cuts must be generated with $\Delta = 0$, but it may be useful to generate additional cuts using values of $\Delta > 0$.

If the original problem (3.84) is a linear 0-1 programming problem, it is useful to linearize the cut. Suppose that (3.84) minimizes $dx + cy$ subject to $Dx + Ay \geq b$, where each $x_j, y_j \in \{0,1\}$. The cut (3.88) can be written

$$v \geq v_{k+1} - \Delta - M\left(|L_1 \cup L_2| - \sum_{\ell \in L_1 \cup L_2} \delta_\ell\right) \\ u^\ell Dx \geq u^\ell b - B_\ell - M\delta_\ell, \text{ all } \ell \in L_1 \\ u^\ell Dx > u^\ell b - B_\ell - v_{k+1} + f(x^k) + \Delta - M\delta_\ell, \text{ all } \ell \in L_2 \tag{3.89}$$

where M is a large number and each δ_ℓ is a 0-1 variable.

Suppose, for example, a Benders method is applied to the problem

$$\min 3x_1 + 4x_2 + 5y_1 + 6y_2 + 7y_3 \\ x_1 + x_2 + 4y_1 + 3y_2 - y_3 \geq 2 \\ x_1 - x_2 - y_1 + y_2 + 4y_3 \geq 3 \\ x_j, y_j \in \{0, 1\} \tag{3.90}$$

in which x_1, x_2 are the master problem variables. The initial master problem has no constraints, and one can arbitrarily choose $x^0 = (0,0)$

Integer Linear Inequalities 181

as the solution. The resulting subproblem (3.85) is the problem (3.75) studied in the previous section. Recall that Leaf Nodes 1 and 3 of the branch-and-relax tree are infeasible, while Leaf Nodes 2 and 4 are feasible. One can check that $(B_1, B_3) = (0, 1)$ and $B_2, B_4 = (-6\frac{3}{4}, -13)$. The Benders cut (3.89) for a given $\Delta \geq 0$ is

$$\begin{aligned} v &\geq 12 - \Delta - M(4 - \delta_1 - \delta_2 - \delta_3 - \delta_4) \\ 5x_1 + 3x_2 &\geq 11 - M\delta_1 \\ x_1 - x_2 &\geq 2 - M\delta_3 \\ \tfrac{7}{4}x_1 - \tfrac{7}{4}x_2 &\geq \Delta - M\delta_2 + \epsilon \\ 0 &\geq 1 + \Delta - M\delta_4 + \epsilon \end{aligned} \qquad (3.91)$$

Note that the strict inequalities are implemented by adding ϵ to the right-hand side. The next master problem minimizes v subject to (3.91). If $\Delta = 0$ in the Benders cut, the optimal solution is $(x_1, x_2) = (1, 0)$ with $(\delta_1, \ldots, \delta_4) = (1, 1, 0, 1)$. Setting $x^1 = (1, 0)$ defines the subproblem for the next iteration.

3.7.4 Exercises

1 Prove Corollary 3.33.

2 Consider the integer programming problem (3.37), which has optimal value 10. The bound $2x_1 + 3x_2 \geq 10$ can be obtained by first taking a linear combination of the constraints with multipliers $\frac{6}{7}$ and $\frac{3}{7}$ and rounding up to obtain a new inequality, then taking a linear combination of the first constraint and the new inequality with multipliers $\frac{1}{4}$ and $\frac{1}{4}$ and rounding up to obtain a second new inequality, and finally taking a linear combination of the two new inequalities with multipliers 3 and 1. Write the corresponding Chvátal function in the form $h(x) = u\lceil M_2\lceil M_1 x\rceil\rceil$. Write an expression for a lower bound on the optimal value if the right-hand side is perturbed to $(0 + \Delta_1, 5 + \Delta_2)$.

3 Construct a branching tree that solves the 0-1 programming problem

$$\begin{aligned} \min\ & 3x_1 + 3x_2 + 5x_3 \\ & 3x_1 - x_2 + 4x_3 \geq 2 \\ & x_1 + 2x_2 - x_3 \geq 1 \\ & x_1, x_2, x_3 \in \{0, 1\} \end{aligned}$$

Write a set of inequalities (one for each leaf node) whose satisfaction by a perturbed problem is sufficient to ensure that the perturbations reduce the optimal value no more than Δ.

4 Indicate how sensitivity analysis based on a branching tree can be extended to general integer programming. Assume that branching occurs in the usual fashion by splitting the domain of a variable.

5 Suppose that the problem

$$\min x_1 + x_2 + 3y_1 + 4y_2$$
$$2x_1 + x_2 - y_1 + 3y_2 \geq 0$$
$$x_1 + 2x_2 + 2y_1 + y_2 \geq 5$$
$$x_1, x_2, y_1, y_2 \geq 0 \text{ and integral}$$

is to be solved by Benders decomposition with x_1, x_2 in the master problem. Suppose that the initial solution of the master problem is $(x_1, x_2) = (0, 0)$, so that the initial subproblem is (3.74) but with variables y_1, y_2. Write a Benders cut based on the subadditive dual.

6 Complete the solution of (3.90) by Benders decomposition based on the branching dual.

3.8 The Element Constraint

The element constraint implements variable indexing, which is a central modeling feature of constraint programming and integrated problem solving. A variable index is an index or subscript whose value depends on one or more variables. The existence of effective filters and good relaxations for the element constraint contributes substantially to the efficiency of integrated algorithms.

The simplest form of the element constraint is

$$\text{element}(y, z \,|\, (a_1, \ldots, a_m)) \tag{3.92}$$

where y is a discrete variable and each a_i is a constant. The constraint says that z must take the yth value in the list $a_1, \ldots, a_m$. Expressions of the form a_y in an integrated model can be implemented by replacing each occurrence of a_y with z and adding (3.92) to the constraint set. One can then apply filtering algorithms and generate relaxations for (3.92).

Another form of the element constraint is

$$\text{element}(y, (x_1, \ldots, x_m), z) \tag{3.93}$$

where each x_i is a variable. It sets z equal to the yth variable in the list $x_1, \ldots, x_m$. This constraint implements an expression of the form

The Element Constraint

x_y, which is common in *channeling* constraints that connect two formulations of the same problem (Section 2.2.5).

Element constraints can also be multidimensional. For example,

$$\text{element}((y_1, y_2), z \,|\, A) \qquad (3.94)$$

selects the element in row y_1 and column y_2 of matrix A and assigns that value to z. For purposes of domain filtering, a constraint of this sort can be treated as a single-dimensional constraint. Thus, if A is $m \times n$, (3.94) can be converted to $\text{element}(y, z \,|\, a)$ and filtered in that form, where $y = m(y_1 - 1) + y_2$ and $a = (A_1, \ldots, A_n)$. Here A_i is a tuple of the elements in row i of A.

A constraint like (3.94) implements doubly-indexed expressions, as in the traveling salesman problem, for example. If A_{ij} is the cost of traveling from point i to point j, the salesman's objective is to minimize $\sum_{ij} A_{y_i y_{i+1}}$ subject to the constraint that $y_1, \ldots, y_n$ take different values (and where y_{n+1} is identified with y_1).

It is sometimes advantageous to analyze a specially structured element constraint. For example, variably indexed coefficients in expressions of the form $a_y x$ are quite common and are used, for example, in Sections 2.2.7 and 2.3.7. They are implemented by the *indexed linear element* constraint

$$\text{element}(y, x, z \,|\, (a_1, \ldots, a_m)) \qquad (3.95)$$

where x is now a single variable. The constraint sets z equal to the yth term in the list $a_1 x, \ldots, a_m x$. It is possible to exploit the structure of (3.95) when designing filters and relaxations. This is more efficient than implementing x_y with (3.93) and then adding the constraints $x_i = a_i x$ for $i = 1, \ldots, m$. There is also a vector-valued version

$$\text{element}(y, x, z \,|\, (A_1, \ldots, A_m))$$

in which z and each A_k are tuples of the same length.

3.8.1 Domain Completeness

Filtering algorithms for element constraints can achieve full domain completeness with a modest amount of computation.

It is a trivial matter to filter domains for an element constraint (3.92) that contains a list of constants. Let D_z, D_y be the current domains of z and y, respectively, and let D'_z, D'_y be the new, reduced domains. It is assumed that D_y and D_z are finite. Then D_z can be restricted to a_is whose indices are in D_y:

$$D'_z = D_z \cap \{a_i \,|\, i \in D_y\}. \qquad (3.96)$$

D_y can now be restricted to indices of the a_is that remain in D'_z:

$$D'_y = \{i \in D_y \mid a_i \in D'_z\} \tag{3.97}$$

This achieves domain completeness.

For example, consider the element constraint

$$\text{element}(y, z \mid (20, 30, 60, 60))$$

where initially the domains are

$$D_z = \{20, 40, 60, 80, 90\}, \quad D_y = \{1, 2, 4\}$$

Rules (3.96)–(3.97) yield the filtered domains

$$D'_z = \{20, 40, 60, 80, 90\} \cap \{20, 30, 60\} = \{20, 60\}$$
$$D'_y = \{1, 2, 4\} \cap \{1, 3, 4\} = \{1, 4\}$$

Achieving domain completeness for an element constraint with variables (3.93) is more interesting. D_y is assumed finite, but since $x_1, \ldots, x_m$ and z may be continuous variables, their domains may be infinite.

First, D_z must be a subset of the combined domains of the variables x_i whose indices are in D_y:

$$D'_z = D_z \cap \bigcup_{i \in D_y} D_{x_i} \tag{3.98}$$

D_y can now be restricted to indices i for which D'_z intersects D_{x_i}:

$$D'_y = \{i \in D_y \mid D'_z \cap D_{x_i} \neq \emptyset\} \tag{3.99}$$

Finally, D_{x_i} can be restricted if i is the only index in D'_y:

$$D'_{x_i} = \begin{cases} D'_z & \text{if } D'_y = \{i\} \\ D_{x_i} & \text{otherwise.} \end{cases} \tag{3.100}$$

This achieves domain completeness.

Consider, for example, the element constraint

$$\text{element}(y, (x_1, x_2, x_3), z)$$

where initially the domains are

$$D_z = [50, 80], \quad D_y = \{1, 3\}$$
$$D_{x_1} = [0, 20], \quad D_{x_2} = [10, 40], \quad D_{x_3} = [30, 60] \cup [70, 90]$$

Rules (3.98)–(3.100) imply that the reduced domains are

$$D'_z = [50, 80] \cap ([0, 20] \cup [30, 60] \cup [70, 90]) = [50, 60] \cup [70, 80]$$
$$D'_y = \{3\}, \quad D'_{x_1} = D_{x_1}, \quad D'_{x_2} = D_{x_2}, \quad D'_{x_3} = D'_z$$

Thus, y is fixed to 3, which means that $x_3 = z$. The common domain of x_3 and z is the intersection of their original domains.

The indexed linear element constraint (3.95) is processed as follows. D_z can be reduced to its intersection with the set of values that can be obtained by multiplying an a_i by a possible value of x:

$$D'_z = D_z \cap \bigcup_{i \in D_y} \{a_i v \mid v \in D_x\} \tag{3.101}$$

D'_y is the set of indices i for which D'_z contains $a_i v$ for some possible value v of x:

$$D'_y = \{i \in D_y \mid D'_z \cap \{a_i v \mid v \in D_x\} \neq \emptyset\} \tag{3.102}$$

Finally, D'_x can be reduced to values whose multiples by some a_i belong to D'_z:

$$D'_x = \bigcup_{i \in D'_y} \{v \in D_x \mid a_i v \in D'_z\} \tag{3.103}$$

The filtering procedure can be extended to the vector-valued constraint. Suppose, for example, that the variables in constraint

$$\text{element}(y, x, z \mid (1, 2, 3))$$

have initial domains

$$D_z = [36, 48], \quad D_y = \{1, 2, 3\}, \quad D_x = [10, 30]$$

Applying (3.101)–(3.103):

$$D'_z = [36, 48] \cap ([10, 30] \cup [20, 60] \cup [30, 90]) = [36, 48]$$
$$D'_y = \{2, 3\}$$
$$D'_x = \emptyset \cup [\tfrac{36}{2}, \tfrac{48}{2}] \cup [\tfrac{36}{3}, \tfrac{48}{3}] = [12, 16] \cup [18, 24]$$

3.8.2 Bounds Completeness

Bounds completeness for an element constraint (3.92) with constants is trivial to achieve. Let $I_z = [L_z, U_z]$ be the current interval domain of z. The updated domain is $[L'_z, U'_z]$, where

$$L'_z = \max\left\{L_z, \min_{i \in D_y}\{a_i\}\right\}, \quad U'_z = \min\left\{U_z, \max_{i \in D_y}\{a_i\}\right\}$$

D_y can also be filtered:
$$D'_y = \{i \in D_y \mid a_i \in [L'_z, U'_z]\}$$

To tighten bounds for an element constraint (3.93) with variables, let $[L_{x_i}, U_{x_i}]$ be the current interval domain of x_i. The domains of z, y can be updated by setting

$$L'_z = \max\left\{L_z, \min_{i \in D_y}\{L_{x_i}\}\right\}, \quad U'_z = \min\left\{U_z, \max_{i \in D_y}\{U_{x_i}\}\right\}$$
$$D'_y = \{i \in D_y \mid [L_{x_i}, U_{x_i}] \cap [L'_z, U'_z] \neq \emptyset\}$$

If D'_y is a singleton $\{i\}$, the bounds on x_i can be updated:
$$L'_{x_i} = \max\{L_{x_i}, L'_z\}, \quad U'_{x_i} = \min\{U_{x_i}, U'_z\},$$

This procedure achieves bounds completeness if the variables (other than y) have interval domains, but otherwise need not do so. For example, if
$$D_z = [20, 30], \quad D_y = \{1\}, \quad D_{x_1} = [0, 10] \cup [40, 50]$$
then applying the procedure to element$(y, (x_1), z)$ reduces the bounds for x_1 to $[20, 30]$ and has no effect on the bounds for z. However, the constraint is infeasible, and bounds completeness is achieved only by reducing the domain of x_1 and z to an empty interval.

Bounds can be updated for element$(y, x, z \mid (a_1, \ldots, a_m))$ by setting

$$L'_z = \max\left\{L_z, \min_{i \in D_y}\{\min\{a_i U_x, a_i L_x\}\}\right\}$$
$$U'_z = \min\left\{U_z, \max_{i \in D_y}\{\min\{a_i L_x, a_i U_x\}\}\right\}$$

and

$$L'_x = \max\left\{L_x, \min_{i \in D_y}\left\{\min\left\{\frac{U'_z}{a_i}, \frac{L'_z}{a_i}\right\}\right\}\right\}$$
$$U'_x = \min\left\{U_x, \max_{i \in D_y}\left\{\min\left\{\frac{L'_z}{a_i}, \frac{U'_z}{a_i}\right\}\right\}\right\}$$

An index i can be deleted from D_y if the reduced interval for z does not overlap the reduced interval for $a_i x$:
$$D'_y = \{i \in D_y \mid [\min\{a_i L'_x, a_i U'_x\}, \max\{a_i L'_x, a_i U'_x\}] \cap [L'_z, U'_z] \neq \emptyset\}$$

This procedure achieves bounds completeness if the variables other than y have interval domains. For example, if
$$D_z = [30, 40], \quad D_y = \{1, 2\}, \quad D_x = [-10, 10]$$
then the updated bounds for element$(y, x, z \mid (-3, 5))$ are $D'_z = [30, 40]$ and $D'_x = [-5, 5]$. Also, D_y is reduced to $\{2\}$.

3.8.3 Exercises

1. Show that formulas (3.96)–(3.97) achieve domain completeness for element$(y, z \,|\, a)$.

2. Show that formulas (3.98)–(3.100) achieve domain completeness for element(y, x, z).

3. Suppose one is given initial domains $y \in \{1, 2, 3\}$, $x \in \{-1\} \cup [1, 3]$, and $z \in [-2, -1] \cup [1, 2]$. Reduce domains to achieve domain consistency for element$(y, x, z \,|\, (-1, 0, 2))$. Also, reduce domains to achieve bounds completeness.

4. Extend the filtering procedure (3.98)–(3.100) to the vector-valued indexed linear element constraint.

5. Interpret the following expression using the appropriate element constraints, where the variables are y, y_i, x_i, and $w_{i1}, \ldots, w_{im}$ for $i = 1, \ldots, n$:

$$\sum_{i=1}^{n} a_{iy_i} x_i = \sum_{i=1}^{n} b_i w_{iy}$$

3.9 The All-Different Constraint

The all-different (alldiff) constraint is one of the most frequently used global constraints in constraint programming models, since it arises in assignment and sequencing problems for which a constraint programming approach is particularly well suited. There are efficient filtering algorithms for achieving domain completeness and bounds completeness. In fact, the matching algorithm commonly used to filter alldiff has inspired a number of similar filtering techniques for other global constraints. There are also continuous relaxations for alldiff, which will be discussed in Section 4.12.

The all-different constraint

$$\text{alldiff}(x_1, \ldots, x_n)$$

requires that the variables $x_1, \ldots, x_n$ all take distinct values, where the domain D_{x_i} of each x_i is finite.

A simple assignment problem involving alldiff might require that each of five workers be assigned one job, and each of six available jobs be assigned to at most one worker. Each worker has the necessary skills to do only certain jobs. The problem is to find a feasible solution for

$$\text{alldiff}(x_1, \ldots, x_5)$$
$$\begin{aligned} D_{x_1} &= \{1\} & D_{x_4} &= \{1, 5\} \\ D_{x_2} &= \{2, 3, 5\} & D_{x_5} &= \{1, 3, 4, 5, 6\} \\ D_{x_3} &= \{1, 2, 3, 5\} & & \end{aligned} \quad (3.104)$$

Thus, Worker 1 can do only Job 1, Worker 2 can do Job 2, 3 or 5, and so forth. One can see right away that Job 1 must be assigned to worker 1 and can therefore be removed from the domains of x_3, x_4 and x_5. In fact, the domains can be reduced considerably more than this.

Some graph theoretic terminology will be useful in discussions to follow. A *path* of a graph G is a sequence of edges

$$(v_1, v_2), (v_2, v_3), \ldots, (v_{k-1}, v_k)$$

in the graph, and such a path *connects* v_1 with v_k. G is *connected* if any two vertices in G are connected by a path. A subset S of vertices of G *induces* the subgraph of G that consists of the vertices in S and all edges (i, j) of G for which $i, j \in S$. If S induces a connected subgraph G' of G, and no proper superset of S does so, then G' is a *connected component* of G. If G is a directed graph, then G is *strongly connected* if, for every pair i, j of vertices in G, some path of G connects i with j and some path connects j with i. A *strongly connected component* is defined in analogy with a connected component.

3.9.1 Bipartite Matching

It is helpful to view the alldiff constraint as posing a matching problem on a graph. Given the constraint alldiff$(x_1, \ldots, x_n)$, construct a bipartite graph with vertices on one side that correspond to variables $x_1, \ldots, x_n$, and vertices on the other side that correspond to elements in the variable domains. The graph contains edge (x_j, i) whenever $i \in D_{x_j}$. A *matching* on the graph is a subset of the edges such that each vertex is incident to at most one edge in the matching. The alldiff constraint is feasible if and only if some matching *covers* the vertices $x_1, \ldots, x_n$; that is, each x_j is incident to an edge in the matching.

The example (3.104) corresponds to the bipartite graph in Figure 3.9. A matching that covers $x_1, \ldots, x_5$ is shown by heavy lines. It corresponds to a solution of the alldiff constraint, namely $(x_1, \ldots, x_5) = (1, 2, 3, 5, 4)$.

An easy way to check whether an alldiff constraint is feasible is to find a *maximum cardinality matching* on the associated graph G, which is a matching that contains a maximum number of edges. The alldiff is feasible if and only if a maximum cardinality matching covers $x_1, \ldots, x_n$.

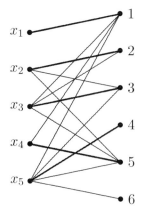

Figure 3.9. Bipartite graph corresponding to an alldiff constraint. The heavy lines indicate a maximum cardinality matching.

A simple algorithm for finding a maximum cardinality matching is based on the idea of an *alternating path*. For a given matching, a path in G is alternating when every other edge in the path belongs to the matching. It is an *augmenting path* when neither endpoint of the path is covered by the matching. The following is a standard result of graph theory and applies to any graph, bipartite or otherwise. It is a special case of the optimality condition for maximum flow that will appear in Theorem 3.39.

THEOREM 3.35 *A matching on a graph G is a maximum cardinality matching if and only if there is no augmenting path in G.*

To find a maximum cardinality matching, start with an arbitrary matching. If there is an augmenting path P, let P' be the edges in P that belong to the matching. Remove the edges in P' from the matching and add the edges in $P \setminus P'$ to the matching. This creates a new matching that contains one more edge than the original matching. Continue until no augmenting path exists.

3.9.2 Domain Completeness

A domain filtering method for alldiff can be derived from properties of a maximum cardinality matching. It is based on the fact that x_j can take the value i if and only the edge (x_j, i) is part of some maximum cardinality matching that covers $x_1, \ldots, x_n$. This can, in turn, be checked by applying the following graph theoretic result. An alternating cycle is a cycle in which every other edge belongs to a given matching.

THEOREM 3.36 *Consider any maximum cardinality matching of a graph G and an edge e that does not belong to the matching. Then e belongs to some maximum cardinality matching if any only if it is part of (a) an alternating cycle, or (b) an even alternating path, one end of which is a vertex that is incident to no edge in the matching.*

Variable domains for alldiff can be now be filtered as follows. Find a maximum cardinality matching as above on the associated bipartite graph G. If the matching fails to cover the vertices $x_1, \ldots, x_n$ on one side of G, then the alldiff is infeasible. Otherwise, for each uncovered vertex on the other side, mark all edges that are part of an alternating path that starts at that vertex. Also, mark every edge that belongs to some alternating cycle, for the same reason. Now delete all unmarked edges that are not part of the matching. The remaining edges correspond to elements of the filtered domains.

In the graph of Figure 3.9, only the vertex labeled 6 is uncovered. The only alternating path starting at that vertex contains also vertices x_5 and 4. Edges $(x_5, 6)$ and $(x_6, 4)$ are therefore marked. The only alternating cycle contains vertices $x_2, 2, x_3$ and 3. Its edges are also marked. One can now delete all edges that are not in the matching except the marked edges $(x_5, 6)$, $(x_2, 3)$ and $(x_3, 2)$. This yields the graph in Figure 3.10. The domains are therefore reduced to

$$D_{x_1} = \{1\} \quad D_{x_4} = \{5\}$$
$$D_{x_2} = \{2, 3\} \quad D_{x_5} = \{4, 6\}$$
$$D_{x_3} = \{2, 3\}$$

The marked edges can be found algorithmically by modifying the given bipartite graph G to obtain a directed graph G' on the same vertices.

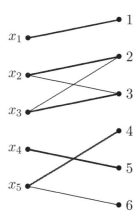

Figure 3.10.   Results of domain filtering for an alldiff constraint.

The All-Different Constraint

Directed edge (x_j, i) belongs to G' if the same (undirected) edge is part of the initial maximum cardinality matching on G. Also (i, x_j) belongs to G' if it belongs to G but is not part of the matching. Identify the strongly connected components of G'. This can be done in time proportional to $m + n$, where m is the number of edges and n the number of vertices. Mark all edges of G' that are part of the matching or lie on directed paths starting at vertices that are uncovered by the matching. Mark all edges that connect vertices in strongly connected components of G'. The unmarked edges can be deleted.

Figure 3.11 shows G' for the graph G of Figure 3.9. Vertices x_2, x_3, 2, and 3 belong to a strongly connected component, and the connecting edges are marked. Also edges on the path from 6 to x_5 to 4 are marked.

3.9.3 Bounds Completeness

Bounds completeness makes sense for the alldiff constraint when the domain elements have a natural ordering, as for example when they are integers. The matching model achieves bounds completeness more rapidly than domain completeness, because the bipartite graph that represents bounds has a convexity property that allows for faster identification of a maximum cardinality matching.

Let $L_j = \min D_{x_j}$ and $U_j = \max D_{x_j}$ be the endpoints of x_j's domain. Alldiff$(x_1, \ldots, x_n)$ is bounds complete if for any variable x_k, $x_k = L_k$ in some solution of the alldiff for which each x_j lies in the interval

$$I_{x_j} = \{L_j, \ldots, U_j\}$$

and $x_k = U_k$ in some solution for which each $x_j \in I_{x_j}$.

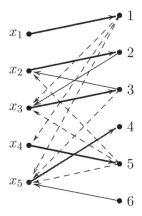

Figure 3.11. Directed bipartite graph corresponding to an alldiff constraint. Dashed edges can be eliminated.

The bipartite graph is constructed as before, except that the interval I_j is treated as the domain of x_j. Consider the example on the left below, where the intervals I_{x_j} are shown on the right.

$$\begin{array}{ll} \text{element}(x_1,\ldots,x_5) & \\ D_{x_1} = \{\,1,2,\quad 4\quad\,\} & I_{x_1} = \{\,1,2,3,4\quad\,\} \\ D_{x_2} = \{\quad 2,3,\quad 6\,\} & I_{x_2} = \{\quad 2,3,4,5,6\,\} \\ D_{x_3} = \{\quad\quad 3,\quad 5\quad\,\} & I_{x_3} = \{\quad\quad 3,4,5\quad\,\} \\ D_{x_4} = \{\quad\quad 3,4\quad\quad\} & I_{x_4} = \{\quad\quad 3,4\quad\quad\} \\ D_{x_5} = \{\quad\quad\quad 4,5\,\} & I_{x_5} = \{\quad\quad\quad 4,5\,\} \end{array} \quad (3.105)$$

The corresponding graph appears in Figure 3.12.

In general, the bipartite graph representing interval domains is *convex*, meaning that if x_j is linked to domain elements i and k for $i < k$, then it is linked to all the elements $i,\ldots,k$. The maximum cardinality matching problem can be solved on a convex graph in linear time relative to the number of variables. This is accomplished as follows. For each vertex $i = 1,\ldots,m$ on the right, let an (x_j, i) be an edge in the matching, where j is the index that minimizes U_j subject to the condition that (x_j, i) is an edge of the graph and x_j is not already covered. If there is no such edge (x_j, i), then i is left uncovered. The matching for problem (3.105) is shown in Figure 3.12.

The matching obtained as just described covers all the variables x_j if any matching does. This is implied by the following result of graph theory. Let a *perfect* matching be one that covers all vertices.

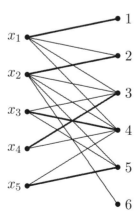

Figure 3.12. Bipartite graph used to achieve bounds consistency for an all-different constraint. The heavy lines indicate a matching that covers all x_j and therefore satisfies alldiff.

THEOREM 3.37 *If G is a convex bipartite graph, the above algorithm finds a perfect matching for G, if one exists.*

The theorem can be specialized to the present situation as follows.

COROLLARY 3.38 *If G is a convex graph with vertices $x_1, \ldots, x_n$ on the left, and $m \geq n$ vertices on the right, the above algorithm finds a matching for G that covers $x_1, \ldots, x_n$, if one exists.*

Proof. Construct graph G' from G by adding vertices $x_{n+1}, \ldots, x_m$ to the left and edges from each new vertex to all vertices on the right. If G has a matching that covers $x_1, \ldots, x_n$, then G' has a perfect matching. By Theorem 3.37, the algorithm finds such a matching when applied to G'. Modify the algorithm so that (i) when two or more edges (x_j, i) in G' minimize U_j, the edge with the smaller index j is added to the matching, and (ii) all edges in the matching that are incident to $x_{n+1}, \ldots, x_m$ are dropped. The modified algorithm is the original algorithm applied to G, and it produces a matching that covers $x_1, \ldots, x_n$. □

Once a matching that covers $x_1, \ldots, x_n$ is found, one can mark edges as described in the previous section. For each unmarked edge (x_j, i) that is not part of the matching, remove i from D_{x_j} if $i \in D_{x_j}$. Let D'_{x_j} be the updated domain for each j. Now, update the bounds by setting $L_j = \min D'_{x_j}$ and $U_j = \max D'_{x_j}$ for each j. This achieves bounds completeness.

In the example, edges from x_1 to 3 and 4 are removed, as are edges from x_2 to 3, 4 and 5. This reduces D_{x_1} from $\{1, 2, 4\}$ to $D'_{x_1} = \{1, 2\}$ and reduces D_{x_2} from $\{2, 3, 6\}$ to $\{2, 6\}$, while leaving the other domains unchanged. So, the bounds for x_1 are updated from $(L_1, U_1) = (1, 4)$ to $(1, 2)$, and the remaining bounds are unaffected.

Again, the marked edges can be found by forming the directed graph described in the previous section and identifying edges that (i) lie on directed paths from uncovered vertices, or (ii) connect vertices in strongly connected components. If properly implemented, the running time is linear in n except for the time required to sort the interval endpoints in increasing order.

3.9.4 Exercises

1 Use the bipartite matching algorithm to achieve domain consistency for alldiff$(x_1, \ldots, x_5)$, where $x_1 \in \{1, 4\}$, $x_2 \in \{1, 3\}$, $x_3 \in \{3, 6\}$, $x_4 \in \{2, 3, 5\}$, $x_5 \in \{1, 2, 3, 4, 5, 6, 7\}$. Do it once by identifying even alternating cycles and even alternating paths on the graph. Then do it again by identifying strongly connected components on the associated directed graph.

2 Use the bipartite matching algorithm for convex graphs to achieve bounds consistency for the alldiff constraint in the previous exercise.

3 The alldiff-except-0(x) constraint requires the variables $x_1, \ldots, x_n$ with nonzero values to take different values. Indicate how to obtain domain completeness for this constraint with a bipartite matching model.

3.10 The Cardinality and Nvalues Constraints

The *cardinality* constraint is an extension of the alldiff constraint that counts how many variables take each of a given set of values. It is also known as the *distribute*, *gcc*, or *generalized cardinality* constraint. The *nvalues* constraint is another extension of alldiff that counts how many different values are assumed by a set of variables. Both are versatile constraints, and both can be efficiently filtered using a network flow model that generalizes the matching model used in the previous section to filter alldiff.

3.10.1 The Cardinality Constraint

The cardinality constraint is illustrated in Section 2.2.5. It can be written

$$\text{cardinality}(x \mid v, \ell, u) \tag{3.106}$$

where $x = (x_1, \ldots, x_n)$ is a tuple of variables, $v = (v_1, \ldots, v_m)$ a tuple of values, $\ell = (\ell_1, \ldots, \ell_m)$ a tuple of lower bounds, and $u = (u_1, \ldots, u_m)$ a tuple of upper bounds. The constraint is satisfied when, for each $j \in \{1, \ldots, m\}$, at least ℓ_j and at most u_j variables x_i assume the value v_j.

Consider for example the constraint

$$\text{cardinality}((x_1, x_2, x_3, x_4) \mid (a, b, c), (1, 1, 0), (2, 3, 2)) \tag{3.107}$$

with domains $D_{x_1} = D_{x_3} = \{a\}$, $D_{x_2} = \{a, b, c\}$, $D_{x_4} = \{b, c\}$. The constraint requires that at least one, and at most two, of the variables $x_1, \ldots, x_4$ take the value a, and analogously for values b and c. Obviously a must be assigned to x_1 and x_3, which means that a cannot be used again and therefore can be removed from the domain of x_2. It will be seen shortly that no other values can be removed from domains.

The alldiff constraint is a special case in which the domain of each x_i is a subset of $\{v_1, \ldots, v_m\}$, $\ell = (0, \ldots, 0)$, and $u = (1, \ldots, 1)$.

3.10.2 Network Flow Model

The network flow model for the cardinality constraint (3.106) associates each variable x_i and each value v_j with a node of a directed graph G. There is a directed edge from v_j to x_j in G when v_j belongs to the domain D_{x_j} of x_j. Directed edges also run from a source vertex s to each value v_j, from each variable x_i to a sink vertex t, and from t to s. The graph corresponding to the example (3.107) appears in Figure 3.13.

Each edge of G has capacity bounds $[\ell, u]$, where ℓ is a lower bound on how much flow the edge can carry, and u an upper bound. A *feasible flow* on G assigns a flow volume to each edge in such a way that the capacity bounds are observed and flow is conserved at each vertex; that is, the total flow into a vertex equals the total flow out. The capacity bounds in the present case are $[\ell_j, u_j]$ for each edge (s, v_j), $[0, 1]$ for each edge (v_j, x_i) and (x_i, t), and $[0, n]$ for (t, s). If there are values in the domains that are not among $\{v_1, \ldots, v_m\}$, they can be represented by a single additional value v in the graph, with capacity $[0, n]$ on the arc (s, v).

A flow of 1 from v_j to x_i is interpreted as assigning value v_j to x_i. Clearly any feasible flow assigns at most one value to each variable and ensures that each value v_j is assigned to at least ℓ_j, and at most u_j, variables. The cardinality constraint (3.106) is feasible if and only if the maximum feasible flow from t to s is n. One maximum flow on the graph corresponding to the cardinality constraint (3.107) appears in

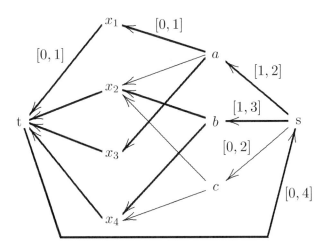

Figure 3.13. Flow model of a cardinality constraint. One feasible flow pattern is indicated by heavy arrows. The positive flow volumes are all 1 except on edges (s, a) and (s, b), where the flow volume is 2.

Figure 3.13. It assigns the values $(x_1, \ldots, x_4) = (a, b, a, b)$, which satisfy the cardinality constraint.

There are fast (polynomial time) algorithms for maximizing flow on any given edge. Since these algorithms require a feasible initial flow, one can begin with all the bounds on edges (s, v_j) set to $[0, u_j]$, so that an initial flow of zero on all edges is feasible. For $j = 1, \ldots, m$ one can now do the following. If the current flow on (s, v_j) is less than ℓ_j, maximize the flow on (s, v_j), restore the bounds to $[\ell_j, u_j]$, and use this flow as a starting feasible solution for the next max flow problem. If at any point maximizing flow on (s, v_j) yields a flow less than ℓ_j, there is no feasible flow. If the capacity bounds are integral, there is a maximum flow in which the flow on every edge is integral.

A maximum flow algorithm can be used to check whether x_j can take a given value v in its domain. Suppose a max flow of n from t to s is found for G. If the flow on (v, x_j) is 1, then obviously x_j can take the value v. If the flow on (v, x_j) is zero, then one can compute the max flow on (v, x_j) in a graph G' that is the same as G except that the flow on (t, s) is fixed to n. If the max flow on (v, x_j) is zero, then x_j cannot take the value v and v can be removed from D_{x_j}.

One can determine quickly whether zero is the max flow on (v, x_j) in G' by checking whether it satisfies the optimality condition for a max flow. This condition is defined on the *residual graph* of G'. Let each edge (i, j) of G' have capacity $[\ell_{ij}, u_{ij}]$, and suppose that a feasible flow f on G' assigns a flow volume f_{ij} to each (i, j). Define the residual graph $R(f)$ for f to be the graph on the same vertices as G' that contains an edge (i, j) with capacity $[0, u_{ij} - f_{ij}]$ whenever $f_{ij} < u_{ij}$ and an edge (j, i) with capacity $[0, f_{ij} - \ell_{ij}]$ whenever $f_{ij} > \ell_{ij}$. Figure 3.14 displays the residual graph $R(f)$ for the flow f shown in Figure 3.13.

It is a standard result of network flow theory that the flow f_{ij} on an edge (i, j) can be increased if and only if there is an *augmenting path* from j to i. This is a path from j to i in $R(f)$ that does not include edge (i, j). In particular, the flow on (i, j) can be increased at least by an amount equal to $\min\{u_{ij}, \min_{i'j'}\{u_{i'j'}\}\}$, where i', j' range over the edges (i', j') on the augmenting path.

THEOREM 3.39 *A given feasible flow f on a graph maximizes the flow on (i, j) if and only if there is no augmenting path from j to i in the residual graph $R(f)$.*

For example, the flow of zero on edge (c, x_2) of the graph G shown in Figure 3.13 can be increased without changing the flow on (t, s). This is because there is an augmenting path from x_2 to c (through vertices b, s) in the residual graph of G', shown in Figure 3.14. In particular, the flow

The Cardinality and Nvalues Constraints 197

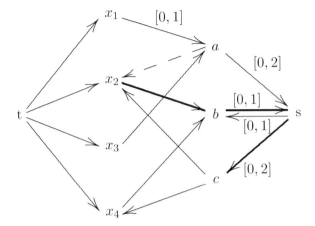

Figure 3.14. Residual graph for the flow indicated in Figure 3.13, assuming the flow on (t, s) is fixed to 4. The heavy arrows show an augmenting path from x_2 to c, which indicates that the flow on (c, x_2) can be increased to 1. The dashed edge represents a flow that must be zero, which means a can be eliminated from the domain of x_2. No other values can be eliminated from domains.

can be increased to 1 because the maximum capacity along the cycle is $\max\{1, 1, 2, 1\} = 1$. The flow of zero on (a, x_2) cannot be increased, however, because there is no augmenting path from x_2 to a.

3.10.3 Domain Completeness for Cardinality

The cardinality constraint (3.106) can be used to eliminate an element v_j from D_{x_i} if the flow volume on (v_j, x_i) cannot be 1 in any feasible flow in the associated graph G. Thus, if a feasible flow f on G places a flow of zero on (v_j, x_i), one can check whether zero is the maximum flow on this edge by applying Theorem 3.39. The theorem says that zero is the maximum flow if and only if there is no augmenting path from x_i to v_j. This, in turn, is true if and only if (v_j, x_i) belongs to no directed cycle of the residual graph $R(f)$. But this is the case if and only if v_j and x_i belong to different strongly connected components of $R(f)$. This demonstrates the following theorem:

THEOREM 3.40 *Let f be a feasible flow in the graph G associated with a cardinality constraint (3.106). An element v_j can be eliminated from the domain of x_i if and only if v_j and x_i belong to different strongly connected components of the residual graph $R(f)$ of G.*

Thus, one can achieve domain completeness for a cardinality constraint (3.106) by the following procedure:

(a) Apply a maximum flow algorithm to find a flow f on G that maximizes the flow volume on (t, s). This can be done in time proportional to n and the number of edges of G. If the maximum flow on (t, s) is less than n, then (3.106) is infeasible.

(b) Identify the strongly connected components of the residual graph $R(f)$ of G'. This can be done in time that is proportional to $m + n$.

(c) For every value $v_j \in D_{x_i}$, eliminate v_j from D_{x_i} if f places a flow of zero on edge (v_j, x_i) of $R(f)$ and the two vertices v_j, x_i belong to different strongly connected components of $R(f)$.

In the example of Figure 3.14, the vertices t and x_1 respectively belong to two strongly connected components, and the remaining vertices belong to a third component. Thus, of the three edges (a, x_2), (c, x_2), and (c, x_3) with zero flow volume in f, only (a, x_2) necessarily carries zero flow, because only this edge has endpoints in different strongly connected components. Thus D_{x_2} can be reduced from $\{a, b, c\}$ to $\{b, c\}$, and domain completeness is achieved.

The flow model for the cardinality constraint becomes a flow model for the alldiff constraint when all the edges (s, v_j) have a capacity range of $[0, 1]$. The matching method presented for filtering alldiff can be understood as a specialization of the max flow method for filtering cardinality. In particular, Theorem 3.36 follows from the optimality conditions for a max flow given in Theorem 3.39.

3.10.4 The Nvalues Constraint

The *nvalues* constraint is written nvalues$(x \mid \ell, u)$, where $x = \{x_1, \ldots, x_n\}$ is a set of variables, ℓ a lower bound, and ℓ an upper bound. It is illustrated in Section 2.2.5. The network flow model presented above is easily modified to provide filtering for the nvalues constraint. This is left as an exercise.

Filtering algorithms have also been developed for versions of nvalues in which ℓ or u is a variable and must be filtered along with x. A max flow model can be adapted to achieve domain completeness for nvalues$(x, \ell \mid u)$, in which ℓ is a variable. Achieving domain completeness for nvalues$(x, u \mid \ell)$, in which u is a variable, is NP-hard, but incomplete polynomial-time filters have been proposed.

3.10.5 Exercises

1 Use the max flow model to filter the constraint
$$\text{cardinality}((x_1, \ldots, x_5) \mid (a, b, c), (1, 1, 1), (2, 2, 2))$$
where $x_1 \in \{a\}$, $x_2 \in \{b\}$, $x_3 \in \{a, c\}$, $x_4 \in \{b\}$, $x_5 \in \{b, c\}$.

The Circuit Constraint

2. Prove Theorem 3.35 as a corollary of Theorem 3.39. Hint: add a source and sink and view the matching problem as a flow problem.

3. Prove Theorem 3.36 as a corollary of Theorem 3.39. Hint: look at the conditions for whether the current flow of zero on edge e is maximum.

4. Formulate a flow model that can be used to achieve domain consistency for the constraint nvalues$(x \mid \ell, u)$. Hint: some variables may not receive flow in a feasible solution.

3.11 The Circuit Constraint

The circuit constraint is similar to the alldiff constraint in that it requires a set of variables to indicate a permutation, but the variables encode the permutation in a different way. In the constraint alldiff$(x_1, \ldots, x_n)$, each variable x_i indicates the ith item in a permutation of $1, \ldots, n$ when the domain of each x_i is a subset of $\{1, \ldots, n\}$. In the circuit constraint

$$\text{circuit}(x_1, \ldots, x_n) \tag{3.108}$$

each variable x_i denotes which item follows i in the permutation, where again the domain of each x_i is a subset of $\{1, \ldots, n\}$. Thus, (3.108) requires that $y_1, \ldots, y_n$ be a permutation of $1, \ldots, n$, where each $y_{i+1} = x_{y_i}$ (and y_{n+1} is identified with y_1).

The circuit constraint can be interpreted as describing a *hamiltonian cycle* on a directed graph. The elements $1, \ldots, n$ may be viewed as vertices of a directed graph G that contains an edge (i, j) whenever $j \in D_{x_i}$. An edge (i, j) is selected when $x_i = j$, and (3.108) requires that the selected edges form a hamiltonian cycle, which is a path through vertices $y_1, \ldots, y_n, y_1$ for which $y_1, \ldots, y_n$ are all distinct. An edge is hamiltonian when it is part of some hamiltonian cycle. An element j can be deleted from D_{x_i} if and only if (i, j) is a nonhamiltonian edge, which means that domain completeness can be achieved by identifying all nonhamiltonian edges.

Achieving domain completeness is more difficult for the circuit constraint than for alldiff and is in fact an NP-hard problem. On the other hand, there may be more potential for domain reduction because the circuit constraint is stronger than alldiff when each variable domain is $\{1, \ldots, n\}$. Any $x = (x_1, \ldots, x_n)$ that satisfies (3.108) also satisfies alldiff$(x_1, \ldots, x_n)$, whereas the reverse is not true. There are also strong relaxations for circuit, based on cutting planes that have been developed for the traveling salesman problem (Section 4.14).

3.11.1 Modeling with Circuit

The circuit constraint is naturally suited to modeling situations in which costs or constraints depend on which item immediately follows another in a permutation. This occurs, for instance, in the traveling salesman problem, in which a salesman wishes to visit each of n cities once and return home while minimizing the distance traveled. If c_{ij} is the distance from city i to city j, the problem is

$$\min \sum_i c_{ix_i} \tag{3.109}$$
$$\text{circuit}(x_1, \ldots, x_n)$$

where x_i is the city visited immediately after city i. If certain cities cannot be visited immediately after city i, this can be reflected in the domain of x_i.

On the other hand, if one wishes to assign workers to jobs, and the cost of assigning worker i to job j is c_{ij}, then the alldiff constraint provides the natural formulation:

$$\min \sum_i c_{iy_i}$$
$$\text{alldiff}(y_1, \ldots, y_n)$$

Here, y_i is the job assigned to worker i, and the objective is to minimize total cost. This is the classical assignment problem. If there is a restriction on which jobs may be assigned to worker i, this can be reflected in the initial domain of y_i.

The traveling salesman problem can be modeled with alldiff as well as circuit, but the alldiff formulation provides no natural way to constrain which cities may follow a given city in the salesman's tour. The alldiff model is

$$\min \sum_i c_{y_i y_{i+1}}$$
$$\text{alldiff}(y_1, \ldots, y_n)$$

where y_i is the ith city visited and city $n+1$ is identified with city 1. If the salesman is allowed to visit only cities in D_i after visiting i, then one must add the *channeling* constraints

$$\begin{aligned} y_{i+1} &= x_{y_i}, \quad i = 1, \ldots, n-1 \\ y_1 &= x_{y_n} \end{aligned} \tag{3.110}$$

Now the city immediately following city i can be constrained by specifying the domain of x_i.

The Circuit Constraint 201

The assignment problem can be modeled with circuit as well as alldiff, but only by introducing variables y_i that indicate the job assigned to worker i:

$$\min \sum_i c_{iy_i}$$

$$\text{circuit}(x_1, \ldots, x_n)$$

constraints (3.110)

The domain of y_i can be restricted to indicate which cities may be assigned to salesman i.

The circuit constraint is well suited for scheduling problems that involve sequence-dependent setup times or costs. Suppose, for example, that when job i immediately precedes job j, the time required to process job i and then set up job j is c_{ij}. The domain of each x_i contains the jobs that can immediately follow job i. If the jobs are to be sequenced so as to minimize makespan (total time required to complete the jobs), the problem can be written as (3.109). This formulation is valid if job n is interpreted as a dummy job that immediately follows the last job processed. Then $c_{in} = c_{ni} = 0$ for $i = 1, \ldots, n-1$, and x_n is the first job to run.

3.11.2 Elementary Filtering Methods

Checking a circuit constraint for feasibility is equivalent to checking whether a directed graph has a hamiltonian cycle, which is an NP-hard problem. Achieving domain completeness for circuit is therefore NP-hard. There are useful incomplete filtering methods, however, that run in polynomial time.

Two elementary filtering methods for circuit are based on alldiff filtering and vertex-degree filtering. The alldiff filtering methods of Section 3.9 can be applied because the variables in $\text{circuit}(x_1, \ldots, x_n)$ must take different values.

Vertex-degree filtering is based on the fact that the in-degree and out-degree of every vertex in a hamiltonian cycle is one. Let the inverse domain of j be $D_j^{-1} = \{i \mid j \in D_{x_i}\}$, which is the index set of all variables whose domain contains j. The vertex-degree filtering algorithm cycles through two steps until no further domain reduction is possible:

1. If D_{x_i} is a singleton $\{j\}$ for some i, remove j from D_{x_k} for all $k \neq i$.
2. If D_j^{-1} is a singleton $\{i\}$ for some j, reduce D_{x_i} to $\{j\}$.

Neither filtering method is redundant of the other, as can be seen in the two examples (a) and (b) of Figure 3.15. Since $D_{x_1} = D_{x_2} = \{3, 4\}$ in example (a), alldiff filtering removes 3 and 4 from the other

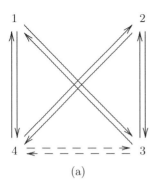 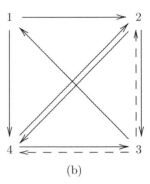

Figure 3.15. (a) Alldiff filtering removes the dashed edges, but vertex-degree filtering has no effect. (b) Vertex-degree filtering removes the dashed edges, but alldiff filtering has no effect.

two domains. Vertex-degree filtering has no effect, however, since the in-degree and out-degree of every vertex is 2. Conversely, vertex-degree filtering removes 2 and 4 from $D_{x_3} = \{1, 2, 4\}$ in example (b), but alldiff filtering has no effect.

3.11.3 Filtering Based on Separators

A more thorough filtering method than alldiff or vertex-degree filtering can be obtained by identifying one or more vertex separators of the associated graph G. A vertex separator is a set of vertices that, when removed, separate G into two or more connected components. By defining a certain kind of labeled graph on a vertex separator, one can state a necessary condition for an edge of that graph to be hamiltonian in G. This allows one to filter domains by analyzing a graph that may be much smaller than G.

Let $G = (V, E)$ be a directed graph with vertex set V and edge set E. A subset S of V is a *(vertex) separator* of G if $V \setminus S$ induces a subgraph with two or more connected components.

The *separator graph* for a separator S of G is a graph $G_S = (S, E_S)$ in which E_S contains *labeled* and *unlabeled* edges. Edge (i, j) is an unlabeled edge of G_S if $(i, j) \in G$. Let C be any connected component of the subgraph of G induced by $V \setminus S$. Edge (i, j) is an edge of G_S with *label* C if (i, c_1) and (c_2, j) are edges of G for some pair of vertices c_1, c_2 of C (possibly $c_1 = c_2$).

Consider, for example, the graph G of Figure 3.16. Vertex set $S = \{1, 2, 3\}$ separates G into three connected components that may be labeled A, B and C, each of which contains only one vertex. The separator graph G_S contains the three edges that connect its vertices in G plus four labeled edges. For example, there is an edge $(1, 2)$ labeled A, which can

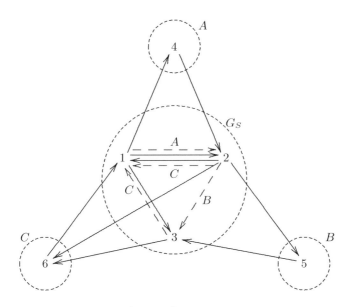

Figure 3.16. Graph G on vertices $\{1, \ldots, 6\}$ contains the solid edges, and the separator graph G_S on $S = \{1, 2, 3\}$ contains the solid (unlabeled) edges and dashed (labeled) edges within the larger circle. The small circles surround connected components of the separated graph.

be denoted $(1, 2)^A$, because there is a directed path from some vertex in component A, through $(1, 2)$, back to a vertex of component A.

A hamiltonian cycle of the separator graph G_S is *permissible* if it contains at least one edge bearing each label. Thus, the edges $(1, 2)^A$, $(2, 3)^B$, and $(3, 1)^C$ form a permissible hamiltonian cycle in Figure 3.15.

THEOREM 3.41 *If S is a separator of directed graph G, then G contains a hamiltonian cycle only if G_S contains a permissible hamiltonian cycle. Furthermore, an edge of G connecting vertices in S is hamiltonian only if it is part of a permissible hamiltonian cycle of G_S.*

Proof. The first task is to show that if H is an arbitrary hamiltonian cycle of $G = (V, E)$, then one can construct a permissible hamiltonian cycle H_S for G_S. Consider the sequence of vertices in H and remove those that are not in S. Let $v_1, \ldots, v_m, v_1$ be the remaining sequence of vertices. H_S can be constructed on these vertices as follows. For any pair v_i, v_{i+1} (where v_{m+1} is identified with v_1), if they are adjacent in H then (v_i, v_{i+1}) is an unlabeled edge of G_S and connects v_i and v_{i+1} in H_S. If v_i, v_{i+1} are not adjacent in H, then all vertices in H between v_i and v_{i+1} lie in the same connected component C of the subgraph of G induced by $V \setminus S$. This means (v_i, v_{i+1}) is an edge of G_S with label

C, and $(v_i, v_{i+1})^C$ connects v_i and v_{i+1} in H_S. Since H passes through all connected components, every label must occur on some edge of H_S, and H_S is permissible.

The second task is to show that if (i,j) with $i,j \in S$ is an edge of a hamiltonian cycle H of G, then (i,j) is an edge of a permissible hamiltonian cycle of G_S. But in this case (i,j) is an unlabeled edge of G_S and, by the above construction, (i,j) is part of H_S. $\square$

As noted earlier, the separator graph G_S in Figure 3.16 contains a permissible hamiltonian cycle with edges $(1,2)^A$, $(2,3)^B$, and $(3,1)^C$. Since this is the only permissible hamiltonian cycle, none of the unlabeled edges $(1,2)$, $(2,1)$, and $(1,3)$ are part of a permissible hamiltonian cycle. Thus, by Theorem 3.41, these edges are nonhamiltonian in the original graph G. The domain D_{x_1} can be reduced from $\{2,3,4\}$ to $\{4\}$, and D_{x_2} from $\{1,5,6\}$ to $\{5,6\}$.

The application of Theorem 3.41 requires that one find one or more separators of the original graph G. One way to find them is to use a simple breadth-first-search heuristic. Let vertices i,j be *neighbors* if (i,j) or (j,i) is an edge of G. Arrange the vertices of G in levels as follows. Arbitrarily select a vertex i of G as a *seed* and let Level 0 contain i alone. Let Level 1 contain all neighbors of i in G. Let level k (for $k \geq 2$) contain all vertices j of G such that (a) j is a neighbor of some vertex on level $k-1$, and (b) j does not occur in levels 0 through $k-1$. If $m \geq 2$, the vertices on any given level k ($0 < k < m$) form a separator of G. Thus the heuristic yields $m-1$ separators.

The heuristic can be run several times as desired, each time beginning with a different vertex on Level 0. In the example, using Vertex 6 as a seed produces the separator of Figure 3.16 on Vertices 1, 2, and 3. Using Vertex 4 as a seed, however, yields a separator graph on Vertices 1 and 2 only, and so forth.

3.11.4 Network Flow Model

The task that remains is to identify edges of G_S that are part of no permissible hamiltonian cycle in G_S, or *nonpermissible* edges for short. Rather than attempt to find all nonpermissible edges, which could be computationally expensive even for relatively small separator graphs, one can identify edges that satisfy a weaker condition. One approach is to construct a network flow model that enforces a relaxation of the condition in Theorem 3.41, as well as a vertex-degree constraint. The flow model is similar to that presented for the cardinality constraint in Section 3.10.

The Circuit Constraint

First, construct a network $N(G, S)$ as follows. There is a source node s, a node C for every label, a node U, a node for every ordered pair (i,j) for which at least one edge connects i to j in G_S, a node for every vertex of G_S, and a sink node t. The arcs of $N(G, S)$ are as follows:

- an arc (s, C) with capacity range $[1, \infty)$ for every label C
- an arc (s, U) with capacity range $[0, \infty)$
- an arc $(C, (i,j))$ with capacity range $[0, 1]$ for every edge $(i,j)^C$ of G_S
- an arc $(U, (i,j))$ with capacity range $[0, 1]$ for every unlabeled edge (i,j) of G_S
- an arc $((i,j), i)$ with capacity range $[0, 1]$ for every node (i,j) of $N(G, S)$
- an arc (i, t) with capacity range $[0, 1]$ for every vertex i if G_s
- a return arc (t, s) with capacity range $[m, m]$.

The network $N(G, S)$ for the separator graph G_S of Figure 3.16 appears in Figure 3.17.

Every permissible hamiltonian cycle on G_S describes a flow pattern on $N(G, S)$, in which a flow of 1 from C to (i,j) indicates that $(i,j)^C$ is part of the cycle, and a flow of 1 from U to (i,j) means that the unlabeled edge (i,j) is part of the cycle. Since the cycle must contain

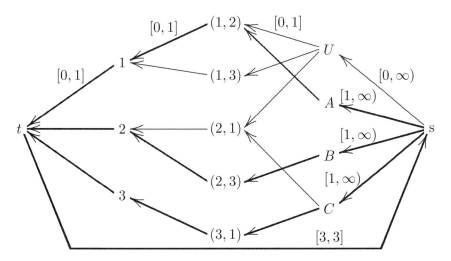

Figure 3.17. Flow model for simultaneous cardinality and out-degree filtering of non-hamiltonian edges. Heavy lines show the only feasible flow. Since the maximum flow on edges (U, z_{12}), (U, z_{13}), and (U, z_{21}) is zero, edges $(1,2)$, $(1,3)$, and $(2,1)$ in the original graph are nonhamiltonian.

all m vertices of G_S, the flow on the return arc must be m. Since each label must occur on the cycle, the flow on each arc (s, C) must be at least 1. Since the out-degree of each vertex i of the cycle is 1, the flow on each (i, t) can be at most 1. In addition,

THEOREM 3.42 *An edge (i, j) of G is nonhamiltonian if there is a separator S of G for which the maximum flow on edge $(U, (i, j))$ of $N(G, S)$ is zero.*

As in Section 3.10, this can be checked by computing a feasible flow f on $N(G, S)$. The maximum flow on $(U, (i, j))$ is zero if f places zero flow on $(U, (i, j))$, and zero flow on this edge satisfies the optimality condition of Theorem 3.39.

For example, the zero flow on edges $(U, (1, 2))$, $(U, (1, 3))$ and $(U, (2, 1))$ of Figure 3.17 is maximum in each case. So, the three edges $(1, 2)$, $(1, 3)$, and $(2, 1)$ of G_S are nonhamiltonian.

A similar test can be devised to combine cardinality filtering with in-degree filtering. It is unclear how to combine cardinality filtering with both out-degree and in-degree filtering in the same network model.

3.11.5 Exercises

1 Assume there is an oracle that can quickly tell whether a graph is hamiltonian. Describe how to use this oracle to check quickly whether a particular edge in a graph is hamiltonian.

2 Apply Theorem 3.41 to the graph in Figure 3.16 when the separator is $S = \{1, 2\}$ and when the separator is $S = \{1, 2, 3, 4\}$. Does the theorem identify all nonhamiltonian edges in the subgraph induced by S? For each separator graph use the flow model to detect nonpermissible edges. Does it identify all nonpermissible edges?

3 Consider the graph with directed edges $(1, 2), (1, 4), (1, 5), (2, 3), (2, 7)$, $(3, 4), (3, 7), (3, 8), (4, 1), (4, 5), (5, 6), (6, 2), (6, 5), (7, 3), (8, 1), (8, 4)$ and the separator $S = \{1, 2, 3, 4\}$. Use Theorem 3.41 to filter as many edges as possible. Does this filter remove any edges that are not removed by vertex-degree filtering? Does it remove any that are not removed by alldiff filtering? Does it remove all nonhamiltonian edges connecting vertices of S?

4 Use the flow model to detect nonpermissible edges in the separator graph constructed in Exercise 3. Does it identify all nonpermissible edges?

5 Show by counterexample that filtering based on Theorem 3.41 is incomplete, even when all separators are used.

3.12 The Stretch Constraint

The stretch constraint was originally designed for scheduling workers in shifts. It is illustrated in Section 2.2.5. Analysis of this constraint provides an opportunity to show how dynamic programming can contribute to filtering.

The stretch constraint is written

$$\text{stretch}(x \mid v, \ell, u, P)$$

where x is a tuple $(x_1, \ldots, x_n)$ of variables. In typical applications, x_i represents the shift that a given employee will work on day i. Also, v is an m-tuple of possible values of the variables, ℓ an m-tuple of lower bounds, and u an m-tuple of upper bounds. A *stretch* is a maximal sequence of consecutive variables that take the same value. Thus, $x_j, \ldots, x_k$ is a stretch if for some value v, $x_j, \ldots, x_k = v$, $x_{j-1} \neq v$ (or $j = 1$), and $x_{k+1} \neq v$ (or $k = n$).

The stretch constraint requires that for each $j \in \{1, \ldots, m\}$, any stretch of value v_j in x have length at least ℓ_j and at most u_j. In addition P is a set of *patterns*, which are pairs of values $(v_j, v_{j'})$. The constraint requires that when a stretch of value v_j immediately precedes a stretch of value $v_{j'}$, the pair $(v_j, v_{j'})$ must be in P. Thus the constraint puts bounds on how many consecutive days the employee can work each shift, and which shifts can immediately follow another. For instance, one of the shifts may represent a day off, and it is common to require that the employee never work two different shifts without at least one intervening day off.

Consider for example a shift scheduling problem in which there are three shifts (a, b, and c) and the domain of each x_j is listed beneath each variable in Table 3.2. A stretch constraint for these variables might be

$$\text{stretch}(x \mid (a, b, c), (2, 2, 2), (3, 3, 3), P) \quad (3.111)$$

where P contains the pairs (a, b), (b, a), (b, c), and (c, b). Thus a stretch can be at least two and at most three days long, and the worker cannot

Table 3.2. Variable domains in a shift scheduling problem.

x_1	x_2	x_3	x_4	x_5	x_6	x_7
a	a	a		a	a	a
	b	b	b		b	b
c	c		c	c		

change directly between shifts a and c. The constraint allows only two feasible schedules: *aabbaaa* and *ccbbaaa*.

There is a cyclic version of the constraint, *stretch-cycle*, which recognizes stretches that continue from x_n to x_1. It can be used when one wants to have the same schedule every week and allows stretches to extend through the weekend into the next week. The filter described below can be modified for the cyclic constraint.

3.12.1 Dynamic Programming Model

A schedule x can be viewed as taking an employee through a sequence of *states*. An employee who is just starting to work a stretch of shift v_j on day i ($x_i = j$) is in state (i, v_j). Stretches take the employee from one state to another. Thus, if the employee is in state (i, v_j), a stretch of value v_j that starts on day i can take the employee to a state of the form $(i + \delta, v_{j'})$, where δ is the length of the stretch, and $v_{j'}$ is a shift that can immediately follow v_j. This is called a feasible *state transition*. A feasible backward transition can similarly bring an employee from state (i, v_j) to $(i - \delta, v_{j'})$ when $(v_{j'}, v_j) \in P$.

A state (i, v_j) is *forward reachable* if there is some feasible sequence of forward state transitions that brings the employee to (i, v_i) from a feasible state on Day 1. It is *backward reachable* if there is a feasible sequence of backward transitions that brings the employee to (i, v_j) from a feasible ending state. So, $x_i = v_j$ in some feasible solution of the stretch constraint if it is part of a stretch whose starting point is forward reachable and whose endpoint is backward reachable.

In the example of Table 3.2, $x_3 = b$ is part of a feasible solution because the stretch bb can occur on Days 3 and 4. It can occur because the first b is forward reachable via stretch aa (or cc) and the second b is backward reachable via stretch aaa.

The stretch constraint has a *Markovian* property that allows the reachable states to be computed efficiently in a recursive fashion—that is, by *dynamic programming*. The Markovian property is simply that the feasible transitions from a given state (i, v_j) depend only on the state itself, and not how the employee reached the state. The bounds on the length of a stretch of shift v_j, for example, depend only on v_j and not on what day it is. The shifts to which the employee is allowed to transition likewise depend only on v_j, because the pattern set P contains only pairs. If triples of consecutive stretches could be constrained, the Markovian property would disappear.

Forward reachability is efficiently computed by defining a function $f(i, v_j)$ that denotes the number of days on or before i on which shift v_j is forward reachable. Such a function is useful because shift v_j is reachable during a given period if and only if $f(i, v_j)$ increases as i moves through the period. That is, v_j is forward reachable during the period $k, \ldots, k'$ if and only if $f(k', v_j) - f(k-1, v_j) > 0$.

The forward reachability function can be computed recursively because (i, v_k) is forward reachable when a shift v_j that can precede v_k is forward reachable on a day i' prior to i such that (a) i' is at most u_j and at least ℓ_j days before i, and (b) v_j is in x_i's domain on every day between i' and i. Condition (b) can be checked by precomputing a run length function r_{ij} that measures the longest stretch ending on day i that shift v_j can run. The recursion is

$$f(i+1, v_k) = \begin{cases} f(i, v_k) + 1 & \text{if } f(i_{\max}, v_j) - f(i_{\min}, v_j) > 0 \text{ for some } v_j \\ & \text{for which } i_{\max} \geq i_{\min} \text{ and } (v_j, v_k) \in P \\ f(i, v_k) & \text{otherwise} \end{cases}$$

for $i = 1, \ldots, n$, where

$$i_{\max} = i - \ell_j + 1, \qquad i_{\min} = i - \min\{u_j, r_{ij}\}$$

The boundary conditions are set to get the recursion started correctly:

$$f(0, v_j) = 0, \ f(1, v_j) = 1, \text{ all } j$$

The run length can itself be computed by a simple recursion,

$$r_{ij} = \begin{cases} r_{i-1,j} + 1 & \text{if } v_j \in D_{x_i} \\ 0 & \text{otherwise,} \end{cases} \quad \text{all } j \text{ and } i = 1, \ldots n$$

where $r_{0j} = 0$ for all j, and D_{x_i} is the domain of x_i. The values of $f(i, v_i)$ for the example of Table 3.2 are displayed in Table 3.3.

A similar backward reachability function $b(i, v_j)$ denotes the number of days on or after day i on which shift v_j is backward reachable. It is recursively computed

$$b(i-1, v_k) = \begin{cases} b(i, v_k) + 1 & \text{if } b(i_{\min}, v_j) - b(i_{\max}, v_j) > 0 \text{ for some } v_j \\ & \text{for which } i_{\min} \leq i_{\max} \text{ and } (v_k, v_j) \in P \\ b(i, v_k) & \text{otherwise} \end{cases}$$

for $i = n, \ldots, 1$, where

$$i_{\min} = i + \ell_j - 1, \qquad i_{\max} = i + \min\{u_j, \bar{r}_{ij}\}$$

with boundary conditions

$$b(n, v_j) = 1, \ b(n+1, v_j) = 0, \text{ all } j$$

Table 3.3. Forward and backward reachability functions for filtering a stretch constraint.

v_j	$i=0$	1	2	3	4	5	6	7	8
a	0	1	1	1	1	2	2	2	2
b	0	1	1	2	3	3	4	5	6
c	0	1	1	1	1	2	2	2	2

$f(i, v_j)$:

v_j	$i=0$	1	2	3	4	5	6	7	8
a	4	4	3	2	2	2	1	1	0
b	5	4	4	4	3	2	1	1	0
c	4	4	3	2	2	2	1	1	0

$b(i, v_j)$:

The run length is measured in a backwards direction,

$$\bar{r}_{ij} = \begin{cases} \bar{r}_{i+1,j} + 1 & \text{if } v_j \in D_{x_i} \\ 0 & \text{otherwise,} \end{cases} \quad \text{all } j \text{ and } i = n, \ldots 1$$

where $\bar{r}_{n+1,j} = 0$ for all j. The function values for the example appear in Table 3.3.

3.12.2 Domain Completeness

Possible values of each x_i can now be computed by examining the forward and backward reachability functions. Consider each possible stretch of each shift that is within the length bounds and compatible with variable domains. For each day i in a stretch of shift v_j, mark v_j as a possible value of x_i if it is forward reachable on the first day of the stretch and backward reachable on the last day of the stretch. All assignments to x_i that remain unmarked after all possible stretches are considered can be deleted from the domain of x_i. This process achieves domain completeness.

More precisely, for each v_j consider every possible stretch $x_k, \ldots, x_{k'}$ of shift v_j for which $\ell_j \leq k' - k + 1 \leq u_j$ and $v_j \in D_{x_i}$ for $i = k, \ldots, k'$. If shift v_j is forward reachable on day k and backward reachable on day k', mark v_j as a possible value of x_i for $i = k, \ldots, k'$. Shift v_j is forward reachable on day k if $f(k, v_j) - f(k-1, v_j) > 0$ and backward reachable on day k' if $b(k', v_j) - b(k'+1, v_j) > 0$. Each v_j that remains unmarked as a possible value of x_i can be removed from D_{x_i}.

For example, a stretch bb on Days 3 and 4 is compatible with the domains shown in Table 3.2. Shift b is forward reachable on Day 3 because $f(4, b) - f(3, b) > 0$ and backward reachable on Day 4 because $b(4, b) - b(5, b) > 0$. So, value b is marked to remain in the domains of x_3

Table 3.4. Reduced variable domains in a shift scheduling problem.

x_1	x_2	x_3	x_4	x_5	x_6	x_7
a	a			a	a	a
		b	b			
c	c					

and x_4. The reduced domains after completion of the filtering algorithm appear in Table 3.4.

The filtering process can be completed in time proportional to nm^2 by maintaining, for each shift, a queue of days that are candidates for the end of a stretch. While examining each day i in reverse order, add i to the back of the queue if i is backward reachable. If i is forward reachable, remove days from the front of the queue that cannot be part of a stretch starting at i, either because the stretch would have the wrong length, or the run length at the front of the queue is too short to reach back to i. Then, add the stretch from i to the front of the queue to a list of possible stretches. The algorithm appears in Figure 3.18.

3.12.3 Exercises

1 Let G be a directed acyclic graph on n vertices, in which each edge (i, j) has length c_{ij}. State a dynamic programming recursion that calculates the shortest distance from every vertex to a terminal vertex

---

Let Q be a queue and front(Q) the element at the front of Q.
Let L be a list of possible stretches.
For $j = 1, \ldots, m$:
    Let $Q, L = \emptyset$.
    For $i = n, \ldots, 1$:
        If $b(i, v_j) - b(i+1, v_j) > 0$ then add i to the back of Q.
        If $f(i, v_j) - f(i-1, v_j) > 0$ then
            Repeat while $Q \neq \emptyset$ and $\min\{u_j, r_{ij}\} < \text{front}(Q) - i$:
                Remove front(Q) from Q.
            If $Q \neq \emptyset$ and $\ell_j \leq \text{front}(Q) - i$ then add $[i, \text{front}(Q)]$ to L.
    For each $[k, k'] \in L$:
        Mark v_j as a feasible value of x_i for $i = k, \ldots, k'$.
For $i = 1, \ldots, n$:
    If no v_j is marked as a feasible value of x_i then remove v_j from D_{x_i}.

Figure 3.18. Algorithm that achieves domain completeness for a stretch constraint.

t. Hints: Let $f_k(i)$ denote the length of the shortest path from i to t that contains k or fewer edges, and define $f_k(\cdot)$ recursively in terms of $f_{k-1}(\cdot)$. The boundary conditions are $f_0(i) = \infty$ for $i \neq t$ and $f_0(t) = 0$.

2 Use the algorithm of Figure 3.18 to reduce the domains in Table 3.2 on the basis of the recursive function values in Table 3.3.

3 Suppose that the domains of $x_1, \ldots, x_7$ are as given beneath each variable below:

x_1	x_2	x_3	x_4	x_5	x_6	x_7
a	a	a		a	a	a
b		b	b	b		b
c	c		c	c	c	c

The three feasible solutions of

$$\text{stretch}(x \mid (a, b, c), (2, 2, 2), (7, 7, 7), P)$$

where $P = \{(a, b), (b, c)\}$ should be evident upon inspection. Use the dynamic programming recursion to filter the domains.

3.13 Disjunctive Scheduling

Disjunctive scheduling is the problem of scheduling jobs that must run one at a time, subject to a release time and deadline for each job. The processing time of each job is fixed. In *preemptive scheduling*, one job can be interrupted to start processing another job, while this is not allowed in *nonpreemptive scheduling*. The focus here is on nonpreemptive scheduling.

The basic disjunctive scheduling constraint is

$$\text{disjunctive}(s \mid p)$$

where $s = (s_1, \ldots, s_n)$ is a tuple of variables s_j indicating the start time of job j. The parameter $p = (p_1, \ldots, p_n)$ is a tuple of processing times p_j for each job. The constraint requires that for any pair of jobs, one must finish before the other starts. That is, it enforces the disjunctive condition

$$(s_i + p_i \leq s_j) \vee (s_j + p_j \leq s_i)$$

for all i, j with $i \neq j$. The constraint programming literature often refers to the disjunctive constraint as a *unary resource* constraint because each job requires one unit of resource while it is running, and only one unit of resource is available at any one time.

Disjunctive Scheduling

Each job j is associated with an earliest start time E_j and a latest completion time L_j. Initially, these are the release time and deadline of the job, respectively, but they may be updated in the course of the solution algorithm. Thus, the current domain of s_j is the interval $[E_j, L_j - p_j]$, and the release time and deadline of each job is indicated by the initial domain of s_j. Constraint programming systems treat the domain of s_j as a sequence of consecutive integers, but none of the techniques described here presuppose that the domain elements be integral. If all the problem data are integral, however, E_j should always be rounded up and L_j rounded down.

The filtering task for the disjunctive constraint is to reduce the domains of the s_js as much as possible. Achieving full bounds completeness is an NP-hard problem, since checking for the existence of a feasible schedule is NP-hard. Yet filtering algorithms that stop short of full bounds completeness can be very valuable in practice.

The most popular filtering methods are based on the edge-finding principle and the not-first/not-last principle. The former finds jobs that must precede or follow others, and the latter finds jobs that cannot be first or last in a given subset of jobs. Either can allow one to shrink some of the time windows $[E_j, L_j]$.

3.13.1 Edge Finding

Edge finding is the best-known filtering method for disjunctive scheduling, and it plays an important role in practical solvers. It identifies subsets of jobs that must all precede, or all follow, a particular job. The name *edge finding* derives from the fact that the procedure finds new edges for the precedence graph, which is a graph in which directed edges indicate which jobs must precede other jobs.

It is helpful to introduce some notation that is specialized to scheduling analysis. For a subset J of jobs, let $E_J = \min_{j \in J}\{E_j\}$, $L_J = \max_{j \in J}\{L_j\}$, and $p_J = \sum_{j \in J} p_j$. Also, $i \gg J$ means that job i starts after every job in J has finished, and $i \ll J$ means that job i finishes before any job in J starts.

The edge-finding principle can be expressed in two symmetrical rules:

$$\begin{aligned} &\text{If } L_J - E_{J \cup \{i\}} < p_i + p_J, \text{ then } i \gg J \quad (a) \\ &\text{If } L_{J \cup \{i\}} - E_J < p_i + p_J, \text{ then } i \ll J \quad (b) \end{aligned} \quad (3.112)$$

Rule (a) is based on the fact that if job i does not start after all the jobs in J, then some job in J must run after job i finishes. This means that all the jobs in $J \cup \{i\}$ must be performed between their earliest start time $E_{J \cup \{i\}}$ and the latest finish time L_J for jobs in J alone. If the total time $p_i + p_J$ required for this will not fit in this interval, then job i must start after all the jobs in J. Rule (b) is based on the same reasoning in reverse direction.

If it is found that job i must follow the jobs in J, then job i cannot start until all the jobs in J finish. A lower bound on this start time is the maximum of $E_{J'} + p_{J'}$ over all subsets J' of J. Similarly, if job i must precede the jobs in J, then i must finish before any of the jobs in J start. An upper bound on this finish time is the minimum of $L_{J'} - p_{J'}$ over all $J' \subset J$. In summary,

$$\begin{aligned}&\text{If } i \gg J, \text{ then update } E_i \text{ to } \max\left\{E_i, \max_{J' \subset J}\{E_{J'} + p_{J'}\}\right\}. \\ &\text{If } i \ll J, \text{ then update } L_i \text{ to } \min\left\{L_i, \min_{J' \subset J}\{L_{J'} - p_{J'}\}\right\}.\end{aligned} \quad (3.113)$$

As an example consider the four-job scheduling problem described in Table 3.5 and illustrated in Figure 3.19. Since

$$L_{\{1,2\}} - E_{\{1,2,4\}} = 6 - 0 < 3 + (3+1) = p_4 + (p_1 + p_2)$$

rule (3.112a) implies that Job 4 must follow Jobs 1 and 2, or $4 \gg \{1, 2\}$. So, E_4 is updated to

$$\max\{E_4, \max\{E_{\{1,2\}} + p_{\{1,2\}}, E_1 + p_1, E_2 + p_2\}\} \\ = \max\{0, \max\{1+4, 2+1, 1+3\}\} = 5$$

Note that, although $2 \ll \{3\}$, edge finding does not deduce this fact.

The practicality of edge finding rests on the fact that one need not examine all subsets J of jobs to find all the bound updates that can be established by the edge-finding rules. In fact, the following polynomial-time algorithm suffices. It runs in time proportion to n^2, where n is the number of jobs. The fastest known algorithm runs in time proportional to $n \log n$, but it requires much more complex data structures.

The first step of the algorithm is to compute the *Jackson preemptive schedule* (JPS) for the given instance. Let us say that a job j is *available* at a given time t if $t \in [E_j, L_j]$ and job j has not completed. Then, as

Table 3.5. Data for a 4-job disjunctive scheduling problem.

j	p_j	E_j	L_j
1	1	2	5
2	3	1	6
3	1	3	8
4	3	0	9

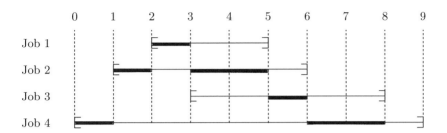

Figure 3.19. Time windows (horizontal lines) for the four-job disjunctive scheduling problem of Table 3.5. The heavy lines show the Jackson preemptive schedule.

one moves forward in time, the job in process at each time t should be the job j that has the smallest L_j among the jobs available (if any) at t. A JPS is illustrated in Figure 3.19.

Now, for each job i, do the following. Let J_i be the set of jobs that are not finished at time E_i in the JPS. Let $\bar{p}_j$ be the processing time left for job j at time E_i in the JPS. Finally, let J_{ik} be the jobs in J_i, other than i, that have deadlines at or before job k's deadline:

$$J_{ik} = \{j \in J_i \setminus \{i\} \mid L_j \leq L_k\}$$

Examine the jobs $k \in J_i$ ($k \neq i$) in decreasing order of deadline L_k, and select the first job for which

$$L_k - E_i < p_i + \bar{p}_{J_{ik}} \tag{3.114}$$

Then conclude that $i \gg J_{ik}$ and update E_i to JPS(i,k), which is the latest completion time in the JPS of the jobs in J_{ik}.

This algorithm updates the earliest start times E_i. The same algorithm is run with the direction of time reversed to update the latest completion times L_i. Table 3.6 details the execution of the algorithm for the example of Figure 3.19. It identifies the one precedence that is established by edge finding, $4 \gg \{1, 2\}$, and updates E_1 from 0 to 5, which is the latest finish time of Jobs 2 and 3 in the JPS shown in Figure 3.19.

THEOREM 3.43 *The above edge-finding algorithm is valid and identifies all updated domains that can be deduced from the edge-finding rules (3.112)–(3.113).*

Proof. First show that the algorithm is valid. It suffices to show that any precedence $i \gg J_{ik}$ discovered by the algorithm is valid, because once it is given that $i \gg J_{ik}$, updating E_i to JPS(i, k) is clearly valid (the

Table 3.6. Execution of an algorithm for edge finding.

i	J_i	$\bar{p}$	k	J_{ik}	$L_k - E_i$	$p_i + \bar{p}_{J_{ik}}$
1	$\{1,2,3,4\}$	$(1,2,1,2)$	4	$\{2,3,4\}$	$9-2$	$1+5$
			3	$\{2,3\}$	$8-2$	$1+3$
			2	$\{2\}$	$6-2$	$1+3$
2	$\{1,2,3,4\}$	$(1,3,1,2)$	4	$\{1,3,4\}$	$9-1$	$3+4$
			3	$\{1,3\}$	$8-1$	$3+2$
			1	$\{3\}$	$5-1$	$3+1$
3	$\{2,3,4\}$	$(0,2,1,2)$	4	$\{2,4\}$	$9-3$	$1+4$
			2	$\{2\}$	$6-3$	$1+2$
4	$\{1,2,3,4\}$	$(1,3,1,3)$	3	$\{1,2,3\}$	$8-0$	$3+5$
			2	$\{1,2\}$	$6-0$	$3+4$

Conclude that $4 \gg \{1,2\}$ and update E_4 from 0 to 5

argument is similar for updating L_i). Suppose, then, that the algorithm derives that $i \gg J_{ik}$, which means that J_{ik} satisfies (3.114). To verify that $i \gg J_{ik}$, it suffices to show that J_{ik} satisfies the edge-finding rule (3.112a):

$$L_{J_{ik}} - E_{J_{ik} \cup \{i\}} < p_i + p_{J_{ik}} \qquad (3.115)$$

This can be deduced as follows from (3.114). By definition of J_{ik}, $L_k = L_{J_{ik}}$. Thus, it suffices to show $E_i - E_{J_{ik} \cup \{i\}} = p_{J_{ik}} - \bar{p}_{J_{ik}}$. But this follows from the fact that none of the jobs in J_{ik} are finished at time E_i in the JPS.

Now show that for any valid update E_i' that can be obtained from the edge-finding rules (3.112)–(3.113), the algorithm obtains an update $E_i'' \geq E_i'$ (the argument is similar for any valid update L_i'). It is given that

$$L_J - E_{J \cup \{i\}} < p_i + p_J \qquad (3.116)$$

for some J and

$$E_i' = E_{J'} + p_{J'} \qquad (3.117)$$

for some $J' \subset J$. Let k be a job in J with the largest L_k. It will first be shown that

$$L_k - E_i < p_i + \bar{p}_{J_{ik}} \qquad (3.118)$$

Clearly, $L_k = L_J$. So, if $\Delta = E_i - E_{J \cup \{i\}}$, then by (3.116) it suffices to show

$$p_J - \bar{p}_{J_{ik}} \leq \Delta \qquad (3.119)$$

Let p^* be the total JPS processing time of jobs in J between $E_{J_{ik} \cup \{i\}}$ and E_i. Also let $\bar{p}_{J_{ik} \setminus J}$ be the total JPS processing time that remains

Disjunctive Scheduling 217

for jobs in $J_{ik} \setminus J$ at time E_i. Then,

$$\bar{p}_{J_{ik}} = p_J + \bar{p}_{J_{ik}\setminus J} - p^* \qquad (3.120)$$

But, since the jobs running between $E_{J_{ik}\cup\{i\}}$ and E_i run one at a time, $p^* \leq \Delta$. Thus, (3.120) implies $\bar{p}_{J_{ik}} \geq p_J - \Delta$, which implies (3.119).

Since (3.118) holds, the algorithm will discover the precedence $i \gg J_{ik'}$ for some k' for which $L_k \leq L_{k'}$ (perhaps $k = k'$), and therefore for which $J_{ik} \subset J_{ik'}$. The algorithm therefore obtains the update $E_i'' = \text{JPS}(i, k')$. Thus

$$E_i'' \geq E_i + \bar{p}_{J_{ik'}} \geq E_i + \bar{p}_{J'} \geq E_{J'} + \hat{p} + \bar{p}_{J'} = E_{J'} + p_{J'} = E_i' \qquad (3.121)$$

The first inequality is due to the fact that the time interval between E_i and $\text{JPS}(i, k')$ is at least the total processing time $\bar{p}_{J_{ik'}}$ that remains for the jobs in $J_{ik'}$. The second inequality is due to the fact that all the jobs in J' that are unfinished in the JPS at time E_i belong to $J_{ik} \subset J_{ik'}$. The third inequality holds if $\hat{p}$ is defined to be the total JPS processing time between $E_{J'}$ and E_i of jobs in J', since these jobs must run one at a time. The first equation is due to the definition of $\bar{p}_{J'}$. Thus, $E_i'' \geq E_i'$, which completes the proof. $\square$

3.13.2 Not-First, Not-Last Rules

Edge finding identifies jobs i that must occur first or last in a set $J \cup \{i\}$ of jobs. A complementary type of rule identifies jobs i that cannot occur first or cannot occur last in $J \cup \{i\}$:

$$\begin{aligned} &\text{If } L_J - E_i < p_i + p_J, \text{ then } \neg(i \ll J). \quad (a) \\ &\text{If } L_i - E_J < p_i + p_J, \text{ then } \neg(i \gg J). \quad (b) \end{aligned} \qquad (3.122)$$

Rule (a) is based on the fact that if job i occurs first in $J \cup \{i\}$, then there must be enough time between E_i and L_J to run all the jobs. If there is not enough time, then i cannot be first. Similarly, rule (b) is based on the fact that if job i occurs last, then there must be enough time between E_J and L_i to run all of the jobs. If there is not, then i cannot be last.

If a not-first or not-last position is established, then the updating of bounds is simpler than in the case of edge finding:

$$\begin{aligned} &\text{If } \neg(i \ll J), \text{ then update } E_i \text{ to } \max\left\{E_i, \min_{j \in J}\{E_j + p_j\}\right\} \quad (a) \\ &\text{If } \neg(i \gg J), \text{ then update } L_i \text{ to } \min\left\{L_i, \max_{j \in J}\{L_j - p_j\}\right\} \quad (b) \end{aligned}$$

$$(3.123)$$

It is simpler because the inner min or max is over all jobs in J rather than all subsets of J.

Returning to the example of Figure 3.19, Rule (3.122a) deduces the fact that $\neg(4 \ll \{1,2\})$ because

$$L_{\{1,2\}} - E_4 = 6 - 0 < 3 + (1+3) = p_4 + (p_1 + p_2)$$

This allows E_4 to be updated from 0 to $\min\{E_1 + p_1, E_2 + p_2\} = \min\{3,4\} = 3$. Actually, this result is dominated by the fact that $4 \gg \{1,2\}$, discovered by edge finding, which updates E_4 to 5. Rule (3.122a) implies that $\neg(3 \ll \{2\})$, however, and rule (3.122b) implies that $\neg(2 \gg \{3\})$. That is, $2 \gg \{3\}$, a fact not deduced by edge finding. The conclusion that $\neg(3 \ll \{2\})$ allows E_3 to be updated from 3 to 4, although $\neg(2 \gg \{3\})$ has no effect on L_2.

As in the case of edge finding, the not-first/not-last rules can be applied without examining all subsets J of jobs. The following algorithm runs in time proportional to n^2, where n is the number of jobs. One part of the algorithm identifies all updates that result from the not-first rule (3.122a)–(3.123a); the procedure for the not-last rule is similar. A running time proportional to $n \log n$ can be obtained by introducing a binary tree data structure.

It is assumed that the jobs are indexed in nondecreasing order of deadlines, so that $j \leq k$ implies $L_j \leq L_k$. Let $\bar{J}_{jk}$ be the set of jobs with deadlines no later than L_k whose earliest finish time is no earlier than job j's earliest finish time. That is,

$$\bar{J}_{jk} = \{\ell \leq k \mid E_j + p_j \leq E_\ell + p_\ell\}$$

Also let LST_{jk} be the following upper bound on the latest time at which the jobs in $\bar{J}_{jk}$ can start,

$$\text{LST}_{jk} = \min_{\ell \leq k} \left\{ L_\ell - p_{\bar{J}_{j\ell}} \right\}$$

where, by default, $p_\emptyset = -\infty$. For each j, the quantity LST_{jk} can be computed recursively for $k = 1, \ldots n$.

The not-first part of the algorithm goes as follows. Let E'_i be the updated release time for each job i, which is initialized to E_i. For each job j do the following. For $i = 1, \ldots, n$ ($i \neq j$) let $E'_i = \max\{E'_i, E_j + p_j\}$ if one of the two conditions below is satisfied:

(i) $E_i + p_i < E_j + p_j$ and $E_i + p_i > \text{LST}_{jn}$
(ii) $E_i + p_i \geq E_j + p_j$ and either $E_i + p_i > \text{LST}_{j,i-1}$ or $E_i > \text{LST}_{jn}$

Disjunctive Scheduling

Table 3.7 shows how the not-first algorithm is applied to the example of Figure 3.19. Supporting data appear in Table 3.8. Note in Table 3.7 that entries appear in the column labeled $E_i + p_i < E_j + p_j$ when this condition is satisfied, and they otherwise appear in the columns headed $E_i + p_i \geq E_j + p_j$. The algorithm updates E_3 from 3 to 4, and E_4 from 0 to 3.

Table 3.7. Execution of a not-first algorithm.

j	$E_j + p_j$	LST_{j4}	i	$E_i + p_i < E_j + p_j$ $E_i + p_i > \text{LST}_{j4}?$	$E_i + p_i \geq E_j + p_j$ $\text{LST}_{j,i-1}$	$E_i + p_i >$ $\text{LST}_{j,i-1}?$	$E_i >$ $\text{LST}_{j4}?$	Update
1	3	1	2		4	no	no	
			3		2	yes	yes	$E'_3 = 3$
			4		2	yes	no	$E'_4 = 3$
2	4	3	1	no				
			3		3	yes	no	$E'_3 = 4$
			4	no				
3	4	7	1	no				
			2		∞	no	no	
			4	no				
4	3	6	1				no	
			2	no				
			3	no				

Table 3.8. Data for execution of the not-first algorithm shown in Table 3.7.

i	E_i	$E_i + p_i$
1	2	3
2	1	4
3	3	4
4	0	3

$\bar{J}_{jk}$	$k = 1$	2	3	4
$j = 1$	$\{1\}$	$\{1, 2\}$	$\{1, 2, 3\}$	$\{1, 2, 3, 4\}$
2	$\emptyset$	$\{2\}$	$\{2, 3\}$	$\{2, 3\}$
3	$\emptyset$	$\emptyset$	$\{3\}$	$\{3\}$
4	$\emptyset$	$\emptyset$	$\emptyset$	$\{4\}$

$L_k - p_{\bar{J}_{jk}}$	$k = 1$	2	3	4
$j = 1$	4	2	3	1
2	∞	3	4	5
3	∞	∞	7	8
4	∞	∞	∞	6

LST_{jk}	$k = 1$	2	3	4
$j = 1$	4	2	2	1
2	∞	3	3	3
3	∞	∞	7	7
4	∞	∞	∞	6

It may be assumed that $E_\ell + p_\ell \leq L_\ell$ for all ℓ, because otherwise it is trivial to check that there is no feasible schedule.

THEOREM 3.44 *The above not-first/not-last algorithm is valid and identifies all updated domains that can be deduced from the not-first/not-last rules (3.122)–(3.123).*

The not-first part of the theorem follows from Lemmas 3.46 and 3.47 below, which in turn rely on Lemma 3.45. The argument is similar for the not-last part of the theorem.

LEMMA 3.45 *If the not-first rule of (3.122a)–(3.123a) updates E_i to $E_j + p_j$ for some set J, then for some $k \geq j$ the rule yields this same update when $J = \bar{J}_{jk} \setminus \{i\}$.*

Proof. It is given that
$$L_J - E_i < p_i + p_J \tag{3.124}$$
and that
$$E_i' = E_j + p_j = \min_{\ell \in J} \{E_\ell + p_\ell\} \tag{3.125}$$
Let k be the largest index in J. It suffices to show
$$L_{\bar{J}_{jk} \setminus \{i\}} - E_i < p_i + p_{\bar{J}_{jk} \setminus \{i\}} \tag{3.126}$$
and
$$E_j + p_j = \min_{\ell \in \bar{J}_{jk} \setminus \{i\}} \{E_\ell + p_\ell\} \tag{3.127}$$
Since the jobs are indexed in nondecreasing order of deadline, $L_{\bar{J}_{jk}} = L_J$ by definition of k. Also, $J \subset \bar{J}_{jk} \setminus \{i\}$ due to (3.125) and the fact that $i \notin J$. This means $p_J \leq p_{\bar{J}_{jk} \setminus \{i\}}$, and so (3.126) follows from (3.124). Finally, (3.127) is true by definition of $\bar{J}_{jk}$ and the fact that $i \neq j$. □

LEMMA 3.46 *If $E_i + p_i < E_j + p_j$, the not-first rule updates E_i to $E_j + p_j$ if and only if $E_i + p_i > \mathrm{LST}_{jn}$.*

Proof. Suppose first that the not-first rule updates E_i to $E_j + p_j$. Then, by Lemma 3.45, there is a $k \geq j$ for which (3.126) holds, which implies
$$L_k - E_i < p_i + p_{\bar{J}_{jk} \setminus \{i\}} \tag{3.128}$$
since $L_k = L_{\bar{J}_{jk} \setminus \{i\}}$. First, note that $E_i + p_i < E_j + p_j$ implies $i \notin \bar{J}_{jk}$. Now,
$$E_i + p_i > L_k - p_{\bar{J}_{jk} \setminus \{i\}} = L_k - p_{\bar{J}_{jk}} \geq \mathrm{LST}_{jn}$$

where the first inequality is due to (3.128), the equation is due to the fact that $i \notin \bar{J}_{jk}$, and the last inequality is due to the definition of LST_{jn}.

Now suppose that $E_i + p_i > \text{LST}_{jn}$. So, $\text{LST}_{jn} < \infty$, which means $\text{LST}_{jn} = L_k - p_{\bar{J}_{jk}}$ for some $k \geq j$ for which $\bar{J}_{jk}$ is nonempty. Now,

$$E_i + p_i > \text{LST}_{jn} = L_k - p_{\bar{J}_{jk}} = L_k - p_{\bar{J}_{jk}\setminus\{i\}} \qquad (3.129)$$

where the second equation is due to the fact that $i \notin \bar{J}_{jk}$. But (3.129) implies (3.128), and so the not-first rule updates E_i to $E_j + p_j$. □

LEMMA 3.47 *If* $E_i + p_i \geq E_j + p_j$, *the not-first rule updates* E_i *to* $E_j + p_j$ *if and only if either* $E_i + p_i > \text{LST}_{j,i-1}$ *or* $E_i > \text{LST}_{jn}$.

Proof. First suppose that the not-first rule updates E_i to $E_j + p_j$. Then, by Lemma 3.45 there is a $k \geq j$ for which (3.128) holds. There are two cases, corresponding to $k < i$ and $i < k$. If $k < i$, then $i \notin \bar{J}_{jk}$ and from (3.128),

$$E_i + p_i > L_k - p_{\bar{J}_{jk}\setminus\{i\}} \geq \text{LST}_{j,i-1}$$

If $i < k$, then $i \in \bar{J}_{kj}$ and $p_{\bar{J}_{jk}} = p_i + p_{\bar{J}_{jk}\setminus\{i\}}$. So it follows from (3.128) that $L_k - E_i < p_{\bar{J}_{jk}}$, which implies

$$E_i > L_k - p_{\bar{J}_{jk}} \geq \text{LST}_{jn}$$

For the converse, first suppose that $E_i + p_i > \text{LST}_{j,i-1}$. Then, since $\text{LST}_{j,i-1} < \infty$, there is a $k \leq i$ for which $\text{LST}_{j,i-1} = L_k - p_{\bar{J}_{jk}}$ and $\bar{J}_{jk}$ is nonempty. So,

$$E_i + p_i > \text{LST}_{j,i-1} = L_k - p_{\bar{J}_{jk}} = L_k - p_{\bar{J}_{jk}\setminus\{i\}}$$

where the second equation is due to $i \notin \bar{J}_{jk}$. But this implies (3.128), which means that the not-first rule updates E_i to $E_j + p_j$. Now suppose that $E_i + p_i \leq \text{LST}_{j,i-1}$ and $E_i > \text{LST}_{jn}$. Then there is a $k \geq j$ for which $\text{LST}_{jn} = L_k - p_{\bar{J}_{jk}}$. Also $k \geq i$, since otherwise $\text{LST}_{jn} = \text{LST}_{j,i-1} < E_i$, which contradicts the assumption that $E_i + p_i \leq \text{LST}_{k,i-1}$. But $k \geq i$ implies $i \in \bar{J}_{jk}$. However, $\bar{J}_{jk} \neq \{i\}$, because otherwise

$$E_i > \text{LST}_{jn} = L_k - p_{\bar{J}_{jk}} = L_k - p_i \geq L_i - p_i$$

which contradicts the assumption that $E_\ell + p_\ell \leq L_\ell$ for all ℓ. So $\bar{J}_{jk} \setminus \{i\}$ is nonempty and satisfies (3.128), which means that the not-first rule updates E_i to $E_j + p_j$. □

3.13.3 Benders Cuts

Disjunctive scheduling constraints commonly arise in the context of planning and scheduling problems. For instance, it may be necessary to assign jobs to facilities as well as schedule the jobs on the facilities to which they are assigned. One such problem is discussed in Section 2.3.7.

Problems of this kind are often suitable for logic-based Benders decomposition. A master problem assigns jobs to facilities, and separable subproblems schedule the jobs assigned to each facility. Generic Benders cuts can be developed for these problems, based on the nature of the objective function. The cuts can normally be strengthened when information from the scheduling algorithm is available, or when the subproblem is re-solved a few times with different job assignments. Section 2.3.7, for example, illustrates how the cuts can exploit information obtained from edge-finding and branching procedures.

Minimizing Cost

The simplest type of objective minimizes the fixed cost of assigning jobs to facilities. Thus, if f_{ij} is the cost of assigning job j to facility i, and variable x_j is the facility assigned to job j, the basic planning and scheduling problem may be written

$$\min \sum_j f_{x_j j}$$
$$s_j + p_{x_j j} \leq d_j, \text{ all } j \qquad (3.130)$$
$$\text{disjunctive}\left((s_j \mid x_j = i) \mid (p_{ij} \mid x_j = i)\right), \text{ all } i$$
$$x_j \in \{1, \ldots, n\}, \; s_j \in [r_j, \infty), \text{ all } j$$

Here, r_j and d_j are the release time and deadline for job j, and variable s_j is the job's start time. The notation $(s_j \mid x_j = i)$ refers to the tuple of start times s_j such that $x_j = i$, and similarly for $(p_{ij} \mid x_j = i)$. The disjunctive constraints therefore require that the jobs on each facility be scheduled sequentially. The disjunctive constraint used here is an extension of the standard one, because the list of variables in the argument depends on the value of the variable x. The standard disjunctive constraint, however, will be used in solving the Benders subproblems.

The master problem can be written

$$\min \sum_j f_{x_j j}$$
$$\text{optional relaxation of subproblem} \qquad (3.131)$$
$$\text{Benders cuts}$$
$$x_j \in \{1, \ldots, n\}, \text{ all } j$$

Section 4.16.3 discusses how the subproblem can be relaxed. If the solution of (3.131) in iteration k assigns each job j to machine x_j^k, the subproblem separates into the following scheduling problem for each facility i:

$$\min \sum_{\substack{j \\ x_j^k = i}} f_{ij}$$

$$s_j + p_{ij} \leq d_j, \text{ all } j \text{ with } x_j^k = i \quad (3.132)$$

$$\text{disjunctive}\Big((s_j \mid x_j^k = i) \mid (p_{ij} \mid x_j^k = i) \Big)$$

$$s_j \in [r_j, \infty), \text{ all } j \text{ with } x_j^k = i$$

Note that the objective function is a constant. If the scheduling problem on some facility i is infeasible, a Benders cut can be generated to rule out assigning those jobs to machine i again:

$$\bigvee_{j \in J_{ik}} x_j \neq i \quad (3.133)$$

Here, J_{ik} is the set of jobs assigned to facility i in iteration k, so that $J_{ik} = \{j \mid x_j^k = i\}$. The cut (3.133) is added to the master problem for facility i on which the scheduling problem is infeasible.

The cut (3.133) can be strengthened by identifying a proper subset of the jobs assigned to facility i that suffice to create infeasibility. The set J_{ik} can then be this proper subset. A smaller J_{ik} can be identified through an analysis of the scheduling algorithm (i.e., and analysis of the solution of the inference dual), as suggested in Section 2.3.7, or by a sampling procedure that re-solves the scheduling problem several times with different sets of jobs assigned to facility i. One simple procedure goes as follows. Initially, let $J_{ik} = \{j_1, \ldots, j_p\}$ contain all the jobs assigned to facility i. For $\ell = 1, \ldots, p$ do the following: try to schedule the tasks in $J_{ik} \setminus \{j_\ell\}$ on facility i, and if there is no feasible schedule, remove j_ℓ from J_{ik}. It can be well worth the effort of re-solving the subproblem a few times in order to obtain a stronger cut.

In practice, the master problem is often formulated as a 0-1 programming problem in which the decision variables x_{ij} are 1 when job i is assigned to machine j, and 0 otherwise. In this case, the master problem (3.131) becomes

$$\min \sum_{ij} f_{ij} x_{ij}$$

optional relaxation of subproblem

Benders cuts

$$x_{ij} \in \{0, 1\}, \text{ all } j$$

The Benders cut (3.133) must now be formulated as a 0-1 knapsack inequality:

$$\sum_{j \in J_{ik}} (1 - x_{ij}) \geq 1 \tag{3.134}$$

Minimizing Makespan

The Benders cuts are less straightforward when the objective is to minimize makespan. In this case, the problem is

$$\begin{aligned} &\min M \\ &M \geq s_j + p_{x_j j}, \text{ all } j \\ &s_j + p_{x_j j} \leq d_j, \text{ all } j \\ &\text{disjunctive}\left((s_j \mid x_j = i) \mid (p_{ij} \mid x_j = i)\right), \text{ all } i \\ &x_j \in \{1, \ldots, n\}, \; s_j \in [r_j, \infty), \text{ all } j \end{aligned} \tag{3.135}$$

The master problem is

$$\begin{aligned} &\min v \\ &\text{optional relaxation of subproblem} \\ &\text{Benders cuts} \\ &x_j \in \{1, \ldots, n\}, \text{ all } j \end{aligned} \tag{3.136}$$

Given a solution x^k of the master problem in iteration k, the subproblem separates into the following min makespan problem for each machine i:

$$\begin{aligned} &\min M_i \\ &M_i \geq s_j + p_{ij}, \; s_j + p_{ij} \leq d_j, \text{ all } j \text{ with } x_j^k = i \\ &\text{disjunctive}\left((s_j \mid x_j^k = i) \mid (p_{ij} \mid x_j^k = i)\right) \\ &s_j \in [r_j, \infty), \text{ all } j \text{ with } x_j^k = i \end{aligned}$$

If M_{ik}^* is the minimum makespan on machine i, the overall minimum makespan is $\max_i\{M_{ik}^*\}$.

The most obvious Benders cut is the one presented in Section 2.3.7. Again, let J_{ik} be the set of jobs assigned to facility i in iteration k. Then a simple Benders cut requires the makespan to be at least M_{ik}^* whenever the jobs in J_{ik} are assigned to facility i:

$$(\{j \mid x_j = i\} \subset J_{ik}) \to (v \geq M_{ik}^*) \tag{3.137}$$

Disjunctive Scheduling

The cut can be linearized as follows when the master problem is a 0-1 programming problem:

$$v \geq M_{ik}^* \left(\sum_{j \in J_{ik}} x_{ij} - |J_{ik}| + 1 \right) \quad (3.138)$$

The cut (3.137) or (3.138) is added to the master problem for every facility i. As before, one may be able to identify a smaller set J_{ik} for which these cuts remain valid, either by an analysis of the scheduling algorithm or a sampling procedure.

A difficulty with the cuts (3.137) and (3.138) is that they become useless when even one of the jobs in J_{ik} is not assigned to facility i. The cuts can be improved by considering the effect on minimum makespan when jobs are removed. The simplest case is that in which all the release times r_j are the same. The improved cuts are based on a simple fact.

LEMMA 3.48 *Consider a minimum makespan problem in which jobs $1, \ldots, n$ with identical release times are scheduled in facility i. If M^* is the minimum makespan, then the minimum makespan for the same problem with jobs $1, \ldots, s$ removed is*

$$M^* - \sum_{j=1}^{s} p_{ij}$$

A valid Benders cut is therefore

$$v \geq M_{ik}^* - \sum_{\substack{j \in J_{ik} \\ x_j \neq i}} p_{ij} \quad (3.139)$$

The cut is easily linearized:

$$v \geq M_{ik}^* - \sum_{j \in J_{ik}} (1 - x_{ij}) p_{ij} \quad (3.140)$$

Minimizing Tardiness

Tardiness can be measured by the number of late jobs or the total time by which the completion times exceed the corresponding deadlines. A model that minimizes the number of late jobs is

$$
\begin{aligned}
&\text{Linear: } \min \sum_j \delta_j \\
&\text{Conditional: } (T_j > 0) \rightarrow (\delta_j = 1), \text{ all } j \\
&\text{Indexed linear: } T_j \geq s_j + p_{x_j j}, \text{ all } j \\
&\text{Disjunctive: } \left((s_j \mid x_j = i) \mid (p_{ij} \mid x_j = i) \right), \text{ all } i \\
&\text{Domains: } x_j \in \{1, \ldots, n\},\ s_j \in [r_j, \infty),\ \delta_j \in \{0,1\}, \text{ all } j
\end{aligned}
\qquad (3.141)
$$

Again, the master problem is (3.136). Given a solution x^k of the master problem, the subproblem on each facility i is

$$
\begin{aligned}
&\min \sum_{\substack{j \\ x_j^k = i}} \delta_j \\
&(T_j > 0) \rightarrow (\delta_j = 1),\ T_j \geq s_j + p_{ij}, \text{ all } j \text{ with } x_j^k = i \\
&\text{disjunctive}\left((s_j \mid x_j^k = i) \mid (p_{ij} \mid x_j^k = i) \right), \text{ all } i \\
&s_j \in [r_j, \infty),\ \delta_j \in \{0,1\}, \text{ all } j \text{ with } x_j^k = i
\end{aligned}
\qquad (3.142)
$$

If L_{ik}^* is the minimum number of late jobs on facility i, then $\sum_i L_{ik}^*$ is the minimum overall.

A trivial Benders cut forces the number of late jobs to be at least L_{ik}^* on each facility i to which all jobs in J_{ik} are assigned. Thus there is a single Benders cut per iteration that consists of several constraints:

$$
\begin{aligned}
&v \geq \sum_i L_i \\
&L_i \geq 0,\ (\{j \mid x_j = i\} \subset J_{ik}) \rightarrow (L_i \geq L_{ik}^*), \text{ all } i
\end{aligned}
\qquad (3.143)
$$

The linearization is:

$$
\begin{aligned}
&v \geq \sum_i L_i \\
&L_i \geq 0,\ L_i \geq L_{ik}^* - L_{ik}^* \sum_{j \in J_{ik}} (1 - x_{ij}), \text{ all } i
\end{aligned}
\qquad (3.144)
$$

If tardiness is measured by the total amount by which finish times exceed deadlines, the objective function in (3.141) becomes $\sum_j T_j$ and the conditional constraint is dropped. The subproblem (3.142) is similarly modified. If T_{ik}^* is the minimum tardiness on facility i, then $\sum_i T_{ik}^*$ is the overall minimum.

Disjunctive Scheduling

A trivial Benders cut analogous to (3.137) or (3.138) can again be written, but a sampling procedure can strengthen such a cut significantly. One procedure goes as follows. Let $T_i(J)$ be the minimum tardiness on facility i that results when the jobs in J are assigned to the facility, so that $T_i(J_{ik}) = T_{ik}^*$. Let Z_{ik} be the set of jobs in J_{ik} that can be removed, one at a time, without reducing the minimum tardiness:

$$Z_{ik} = \{j \in J_{ik} \mid T_i(J_{ik} \setminus \{j\}) = T_{ik}^*\}$$

Now let T_{ik}^0 be the minimum tardiness that results from removing all the jobs in Z_{ik} at once, so that $T_{ik}^0 = T_i(J_{ik} \setminus Z_{ik})$. This leads to the Benders cut consisting of

$$v \geq \sum_i T_i \qquad (3.145)$$
$$T_i \geq 0, \text{ all } i$$

and the conditional constraints

$$\begin{aligned}(\{j \in J_{ik} \mid x_j \neq i\} = \emptyset) &\to (v \geq T_{ik}^*), \text{ all } i \\ (\{j \in J_{ik} \mid x_j \neq i\} \subset Z_{ik}) &\to (v \geq T_{ik}^0), \text{ all } i\end{aligned} \qquad (3.146)$$

The second conditional constraint can be omitted for facility i when $T_{ik}^0 = T_{ik}^*$. The constraints (3.146) are linearized as follows:

$$\begin{aligned}v &\geq T_{ik}^* - T_{ik}^* \sum_{j \in J_{ik}}(1 - x_{ij}), \text{ all } i \\ v &\geq T_{ik}^0 - T_{ik}^* \sum_{j \in J_{ik} \setminus Z_{ik}}(1 - x_{ij}), \text{ all } i\end{aligned} \qquad (3.147)$$

Again, the second constraint can be dropped when $T_{ik}^* = T_{ik}^0$. Generation of these cuts requires n additional calls to the scheduler in each Benders iteration, where n is the number of jobs.

When all the release times are equal, a second type of Benders cut can be derived from the following lemma.

LEMMA 3.49 *Consider a minimum tardiness problem P in which tasks $1, \ldots, n$ have release time 0, due dates $d_1, \ldots, d_n$, and are to be scheduled on a single facility i. Let T^* be the minimum tardiness for this problem, and $\hat{T}$ the minimum tardiness for the problem $\hat{P}$ that is identical to P except that tasks $1, \ldots, s$ are removed. Then,*

$$T^* - \hat{T} \leq \sum_{j=1}^{s}\left(\sum_{\ell=1}^{n} p_{i\ell} - d_j\right)^+ \qquad (3.148)$$

Proof. Consider any optimal solution $\hat{S}$ of $\hat{P}$. One may assume that the makespan of $\hat{S}$ is at most $M = \sum_{\ell=s+1}^{n} p_{i\ell}$, because if it is greater than M, at least one task can be moved to an earlier time without increasing the total tardiness of $\hat{S}$. To obtain a feasible solution S for P, schedule tasks $1, \ldots, s$ sequentially after M. That is, for $j = 1, \ldots, s$ let task j start at time $M + \sum_{\ell=1}^{j-1} p_{i\ell}$. The tardiness of task j in S is at most

$$\left(M + \sum_{\ell=1}^{s} p_{i\ell} - d_j\right)^+ = \left(\sum_{\ell=1}^{n} p_{i\ell} - d_j\right)^+$$

The total tardiness of S is therefore at most

$$\hat{T} + \sum_{k=1}^{s} \left(\sum_{\ell=1}^{n} p_{i\ell} - d_j\right)^+$$

from which (3.148) follows. □

This leads to the Benders cut

$$v \geq \sum_{i} T'_i$$
$$T'_i \geq 0, \quad T'_i \geq T^*_{ik} - \sum_{\substack{j \in J_{ik} \\ x_j \neq i}} \left(\sum_{\ell \in J_{ik}} p_{i\ell} - d_j\right)^+, \quad \text{all } i \tag{3.149}$$

which has the linearization

$$v \geq \sum_{i} T'_i$$
$$T'_i \geq 0, \quad T'_i \geq T^*_{ik} - \sum_{j \in J_{ik}} \left(\sum_{\ell \in J_{ik}} p_{i\ell} - d_j\right)^+ (1 - x_{ij}), \quad \text{all } i \tag{3.150}$$

3.13.4 Exercises

1 In the example of Table 3.5, verify that edge finding does not deduce the valid precedence $2 \gg \{3\}$.

2 Consider the 3-machine disjunctive scheduling problem in which

$$(E_1, E_2, E_3) = (2, 1, 0), \quad (L_1, L_2, L_3) = (6, 5, 8), \quad p = (3, 2, 2)$$

Use the edge-finding conditions (3.112) to check for valid precedences, and update the bounds accordingly. Does edge finding identify all valid precedences?

Disjunctive Scheduling 229

3. Apply the polynomial-time edge-finding algorithm to the problem of Exercise 2.

4. Write the polynomial-time edge-finding algorithm for updating E_is in pseudocode.

5. State a polynomial-time edge-finding algorithm for updating L_is and apply it to the problem of Table 3.5.

6. The proof of Theorem 3.43 omits the updating of L_is. State the argument for this case.

7. Apply the not-first and not-last rules to the example of Exercise 2 and update the bounds accordingly.

8. Apply the polynomial-time not-first algorithm to the example of Exercise 2.

9. State a polynomial-time not-last algorithm and apply it to the problem of Table 3.5.

10. Suppose that a minimum makespan planning and scheduling problem is to be solved by logic-based Benders decomposition. In the first iteration, Jobs 1, 2 and 3 are assigned to Machine A, on which $(p_{A1}, p_{A2}, p_{A3}) = (2, 3, 2)$. Use release times $(E_1, E_2, E_3) = (0, 2, 2)$ and deadlines $(L_1, L_2, L_3) = (3, 5, 7)$. Write the appropriate Benders cut (3.138).

11. Suppose in the problem Exercise 10 that all the release times are zero. Write the Benders cut (3.140).

12. Prove Lemma 3.48.

13. Give a counterexample to show that (3.140) may not be a valid cut if the release times are different.

14. Suppose that a Benders method is applied to a minimum total tardiness planning and scheduling problem. In the first iteration, Jobs 1, 2, 3 and 4 are assigned to Machine A, on which $(p_{A1}, p_{A2}, p_{A3}, p_{A4}) = (1, 1, 2, 2)$. The release times are $(E_1, \ldots, E_4) = (0, 0, 2, 0)$ and the due dates are $(L_1, \ldots, L_4) = (2, 2, 3, 5)$. Write the resulting Benders cut (3.147).

15. Show by example that $T_i(J_{ik} \setminus Z_{ik}) < T_i(J_{ik})$ is possible in a disjunctive scheduling problem. Hint: see Exercise 14.

16. Show that $T_i(J_{ik} \setminus Z_{ik}) = T_i(J_{ik})$ in a disjunctive scheduling problem when all the release dates are the same.

3.14 Cumulative Scheduling

Cumulative scheduling differs from disjunctive scheduling in that several jobs may run simultaneously. Each job consumes a certain amount of resource, however, and the rate of resource consumption at any one time must not exceed a limit. Cumulative scheduling may therefore be seen as a form of resource-constrained scheduling.

There may be one resource, or multiple resources with a different limit for each one. A resource limit may be constant or variable over time, and schedules may be preemptive or nonpreemptive. Disjunctive scheduling is a special case of cumulative scheduling in which there is one resource, each job consumes one unit of it, and the limit is always one.

Cumulative scheduling is one of the more successful application areas for constraint programming, due in part to sophisticated filtering algorithms that have been developed for the associated constraints. The most important filtering methods include edge finding, extended edge finding, not-first/not-last rules, and energetic reasoning.

Attention is restricted here to the most widely used cumulative scheduling constraint, which schedules nonpreemptively subject to a single resource limit that is constant over time. The constraint is written

$$\text{cumulative}(s \mid p, c, C) \qquad (3.151)$$

where $s = (s_1, \ldots, s_n)$ is a tuple of variables s_j representing the start time of job j. The remaining arguments are parameters: $p = (p_1, \ldots, p_n)$ is a tuple of processing times for each job, $c = (c_1, \ldots, c_n)$ is a tuple of resource consumption rates, and C is the limit on total resource consumption at any one time. The constraint requires the following:

$$\sum_{\substack{j \\ s_j \leq t \leq s_j + p_j}} c_j \leq C, \quad \text{for all times } t$$

That is, the total rate of resource consumption of the jobs underway at any time t is at most C.

As with disjunctive scheduling, each job j is associated with an earliest start time E_j and a latest completion time L_j. Initially, these are the release time and deadline of the job, respectively, but they may be updated in the course of the solution algorithm. Thus, the current domain of s_j is the interval $[E_j, L_j - p_j]$, and the release time and deadline of each job is indicated by the initial domain of s_j. If all the problem data are integral, E_j should always be rounded up and L_j rounded down.

A small instance of a cumulative scheduling problem is presented in Table 3.9, and a feasible schedule is illustrated in Figure 3.20.

Cumulative Scheduling

Table 3.9. A small instance of a cumulative scheduling problem. The resource limit is $C = 4$.

j	p_j	c_j	E_j	L_j
1	5	1	0	5
2	3	3	0	5
3	4	2	1	7

3.14.1 Edge Finding

Edge finding for disjunctive scheduling can be generalized to cumulative scheduling. The key concept that makes generalization possible is the *energy* e_j of a job j, which is the product $p_j c_j$ of the processing time and resource consumption rate. In Figure 3.20, the energy of each job is the area it consumes in the chart. While disjunctive edge finding checks whether the total processing time of a set of jobs fits within a given time interval, cumulative edge finding checks whether the total energy demand exceeds the supply, which is the product of the time interval and the resource limit C.

Unlike disjunctive edge finding, cumulative edge finding does not deduce that one job must start after certain others finish. Rather, it deduces that it must *finish* after the others finish. Nonetheless, cumulative edge-finding algorithms are quite useful for reducing domains.

The notation is similar to that of the previous section. For a set J of jobs, let $p_J = \sum_{j \in J} p_j$, $e_J = \sum_{i \in J} e_j$, $E_J = \min_{j \in J}\{E_j\}$, and

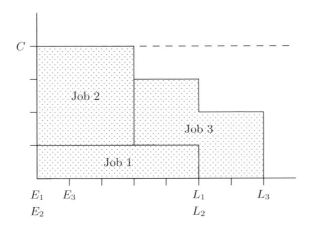

Figure 3.20. A feasible solution of a small cumulative scheduling problem. The horizontal axis is time and the vertical axis is resource consumption.

$L_J = \max_{j \in J}\{L_j\}$. The notation $i > J$ indicates that job i must finish after all the jobs in J finish, and $i < J$ indicates that job i must start before any job in J starts.

The edge-finding rules are based on the principle that if a set J of jobs requires total energy e_J, then the time interval $[t_1, t_2]$ in which they are scheduled must have a length of at least e_J/C. Thus, one must have $e_J \leq C \cdot (t_2 - t_1)$. In Figure 3.20, the jobs have a total energy of 22 and therefore require a time interval of at least $\lceil 22/4 \rceil = \lceil 5.5 \rceil = 6$ (rounding up because the data are integral). In fact, they require a time interval of 7 because the jobs cannot be packed perfectly into a rectangular area.

The edge-finding rules may be stated

$$\begin{aligned} \text{If } e_i + e_J &> C \cdot (L_J - E_{J \cup \{i\}}), \text{ then } i > J. \quad (a) \\ \text{If } e_i + e_J &> C \cdot (L_{J \cup \{i\}} - E_J), \text{ then } i < J. \quad (b) \end{aligned} \qquad (3.152)$$

Rule (a) is based on the fact that if job i does not finish after all the jobs in J finish, then the time interval from $E_{J \cup \{i\}}$ to L_J must cover the energy demand $e_i + e_J$ of all the jobs. If there is not enough energy, then job i must finish after the other jobs finish. The reasoning for rule (b) is analogous. In Figure 3.20, job 3 must finish after the other jobs finish (i.e., $3 > \{1, 2\}$) because

$$e_3 + e_{\{1,2\}} = 22 > 4 \cdot (5 - 0) = C \cdot (L_{\{1,2\}} - E_{\{1,2,3\}})$$

Once it is established that $i > J$, it may be possible to update the release time E_i of job i, as illustrated in Figure 3.21. If the total energy e_J of the jobs in J exceeds the energy available between E_J and L_J within a resource limit of $C - c_i$, then at some time in the schedule, the jobs in J must consume more resource than $C - c_i$. Since the excess energy is

$$R(J, c_i) = e_J - (C - c_i)(L_J - E_J)$$

the jobs in J must consume more resource than $C - c_i$ for a period of at least $R(J, c_i)/c_i$. None of this excess resource is consumed after job i finishes, because $i > J$. It must be consumed before job i starts, which means that job i can start no earlier than $E_J + R(J, c_i)/c_i$.

On the other hand, suppose that job i must start before the jobs in J start (i.e., $i < J$). The excess resource $R(J, c_i)$ cannot be consumed before job i starts. It must be consumed after job i finishes, which means that job i can finish no sooner than $E_J - R(J, c_i)/c_i$. Thus the reasoning

Cumulative Scheduling

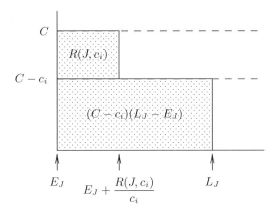

Figure 3.21. Updating of E_i to $E'_i = E_J + R(J, c_i)/c_i$. The area of each rectangle is indicated. The entire shaded area has area e_J.

is similar when $i < J$. This leads to the update rules:

If $i > J$ and $R(J, c_i) > 0$, update E_i to $\max\left\{E_i, E_J + \dfrac{R(J, c_i)}{c_i}\right\}$.

If $i < J$ and $R(J, c_i) > 0$, update L_i to $\min\left\{L_i, L_J - \dfrac{R(J, c_i)}{c_i}\right\}$.

The domains can in general be reduced further by doing a similar update for all subsets J' of J, because $i > J$ implies $i > J'$ and $i < J$ implies $i < J'$. Thus if $i > J$, E_i can be updated to $\max\{E_i, E'_i(J)\}$, where

$$E'_i(J) = \max_{\substack{J' \subset J \\ R(J', c_i) > 0}} \left\{E_{J'} + \frac{R(J', c_i)}{c_i}\right\} \quad (3.153)$$

and if $i < J$, L_i can be updated to $\min\{L_i, L'_i\}$, where

$$L'_i(J) = \min_{\substack{J' \subset J \\ R(J', c_i) > 0}} \left\{L_{J'} - \frac{R(J', c_i)}{c_i}\right\} \quad (3.154)$$

If this is done for all subsets J of jobs, one can update E_i to $\max\{E_i, E'_i\}$, where

$$E'_i = \min_{\substack{J \\ i \notin J \\ C \cdot (L_J - E_{J \cup \{i\}}) < e_i + e_J}} \{E'_i(J)\} \quad (3.155)$$

and update L_i to $\min\{L_i, L'_i\}$, where

$$L'_i = \max_{\substack{J \\ i \notin J \\ C \cdot (L_J - E_{J \cup \{i\}}) < e_i + e_J}} \{L'_i(J)\} \tag{3.156}$$

In the example of Figure 3.20, the only precedence discovered by the edge-finding rule (3.152a) is $3 > \{1, 2\}$. Since $R(\{1\}, c_3) = -5$, $R(\{2\}, c_3) = -1$, and $R(\{1, 2\}, c_3) = 4$, one has

$$E'_3(\{1,2\}) = E_{\{1,2\}} + \frac{R(\{1,2\}, c_3)}{c_3} = 2$$

Thus $E_3 = 1$ can be updated to $E'_3 = 2$.

Figure 3.22 presents an algorithm that computes updates (3.155) in time proportional to n^2. The algorithm assumes that the jobs are indexed in nondecreasing order of release times, so that $E_i \leq E_j$ when $i < j$. For each value of k in the outer loop, the algorithm identifies jobs i for which $i > J_k$, where $J_k = \{j \mid L_j \leq L_k\}$ is the subset of jobs with deadlines no later than job k.

Let $E'_j = E_j$ for all j.
For $k = 1, \ldots, n$:
    Let $e_J = 0$, $e^*_J = \infty$, $\Delta e = \infty$, and $E^* = L_k$.
    For $i = n, n-1, \ldots, 1$:
        If $L_i \leq L_k$ then
            Let $e_J = e_J + e_i$.
            If $e_J/C > L_k - E_i$ then exit (no feasible schedule).
            If $e_J + CE_i > e^*_J + CE^*$ then let $e^*_J = e_J$ and $E^* = E_i$.
        Else
            Let $R = e^*_J - (C - c_i)(L_k - E^*)$.
            If $R > 0$ then let $E''_i = E^* + R/c_i$.
            Else let $E''_i = -\infty$.
    For $i = 1, \ldots, n$:
        If $L_i \leq L_k$ then
            Let $\Delta e = \min\{\Delta e, C \cdot (L_k - E_i) - e_J\}$.
            Let $e_J = e_J - e_i$.
        Else
            If $C \cdot (L_k - E_i) < e_J + e_i$ then let $E'_i = \max\{E'_i, E''_i\}$.
            If $\Delta e < e_i$ then
                Let $R = e^*_J - (C - c_i)(L_k - E^*)$.
                If $R > 0$ then let $E'_i = \max\{E'_i, E^* + R/c_i\}$.

Figure 3.22. Edge-finding algorithm for computing updated release times E'_i in time proportional to n^2. The algorithm assumes that jobs are indexed in nondecreasing order of release time E_j.

The task of the first i-loop is to compute updates E_i''' based on subsets of J_k in which the jobs have release dates no earlier than E_i. Define $J_{ki} = \{j \in J_k \mid j \geq i\}$, recalling that $j \geq i$ implies $E_j \geq E_i$. Then it can be shown that E_i''', as computed by the algorithm, is equal to

$$\max_{\substack{J' \subset J_{ki} \\ R(J', c_i) > 0}} \left\{ E_{J'} + \frac{R(J', c_i)}{c_i} \right\} \tag{3.157}$$

To see this, let j^* be the index that maximizes $e_{J_{kj}} + CE_j$ over $j \in J_{ki}$. Then, e_j^* as computed in the algorithm is $e_{J_{kj^*}}$, and E^* is E_{j^*}. Also, if $e_{J_{kj^*}} - (C - c_i)(L_k - E_{j^*}) > 0$, then E_i''' as computed is

$$E_i''' = E_{j^*} + \frac{e_{J_{kj^*}} - (C - c_i)(L_k - E_{j^*})}{c_i}$$

Because $j = j^*$ maximizes $e_{J_{kj}} + CE_j$ over $j \in J_{ki}$, however, it also maximizes

$$E_j + \frac{e_{J_{kj}} - (C - c_i)(L_k - E_j)}{c_i} = \frac{e_{J_{kj}} + CE_j}{c_i} + \text{constant}$$

over $j \in J_{ki}$, which implies that E_i''' is equal to (3.157).

The *else* portion of the second i-loop has two functions. The first line updates E_i' to E_i''' when $i > J_{ki}$. The remaining lines identify precedences $i > J$ for subsets J of J_k containing at least one job with release date before E_i. To accomplish this, the *if* part of the second i-loop computes

$$\Delta e = \min_{\substack{j \in J_k \\ j < i}} \{C \cdot (L_k - E_j) - e_{J_{kj}}\} \tag{3.158}$$

Thus, Δe is also the minimum of $C \cdot (L_k - E_J) - e_J$ over all subsets J of J_k containing at least one j with $E_j < E_i$. If j' is the minimizing index in (3.158), then $\Delta e \leq e_i$ implies that

$$e_i + e_{J_{kj'}} > C \cdot (L_J - E_{J_{kj'} \cup \{i\}})$$

which implies $i > J_{kj'}$ by the edge-finding rule. Now E_i can be updated to

$$\max \left\{ E_i', E_{j'} + \frac{e_{J_{kj'}} - (C - c_i)(L_k - E_{j'})}{c_i} \right\}$$

But since j^* maximizes

$$E_j + \frac{e_{J_{kj}} - (C - c_i)(L_k - E_j)}{c_i}$$

over $j \in J_{ki} \subset J_{kj'}$, E_i can be updated to

$$\max\left\{E'_i, E_{j^*} + \frac{e_{J_{kj^*}} - (C - c_i)(L_k - E_{j^*})}{c_i}\right\}$$
$$= \max\left\{E'_i, E^* + \frac{e^*_J - (C - c_i)(L_k - E^*)}{c_i}\right\}$$

when $E^* + (e^*_J - (C - c_i)(L_k - E^*))/c_i > 0$, as is done in the algorithm.
Based in part on the above reasoning, one can conclude the following.

THEOREM 3.50 *The algorithm of Figure 3.22 computes all updates of the form (3.155), and a similar algorithm computes all updates of the form (3.156).*

When the algorithm is applied to the problem of Figure 3.20, the update $E'_3 = 2$ is discovered when $k = 1$ and $i = 3$ in the second i-loop. The first i-loop has already computed $e^*_J = 14$ and $E^* = 0$, because the index j that maximizes $e_{J_{kj}} + CE_j$ over $j \in J_{ki} = J_{13} = \{1,2\}$ is $j^* = 1$. So $e^*_J = e_{J_{kj^*}} = e_{\{1,2\}} = 14$, and $E^* = E_{j^*} = 0$. Also, $\Delta e = 6$, because the index j that minimizes $C \cdot (L_k - e_j) - e_{J_{kj}}$ over $j \in J_k$ and $j < i$ (i.e., over $j = 1, 2$) is $j' = 1$, so that $\Delta e = C \cdot (L_k - E_{j'}) - e_{J_{kj'}} = 6$. Since $\Delta e < e_i$, the algorithm has established that $i > J_{kj'}$ (i.e., $3 > \{1,2\}$). Now since $R = e^*_J - (C - c_i)(L_k - E^*) = 4 > 0$, it updates $E'_3 = 1$ to $E^* + R/c_3 = 2$.

3.14.2 Extended Edge Finding

A weakness of the edge-finding rules is that they may fail to detect that job i must finish after the jobs in J finish when job i has an earlier release time than the other jobs. In such cases, the total time available $L_J - E_{J \cup \{i\}}$ may provide the required energy if it were available to all the jobs, but the period between $E_{J \cup \{i\}}$ and E_J is not available to the jobs in J. If job i cannot finish by E_J, it may have to finish after the other jobs finish. A similar situation can occur when job i has a later deadline than the other jobs. An extended version of edge finding corrects this problem, albeit at the cost of more computation.

Edge finding is extended by adding the rules below:

If $E_i \leq E_J < E_i + p_i$ and
$\quad e_J + c_i(E_i + p_i - E_J) > C \cdot (L_J - E_J)$, then $i > J$ $\quad$ (a)

If $L_i - p_i < L_J \leq L_i$ and
$\quad e_J + c_i(L_J - L_i + p_i) > C \cdot (L_J - E_J)$, then $i < J$ $\quad$ (b)

$$\quad (3.159)$$

Cumulative Scheduling

The reasoning behind Rule (a) is as follows. It is supposed that job i has release time no later than E_J but cannot finish by E_J. If job i finishes before some job in J finishes (or at the same time), then it must finish by L_J. Thus, the energy required between E_J and L_J is at least the energy e_J of the jobs in J and the energy $c_i(E_i + p_i - E_J)$ of the portion of job i that must run during this period. If this exceeds the available energy during the period, then job i must finish after all the jobs in J finish. The reasoning is similar for Rule (b).

Rule (a) is illustrated in the example of Figure 3.23. Here, job 4 must finish after the others finish, because $E_4 < E_{\{1,2,3\}} < E_4 + p_4$ and

$$e_{\{1,2,3\}} + c_4(E_4 + p_4 - E_{\{1,2,3\}}) = 10 + 1 \cdot (4-1) = 13$$
$$> 12 = 2 \cdot (7-1) = C \cdot (L_{\{1,2,3\}} - E_{\{1,2,3\}})$$

Note that ordinary edge finding does not deduce that $4 > \{1,2,3\}$, because

$$e_4 + e_{\{1,2,3\}} = 14 \leq 2 \cdot (7-0) = C \cdot (L_{\{1,2,3\}} - E_{\{1,2,3,4\}})$$

If $i > J$, then as with ordinary edge finding, E_i can be updated to $\max\{E_i, E'_i(J)\}$, where $E'_i(J)$ is given by (3.153). Similarly, if $i < J$ then L_i can be updated to $\min\{L_i, L'_i(J)\}$. Thus, E_i can be updated to the maximum of $E'_i(J)$ over all J satisfying the extended edge-finding rule (3.159a), and L_i can be updated to the minimum of $L'_i(J)$ over all J satisfying (3.159b). The updates appear to require more computation than the original edge-finding updates. At least one algorithm, not presented here, computes them in time proportional to n^3.

In the example, $R(J, c_4) > 0$ when J is $\{1,2\}$, $\{2,3\}$ or $\{1,2,3\}$. Thus $E_4 = 0$ is updated to $E'_4(\{1,2,3\}) = \max\{1 + \frac{1}{2}, 1 + \frac{1}{2}, 1 + \frac{4}{2}\} = 3$.

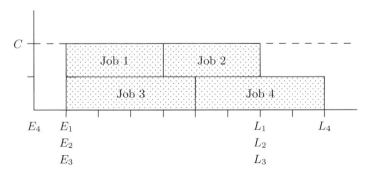

Figure 3.23. A feasible solution of a small cumulative scheduling problem. Extended edge finding deduces that $4 > \{1, 2, 3\}$, but ordinary edge finding does not.

3.14.3 Not-First, Not-Last Rules

As in the case of disjunctive scheduling, one can sometimes deduce in a cumulative scheduling context that a job i cannot be scheduled first, or cannot be scheduled last, in a set $J \cup \{i\}$ of jobs. One must be careful about what *not first* and *not last* mean, however. In disjunctive scheduling, a job i is *not first* when $\neg(i \ll J)$, but in cumulative scheduling *not first* does not mean $\neg(j < J)$. Rather, it means the same thing as in disjunctive scheduling: job i starts after some job in J finishes. Similarly, job i is not last when it finishes before some job in J starts.

Let $F_J = \min_{j \in J}\{E_j + p_j\}$ be the minimum earliest finish time of the jobs in J, and let $S_J = \max_{j \in J}\{L_j - p_j\}$ be the maximum latest start time of the jobs in J. If job i is not first, then it cannot start earlier than F_J, which means that E_i can be updated to $\max\{E_i, F_J\}$. If job i is not last, then it cannot finish later than S_J, and L_i can be updated to $\min\{L_i, S_J\}$. In view of this, the not-first/not last rules can be stated:

$$
\begin{aligned}
&\text{If } E_J \leq E_i < F_J \text{ and} \\
&e_J + c_i \left(\min\{E_i + p_i, L_J\} - E_J\right) > C \cdot (L_J - E_J), \quad (a) \\
&\quad \text{then update } E_i \text{ to } \max\{E_i, F_J\} \\
\\
&\text{If } S_J < L_i \leq L_J \text{ and} \\
&e_J + c_i \left(L_J - \max\{L_i - p_i, E_J\}\right) > C \cdot (L_J - E_J), \quad (b) \\
&\quad \text{then update } L_i \text{ to } \min\{L_i, S_J\}
\end{aligned}
\tag{3.160}
$$

Rule (a) can be proved as follows. Suppose job i and set J satisfy the conditions of the rule but job i starts at some time t before F_J. Thus, $E_i \leq t < F_J$. Since, by hypothesis, no job in J can finish by t, a resource capacity of c_i must remain unused during the period from E_J to t. This means that the total energy required between E_J and L_J by the jobs in $J \cup \{i\}$, including the energy that cannot be used, is

$$e_J + c_i \left(\min\{t + p_i, L_J\} - E_J\right)$$

But since $t > E_i$, this quantity is greater than or equal to

$$e_J + c_i \left(\min\{E_i + p_i, L_J\} - E_J\right)$$

which, by hypothesis, exceeds the available energy $C \cdot (L_J - E_J)$. So job i can start no earlier than F_J. The argument for Rule (b) is analogous.

The example of Figure 3.20 is modified in Figure 3.24 to illustrate the not-first principle. Here, $F_{\{1,2\}} = \min\{0+3, 0+6\} = 3$. One can deduce that job 3 is not first because $E_{\{1,2\}} \leq E_3 < F_{\{1,2\}}$ and

$$
\begin{aligned}
&e_{\{1,2\}} + c_3 \cdot \left(\min\{e_3 + p_3, L_{\{1,2\}}\} - E_{\{1,2\}}\right) = \\
&\quad 15 + 2 \cdot (\min\{5, 6\} - 0) = 25 > 4 \cdot (6 - 0) = C \cdot (L_{\{1,2\}} - E_{\{1,2\}})
\end{aligned}
$$

Cumulative Scheduling

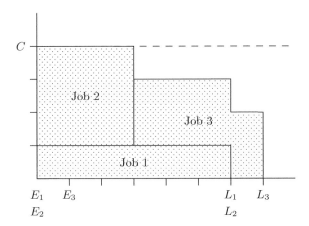

Figure 3.24. A feasible solution of a small cumulative scheduling problem. It can be deduced that job 3 is not first, and E_3 can therefore be updated to 3. Neither edge finding nor extended edge finding can update E_3.

E_3 is therefore updated to $\max\{E_3, F_{\{1,2\}}\} = \max\{1, 3\} = 3$. Note that although $3 \gg \{1, 2\}$, this cannot be deduced by any of the edge-finding rules discussed earlier. Thus, the not-first rule discovers an update that is missed by edge finding, even when extended.

The not-first/not-last rules (3.160) can be used to update each E_i to the maximum of $\max\{E_i, F_J\}$ over all sets J satisfying the conditions in rule (a), and to update L_i to the minimum of $\min\{L_i, S_J\}$ over all sets J satisfying the conditions in (b). It is possible to compute these updates in time proportional to n^3.

3.14.4 Energetic Reasoning

The concept of energy plays a role in all of the filtering methods presented here for the cumulative scheduling constraint. A sharper analysis of the energy required by a set of jobs in a given time interval $[t_1, t_2]$, however, provides an additional technique for reducing domains.

The analysis is based on *left-shifting* and *right-shifting* jobs. A job i is left-shifted if it is scheduled as early as possible (i.e., if it starts at E_i) and is right-shifted if it is scheduled as late as possible (i.e., starts at $L_i - p_i$). Suppose that job i runs $p_i^\ell(t_1)$ time units past t_1 when it is left-shifted and starts $p_i^r(t_2)$ time units before time t_2 when right-shifted. Thus

$$p_i^\ell(t_1) = \max\{0, p_i - \max\{0, t_1 - E_i\}\}$$
$$p_i^r(t_2) = \max\{0, p_i - \max\{0, L_i - t_2\}\}$$

The amount of time job i runs during $[t_1, t_2]$ is at least the minimum of $p_i^\ell(t_1)$, $p_i^r(t_2)$, and $t_2 - t_1$. So the minimum energy consumption of job i during $[t_1, t_2]$ is

$$e_i(t_1, t_2) = c_i \min\left\{p_i^\ell(t_1), p_i^r(t_2), t_2 - t_1\right\}$$

The minimum total energy consumption during interval $[t_1, t_2]$ is therefore $e(t_1, t_2) = \sum_i e_i(t_1, t_2)$, where the sum is taken over all jobs.

There is clearly no feasible schedule if $e(t_1, t_2) > C \cdot (t_2 - t_1)$ for some interval $[t_1, t_2]$. Fortunately, it is not necessary to examine every interval to check whether such an interval exists. To see this, let

$$\Delta e(t_1, t_2) = C \cdot (t_2 - t_1) - e(t_1, t_2)$$

be the excess energy capacity during $[t_1, t_2]$. Then there is no feasible schedule if $\Delta e(t_1, t_2) < 0$ for some interval $[t_1, t_2]$. If $F_i = E_i + p_i$ is the earliest finish time of job i, and $S_i = L_i - p_i$ the latest start time, the following can be shown:

LEMMA 3.51 *All local minima of the function $\Delta e(t_1, t_2)$ occur at points (t_1, t_2) belonging to the set T^* defined to be the union of the following sets:*

$$\{(t_1, t_2) \mid t_1 \in T_1,\ t_2 \in T_2,\ t_1 > t_2\}$$
$$\{(t_1, t_2) \mid t_1 \in T_1,\ t_2 \in T(t_1),\ t_1 > t_2\}$$
$$\{(t_1, t_2) \mid t_2 \in T_1,\ t_1 \in T(t_2),\ t_1 > t_2\}$$

where

$$T_1 = \{E_i, F_i, S_i \mid i = 1, \ldots, n\}$$
$$T_2 = \{F_i, S_i, L_i \mid i = 1, \ldots, n\}$$
$$T(t) = \{E_i + L_i - t \mid i = 1, \ldots, n\}$$

Due to this fact, one can check whether $\Delta e(t_1, t_2)$ goes negative for some interval by evaluating it only for $(t_1, t_2) \in T^*$.

Energy reasoning can be extended to domain reduction, since even when $\Delta e(t_1, t_2)$ is nonnegative, it may be small enough to restrict when some jobs can run. The set of all jobs other than i requires energy of at least $e(t_1, t_2) - e_i(t_1, t_2)$ in the interval $[t_1, t_2]$. Thus, the energy available to run job i in the interval $[t_1, t_2]$ is at most

$$\Delta e_i(t_1, t_2) = C \cdot (t_2 - t_1) - e(t_1, t_2) + e_i(t_1, t_2)$$

If job i is left-shifted, it requires energy $c_i p_i^\ell(t_1)$ after t_1. If $c_i p_i^\ell(t_1) > \Delta e_i(t_1, t_2)$, then

$$c_i p_i^\ell(t_1) - \Delta e_i(t_1, t_2)$$

Cumulative Scheduling 241

is the amount of job i's energy consumption that must be moved outside $[t_1, t_2]$ to the right, beyond t_2. This means that job i must run for a time of at least $p_i^\ell(t_1) - \Delta e_i(t_1, t_2)/c_i$ past t_2, and job i's earliest finish time F_i can be updated to

$$\max\left\{F_i,\ t_2 + p_i^\ell(t_1) - \frac{\Delta e_i(t_1, t_2)}{c_i}\right\}$$

This implies the following:

THEOREM 3.52 *If $c_i p_i^\ell(t_1) > \Delta e_i(t_1, t_2)$, then E_i can be updated to*

$$\max\left\{E_i,\ t_2 + p_i^\ell(t_1) - p_i - \frac{\Delta e_i(t_1, t_2)}{c_i}\right\}$$

Similarly, if $c_i p_i^r(t_2) > \Delta e_i(t_1, t_2)$, then L_i can be updated to

$$\min\left\{L_i,\ t_1 - p_i^r(t_2) + p_i - \frac{\Delta e_i(t_1, t_2)}{c_i}\right\}$$

One can apply Theorem 3.52 to all jobs i and all pairs $(t_1, t_2) \in T^*$ in time proportional to n^3. It is apparently an open question whether this necessarily detects all updates that can be obtained from the theorem.

3.14.5 Benders Cuts

Like the disjunctive constraint, the cumulative constraint commonly occurs in planning and scheduling problems are amenable to the logic-based Benders decomposition. The Benders cuts are similar to those developed in Section 3.13.3 for disjunctive scheduling.

Minimizing Cost

Consider first a planning and cumulative scheduling problem that minimizes the fixed cost of assigning jobs to facilities. The model is identical to the model (3.130) stated earlier for planning and disjunctive scheduling, except that the disjunctive constraint is replaced by

$$\text{cumulative}((s_j \mid x_j = i) \mid (p_{ij} \mid x_j = i), (c_{ij} \mid x_j = i), C_i),\ \text{all } i$$

where c_{ij} the rate of resource consumption of job j in facility i, and C_i is the maximum resource consumption rate in facility i.

The Benders master problem is identical to the master problem (3.131) for disjunctive scheduling, aside from the subproblem relaxation, which becomes more complicated for minimum cost as well as other objectives when one moves to cumulative scheduling. Relaxations for the cumulative scheduling subproblem are presented in Section 4.16.3. The subproblem is the same as before, except that the disjunctive constraint in (3.132) is replaced with a cumulative constraint. The Benders cuts are again (3.133) and have the linearized version (3.133). As before, the cuts can be strengthened by analyzing the solution of the scheduling subproblem (or more precisely, the solution of its inference dual) and identifying which jobs play a role in the proof of infeasibility.

Minimizing Makespan

If the objective is to minimize makespan, the model is (3.135) with the disjunctive constraint replaced by a cumulative constraint. The simple Benders cut (3.138) is still valid. However, Lemma 3.48 does not hold for cumulative scheduling, and the cut (3.140) is not valid. However, a weaker form of the lemma can be proved and a cut written on that basis.

LEMMA 3.53 *Consider a minimum makespan problem P in which jobs $1, \ldots, n$, with release time 0 and deadlines $d_1, \ldots, d_n$, are to be scheduled on a single facility i. Let M^* be the minimum makespan for P, and $\hat{M}$ the minimum makespan for the problem $\hat{P}$ that is identical to P except that jobs $1, \ldots, s$ are removed. Then,*

$$M^* - \hat{M} \leq \Delta + \max_j\{d_j\} - \min_j\{d_j\} \tag{3.161}$$

where $\Delta = \sum_{j=1}^{s} p_{ij}$. In particular, when all the deadlines are the same, $M^* - \hat{M} \leq \Delta$.

Proof. Consider any optimal solution of $\hat{P}$ and extend it to a solution S of P by scheduling jobs $1, \ldots, s$ sequentially after $\hat{M}$. That is, for $k = 1, \ldots, s$ let job k start at time $\hat{M} + \sum_{j=1}^{k-1} p_{ij}$. The makespan of S is $\hat{M} + \Delta$. If $\hat{M} + \Delta \leq \min_j\{d_j\}$, then S is clearly feasible for P, so that $M^* \leq \hat{M} + \Delta$ and the lemma follows. Now, suppose $\hat{M} + \Delta > \min_j\{d_j\}$. This implies

$$\hat{M} + \Delta + \max_j\{d_j\} - \min_j\{d_j\} > \max_j\{d_j\} \tag{3.162}$$

Since $M^* \leq \max_j\{d_j\}$, (3.162) implies (3.161), and again the lemma follows.□

The bound $M^* - \hat{M} \leq \Delta$ need not hold when the deadlines differ. Consider, for example, an instance with three jobs where $(r_1, r_2, r_3) = (0, 0, 0)$, $(d_1, d_2, d_3) = (2, 1, \infty)$, $(p_{i1}, p_{i2}, p_{i3}) = (1, 1, 2)$, $(c_{i1}, c_{i2}, c_{i3}) = (2, 1, 1)$, and $C = 2$. Then, if $s = 1$, $M^* - \hat{M} = 4 - 2 > \Delta = p_{i1} = 1$.

Cumulative Scheduling

To write a Benders cut, suppose first that all the deadlines are the same. Then by Lemma 3.48, each job removed from facility i reduces the minimum makespan at most p_{ij}:

$$v \geq M_{ik}^* - \sum_{\substack{j \in J_{ik} \\ x_j \neq i}} p_{ij} \tag{3.163}$$

The cut is easily linearized:

$$v \geq M_{ik}^* - \sum_{j \in J_{ik}} (1 - x_{ij}) p_{ij} \tag{3.164}$$

When the deadlines differ, the cut (3.163) becomes

$$v \geq M_{ik}^* - \left(\sum_{\substack{j \in J_{ik} \\ x_j \neq i}} p_{ij} + \max_{j \in J_{ik}} \{d_j\} - \min_{j \in J_{ik}} \{d_j\} \right) \tag{3.165}$$

and its linearization is

$$v \geq M_{ik}^* - \left(\sum_{j \in J_{ik}} p_{ij}(1 - x_{ij}) + \max_{j \in J_{ik}} \{d_j\} - \min_{j \in J_{ik}} \{d_j\} \right) \tag{3.166}$$

Minimizing Tardiness

When the objective is to minimize the number of late jobs, the model is (3.141) with the disjunctive constraint replaced by a cumulative constraint. The Benders cuts are again (3.143)–(3.144).

Finally, when minimizing total tardiness, one can again use the Benders cuts (3.145)–(3.147). Lemma 3.49 remains valid, but the resulting Benders cuts (3.149)–(3.150) can be quite weak. The reason is that the proof of Lemma 3.49 assumes that the minimum makespan $\hat{M}$ of problem $\hat{P}$ is at most $M = \sum_{j=s+1}^{n} p_{ij}$. The true makespan is likely to be much less than M because some jobs can be scheduled concurrently.

A generally stronger cut could be obtained by assuming $\hat{M}$ is at most the makespan M^* of problem P's minimum tardiness solution. Unfortunately, this is not a valid assumption, because in exceptional cases $\hat{M} > M^*$ even though $\hat{P}$ has fewer jobs than P. Suppose, for example, that $n = 4$ and $s = 1$, with $r = (0,0,0,0)$, $d = (5,3,3,6)$, $p = (1,2,2,4)$, $c = (2,1,1,1)$, and $C = 2$. An optimal solution of P puts $t = (4,0,2,0)$ with tardiness $T^* = 1$ and makespan $M^* = 5$, but the only optimal solution of the smaller problem $\hat{P}$ puts $(t_2, t_3, t_4) = (0, 0, 2)$ with tardiness $\hat{T} = 0$ and makespan $\hat{M} = 6 > M^*$.

It is therefore probably advisable to use Benders cuts (3.145)–(3.147) rather than (3.149)–(3.150) when the subproblem involves cumulative scheduling.

3.14.6 Exercises

1. Consider a cumulative scheduling problem with four jobs in which $p = (3, 3, 4, 3)$, $c = (2, 2, 1, 1)$, and the resource limit is 3. All release times are zero, and the deadlines are $(L_1, \ldots, L_4) = (6, 6, 6, 7)$. Apply the edge-finding rules and update the bounds accordingly.

2. State the argument for the second edge-finding rule (3.152b).

3. Apply the algorithm of Fig 3.22 to the problem of Exercise 1.

4. State the argument for the second extended edge-finding rule (3.159b).

5. Consider a cumulative scheduling problem with four jobs in which $p = (5, 3, 3, 3)$, $c = (1, 2, 2, 1)$, and the resource limit is 3. The release times are $(E_1, \ldots, E_4) = (0, 2, 2, 2)$, and the deadlines are $(L_1, \ldots, L_4) = (9, 8, 8, 8)$. Apply the extended edge-finding rules and update the bounds accordingly. Does extended edge finding identify a valid precedence that is not identified by ordinary edge finding?

6. State the argument for the not-last rule (3.160b).

7. Consider a cumulative scheduling problem with four jobs in which $p = (3, 3, 4, 3)$, $c = (2, 2, 1, 1)$, and the resource limit is 3. The release times are $(E_1, \ldots, E_4) = (1, 1, 1, 0)$, and all the deadlines are 7. Apply the not-first/not-last rules and update the bounds accordingly. Does this identify any precedences not identified by edge finding?

8. Show by counterexample that Lemma 3.48 is not valid for cumulative scheduling.

9. Suppose that a Benders method is applied to a minimum makespan planning and cumulative scheduling problem. In the first iteration, Jobs 1, 2 and 3 are assigned to machine A, on which the resource capacity is 2, $(p_{A1}, p_{A2}, p_{A3}) = (1, 1, 2)$, and $(c_{A1}, c_{A2}, c_{A3}) = (2, 1, 1)$. All the release times are zero, and all the deadlines are 4. Write the Benders cut (3.164).

10. Change the deadlines in the problem of Exercise 9 to $(L_1, L_2, L_3) = (2, 1, 4)$. Write the Benders cut (3.165). Show that the stronger cut (3.164) is not valid in this case.

11 Show by example that $T_i(J_{ik}\setminus Z_{ik}) < T_i(J_{ik})$ is possible for cumulative scheduling, even when all the release dates are the same.

3.15 Bibliographic Notes

Section 3.1. The concept of domain completeness was originally developed for binary (two-variable) constraints under the name of arc consistency [228, 241], because constraints were seen as relations forming a network. The concept was generalized to multiple-variable constraints in [240] and implicitly in [101]. The concept is still called hyperarc consistency, or generalized arc consistency, in recognition of the network model. The term *completeness* is used here rather than *consistency* because it connects naturally with the completeness of the inference method used to achieve it, and because the word *consistency* suggests feasibility to those outside the constraints community. This usage also reserves the term *consistency* for k-consistency, where it has a slightly different meaning.

Bounds completeness has arisen in many contexts. It is called bounds consistency or interval consistency in the constraints community.

An extension of the idea of completeness to partial assignments yields k-completeness, which is stronger than the concept of k-consistency that originated in the constraints community [134]. The fundamental theorem relating backtracking with strong k-completeness and width of the dependency graph (Theorem 3.1) is due to [135]. This and other forms of consistency are studied in [321, 173].

When the dependency graph has small *induced width*, its structure can be exploited by nonserial dynamic programming, which has been known in operations research for more than 30 years [48]. Essentially the same idea has surfaced in a number of contexts, including Bayesian networks [219], belief logics [302, 305], pseudoboolean optimization [94], location theory [79], k-trees [10, 11], and bucket elimination [105].

Section 3.2. Inference duality is developed in [183, 185, 194, 201]. The connection between inference duality and sensitivity analysis appears in [183] and is applied to mixed integer programming in [103]. The connection between inference duality and constraint-directed search is described in [185, 192, 193, 201].

Section 3.3. The Farkas Lemma (Theorem 3.4) was first proved in [121]. Fourier-Motzkin elimination can be traced to Fourier's work in the early nineteenth century and was used by Boole to solve linear programming problems [62]. Linear programming duality is generally credited to John von Neumann. Good textbook expositions of the linear programming concepts discussed here can be found in [85] (elementary), [325] (intermediate), and [297] (advanced).

Section 3.4. Surrogate duality was introduced by [149]. Lagrangean duality is an old idea that is first applied to integer programming in [170, 171], where the dual is solved by subgradient optimization. A good exposition of the application of Lagrangean duality to integer programming can be found in [125] and to nonlinear programming in [31]. A comparison of several types of integer programming duality appears in [186].

Section 3.5. The resolution method for propositional logic (Theorem 3.19) is due to [275, 276] and is extended to first-order predicate logic in [287]. It is extended to clauses with multi-valued variables in [198]. The connection between Horn clauses and unit resolution (Theorem 3.20) is observed in [111]. The completeness of parallel resoution (Theorem 3.24) as well as its polynomial complexity in partial-order dynamic backtracking are proved in [185].

Section 3.6. The domination theorems for 0-1 inequalities (Theorems 3.25 and 3.26) are proved in [179, 185]). The algorithm for generating all nonredundant clauses implied by a 0-1 linear inequality (Figure 3.7) is generalized to nonlinear 0-1 inequalities in [157, 158]. The implication theorems for cardinality clauses (Theorems 3.28 and 3.29) are proved in [179, 185]. Additional domination results are proved in [29], which states an efficient algorithm for deriving all cardinality clauses implied by a 0-1 inequality. This algorithm is used as the basis for a 0-1 programming solver OPBDP. The completeness proofs for 0-1 resolution (Theorems 3.30 and 3.31) are due to [181].

Section 3.7. The subadditive (superadditive) dual is stated in [207]. Its connection with integer programming is explored in [52], and in particular with Chvátal-Gomory cuts (Corollary 3.33) in [53, 204, 336]. These ideas are presented in Section II.3 of [251].

The branching dual is used in [183] for sensitivity analysis and applied to mixed integer programming for the same purpose in [103]. Another type of branching dual is used in [296] for mixed integer sensitivity analysis (with respect to the right-hand side only) by computing a piecewise linear value function.

Section 3.8. The element constraint was introduced by [174]. Filtering for domain and bounds completeness is straightforward. Filtering algorithms are widely implemented, except those for the multidimensional and specially structured versions of the constraint.

Section 3.9. The augmenting path condition for maximum cardinality bipartite matching (Theorem 3.35) is due to [46], but the basic idea is found in Egerváry's 1931 "Hungarian" algorithm for the assignment

Bibliographic Notes

problem, as interpreted in [217]. A number of fast algorithms for maximum cardinality bipartite matching have been proposed, including one with $O(n^{1/2}m)$ complexity [202], where n is the number of vertices and m the number of edges, and an algorithm with $O(n^{1.5}(m/\log n)^{0.5})$ complexity [6].

The alldiff constraint first appeared in [218]. The matching-based filtering method presented here is due to [93, 281]. Theorem 3.36, from which it is derived, is proved in [47]. The filtering method given here for achieving bounds consistency appeared in [237]. The convex graph result on which it is based, Theorem 3.37, is proved in [147].

Section 3.10. The network flow model given here for filtering the cardinality constraint is due to [282]. The augmenting path theorem on which it is based (Theorem 3.39) was first used in [131, 132] to design maximum flow algorithms. Numerous fast algorithms have since been developed, the best having complexity close to $O(mn)$. For example, the algorithm of [151] has complexity $O(mn \log(n^2/m))$. There is a bounds consistency algorithm based on flows [210] that exploits convexity of the graph. Filters for the nvalues constraint are discussed under that entry in Chapter 5.

Section 3.11. The circuit constraint was first formulated as such by [218], but hamiltonian cycles have been studied by Hamilton, Kirkman, and others at least since the mid-nineteenth century. Elementary filtering methods have been used by [73, 306]. The filtering algorithm based on separators, presented here, appears in [211]. Filtering can also be based on sufficient conditions for nonhamiltonity of a graph, some of which appear in [83, 84, 86]. Given an edge (i, j) in directed graph G, let G_{ij} be the graph that results from inserting a vertex in edge (i, j). Then the edge (i, j) belongs to a hamiltonian path if and only if G_{ij} is hamiltonian. Checking whether nonhamiltonicity conditions are met for each edge may require too much computation for a practical filter, however.

Section 3.12. The stretch constraint originated in [266]. The dynamic-programming-based filter presented here is due to [172]. A good introductory text on dynamic programming is [108]; and a more advanced treatment is [49]. Domain filtering in a dynamic programming context is discussed in [319].

Section 3.13. An early study of disjunctive scheduling is [68], and edge finding is introduced in [69]. The $O(n^2)$ edge-finding algorithm described here is that of [70]. Another $O(n^2)$ algorithm appears in [24, 254, 257]. An algorithm that achieves $O(n \log n)$ complexity with complex data structures is given in [71], and an $O(n^3)$ algorithm that allows incremental updates in [72]. Extensions that take setup times into account are presented in [66, 130]. The propagation algorithm for not-first/not-last rules given here appears in [23], while others appear in [110, 317]. A comprehensive treatment of disjunctive and cumulative scheduling is provided by [24].

The Benders cuts for minimizing cost and makespan are used in [185, 203]. They are strengthened by analyzing the edge-finding process in [190] and by re-solving the subproblem in [196]. The cuts for minimizing the number of late jobs and total tardiness are developed in [191] and strengthened in [196].

Section 3.14. The cumulative scheduling constraint originated in [3]. The $O(n^2)$ edge-finding algorithm presented here, which appears in [24], is an improvement of one in [255, 256]. Another algorithm appears in [72]. The extended edge-finding algorithm given here is based on [24, 254]. Not-first/not-last propagation appears in [255, 256] and energetic reasoning in [116, 117].

The Benders cuts for minimizing cost and makespan appear in [185] and are strengthened in [190, 196]. The cuts for minimizing the number of late jobs and total tardiness are developed in [191] and strengthened in [196].

Chapter 4

RELAXATION

The ideal problem relaxation is both easy to solve and in some sense *tight*, meaning that it closely resembles the original problem. The solution of a tight relaxation is more likely to be feasible in the original problem, or if not, to provide a good bound on the optimal value of the original problem.

The most widely used medium for formulating a relaxation is a system of linear inequalities in continuous variables. This is due largely to the fact that the mathematical programming field has focused on inequality-constrained problems and has made continuous linear relaxation one of its primary tools for solving them. A relaxation of this sort is easy to solve because it is a linear programming problem. It can also be tight if the problem structure is carefully analyzed and the inequalities selected wisely.

There are at least two ways to obtain linear inequality-constrained relaxations. One is to reformulate the constraints with a mixed integer linear programming (MILP) model, and then to take a continuous relaxation of the resulting model. An MILP model consists of linear inequalities with additional 0-1 or integer-valued variables that capture the discrete elements of the problem. Experience has taught that integer variables allow a sufficiently clever modeler to formulate a wide range of constraints. A continuous relaxation is obtained simply by dropping the integrality constraint on the variables, and the relaxation is strengthened by the addition of valid inequalities (cutting planes), an intensely studied topic of mathematical programming.

A second strategy is to design a relaxation directly, without writing an MILP model, using only the original variables and exploiting the specific structure of the constraints. This approach received relatively little attention before the recent rise of interest in integrated methods, because the community most interested in continuous relaxation, the mathematical programming community, normally formulated the problem as an MILP from the start.

The choice between the strategies is not either/or. One can relax some metaconstraints directly and relax others using an MILP model, and then pool the individual relaxations into a single relaxation for the entire problem. The pooled relaxation is solved to obtain a bound on the optimal value of the original problem. This constraint-oriented approach is very much in the spirit of integrated methods, because it allows one to take advantage of the peculiar characteristics of each metaconstraint.

In some cases, it is advantageous to formulate not one relaxation but many. If the relaxations are sufficiently easy to solve, one can examine the space of relaxations in search of a tight one. The problem of finding the tightest relaxation is the *relaxation dual*. One well-known relaxation dual is the Lagrangean dual, which is defined for inequality-constrained problems.

This chapter begins with an introduction to relaxation duality, which finds immediate application in the subsequent discussion of linear inequalities. The basic concepts of linear optimization are presented because one must know how to solve continuous linear relaxations if they are to be useful. There is also a brief description of how to relax semicontinuous piecewise linear functions without using an MILP formulation. Piecewise linear functions not only occur frequently in their own right, but can be used to approximate some nonlinear problems that would otherwise be difficult to relax.

At this point, the relaxation of MILP models is taken up. This requires an extended discussion of cutting plane theory, because cutting planes are often necessary to obtain a tight continuous relaxation. General purpose cutting planes are developed for 0-1, integer, and mixed integer constraint sets. None of this is applicable unless an MILP model can be written in the first place, and a section is therefore devoted to MILP modeling.

Much of the cutting plane discussion is concerned with *separating cuts*, which are generated only after the continuous relaxation is solved. These cutting planes are therefore not part of the original linear relaxation that might be generated for a metaconstraint. They are added to the pooled relaxation only after it is solved.

The remainder of the chapter is devoted to identifying relaxations for specific metaconstraints—disjunctions of linear and nonlinear inequality systems, formulas of propositional logic, and several popular global constraints from the constraint programming field (disjunctions are discussed before MILP modeling in order to provide necessary background). The global constraints canvassed include the element, all-different, cardinality, and circuit constraints, along with disjunctive and cumulative scheduling constraints. Some constraints are relaxed directly, some are given MILP formulations, and some are treated in both ways. Chapter 5 cites relaxations given in the literature for several additional metaconstraints.

Convex nonlinear relaxations can be useful for relaxing continuous nonlinear constraints for which a good linear relaxation is difficult to identify. This possibility is only briefly addressed here by presenting convex relaxations for a disjunction of convex nonlinear systems.

4.1 Relaxation Duality

In many cases, relaxations can be parameterized by a set of *dual variables*, thus providing a choice from an entire family of relaxations. The problem of finding the relaxation that provides the tightest bound can be posed as an optimization problem over the dual variables. This problem is the *relaxation dual*. Various relaxation duals have made enormous contributions to optimization, including the linear programming dual and its special cases, the Lagrangean dual, and to a lesser extent the surrogate and subadditive duals.

Consider an optimization problem

$$\min\ f(x) \\ x \in S \tag{4.1}$$

A parameterized relaxation of (4.1) can be written

$$\min\ f(x, u) \\ x \in S(u) \tag{4.2}$$

where $u \in U$ is a vector of *dual variables*. Each $u \in U$ defines a relaxation of (4.1), in the sense that $S(u) \supset S$, and $f(x, u) \leq f(x)$ for all $x \in S$. Clearly, the optimal value of the relaxation (4.2) is a lower bound on the optimal value of the original problem (4.1). Note that to ensure this property, it is enough that $f(x, u)$ bound $f(x)$ for all $x \in S$, not necessarily for all $x \in S(u)$.

Let $\theta(u)$ be the optimal value of (4.2) for a given u, where $\theta(u)$ is ∞ if (4.2) is infeasible, and $-\infty$ if it is unbounded. Then the *relaxation dual* is the problem of finding the tightest relaxation:

$$\max\ \theta(u) \\ u \in U \tag{4.3}$$

Since $\theta(u)$ is a lower bound on the optimal value of (4.1) for each u, the same is true of the optimal dual value. This property is called *weak duality*.

LEMMA 4.1 (WEAK DUALITY) *If $\theta(u)$ is the optimal value of a relaxation (4.2) of (4.1) for each $u \in U$, then*

$$\min\,\{f(x) \mid x \in S\} \geq \max\,\{\theta(u) \mid u \in U\}$$

Generally there is a gap between the optimal value of the original problem and the optimal value of even the tightest relaxation. It is known as the *duality gap*, which is defined as the difference between the left- and right-hand sides of the inequality in Lemma 4.1. Occasionally, however, there is a *strong duality* property, meaning that there is no duality gap. This occurs in linear programming duality, for example.

The main utility of a relaxation dual is that it can provide a tight bound on the optimal value of the original problem, and may therefore accelerate the search for an optimal solution. An obvious difficulty is that merely evaluating $\theta(u)$ requires the solution of an *inner* optimization problem. Nonetheless, if the dual is cleverly constructed, it may be practical to find a good bound. An optimal dual solution is not essential, because any feasible solution of the dual provides a valid lower bound. A local search algorithm, for example, may be perfectly satisfactory for solving the dual, as in the case of subgradient optimization (Section 4.6.5).

4.2 Linear Inequalities

Linear inequality constraint sets with continuous variables generally do not require relaxation, because the solution technology for them is extremely well developed. Commercial software packages routinely solve linear optimization problems with millions of variables. Nonetheless, it useful to understand the basic concepts of linear optimization, partly due to the importance of the problem and partly because they have application to the relaxation of integer linear inequalities. In addition, the relaxation dual for linear optimization is identical to the inference dual and is useful in many contexts.

4.2.1 Linear Optimization

A linear optimization problem, also known as a linear programming problem, minimizes (or maximizes) a linear function subject to linear

Linear Inequalities

equations and inequalities. As noted in Section 3.3.6, such a problem can always be written

$$\min cx \\ Ax = b, \quad x \geq 0 \qquad (4.4)$$

Recall that a basic solution of (4.4) is one in which at most m variables are positive, where m is the rank of A. A fundamental fact of linear programming is that if (4.4) has a finite optimal solution, then it has an optimal basic solution. Geometrically speaking, each basic feasible solution is a vertex of the polyhedron defined by the constraints of (4.4). Since the objective function is linear, it is intuitively clear that some vertex is an optimal solution.

This fact can be developed algebraically as follows. As in Section 3.3.6, rearrange the columns of A so that $A = [B\ N]$, where the columns of the basis B correspond to the basic variables x_B and the columns of N to the nonbasic variables x_N. The problem (4.4) may be written

$$\min c_B x_B + c_N x_N \\ B x_B + N x_N = b, \quad x \geq 0 \qquad (4.5)$$

and the equality constraint can be solved to obtain $x_B = B^{-1}b - B^{-1}N x_N$. It is convenient to write this

$$x_B = \hat{b} - \hat{N} x_N \qquad (4.6)$$

where $\hat{b} = B^{-1}b$ and $\hat{N} = B^{-1}N$. Thus, every solution (x_B, x_N) of $Ax = b$ has the form $(\hat{b} - \hat{N} x_N, x_N)$ for some x_N, and every feasible solution of (4.5) has this form for some $x_N \geq 0$. If $x_N = 0$, the resulting solution $(\hat{b}, 0)$ is obviously basic, because at most the m components of x_B are positive in the solution. In fact, it is convenient to assume that all m components of x_B are positive, a situation known as *primal nondegeneracy*. It is not hard to extend the argument below to deal with degeneracy, but this is somewhat tedious and will not be undertaken here.

Substituting (4.6) into the objective function of (4.5) allows cost to be expressed as a function of the nonbasic variables x_N:

$$c_B \hat{b} + (c_N - c_B \hat{N}) x_N$$

This can be written $c_B \hat{b} + r x_N$, where the components of the row vector $r = c_N - c_B \hat{N}$ are the reduced costs associated with the nonbasic variables x_N. The basic solution $(\hat{b}, 0)$ has cost $c_B \hat{b}$, which is optimal if all the reduced costs are nonnegative ($r \geq 0$).

It remains to show that if (4.5) has a finite optimal solution, then some basic solution is optimal. This can be shown by using the *simplex algorithm* to find an optimal basic solution. If the basic solution $(\hat{b}, 0)$ is optimal, then an optimal basic solution is already at hand. Otherwise increase a variable x_j with negative reduced cost until some component x_i of x_B hits zero. Since $x_B > 0$ (nondegeneracy), x_j can be increased at least a little without forcing any component of x_B to go negative.

To find which x_i hits zero first, one can use (4.6) to write x_B in terms of x_j only:

$$x_B = \hat{b} - \hat{N}_j x_j \tag{4.7}$$

where $\hat{N}_j$ is the column of $\hat{N}$ that corresponds to x_j. The component x_i of x_B that hits zero first is the one for which the ratio $\hat{b}_i / \hat{N}_{ij}$ is positive and largest (this is known as the *ratio test*). The ratio is positive for at least one i, because otherwise x_j could increase indefinitely and there would be no finite optimal solution.

Setting x_j equal to the largest ratio forces x_i to zero and yields a second basic solution that has a strictly lower objective function value. One can now construct a new basis B' by making x_j basic and x_i nonbasic. Remove from B the column that corresponds to x_i and add the column A_j of N that corresponds to x_j. The new basis B' is, in fact, a basis, since if it were singular, A_j would be a linear combination of the columns of B other than column i. This would imply that $\hat{N}_{ij}$ in (4.7) is zero rather than positive. The new basic solution can be checked for optimality by computing reduced costs as above, again assuming nondegeneracy. The process cannot continue indefinitely, because there are a finite number of possible bases using the columns of A, and a strictly decreasing objective function means that no basis can occur twice. The simplex algorithm therefore terminates with an optimal basic solution.

As an example, consider

$$\begin{aligned} & \min 2x_1 + 3y \\ & x_1 + 2y = -1 \\ & x_1 - 2y \leq 2 \\ & x_1 \geq 0 \end{aligned} \tag{4.8}$$

The problem can be written in the form (4.4) by replacing the unrestricted variable y with the difference $x_2 - x_3$ of nonnegative variables,

Linear Inequalities

and introducing a slack variable x_4 into the second constraint.

$$\begin{aligned} \min\ & 2x_1 + 3x_2 - 3x_3 \\ & x_1 + 2x_2 - 2x_3 = -1 \\ & x_1 - 2x_2 + 2x_3 + x_4 = 2 \\ & x_1, \ldots, x_4 \geq 0 \end{aligned} \quad (4.9)$$

Beginning arbitrarily with basic variables $x_B = (x_3, x_4)$, one obtains

$$B = \begin{bmatrix} -2 & 0 \\ 2 & 1 \end{bmatrix} \quad N = \begin{bmatrix} 1 & 2 \\ 1 & -2 \end{bmatrix} \quad B^{-1} = \begin{bmatrix} -\frac{1}{2} & 0 \\ 1 & 1 \end{bmatrix}$$
$$\hat{b} = \begin{bmatrix} \frac{1}{2} \\ 1 \end{bmatrix} \quad \hat{N} = \begin{bmatrix} -\frac{1}{2} & -1 \\ 2 & 0 \end{bmatrix} \quad r = [-\frac{1}{2}\ 0] \quad (4.10)$$

which corresponds to the feasible basic solution

$$(x_B, x_N) = (x_3, x_4, x_1, x_2) = (\tfrac{1}{2}, 1, 0, 0)$$

with cost $-\frac{3}{2}$. Since $r_1 < 0$, the cost can be reduced further by increasing x_1. The ratios $\hat{b}_i / \hat{N}_{i1}$ are $\frac{1}{2}/(-\frac{1}{2})$ and $1/2$ for $i = 3, 4$. Since $\frac{1}{2}$ is the largest (and only) positive ratio, x_1 is increased to $\frac{1}{2}$ and becomes basic, which causes x_4 to hit zero and become nonbasic. The new basic solution is $(x_{B'}, x_{N'}) = (x_1, x_3, x_2, x_4) = (\tfrac{1}{2}, \tfrac{3}{4}, 0, 0)$, with cost $-\frac{5}{4}$. It can be checked that the reduced costs of x_2, x_4 are 0 and $-\frac{1}{4}$, which means that this solution is not yet optimal. At this point, x_4 is brought into the basis, and the process continues until all reduced costs are nonnegative.

The simplex algorithm has been used to solve linear programming problems since its invention by George Dantzig more than half a century ago. It requires a starting basic solution $(\hat{b}, 0)$ with $\hat{b} \geq 0$. Such a solution can be obtained by using the simplex method to solve a *phase I* problem that minimizes the sum of constraint violations and has an obvious starting solution. *Phase II* then starts with $(\hat{b}, 0)$ and finds an optimal solution. The computational burden of the simplex method is modest because B^{-1} can be easily modified to obtain $(B')^{-1}$ (see Exercises). Recent implementations of the simplex method, as well as interior point methods, are remarkably effective. One rarely encounters a linear programming problem in practice that cannot be easily solved.

4.2.2 Relaxation Dual

A linear programming problem can be relaxed by replacing its constraints with a surrogate. The problem of finding a tightest surrogate relaxation is a relaxation dual that is equivalent to the inference dual defined in Section 3.3.4 for linear programming. The inference dual is itself equivalent to the classical linear programming dual when the primal problem is feasible (Theorem 3.7).

Consider a linear programming problem in inequality form:

$$\begin{aligned} \min \ & cx \\ & Ax \geq b, \quad x \geq 0 \end{aligned} \tag{4.11}$$

A surrogate relaxation of (4.11) replaces $Ax \geq b$ with a single surrogate inequality

$$\begin{aligned} \min \ & cx \\ & uAx \geq ub, \quad x \geq 0 \end{aligned} \tag{4.12}$$

where $u \geq 0$. If $\sigma(u)$ is the optimal value of (4.12), then the relaxation dual problem is to maximize $\sigma(u)$ subject to $u \geq 0$.

THEOREM 4.2 *The relaxation dual of a linear programming problem has the same optimal value as the inference dual. In particular, any finite optimal solution of one is an optimal solution of the other.*

Proof. Let (v^*, u^*) be a finite optimal solution of the inference dual

$$\begin{aligned} \max \ & v \\ & Ax \geq b \overset{Q}{\vdash} (cx \geq v) \\ & Q \in \mathcal{Q} \end{aligned} \tag{4.13}$$

where the proofs in $\mathcal{Q}$ correspond to nonnegative linear combinations and domination. Thus, by definition, (4.13) is equivalent to

$$\max \{v \mid uAx \geq ub \text{ dominates } cx \geq v \text{ for some } u \geq 0\} \tag{4.14}$$

Since v^* is finite, no surrogate of $Ax \geq b$ is infeasible, because an infeasible surrogate dominates all inequalities. Thus,

$$\sigma(u) = \max \{v \mid uAx \geq ub \text{ dominates } cx \geq v\} \tag{4.15}$$

Since u^* solves (4.14), $\sigma(u^*) \geq \sigma(u)$ for all $u \geq 0$, which means that u^* solves the relaxation dual. Also, (4.15) implies $\sigma(u^*) = v^*$, so that the relaxation dual has optimal value v^*.

It is similarly shown that any finite optimal solution of the relaxation dual is an optimal solution of the inference dual with the same optimal value. This is left as an exercise.

Now suppose that the inference dual is infeasible and therefore has optimal value $-\infty$. This means that for any $u \geq 0$, $uAx \geq ub$ dominates

Linear Inequalities 257

$cx \geq v$ for no v. Thus, no surrogate of $Ax \geq b$ is infeasible, and (4.15) holds. This means $\sigma(u) = -\infty$ for all $u \geq 0$, and the relaxation dual has optimal value $-\infty$.

Finally, suppose the inference dual is unbounded and therefore has optimal value ∞. This means that for any v, however large, some surrogate $uAx \geq ub$ dominates $cx \geq v$. Thus, $Ax \geq b, x \geq 0$ is infeasible, which means by Corollary 3.5 that some surrogate $uAx \geq ub$ is an infeasible inequality, and $\sigma(u) = \infty$. □

4.2.3 Exercises

1 Let the linear programming problem (4.4) be nondegenerate, and let $(B^{-1}b, 0)$ be an optimal basic solution. State a necessary and sufficient condition for the existence of another optimal solution. This condition is known as *dual degeneracy*. Hint: look at the reduced costs.

2 In problem (4.9), verify that $(x_1, x_3, x_2, x_4) = (\frac{1}{2}, \frac{1}{4}, 0, 0)$ is the basic solution when x_1, x_3 are basic. Also verify that x_2, x_4 have reduced costs 0 and $-\frac{1}{4}$ by inverting the new basis matrix B':

$$\begin{bmatrix} 2 & 1 \\ -2 & 1 \end{bmatrix}$$

3 In the solution of (4.9), the initial basis matrix B consists of the first two columns on the left below.

$$\begin{bmatrix} -2 & 0 & 1 & 0 \\ 2 & 1 & 0 & 1 \end{bmatrix}$$

By doing row operations to reduce the leftmost two columns to the identity matrix, one in effect multiplies the entire matrix by B^{-1}, and the right two columns are transformed to B^{-1}:

$$\begin{bmatrix} 1 & 0 & -\frac{1}{2} & 0 \\ 0 & 1 & 1 & 1 \end{bmatrix} \quad (4.16)$$

Let B' be the result of replacing column 2 of B with A_1 (the x_1 column of A) as in Exercise 2. To compute $(B')^{-1}$, replace column 2 of (4.16) with A_1

$$\begin{bmatrix} 1 & 1 & -\frac{1}{2} & 0 \\ 0 & 1 & 1 & 1 \end{bmatrix}$$

and do one row operation to restore the identity matrix on the left.

4 The calculation of Exercise 3 can be generalized. Show that if column i of B is replaced by A_j (column j of A) to obtain a new basis matrix B', then $(B')^{-1} = EB^{-1}$, where the *elementary matrix* E is the result of replacing column i of an $m \times m$ identity matrix with

$$\eta = \begin{bmatrix} -\hat{A}_{1j}/\hat{A}_{ij} \\ \vdots \\ -\hat{A}_{i-1,j}/\hat{A}_{ij} \\ 1/\hat{A}_{ij} \\ -\hat{A}_{i+1,j}/\hat{A}_{ij} \\ \vdots \\ -\hat{A}_{mj}/\hat{A}_{ij} \end{bmatrix}$$

where $\hat{A}_j = B^{-1}A_j$. Hint: assume without loss of generality that $i = 1$. Some linear programming solvers obtain the current basis inverse by saving the vectors $\eta_1, \ldots, \eta_k$ from previous iterations $1, \ldots, k$ and computing $E_k \ldots E_1 B^{-1}$, where B is the original basis matrix and $E_1, \ldots, E_k$ are the reconstructed elementary matrices. The basis matrix is periodically reinverted from scratch to avoid accumulated roundoff errors.

5 Show that the formula in the previous exercise correctly obtains the new B^{-1} in Exercise 3. Complete the simplex algorithm for problem (4.9) to obtain $(x_1, \ldots, x_4) = (0, 0, \frac{1}{2}, 1)$.

6 Use the simplex method to minimize $3x_1 + x_2$ subject to $x_1 + x_2 \geq 2$, $x_1 + 3x_2 \geq 3$, and $x_1, x_2 \geq 0$. Add surplus variables x_3, x_4 to convert the constraints to equations, and let them be basic in the initial basic solution.

7 Exhibit a linear programming problem that is unbounded and apply the simplex method to it.

8 If it is given that the optimal value of (4.9) is $-\frac{3}{2}$, show that $(u_1, u_2) = (\frac{3}{2}, 0)$ solves the inference dual of (4.9). Verify that $\theta(\frac{3}{2}, 0) = -\frac{3}{2}$ in the relaxation dual, which implies by weak duality that $u = (\frac{3}{2}, 0)$ solves the relaxation dual. Verify that $u = (\frac{3}{2}, 0)$ yields an objective value of $-\frac{3}{2}$ in the classical dual of (4.9) and therefore solves it.

9 Complete the proof of Theorem 4.2 by showing that any finite optimal solution of the relaxation dual of a linear programming problem is an optimal solution of the inference dual with the same optimal value.

4.3 Semicontinuous Piecewise Linear Functions

A natural generalization of linear inequality constraints is to inequalities or equations with a sum of piecewise linear functions on the left-hand side. Each function takes one variable as an argument.

Constraints of this sort are a highly versatile modeling device, partly because they can approximate separable nonlinear functions. They also represent an excellent application of integrated methods. Although piecewise linear functions can be given mixed integer models using *type II specially ordered sets* of variables, this requires additional variables. By replacing a piecewise linear function with an appropriate metaconstraint, a convex hull relaxation can be quickly generated without adding variables. Branching is also effective and results in substantial propagation. "Gaps" in the function (which result in a semicontinuity) are accommodated without effort, whereas they require still more auxiliary variables in a mixed integer model.

A piecewise linear inequality constraint has the form

$$\sum_j g_j(x_j) \geq b \qquad (4.17)$$

where each g_j is a continuous or semicontinuous piecewise linear function. In general a semicontinuous piecewise linear function $g(x)$ can be written

$$g(x) = \frac{U_k - x}{U_k - L_k} c_k + \frac{x - L_k}{U_k - L_k} d_k \text{ for } x \in [L_k, U_k], \quad k = 1, \ldots, m$$

where intervals $[L_k, U_k]$ intersect in at most a point. A sample function is shown in Figure 4.1.

A separable nonlinear function $\sum_j h_j(x_j)$ can be approximated by a piecewise linear function to a high degree of accuracy by replacing each nonlinear function $h_j(x)$ with a piecewise linear function having sufficiently many linear pieces. The overhead of doing so is small, because a convex hull relaxation can be generated quickly for each function without requiring additional variables.

4.3.1 Convex Hull Relaxation

Figure 4.1 illustrates how a semicontinuous piecewise linear function $g(x)$ can be approximated with a convex hull relaxation. Note that the domain of x can be used to reduce the interval $[L_3, U_3]$ to $[L_3, U_x]$. Techniques from computational geometry use a divide-and-conquer approach to compute the convex hull in time proportional to $m \log m$.

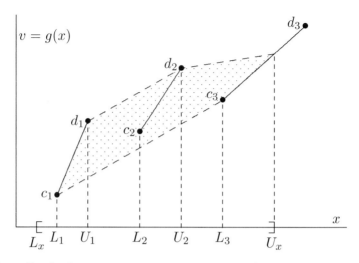

Figure 4.1. Graph of a semicontinuous piecewise linear function and its convex hull (shading). The interval $[L_x, U_x]$ is the current domain of x.

In general, a constraint (4.17) is relaxed by replacing it with $\sum_j v_j \geq b$, and each v_j is constrained by a system of linear inequalities in v_j, x_j that describes the convex hull of the graph of $v_j = g_j(x_j)$.

Branching on x can tighten the domains and relaxations considerably. Branching is called for when the solution value $(\bar{v}, \bar{x})$ of (v, x) in the current relaxation of the problem does not satisfy $g(\bar{x}) = \bar{v}$. If $\bar{x}$ is between U_1 and U_2, for example, the domain of x is split into the intervals $[L_x, U_1]$ and $[L_2, U_x]$, resulting in the convex hull relaxations of Figure 4.2. If $\bar{x}$ lies in the interval $[L_2, U_2]$, one can split the domain of x into three intervals, $[L_x, U_1]$, $[L_2, U_2]$, and $[L_3, U_x]$. In this small example, the resulting convex hull relaxations become exact.

4.3.2 Exercises

1 Let $h(x) = \sum_{j=1}^n h_j(x_j)$ be a separable function, where each x_j has domain D_{x_j}. If

$$S_j = \{(v_j, x_j) \mid v_j = h_j(x_j),\ x_j \in D_{x_j}\}$$

for each j, let $\mathcal{C}_j$ be a constraint set that describes the convex hull of S_j. Show that the feasible set of

$$\mathcal{C} = \left\{ v = \sum_{j=1}^n v_j \right\} \cup \bigcup_{j=1}^n \mathcal{C}_j$$

0-1 Linear Inequalities

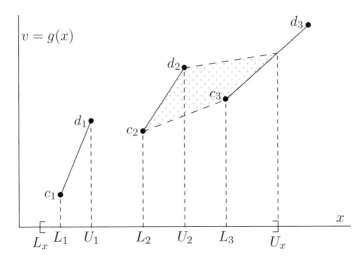

Figure 4.2. Convex hull relaxations after branching on x with a value between U_1 and L_2.

projected onto (v, x) is the convex hull of

$$S = \{(v, x) \mid v = h(x),\ x \in D\}$$

where $D = D_{x_1} \times \cdots \times D_{x_n}$. Thus one can model the convex hull of h by modeling the convex hull of each h_j.

2 Let each $h_j(x_j)$ in Exercise 1 be a fixed-charge function

$$h_j(x_j) = \begin{cases} 0 & \text{if } x_j = 0 \\ f_j + c_j x_j & \text{if } x_j > 0 \end{cases}$$

where each $x_j \in [0, U_j]$. The convex hull of S_j is described by

$$\frac{f_j + c_j U_j}{U_j} x_j \le v_j \le f_j + c_j x_j,\ \ x_j \ge 0$$

and is illustrated by the light shaded area in Figure 4.12(ii) of Section 4.9.2. Describe the convex hull of S using only the variables v and $x = (x_1, \ldots, x_n)$.

4.4 0-1 Linear Inequalities

A *0-1 linear inequality* is a linear inequality with 0-1 variables. It has the form $ax \ge a_0$, where $x = (x_1, \ldots, x_n)$ and $x \in \{0, 1\}^n$. General integer linear inequalities, in which the value of each x_j must be a nonnegative integer, are discussed in Section 4.5. It is assumed throughout this section that the coefficients a_j are integral. This entails negligible loss of generality, because an inequality with rational coefficients can be multiplied by a suitable positive scalar to covert the coefficients to integers.

A continuous relaxation of a system 0-1 linear inequalities

$$Ax \geq b, \quad x \in \{0,1\}^n \tag{4.18}$$

can be obtained by replacing $x_j \in \{0,1\}$ with $0 \leq x_j \leq 1$ for each j. A widely-used technique for strengthening this continuous relaxation is to generate *cutting planes*. These are inequalities that are implied by (4.18) but are *not* implied by the continuous system $Ax \geq b$. That is, if $cx \geq c_0$ is a cutting plane, then any $x \in \{0,1\}^n$ that satisfies $Ax \geq b$ also satisfies $cx \geq c_0$, but this is not true of any $x \in [0,1]^n$. Cutting planes are also known as *valid inequalities* or *valid cuts*.

A 0-1 system (4.18) can therefore be relaxed by combining the continuous linear system $Ax \geq b$ with cutting planes and bounds $0 \leq x_j \leq 1$ for each x_j. Since the cutting planes are not redundant of the *continuous* system $Ax \geq b$, they strengthen the continuous relaxation by *cutting off* points that would otherwise satisfy the relaxation.

Cutting planes are often used to help solve *0-1 linear programming problems*, in which all the constraints are 0-1 linear inequalities. Such problems can be written

$$\begin{aligned} \min \ & cx \\ & Ax \geq b, \quad x \in \{0,1\}^n \end{aligned} \tag{4.19}$$

where $x = (x_1, \ldots, x_n)$. Cutting planes can also be generated for systems with general integer and continuous variables, which are discussed in Section 4.5.

Cutting planes can be generated by a general purpose method, or by a method that is specialized to inequality sets with a particular type of structure. Some general purpose methods are presented in this section, including the Chvátal-Gomory procedure, knapsack cut generation, and lifting. The next section presents Gomory cuts, which are general purpose cuts for general integer as well as 0-1 inequalities, and mixed integer rounding cuts for inequalities with both integral and continuous variables. Specialized cutting planes will be discussed in later sections in connection with specific types of constraints.

4.4.1 Chvátal-Gomory Cuts

A relatively simple algorithm, known as the Chvátal-Gomory procedure, can in principle generate any valid cut for a system of linear inequalities

0-1 Linear Inequalities

with integer-valued variables. Although the procedure is rarely used in practice, it provides a tool for identifying and studying special-purpose cuts. It also reveals an interesting connection between logic and cutting plane theory, because the Chvátal-Gomory procedure for 0-1 linear inequalities is related to the resolution method of inference.

The Chvátal-Gomory procedure can be applied to any system of integer linear inequalities, but it is convenient to begin with a 0-1 system (4.18). $Ax \geq b$ is understood to include bounds $0 \leq x \leq e$, where e is a vector of ones. This imposes no restriction but is convenient for generating cuts.

The procedure is based on the idea of an *integer rounding cut*. Given an inequality $ax \geq a_0$ where x is a nonnegative integer, an integer rounding cut is obtained by rounding up any nonintegers among the coefficients and right-hand side. The cut therefore has the form $\lceil a \rceil x \geq \lceil a_0 \rceil$ and is clearly valid for $ax \geq a_0$. The coefficients a on the left-hand side can be rounded up, without invalidating the inequality, because $x \geq 0$. Now that the left-hand side is integral, the right-hand side a_0 can be rounded up as well.

Each step of the Chvátal-Gomory procedure generates an integer rounding cut for a surrogate of $Ax \geq b$ and adds this cut to the system $Ax \geq b$. Recall from Section 3.3.1 that a surrogate of $Ax \geq b$ is a nonnegative linear combination $uAx \geq ub$. The generated cut therefore has the form $\lceil uA \rceil x \geq \lceil ub \rceil$ where $u \geq 0$. Any inequality generated by a finite number of such steps is a *Chvátal-Gomory cut* for (4.18). A cut that can be generated in k steps, but not $k-1$ steps, is a cut of *rank k*. It is a fundamental result of cutting plane theory that every valid cut for (4.18) is a Chvátal-Gomory cut.

As an example, consider the 0-1 system:

$$x_1 + 3x_2 \geq 2$$
$$6x_1 - 3x_2 \geq -1$$
$$0 \leq x_j \leq 1, \quad j = 1, 2$$
$$x \in \{0, 1\}^2$$

It has the valid rank 2 cut $2x_1 + 3x_2 \geq 5$. There are many Chvátal-Gomory derivations of this cut, such as the following two-step derivation. The cuts $x_1 \geq 1$ and $x_2 \geq 1$ are generated first:

$$
\begin{array}{ll}
x_1 + 3x_2 \geq 2 \quad (\tfrac{1}{7}) & \quad x_1 + 3x_2 \geq 2 \quad (\tfrac{1}{3}) \\
6x_1 - 3x_2 \geq -1 \quad (\tfrac{1}{7}) & \quad -x_1 \quad\quad\quad \geq -1 \quad (\tfrac{1}{3}) \\
\hline
x_1 \quad\quad \geq \tfrac{1}{7} \Rightarrow x_1 \geq 1 & \quad x_2 \geq \tfrac{1}{3} \Rightarrow x_2 \geq 1
\end{array}
$$

The nonzero multipliers u_i are shown in parentheses, and the surrogate $uAx \geq ub$ appears below the line along with the cut $\lceil uA \rceil x \geq \lceil ub \rceil$. The constraint $-x_1 \geq -1$ reflects the bound $x_1 \leq 1$. A linear combination of these two cuts yields $2x_1 + 3x_2 \geq 5$ in the second step.

A longer derivation of $2x_1 + 3x_2 \geq 5$, outlined in Figure 4.3, points the way to a proof of the Chvátal-Gomory theorem. The derivation starts with the weaker inequality $2x_1 + 3x_2 \geq 0$, which can be derived from bounds $x_j \geq 0$. From this it obtains, in a manner described in the proof below, the four Chvátal-Gomory cuts at the leaf nodes of the enumeration tree. These four combine in pairs to obtain the cuts at the middle level. For instance, the first two combine as follows:

$$
\begin{array}{ll}
(2x_1 + 3x_2) + x_1 + & x_2 \geq 1 \quad (\tfrac{1}{2}) \\
(2x_1 + 3x_2) + x_1 + (1 - x_2) \geq 1 \quad (\tfrac{1}{2}) \\
\hline
(2x_1 + 3x_2) + x_1 & \geq \tfrac{1}{2} \Rightarrow (2x_1 + 3x_2) + x_1 \geq 1
\end{array}
$$

This inference is closely parallel to a resolution step (Section 3.5.2). The "logical" parts of the two inequalities are

$$
\begin{aligned}
x_1 + x_2 &\geq 1 \\
x_1 + (1 - x_2) &\geq 1
\end{aligned}
$$

These correspond to logical clauses $x_1 \vee x_2$ and $x_1 \vee \neg x_2$, which have the resolvent x_1 (i.e., $x_1 \geq 1$). So the Chvátal-Gomory operations in Figure 4.3 are parallel to a resolution proof of the empty clause. The conclusion, however, is not the empty clause, but $2x_1 + 3x_2 \geq 1$. This becomes the premise for another round of cuts, which yields $2x_1 + 3x_2 \geq 2$, and so forth, until $2x_1 + 3x_2 \geq 5$ is obtained.

THEOREM 4.3 *Every valid cut for a 0-1 linear system is a Chvátal-Gomory cut.*

Proof. Let $cx \geq c_0$ be a valid cut for (4.18). Let $c_0 - \Delta$ be the smallest possible value of cx in the unit box defined by $0 \leq x \leq e$. Thus, the Chvátal-Gomory cut

$$cx \geq c_0 - \Delta \tag{4.20}$$

is obtained from a linear combination of bounds (namely, $x_j \geq 0$ with weight c_j when $c_j \geq 0$ and $-x_j \geq -1$ with weight $-c_j$ when $c_j < 0$).

There is a sufficiently large number M such that the following is also a Chvátal-Gomory cut

$$cx + M \sum_{j \in P} x_j + M \sum_{j \in N} (1 - x_j) \geq c_0 \tag{4.21}$$

0-1 Linear Inequalities

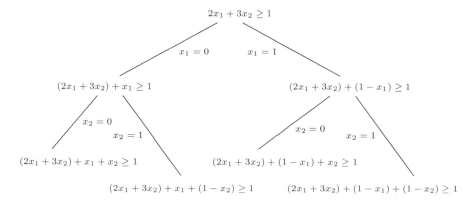

Figure 4.3. Illustration of the proof of the Chvátal-Gomory theorem for 0-1 inequalities.

for any partition P, N of $\{1, \ldots, n\}$. This is shown as follows. First, (4.21) is valid for the polyhedron defined by $Ax \geq b$ and $0 \leq x \leq e$. To see this, it suffices to observe that (4.21) is satisfied at all vertices of this polyhedron. It is satisfied at 0-1 vertices because $cx \geq c_0$ is valid for the 0-1 system. If $\alpha > 0$ is the smallest value of $\sum_{j \in P} x_j + \sum_{j \in N}(1 - x_j)$ over all nonintegral vertices and all partitions, then (4.21) is satisfied at any noninteger vertex when $M = \Delta/\alpha$, due to (4.20). Now, since (4.21) is valid for the polyhedron, by Corollary 3.6 it is dominated by a surrogate of $Ax \geq b$ and $x \leq e$ and is therefore a Chvátal-Gomory cut.

To complete the proof, it suffices to show that if

$$cx \geq c_0 - \delta - 1 \tag{4.22}$$

is a Chvátal-Gomory cut and $cx \geq c_0 - \delta$ is valid, then $cx \geq c_0 - \delta$ is a Chvátal-Gomory cut.

First, for any partition P, N of $\{1, \ldots, n\}$, the inequality

$$cx + \sum_{j \in P} x_j + \sum_{j \in N}(1 - x_j) \geq c_0 - \delta \tag{4.23}$$

is a Chvátal-Gomory cut. This is seen by combining (4.21) and (4.22) with weights $1/M$ and $(M-1)/M$, respectively.

Next, suppose the following are Chvátal-Gomory cuts

$$cx + \sum_{j \in P' \cup \{i\}} x_j + \sum_{j \in N'}(1 - x_j) \geq c_0 - \delta$$

$$cx + \sum_{j \in P'} x_j + \sum_{j \in N' \cup \{i\}}(1 - x_j) \geq c_0 - \delta$$

where P', N' and $\{i\}$ partition a subset of $\{1,\ldots,n\}$. Combining these cuts with a weight of $1/2$ each yields the "resolvent"

$$cx + \sum_{j \in P'} x_j + \sum_{j \in N'} (1 - x_j) \geq c_0 - \delta \tag{4.24}$$

which is therefore a Chvátal-Gomory cut. By applying this operation repeatedly, starting with (4.23), the Chvátal-Gomory cut $cx \geq c_0 - \delta$ is obtained. □

4.4.2 0-1 Knapsack Cuts

It is frequently useful to derive cutting planes for individual 0-1 knapsack inequalities as well as systems of 0-1 inequalities. These cutting planes are known as *knapsack cuts* or *cover inequalities*. Knapsack cuts can often be strengthened by a *lifting* process that adds variables to the cut. Two lifting procedures are described in the following two sections.

Recall that 0-1 knapsack packing inequalities have the form $ax \leq a_0$ and knapsack covering inequalities have the form $ax \geq a_0$, where $a \geq 0$ and each $x_j \in \{0, 1\}$. The nonnegativity restriction on a incurs minimal loss of generality, because one can always obtain $a \geq 0$ by replacing each x_j that has a negative coefficient with $1 - x_j$. Cuts are developed here for packing inequalities $ax \leq a_0$. Covering inequalities can be accommodated by obtaining cuts for $-ax \leq -a_0$.

Define a *cover* for $ax \leq a_0$ to be an index set $J \in \{1,\ldots,n\}$ for which $\sum_{j \in J} a_j > a_0$. A cover is *minimal* if no proper subset is a cover. If J is a cover, the *cover inequality*

$$\sum_{j \in J} x_j \leq |J| - 1 \tag{4.25}$$

is obviously valid for $ax \leq a_0$. Only minimal covers need be considered, because nonminimal cover inequalities are redundant of minimal ones.

For example, $J = \{1, 2, 3, 4\}$ is a minimal cover for the inequality

$$6x_1 + 5x_2 + 5x_3 + 5x_4 + 8x_5 + 3x_6 \leq 17 \tag{4.26}$$

and gives rise to the cover inequality

$$x_1 + x_2 + x_3 + x_4 \leq 3 \tag{4.27}$$

4.4.3 Sequential Lifting

A cover inequality can often be strengthened by adding variables to the left-hand side; that is, by *lifting* the inequality into a higher dimensional

0-1 Linear Inequalities

space. *Sequential lifting*, in which terms are added one at a time, is presented first. The resulting cut depends on the order in which terms are added. There are also techniques for adding several terms simultaneously, and one of these is described in the next section.

Given a cover inequality (4.25), the first step of sequential lifting is to add a term $\pi_k x_k$ to the left-hand side to obtain

$$\sum_{j \in J} x_j + \pi_k x_k \leq |J| - 1 \qquad (4.28)$$

where π_k is the largest coefficient for which (4.28) is still valid. Thus, one can set

$$\pi_k = |J| - 1 - \max_{\substack{x_j \in \{0,1\} \\ \text{for } j \in J}} \left\{ \sum_{j \in J} x_j \;\middle|\; \sum_{j \in J} a_j x_j \leq a_0 - a_k \right\} \qquad (4.29)$$

For example, (4.27) can be lifted to

$$x_1 + x_2 + x_3 + x_4 + 2x_5 \leq 3 \qquad (4.30)$$

because in this case

$$\pi_5 = 3 - \max\{x_1 + x_2 + x_3 + x_4 \mid 6x_1 + 5x_1 + 5x_2 + 5x_4 \leq 17 - 8\} = 2$$

At this point, (4.28) can be lifted further by adding another new term, and so forth. At any stage in this process, the current inequality is

$$\sum_{j \in J} x_j + \sum_{j \in J'} \pi_j x_j \leq |J| - 1$$

The lifting coefficient for the next x_k, where $k \notin L = J \cup J'$, is

$$\pi_k = |J| - 1 - \max_{\substack{x_j \in \{0,1\} \\ \text{for } j \in L}} \left\{ \sum_{j \in J} x_j + \sum_{j \in J'} \pi_j x_j \;\middle|\; \sum_{j \in L} a_j x_j \leq a_0 - a_k \right\} \qquad (4.31)$$

For example, if one attempts to lift (4.30) further by adding x_6, the inequality is unchanged because

$$\pi_6 = 3 - \max\left\{ \sum_{j=1}^{4} x_j + 2x_5 \;\middle|\; 6x_1 + 5x_2 + 5x_3 + 5x_4 + 8x_5 \leq 14 \right\} = 0$$

The order of lifting can affect the outcome; if x_6 is added before x_5, the resulting cut is $x_1 + x_2 + x_3 + x_4 + x_5 + x_6 \leq 3$.

The computation of lifting coefficients π_k by (4.29) in effect requires solving a 0-1 programming problem and can be time consuming. Fortunately, the coefficients can be computed recursively. An algorithm for computing the initial lifting coefficient π_k will be given first, and then a different recursion for computing the remaining coefficients.

This approach actually requires one to compute more than π_k; one must compute the function

$$\pi_J(u) = |J| - 1 - g^*(u)$$

for $u \geq 0$, where

$$g^*(u) = \max_{\substack{x_j \in \{0,1\} \\ \text{for } j \in J}} \left\{ \sum_{j \in J} x_j \ \bigg| \ \sum_{j \in J} a_j x_j \leq a_0 - u \right\} \quad (4.32)$$

Then, in particular, $\pi_k = \pi_J(a_k)$.

The function $\pi_J(u)$ can be computed directly by solving a dynamic programming problem for each value of u, but it is more efficient to solve a dual problem that exchanges the constraint and objective function in (4.32), and then recover $\pi_J(u)$. The dual problem is to compute the following for $t = 0, \ldots, |J|$:

$$h^*(t) = \min \left\{ \sum_{j \in J} a_j x_j \ \bigg| \ \sum_{j \in J} x_j \geq t \right\}$$

Then given any integer $t \in \{0, \ldots, |J|-1\}$, one can deduce that $g^*(u) = t$ for all u satisfying

$$a_0 - h^*(t+1) < u \leq a_0 - h^*(t)$$

Also, $g^*(u) = |J|$ for $u \geq h^*(|J|)$. The function $h^*(t)$ can be computed using the recursion

$$h_k(t) = \min\{a_k + h_{k-1}(t-1), h_{k-1}(t)\}, \quad t = 1, \ldots, |J| \quad (4.33)$$

with the boundary conditions $h_k(0) = 0$ for all k and $h_0(t) = \infty$ for $t > 0$. Now $h^*(t) = h_{|J|}(t)$.

In the example, the first three stages of the recursion (4.33) become

$$h_1(1) = \min\{a_1 + h_0(0), h_0(1)\} = 6$$
$$h_2(1) = \min\{a_2 + h_1(0), h_1(1)\} = 5$$
$$h_2(2) = \min\{a_2 + h_1(1), h_1(2)\} = 11$$
$$h_3(1) = 5, \quad h_3(2) = 10, \quad h_3(3) = 16$$

0-1 Linear Inequalities 269

with $h_k(t) = \infty$ for $t > k$. The fourth and last stage yields

$$\begin{aligned} h^*(0) &= h_4(0) = 0 \\ h^*(1) &= h_4(1) = 5 \\ h^*(2) &= h_4(2) = 10 \\ h^*(3) &= h_4(3) = 15 \\ h^*(4) &= h_4(4) = 21 \end{aligned}$$

From this one can read off the function $g^*(u)$

$$g^*(u) = \begin{cases} 3 & \text{for } 0 \leq u \leq 2 \\ 2 & \text{for } 2 < u \leq 7 \\ 1 & \text{for } 7 < u \leq 12 \\ 0 & \text{for } 12 < u \leq 17 \\ -\infty & \text{for } 17 < u \end{cases}$$

and $\pi_J(u) = 3 - g^*(u)$. Thus, the lifting coefficient for x_5 is $\pi_5 = \pi_J(a_5) = \pi_J(8) = 2$.

The remaining sequential lifting coefficients are determined as follows. If coefficients π_j have been derived for $j \in J'$, then the next coefficient π_k is $\pi_L(a_k)$, where $L = J \cup J'$ and

$$\pi_L(u) = |J| - 1 - \max_{\substack{x_j \in \{0,1\} \\ \text{for } j \in L}} \left\{ \sum_{j \in J} x_j + \sum_{j \in J'} \pi_j x_j \,\middle|\, \sum_{j \in L} a_j x_j \leq a_0 - u \right\}$$

The function $\pi_L(u)$ can be computed recursively when $\pi_J(u)$ is known:

$$\pi_{L \cup \{k\}}(u) = \min\{\pi_L(u), \pi_L(u + a_k) - \pi_L(a_k)\} \quad (4.34)$$

where the first option inside the min corresponds to setting $x_k = 0$, and the second to $x_k = 1$.

In the example, if π_6 is to be computed after π_5, then

$$\pi_6 = \pi_{J \cup \{5\}}(a_6) = \min\{\pi_J(3), \pi_J(6) - \pi_J(3)\} = \min\{1, 1 - 1\} = 0$$

as stated earlier.

The computation of optimal lifting coefficients π_k can be practical when there are not too many variables involved or the problem has special structure. In other cases, one may wish to use a heuristic algorithm to obtain a π_k that may be smaller than the optimal coefficient.

4.4.4 Sequence-Independent Lifting

It is possible to lift all the missing variables in the cover inequality (4.25) simultaneously. This can be done by constructing a function that is analogous to $\pi_J(u)$ in the previous section, but that provides valid lifting coefficients for all the missing variables when they are added as a group. The resulting cut may not be as strong as one obtained by optimal sequential lifting, but it requires much less computation. Sequence-independent lifting has therefore become a standard feature of integer programming solvers.

Recall that a function ρ is superadditive if $\rho(u_1 + u_2) \geq \rho(u_1) + \rho(u_2)$ for all real numbers u_1, u_2. The lifting procedure is based on the following fact.

THEOREM 4.4 *Suppose that $\rho(u) \leq \pi_J(u)$ for all real numbers u, $\rho(u)$ is superadditive, and J is a cover for the 0-1 knapsack inequality $ax \leq a_0$. Then, the following lifted inequality is valid for $ax \leq a_0$:*

$$\sum_{j \in J} x_j + \sum_{j \notin J} \rho(a_j) x_j \leq |J| - 1 \tag{4.35}$$

Proof. It suffices to show that any 0-1 vector $\bar{x}$ satisfying $ax \leq a_0$ also satisfies (4.35). Let $J' = \{j \notin J \mid \bar{x}_j = 1\}$. Then

$$\sum_{j \in J} \bar{x}_j + \sum_{j \notin J} \rho(a_j) \bar{x}_j = \sum_{j \in J} \bar{x}_j + \sum_{j \in J'} \rho(a_j)$$

$$\leq \sum_{j \in J} \bar{x}_j + \rho\left(\sum_{j \in J'} a_j\right) \leq \sum_{j \in J} \bar{x}_j + \pi_J\left(\sum_{j \in J'} a_j\right)$$

where the first inequality is due to the superadditivity of ρ and the second is due to the fact that π_J bounds ρ. It therefore suffices to show that

$$\sum_{j \in J} \bar{x}_j + \pi_J\left(\sum_{j \in J'} a_j\right) \bar{y} \leq |J| - 1 \tag{4.36}$$

with $\bar{y} = 1$. Let $J = \{1, \ldots, p\}$, and note that $(\bar{x}_1, \ldots, \bar{x}_p, \bar{y})$ satisfies

$$\sum_{j \in J} a_j x_j + \left(\sum_{j \in J'} a_j\right) y \leq a_0 \tag{4.37}$$

Since

$$\sum_{j \in J} x_j + \pi_J\left(\sum_{j \in J'} a_j\right) y \leq |J| - 1 \tag{4.38}$$

is a valid lifted cut for (4.37), $(\bar{x}_1, \ldots, \bar{x}_p, \bar{y})$ satisfies (4.38). Therefore, (4.36) holds. $\square$

0-1 Linear Inequalities

A function ρ that satisfies the conditions of Theorem 4.4 can be obtained as follows. Let $\Delta = \sum_{j \in J} a_j - a_0$ and $J = \{1, \ldots, p\}$ with $a_1 \geq \cdots \geq a_p$. Let $A_j = \sum_{k=1}^{j} a_k$ with $A_0 = 0$. Then define

$$\rho(u) = \begin{cases} j & \text{if } A_j \leq u \leq A_{j+1} - \Delta \text{ and } j \in \{0, \ldots, p-1\} \\ j + (u - A_j)/\Delta & \text{if } A_j - \Delta \leq u < A_j \text{ and } j \in \{1, \ldots, p-1\} \\ p + (u - A_p)/\Delta & \text{if } A_p - \Delta \leq u \end{cases} \tag{4.39}$$

It can be shown that $\rho(u)$ is superadditive and bounded above by $\pi_J(u)$. Note that it is easy to compute the lifting coefficients $\rho(a_j)$ using (4.39).

The function $\rho(u)$ for example (4.26) appears in Figure 4.4. Since $\rho(a_5) = \rho(8) = \frac{5}{4}$ and $\rho(a_6) = \rho(3) = \frac{1}{4}$, the result of lifting the cover inequality (4.27) is

$$x_1 + x_2 + x_3 + x_4 + \tfrac{5}{4} x_5 + \tfrac{1}{4} x_6 \leq 3$$

4.4.5 Set Packing Inequalities

The cover inequalities described in the previous three sections are derived from individual knapsack constraints. In many cases, stronger inequalities can be obtained by considering several knapsack constraints simultaneously. One way to do this is to deduce valid inequalities for the collection of cover inequalities derived from several knapsack constraints. Mixed integer solvers typically focus on cover inequalities with a right-hand side of one. A collection of these inequalities define a *set packing* problem, for which several classes of valid inequalities have been studied.

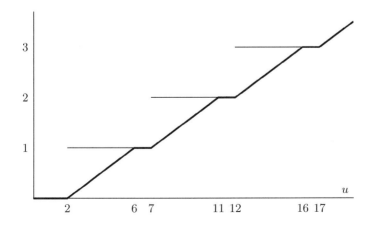

Figure 4.4. Graph of $\pi_J(u)$ (horizontal line segments) and $\rho(u)$ (heavy line) for a knapsack lifting problem.

A set packing problem asks, for a given collection of sets, whether one can choose k sets that are pairwise disjoint. When written in 0-1 form, the problem is to find a feasible solution for

$$Ax \leq e, \quad x_j \in \{0,1\}, \; j = 1, \ldots, n \tag{4.40}$$

and $\sum_j x_j \geq k$, where A is a 0-1 matrix and e is a tuple of ones. The columns of A correspond to sets and the rows to elements of the sets. Set j contains element i if $A_{ij} = 1$, and set j is selected if $x_j = 1$.

Two families of valid inequalities can be derived right away. They are most easily described by defining the *intersection graph* G that corresponds to (4.40). G has a vertex for each variable x_j and an undirected edge (x_j, x_k) whenever $A_{ij} = A_{ik} = 1$ for some row i. An *odd cycle* of G is a cycle with an odd number of edges.

THEOREM 4.5 *Let G be the intersection graph for the set packing problem (4.40). If C is a subset of vertices of G defining an odd cycle, then the* odd cycle inequality

$$\sum_{j \in C} x_j \leq \frac{|C| - 1}{2} \tag{4.41}$$

is valid for (4.40).

It can also be shown that if C defines an odd hole, then (4.41) is facet defining. An odd cycle is an *odd hole* if it has no chords (i.e., no edges other than the edges in the cycle that connect two vertices of the cycle). An inequality is *facet defining* for (4.40) if it describes a facet, or $(n-1)$-dimensional face, of the convex hull of the feasible set. Facet-defining inequalities are desirable as valid inequalities because all valid inequalities are dominated by surrogates of the facet-defining inequalities.

Consider, for example, the small set packing problem

$$\begin{aligned} x_1 + x_2 & \leq 1 \\ x_2 + x_3 & \leq 1 \\ x_3 + x_4 & \leq 1 \\ x_4 + x_5 & \leq 1 \\ x_1 + x_5 & \leq 1 \end{aligned}$$

The intersection graph is a pentagon on the vertices $x_1, \ldots, x_5$. Since this is an odd hole, the valid inequality $x_1 + x_2 + x_3 + x_4 + x_5 \leq 2$ defines a facet of the convex hull of the feasible set.

Clique inequalities are also widely used in practical solvers. A subset C of vertices of a graph G induce a *clique* if every two vertices of C are connected by an edge in G.

THEOREM 4.6 *Let G be the intersection graph for the set packing problem (4.40). If a subset C of vertices of G induce a clique of G, then the clique inequality*

$$\sum_{j \in C} x_j \leq 1 \qquad (4.42)$$

is valid for (4.40).

It can also be shown that (4.42) is facet defining if C defines a maximal clique (i.e., no proper superset of C induces a clique).

For example, the intersection graph of the set packing problem

$$\begin{aligned} x_1 + x_2 + x_3 & \leq 1 \\ x_1 \phantom{{}+x_2+x_3} + x_4 & \leq 1 \\ \phantom{x_1+{}} x_2 \phantom{{}+x_3} + x_4 & \leq 1 \\ \phantom{x_1+x_2+{}} x_3 + x_4 & \leq 1 \end{aligned}$$

is itself a clique. The inequality $x_1 + x_2 + x_3 + x_4 \leq 1$ is therefore valid and facet defining.

4.4.6 Exercises

1 Use the procedure in the proof of Theorem 4.3 to show that the inequality $(2x_1 + 3x_2) + x_1 + x_2 \geq 1$ at the left-most leaf node in Figure 4.3 is a Chvátal-Gomory cut. In particular, what are Δ, α, and M in this case?

2 Identify a minimal cover for (4.26) other than $\{1, 2, 3, 4\}$, and write the corresponding cover inequality.

3 Show that cover inequalities corresponding to nonminimal covers are redundant of those corresponding to minimal covers.

4 Suppose that (4.26) is part of a larger problem whose continuous relaxation has the solution $\bar{x} = (0, 0, \frac{4}{5}, \frac{4}{5}, \frac{7}{8}, \frac{2}{3})$. Find a minimal cover of (4.26) for which $\bar{x}$ violates the corresponding cover inequality. Such an inequality is a *separating cut* that cuts off the solution $\bar{x}$.

5 Suppose that the 0-1 knapsack packing inequality $ax \leq a_0$ is part of a larger problem whose continuous relaxation has the solution $\bar{x}$. Show that a cover inequality $\sum_{j \in J} x_j \leq |J| - 1$ cuts off the solution

$\bar{x}$ if and only if

$$\sum_{j=1}^{n}(1-\bar{x}_j)y_j > 1, \quad \sum_{j\in J} a_j y_j \geq a_0 + 1$$

where $y_j = 1$ when $j \in J$ and $y_j = 0$ otherwise. Describe a 0-1 knapsack problem one can solve to find the cover inequality that is most violated by $\bar{x}$. This is the separation problem for $ax \leq a_0$.

6 The 0-1 knapsack inequality $5x_1 + 4x_2 + 4x_3 + 4x_4 + 9x_5 + 5x_6 \leq 16$ has the minimal cover inequality $x_1 + x_2 + x_3 + x_4 \leq 3$, among others. Use the lifting formula (4.29) to find a lifting coefficient for x_5. Then compute a lifting coefficient for x_6 using (4.31).

7 Compute the lifting coefficient for x_5 in Exercise 6 by recursively computing $h^*(t)$. Then, compute the lifting coefficient for x_6 using the recursive formula (4.34).

8 Show that the recursive formula (4.34) is correct.

9 Plot $\rho(u)$ against u for the cover inequality in Exercise 6 using the formula (4.39). What are the sequence-independent lifting coefficients? Note that the resulting cut is, in this case, the same as the one obtained from sequential lifting.

10 Show that $\rho(u)$ as defined by (4.39) is superadditive and bounded above by $\pi_J(u)$.

11 Derive all valid odd cycle and clique inequalities from the 0-1 system $Ax \leq e$ that are not already in the system, where

$$A = \begin{bmatrix} 1 & 1 & 0 & 0 & 0 \\ 1 & 0 & 1 & 0 & 0 \\ 1 & 1 & 0 & 1 & 0 \\ 0 & 1 & 1 & 0 & 0 \\ 1 & 0 & 1 & 1 & 0 \\ 1 & 0 & 0 & 1 & 1 \end{bmatrix}$$

12 Prove Theorem 4.5.

13 Prove Theorem 4.6.

14 Prove that if C defines a maximal clique in Theorem 4.6, then (4.42) is facet defining. Hints: it suffices to exhibit n affinely independent feasible points that satisfy $\sum_{j\in C} x_j \leq 1$ as an equation. Let e^j be the jth unit vector (i.e., a vector of zeros except for a 1 in the jth place)

and consider the points e^j for $j \in C$, as well as the points $e^j + e^{i_j}$ for $j \notin C$. Here, $i_j \in C$ is selected so that no row of A contains a 1 in columns j and i_j (show that i_j exists).

15 Prove that an odd cycle inequality in Theorem 4.5 that corresponds to an odd hole is facet-defining.

4.5 Integer Linear Inequalities

Perhaps the most popular general purpose cutting planes are Gomory cuts and mixed integer rounding cuts. Gomory cuts can be generated equally well for 0-1 and general integer linear inequalities. Mixed integer rounding cuts are designed for mixed integer/linear systems, in which some of the variables are continuous, and others are required to take 0-1 or general integer values.

The Chvátal-Gomory theorem, proved in the previous section for 0-1 linear inequalities, can also be extended to general integers. In fact, Gomory cuts are actually a special case of rank 1 Chvátal-Gomory cuts, distinguished by the fact that they are *separating cuts*. Suppose that solution $\bar{x}$ of a continuous relaxation of the integer system $Ax \geq b$ is infeasible because it has some nonintegral components. A separating cut $cx \geq c_0$ *cuts off* $\bar{x}$ in the sense that $c\bar{x} < c_0$. The cut is *separating* in that it defines a hyperplane that separates $\bar{x}$ from the convex hull of feasible solutions. There is always a Gomory cut that cuts off $\bar{x}$ when it has some nonintegral components. Typically, one generates several Gomory cuts and adds them to the relaxation to strengthen it.

The motivation for using separating cuts is it provides a principle for selecting relevant cuts. In general, there are a large number of nonredundant valid cuts, and adding them all to the continuous relaxation would be impractical. Separating cuts are relevant in the sense that they exclude the solution of the current relaxation so that when the relaxation is re-solved after adding the cuts, a different solution will be obtained. If desired, one can generate separating cuts for this new solution and add them to the relaxation, and so forth through several rounds.

In some cases, the continuous relaxation of an integer linear system describes an integral polyhedron, meaning that the vertices of the polyhedron have all integral coordinates. In such cases, one can solve the system by finding a vertex solution of its continuous relaxation—for example by using the simplex method. One sufficient condition for an integral polyhedron is that the coefficient matrix of the problem be totally unimodular. Total unimodularity can, in turn, be characterized by

276 Relaxation

a necessary and sufficient condition that is sometimes useful for showing that certain classes of problems have integral polyhedra.

4.5.1 Chvátal-Gomory Cuts

The Chvátal-Gomory procedure is a complete inference method for general integer inequalities as well as for 0-1 inequalities. It is applied to an inequality system

$$Ax \geq b$$
$$x \text{ integral} \tag{4.43}$$

$Ax \geq b$ is understood to include bounds $0 \leq x_j \leq h_j$, which imposes no practical restriction and avoids the necessity of studying the more difficult unbounded case.

In the proof, the resolution pattern is replaced by a more complex combination of inequalities. Suppose again the goal is to derive an inequality $cx \geq c_0$. As in the 0-1 case, a weaker inequality $cx \geq \delta_0$ (for sufficiently small δ_0) can be derived from variable bounds $0 \leq x \leq h$. It therefore suffices to show that whenever $cx - \delta \geq -1$ is a Chvátal-Gomory cut and $cx - \delta \geq 0$ is valid, then $cx - \delta \geq 0$ is a Chvátal-Gomory cut.

The derivation pattern is illustrated in Figure 4.5. It can be assumed without loss of generality that each upper bound h_j is the same, namely h_0. Each node on level k of the enumeration tree corresponds to setting variables $x_1, \ldots, x_k$ to certain values $v_1, \ldots, v_k$. The inequality

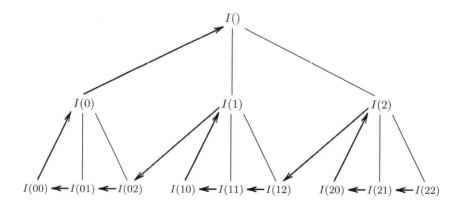

Figure 4.5. Illustration of the proof of the Chvátal-Gomory theorem for general integer inequalities. $I(v_1 v_2)$ stands for the inequality $T(v_1, v_2) \geq 0$.

$T(v_1, \ldots, v_k) \geq 0$ is associated with this node, where

$$T(v_1, \ldots, v_k) = \prod_{i=1}^{k}(h_0 + 1 - v_i)(cx - \delta) + \sum_{i=1}^{k} \prod_{j=i+1}^{k}(h_0 + 1 - v_j)x_i$$

The inequalities associated with the nodes of Figure 4.5, beginning at the lower right, are

$$
\begin{aligned}
T(2,2) &= (cx - \delta) + x_1 + x_2 &\geq 0 \\
T(2,1) &= 2(cx - \delta) + 2x_1 + x_2 &\geq 0 \\
T(2,0) &= 3(cx - \delta) + 3x_1 + x_2 &\geq 0 \\
T(2) &= (cx - \delta) + x_1 &\geq 0 \\
T(1,2) &= 2(cx - \delta) + x_1 + x_2 &\geq 0 \\
T(1,1) &= 4(cx - \delta) + 2x_1 + x_2 &\geq 0 \\
T(1,0) &= 6(cx - \delta) + 4x_1 + x_2 &\geq 0 \\
T(1) &= 2(cx - \delta) + x_1 &\geq 0 \\
T(0,2) &= 3(cx - \delta) + x_1 + x_2 &\geq 0 \\
T(0,1) &= 6(cx - \delta) + 2x_1 + x_2 &\geq 0 \\
T(0,0) &= 9(cx - \delta) + 3x_1 + x_2 &\geq 0 \\
T(0) &= 3(cx - \delta) + x_1 &\geq 0 \\
T() &= (cx - \delta) &\geq 0
\end{aligned}
$$

Each inequality at a nonleaf node is derived (in part) from its left-most immediate successor, as indicated by arrows in the figure. The remaining arrows show (in part) how the inequalities at leaf nodes are derived.

THEOREM 4.7 *Every valid inequality for a bounded integer system is a Chvátal-Gomory cut.*

Proof. It suffices to show that if

$$cx - \delta \geq -1 \tag{4.44}$$

is a Chvátal-Gomory cut, and $cx - \delta \geq 0$ is valid, then $T() \geq 0$ is a Chvátal-Gomory cut. In fact, it will be shown that $T(v_1, \ldots, v_k) \geq 0$ is a Chvátal-Gomory cut for all $v_1, \ldots, v_k$ and all $k \leq n$.

First, it is easy to check that each inequality $T(v_1, \ldots, v_k) \geq 0$ for $k < n$ can be derived from the inequality $T(v_1, \ldots, v_k, 0) \geq 0$ and the bound $-x_{k+1} \geq -1$ by using $1/(h_0 + 1)$ as the multiplier for each.

Now consider the inequalities $T(v_1, \ldots, v_n) \geq 0$. It will be shown inductively that each of these inequalities, beginning with the inequality

$T(h_0, \ldots, h_0) \geq 0$, can be derived from inequalities that have already been shown to be Chvátal-Gomory cuts, namely from

$$cx - \delta \geq -1$$
$$x_i \geq 0 \qquad \text{for all } i = 1, \ldots, n \text{ with } v_i = h_0 \qquad (4.45)$$
$$T(v_1, \ldots, v_{i-1}, v_i + 1) \geq 0 \quad \text{for all } i = 1, \ldots, n \text{ with } v_i < h_0$$

Note first that $T(v_1, \ldots, v_n) \geq -1$ is the sum of the inequalities (4.45). It will be shown that $T(v_1, \ldots, v_n) \geq -1$ cannot be satisfied at equality, which implies that $T(v_1, \ldots, v_n) \geq \epsilon - 1$ is valid for some $\epsilon > 0$. Rounding up the right-hand side yields that $T(v_1, \ldots, v_n) \geq 0$ is a Chvátal-Gomory cut, and the theorem follows.

Suppose then that $T(v_1, \ldots, v_n) \geq -1$ is satisfied at equality. Since each inequality in (4.45) is valid, $T(v_1, \ldots, n_n)$ can be -1 only if $cx - \delta = -1$. But $cx - \delta$ can be -1 and the second inequality in (4.45) satisfied only if $x_1 = v_1$. Given that $x_1 = v_1$, $cx - \delta$ can be -1 and the third inequality in (4.45) satisfied only if $x_2 = v_2$, and so forth. Thus, $(x_1, \ldots, x_n) = (v_1, \ldots, v_n)$ satisfies $cx - \delta = -1$, which means $cx - \delta \geq 0$ is violated by an integer point. This is impossible, because it is given that $cx - \delta \geq 0$ is valid for all integer solutions of $Ax \geq b$. Thus, $T(v_1, \ldots, v_n) \geq -1$ cannot be satisfied at equality. □

4.5.2 Gomory Cuts

There is a systematic and easily implemented method for the generation of separating Chvátal-Gomory cuts. These separating cuts are popularly known as *Gomory cuts* because one of the earliest algorithms for integer programming, invented by Ralph Gomory, is based on them. Gomory originally proposed a pure cutting plane algorithm that repeatedly re-solves the continuous relaxation after adding inequalities that cut off the fractional solution of the last relaxation. Such an algorithm is rarely efficient, but Gomory cuts can be very effective in a branch-and-cut method and are widely used in commercial MILP solvers today.

To generate Gomory cuts for an integer programming problem

$$\begin{aligned} & \min \bar{c}\bar{x} \\ & \bar{A}\bar{x} \geq b \\ & \bar{x} \geq 0 \text{ and integral} \end{aligned} \qquad (4.46)$$

where $\bar{x} = (x_1, \ldots, x_n)$, the problem is first converted to an equality-constrained problem by adding surplus variables $x_{n+1}, \ldots, x_{n+m}$:

$$\begin{aligned} & \min cx \\ & Ax = b, \; x \geq 0 \text{ and integral} \end{aligned} \qquad (4.47)$$

Integer Linear Inequalities

where $x = (x_1, \ldots, x_{n+m})$, $A = [\bar{A} \ -I]$, and $c = [\bar{c} \ 0]$. The surplus variables can be restricted to be integral if it is supposed (with very little loss of generality) that $\bar{A}$ and b have integral components.

Gomory cuts are based on a simple idea from modular arithmetic. Let $frac(\alpha) = \alpha - \lfloor \alpha \rfloor$ be the fractional part of a real number α. Then clearly, $frac(\alpha k) \leq frac(\alpha) k$ for any nonnegative integer k.

Now consider any noninteger basic solution $(x_B, x_N) = (x_B, 0)$ of the continuous relaxation of (4.47). That is, some component x_i of x_B is not an integer. The aim is to find a valid inequality for (4.47) that cuts off this solution.

As indicated in Section 3.3.6, any solution of $Ax = b$ has the form (x_B, x_N) where $x_B = \hat{b} - \hat{N} x_N$ (recall that $\hat{b} = B^{-1}b$ and $\hat{N} = B^{-1}N$). Thus in particular,

$$x_i = \hat{b}_i - \hat{N}_i x_N \tag{4.48}$$

where $\hat{N}_i$ is row i of $\hat{N}$. Since x_i must be integer in any feasible solution of (4.47), the two terms of (4.48) must have the same fractional part:

$$frac(\hat{N}_i x_N) = frac(\hat{b}_i) \tag{4.49}$$

Because x_N consists of nonnegative integers in any feasible solution of (4.47), $frac(\hat{N}_i x_N) \leq frac(\hat{N}_i) x_N$. This and (4.49) imply the *Gomory cut*

$$frac(\hat{N}_i) x_N \geq frac(\hat{b}_i) \tag{4.50}$$

This obviously cuts off the solution $(x_B, x_N) = (x_B, 0)$, since the right-hand side is strictly positive.

The Gomory cut (4.50) can be written

$$(\hat{N}_i - \lfloor \hat{N}_i \rfloor) x_N \geq \hat{b}_i - \lfloor \hat{b}_i \rfloor$$

Subtracting (4.48) from this yields an alternate form of the cut that contains x_i:

$$x_i + \lfloor \hat{N}_i \rfloor x_N \leq \lfloor \hat{b}_i \rfloor \tag{4.51}$$

As an example, consider the following problem, which was discussed in Section 3.7.1:

$$\begin{aligned} \min \ & 2x_1 + 3x_2 \\ & x_1 + 3x_2 \geq 3 \quad (a) \\ & 4x_1 + 3x_2 \geq 6 \quad (b) \\ & x_1, x_2 \geq 0 \text{ and integral} \end{aligned} \tag{4.52}$$

It can be written as an equality-constrained problem by introducing surplus variables x_3, x_4:

$$\begin{aligned}
\min\ & 2x_1 + 3x_2 \\
& x_1 + 3x_2 - x_3 = 3 & (a) \\
& 4x_1 + 3x_2 - x_4 = 6 & (b) \\
& x_1, \ldots, x_4 \geq 0 \text{ and integral}
\end{aligned} \qquad (4.53)$$

The continuous relaxation has the optimal solution

$$F(x_B, x_N) = (x_1, x_2, x_3, x_4) = (1, \tfrac{2}{3}, 0, 0)$$

which is illustrated in Figure 4.6. At this solution,

$$B = \begin{bmatrix} 1 & 3 \\ 4 & 3 \end{bmatrix} \quad B^{-1}N = \hat{N} = \begin{bmatrix} \tfrac{1}{3} & -\tfrac{1}{3} \\ -\tfrac{4}{9} & \tfrac{1}{9} \end{bmatrix} \quad B^{-1}b = \hat{b} = \begin{bmatrix} 1 \\ \tfrac{2}{3} \end{bmatrix}$$

Since x_2 is noninteger, a Gomory cut can be formulated to cut off this solution. The cut (4.50) is

$$frac(-\tfrac{4}{9})x_3 + frac(\tfrac{1}{9})x_4 \geq frac(\tfrac{2}{3}), \quad \text{or} \quad \tfrac{5}{9}x_3 + \tfrac{1}{9}x_4 \geq \tfrac{2}{3} \qquad (4.54)$$

The cut in form (4.51) is

$$x_2 + \lfloor -\tfrac{4}{9} \rfloor x_3 + \lfloor \tfrac{1}{9} \rfloor x_4 \leq \lfloor \tfrac{2}{3} \rfloor, \quad \text{or} \quad x_2 - x_3 \leq 0 \qquad (4.55)$$

Note that either cut excludes the solution $(1, \tfrac{2}{3}, 0, 0)$.

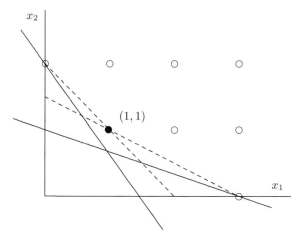

Figure 4.6. An integer programming problem with two Gomory cuts (dashed lines). The small circles show some of the feasible solutions, and the black dot is the optimal solution.

Integer Linear Inequalities

The relaxation is now re-solved with the additional constraint (4.55) written as an equality constraint $-x_2+x_3-x_5 = 0$, where x_5 is a new surplus variable. The solution remains fractional, with $x_B = (x_1, x_2, x_3) = (\frac{3}{5}, \frac{6}{5}, \frac{6}{5})$. Now, the cut (4.50) is

$$\tfrac{3}{5}x_4 + \tfrac{3}{5}x_5 \geq \tfrac{3}{5} \tag{4.56}$$

and in form (4.51) is

$$x_1 - x_4 \leq 0$$

When this is added to the relaxation in equality form $-x_1+x_4-x_6 = 0$, the solution is integral with $(x_1, x_2) = (1, 1)$ and optimal value 5.

Gomory cuts can be written in the original variables $x_1, \ldots, x_n$ by eliminating any surplus variables that occur in cut (4.50). Simply replace any surplus variable x_j in (4.50) with $\bar{A}^{j-n}\bar{x} - b_{j-n}$, where $\bar{A}^{j-n}$ is row $j-n$ of $\bar{A}$. In the example, the two cuts (4.54) and (4.56) can be written in the original variables x_1, x_2 by substituting $x_3 = x_1 + 3x_2 - 3$ and $x_4 = 4x_1 + 3x_2 - 6$. This yields

$$\begin{aligned} x_1 + 2x_2 &\geq 3 \quad (c) \\ 3x_1 + 3x_2 &\geq 6 \quad (d) \end{aligned} \tag{4.57}$$

These cuts are illustrated in Figure 4.6.

The surplus variable coefficients in (4.54) and (4.56) also indicate how to obtain the cuts (4.57) by taking linear combinations and rounding. Constraints (a) and (b) of the original problem (4.52) are first combined with multipliers $\frac{5}{9}$ and $\frac{1}{9}$ to obtain $x_1 + 2x_2 \geq \frac{21}{9}$ which, after rounding, is (c) above. Thus, cut (c) is a rank 1 Chvátal-Gomory cut for the original system (4.52). Next constraints (c), (d), and (e) are combined with coefficients 0, $\frac{3}{5}$, and $\frac{3}{5}$. Note that surplus variable x_3 has coefficient 0 in (4.56). This yields $3x_1 + 3x_2 \geq \frac{27}{5}$ which, after rounding, is cut (d). In general, one can say the following.

THEOREM 4.8 *The Gomory cut (4.50), when expressed in the original variables, is a rank 1 Chvátal-Gomory cut for (4.52).*

Proof. The proof relies on the identity

$$\sum_j frac(a_j)y_j + frac\left(-\sum_j a_j y_j\right) = \left[\sum_j frac(a_j)y_j\right] \tag{4.58}$$

where each y_j is an integer. To prove the theorem, it suffices to exhibit a surrogate of (4.52)'s continuous relaxation that is equivalent to (4.50)

after rounding. Let
$$J_1 = \{j \in \{1,\ldots,n\} \mid x_j \text{ is nonbasic}\}$$
$$J_2 = \{j \in \{n+1,\ldots,n+m\} \mid x_j \text{ is nonbasic}\}$$

Assign multiplier $frac(\hat{N}_{ij})$ to inequality $\bar{A}^{j-n}\bar{x} \geq b_{j-n}$ for each $j \in J_2$ and to bound $x_j \geq 0$ for each $j \in J_1$, and assign a multiplier of zero to all other constraints. The resulting surrogate is

$$\sum_{j \in J_1} frac(\hat{N}_{ij})x_j + \sum_{j \in J_2} frac(\hat{N}_{ij})\bar{A}^{j-n}\bar{x} \geq \sum_{j \in J_2} frac(\hat{N}_{ij})b_{j-n} \quad (4.59)$$

It will be shown that (4.59) with the right-hand side rounded up is equivalent to (4.50). Since $x_j = \bar{A}^{j-n}\bar{x} - b_{j-n}$ for $j \in J_2$, (4.50) expressed in the original variables is

$$\sum_{j \in J_1} frac(\hat{N}_{ij})x_j + \sum_{j \in J_2} frac(\hat{N}_{ij})\bar{A}^{j-n}\bar{x} \geq frac(\hat{b}_i) + \sum_{j \in J_2} frac(\hat{N}_{ij})b_{j-n} \quad (4.60)$$

Since $A = [\bar{A} \ -I]$,

$$\hat{b}_i = (B^{-1})^i b = -\hat{N}^i b = -\sum_{j \in J_2} \hat{N}_{ij} b_{j-n}$$

So, (4.60) can be written

$$\sum_{j \in J_1} frac(\hat{N}_{ij})x_j + \sum_{j \in J_2} frac(\hat{N}_{ij})\bar{A}^{j-n}\bar{x} \geq frac\left(-\sum_{j \in J_2} \hat{N}_{ij} b_{j-n}\right) + \sum_{j \in J_2} frac(\hat{N}_{ij}) b_{j-n}$$

Due to (4.58) and the fact the b is integral, the right-hand side of this inequality is the result of rounding up the right-hand side of the surrogate (4.59). □

It can be shown that the Gomory cutting plane algorithm terminates with an optimal solution after a finite number of steps. Since each step generates a Chvátal-Gomory cut (by Theorem 4.8), the Gomory algorithm provides an alternate proof of Theorem 4.7.

4.5.3 Mixed Integer Rounding Cuts

A *mixed integer/linear inequality* is a linear inequality in which some variables are continuous and other variables take integer values. Inequalities of this kind are widely used in integer programming, partly because disjunctions of linear constraints and other logical conditions

Integer Linear Inequalities

are commonly written in inequality form by introducing 0-1 variables. An integrated solver does not require such formulations in the original model, but mixed integer/linear inequalities may appear in other contexts. For instance, the solver may generate a mixed integer model for a constraint to obtain a continuous relaxation by dropping the integrality requirement. It is therefore useful to know how to generate cuts for mixed integer constraint sets.

The integer rounding cut used in the Chvátal-Gomory procedure can be extended to mixed integer/linear inequalities, resulting in a *mixed integer rounding cut*. It is less obvious how to write a rounding cut in the mixed integer case, but it is derived from a simple observation. Consider the two-variable mixed integer inequality $y+x \geq b$ where $x \geq 0$, b is nonintegral and y is any integer (possibly negative). One can see from Figure 4.7 that the rounding cut

$$y + \frac{x}{\mathrm{frac}(b)} \geq \lceil b \rceil \qquad (4.61)$$

is a valid inequality for these constraints. More formally,

THEOREM 4.9 *The inequality (4.61) is a valid cut for the constraints $y + x \geq b$ and $x \geq 0$, where b is nonintegral and y is integer valued.*

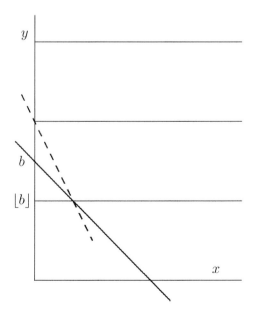

Figure 4.7. A mixed integer cut (dashed line) for the inequality $y+x \geq b$ (solid line).

Proof. Because any point (x, y) satisfying the constraints satisfies $y \geq \lceil b \rceil$ or $y \leq \lceil b \rceil - 1$, it suffices to show that (4.61) holds in either case. If $y \geq \lceil b \rceil$, one can take a linear combination of $y \geq \lceil b \rceil$ with multiplier $frac(b)$ and $x \geq 0$ with multiplier 1 to obtain

$$frac(b)y + x \geq frac(b)\lceil b \rceil \tag{4.62}$$

which is equivalent to (4.61). If $y \leq \lceil b \rceil - 1$, one can take a linear combination of $y \leq \lceil b \rceil - 1$ with multiplier $1 - frac(b)$ and $y + x \geq b$ with multiplier 1 and again obtain (4.62). □

The cut (4.61) is useful because a general mixed integer/linear inequality can be relaxed to an inequality that has the form $y + x \geq b$. Thus, a valid cut for $y + x \geq b$ leads to a general mixed integer rounding cut.

A general mixed integer inequality can be written

$$\sum_{j \in J} a_j y_j + cx \geq b \tag{4.63}$$

where each $y_j \geq 0$ is integer-valued, and each $x_j \geq 0$ is real-valued. Let J_1 be the set of indices j for which $frac(a_j) \geq frac(b)$, and let $J_2 = J \setminus J_1$. Then (4.63) can be written

$$\sum_{j \in J_1} a_j y_j + \sum_{j \in J_2} a_j y_j + z_1 - z_2 \geq b \tag{4.64}$$

where the continuous variable $z_1 \geq 0$ is the sum of all terms $c_j x_j$ with $c_j > 0$, and $z_2 \geq 0$ the negated sum of the remaining terms $c_j x_j$. The reason for the partition of J into J_1 and J_2 will become evident shortly.

The inequality (4.64) remains valid if the coefficients a_j for $j \in J_1$ are rounded up. Thus, since $a_j = \lfloor a_j \rfloor + frac(a_j)$ and $z_2 \geq 0$, (4.64) implies

$$\sum_{j \in J_1} \lceil a_j \rceil y_j + \sum_{j \in J_2} \lfloor a_j \rfloor y_j + \sum_{j \in J_2} frac(a_j) y_j + z_1 \geq b \tag{4.65}$$

Setting y_0 equal to the sum of the first two terms and z_0 equal to the sum of the last two, (4.65) can be written $y_0 + z_0 \geq b$. Because y_0 is a (possibly negative) integer variable and $z_0 \geq 0$, this inequality has the form required by Theorem 4.9. For nonintegral b, it implies the cut $y_0 + z_0/frac(b) \geq \lceil b \rceil$. Restoring the expressions for y_0 and z_0 yields a mixed integer rounding cut for (4.64):

$$\sum_{j \in J_1} \lceil a_j \rceil y_j + \sum_{j \in J_2} \left(\lfloor a_j \rfloor + \frac{frac(a_j)}{frac(b)} \right) y_j + \frac{z_1}{frac(b)} \geq \lceil b \rceil \tag{4.66}$$

Integer Linear Inequalities 285

Note that for $j \in J_2$ the coefficient of y_j is less than $\lceil a_j \rceil$ when a_j is nonintegral, and therefore strengthens the cut. Partitioning J into J_1 and J_2 avoids using a coefficient of this form when it is greater than $\lceil a_j \rceil$ and would weaken the cut. If $c_j^+ = \max\{c_j, 0\}$, the following has been shown.

THEOREM 4.10 *Let $J_1 = \{j \in J \mid \text{frac}(a_j) \geq \text{frac}(b)\}$ and $J_2 = J \setminus J_1$. Then the mixed integer rounding cut*

$$\sum_{j \in J_1} \lceil a_j \rceil y_j + \sum_{j \in J_2} \left(\lfloor a_j \rfloor + \frac{\text{frac}(a_j)}{\text{frac}(b)} \right) y_j + \frac{1}{\text{frac}(b)} \sum_j c_j^+ x_j \geq \lceil b \rceil \quad (4.67)$$

is valid for (4.63) when b is nonintegral, $x_j, y_j \geq 0$, and y_j is integer valued.

For example, the mixed integer/linear inequality

$$\tfrac{5}{3} y_1 - \tfrac{2}{3} y_2 + \tfrac{1}{4} y_3 + 2x_1 + 3x_2 - 4x_3 \geq \tfrac{3}{2}$$

has $J_1 = \{1\}$ and $J_2 = \{2, 3\}$. The mixed integer rounding cut (4.67) is

$$2y_1 - \tfrac{1}{3} y_2 + \tfrac{1}{2} y_3 + 4x_1 + 6x_2 \geq 2$$

4.5.4 Separating Mixed Integer Rounding Cuts

The mixed integer rounding cuts derived in the previous section lead directly to separating cuts for a mixed integer system of the form

$$\begin{aligned} A_1 y + A_2 x = b, \quad x, y \geq 0 \\ y \in \mathcal{Z}^n \end{aligned} \quad (4.68)$$

If some y_i has a nonintegral value in a basic solution of the continuous relaxation of (4.68), a cut violated by this solution can be derived as follows.

The coefficient matrix $[A_1 \ A_2]$ is first partitioned $[B \ N]$, where B contains the basic columns. Some of the columns of B may correspond to y_js and others to x_js, and similarly for N. Let J be the set of indices j for which y_j is nonbasic, and K the set of indices for which x_j is nonbasic. As in Section 4.2.1, one can solve for y_i in terms of the nonbasic variables:

$$y_i = \hat{b}_i - \sum_{j \in J} \hat{N}_{ij} y_j - \sum_{j \in K} \hat{N}_{ij} x_j \quad (4.69)$$

where $\hat{b} = B^{-1} b$ and $\hat{N} = B^{-1} N$. (4.69) implies the inequality

$$y_i + \sum_{j \in J} \hat{N}_{ij} y_j + \sum_{j \in K} \hat{N}_{ij} x_j \geq \hat{b}_i$$

Applying Theorem 4.9 to this inequality yields the desired cut.

THEOREM 4.11 *Let y_i be a noninteger in a basic solution of the continuous relaxation of (4.68), in which N is the matrix of nonbasic columns. Let J be the set of indices j for which y_j is nonbasic, and K the set of indices for which x_j is nonbasic. Then,*

$$y_i + \sum_{j \in J_1} \lceil \hat{N}_{ij} \rceil y_j + \sum_{j \in J_2} \left(\lfloor \hat{N}_{ij} \rfloor + \frac{\mathrm{frac}(\hat{N}_{ij})}{\mathrm{frac}(\hat{b}_i)} \right) y_j + \frac{1}{\mathrm{frac}(\hat{b}_i)} \sum_{j \in K} \hat{N}_{ij}^+ x_j \geq \lceil \hat{b}_i \rceil$$

is a separating mixed integer rounding cut, where the index sets are $J_1 = \{ j \in J \mid \mathrm{frac}(\hat{N}_{ij}) \geq \mathrm{frac}(\hat{b}_{ij}) \}$ and $J_2 = J \setminus J_1$.

The cut is clearly separating because y_i has the noninteger value $\hat{b}_i$ in the given solution, and all the other variables in the cut are nonbasic and equal to zero.

The cut may be illustrated with the constraint set

$$-6y_1 - 4y_2 + 3x_1 + 4x_2 = 1$$
$$-y_1 - y_2 + x_1 + 2x_2 = 3$$
$$y_1, y_2, x_1, x_2 \geq 0, \quad y_1, y_2 \in \mathcal{Z}$$

In one basic solution, the basic variables are $(y_1, x_1) = (\frac{8}{3}, \frac{17}{3})$. The relevant data appear below.

$$B = \begin{bmatrix} -6 & 3 \\ -1 & 1 \end{bmatrix} \quad N = \begin{bmatrix} -4 & 4 \\ -1 & 2 \end{bmatrix} \quad B^{-1} = \begin{bmatrix} -\frac{1}{3} & 1 \\ -\frac{1}{3} & 2 \end{bmatrix}$$

$$\hat{b} = \begin{bmatrix} \frac{8}{3} \\ \frac{17}{3} \end{bmatrix} \quad \hat{N} = \begin{bmatrix} \frac{1}{3} & \frac{2}{3} \\ -\frac{2}{3} & \frac{8}{3} \end{bmatrix}$$

Since y_1 is nonintegral, Theorem 4.11 provides a separating cut:

$$y_1 + \left(\lfloor \tfrac{1}{3} \rfloor + \tfrac{1/3}{2/3} \right) y_2 + \tfrac{1}{2/3} (\tfrac{2}{3})^+ x_2 \geq \lceil \tfrac{8}{3} \rceil$$

or $y_1 + \frac{1}{2} y_2 + x_2 \geq 3$. In this instance, $J_1 = \emptyset$ and $J_2 = \{2\}$.

4.5.5 Integral Polyhedra

A polyhedron is *integral* when every vertex has all integral coordinates. When a set of mixed integer linear inequalities describes an integral polyhedron, it has the same feasible set as its continuous relaxation. In this happy circumstance, one can solve the constraints by solving their continuous relaxation, which is a much easier problem.

Integer Linear Inequalities

There is no easy rule for recognizing when mixed integer inequalities define an integral polyhedron, but there is a well-known sufficient condition that is often useful. Namely, $Ax \geq b$, $x \geq 0$ describes an integral polyhedron if b is integral and the matrix A is *totally unimodular*, meaning that every square submatrix of A has a determinant equal to 0, 1, or -1.

Several properties of total unimodularity follow immediately from its definition. Recall that a unit vector e^i consists of all zeros except for a 1 in the ith place.

THEOREM 4.12 *Matrix A is totally unimodular if and only if A' is totally unimodular, where A' is the matrix obtained by any of the following operations on A: transposition; swapping two columns; negating any column; or adding a unit column e^i.*

The main property is the following:

THEOREM 4.13 *A matrix A with integral components is totally unimodular if and only if $Ax \geq b$, $x \geq 0$ defines an integral polyhedron for any integral b.*

Proof. Suppose A is totally unimodular. Any vertex of the polyhedron $P = \{x \geq 0 \mid Ax \geq b\}$ is a basic feasible solution $(B^{-1}b, 0)$ of $Ax - s = b$, $x, s \geq 0$ for some square submatrix B of $[A \ -I]$. But by Theorem 4.12, $[A \ -I]$ is totally unimodular, which means B^{-1}, and therefore $B^{-1}b$, are integral.

Conversely, suppose P is integral for every integral b, and let $\bar{B}$ be any square submatrix of A. If $\bar{B}$ is singular then $\det \bar{B} = 0$, and so one may assume $\bar{B}$ is nonsingular. Let B be the following square matrix consisting of a subset of the columns of $[A \ -I]$:

$$\begin{bmatrix} \bar{B} & 0 \\ C & -I \end{bmatrix}$$

Note that B is nonsingular. Consider the system $Ax - s = b$, $x, s \geq 0$, where $b = Bz + e^i$ for arbitrary unit vector e^i. Here, z is any integral vector chosen so that $z + (B^{-1})_i \geq 0$, where $(B^{-1})_i$ is column i of B^{-1}. Then $(B^{-1}b, 0) = (z + (B^{-1})_i, 0)$ is a basic feasible solution of $Ax - s = b$, $x, s \geq 0$, which by hypothesis is integral. Since i is arbitrary, B^{-1} and therefore $\bar{B}^{-1}$ are integral. Since $\bar{B}$ is integral (due to the integrality of A), $\det \bar{B}$ and $\det \bar{B}^{-1}$ are integers. Thus, since

$$|\det \bar{B}||\det \bar{B}^{-1}| = |\det(\bar{B}\bar{B}^{-1})| = 1$$

it follows that $|\det \bar{B}| = 1$. □

A necessary and sufficient condition for total unimodularity is the following.

THEOREM 4.14 *The $m \times n$ matrix A is totally unimodular if and only if for every $J \subset \{1, \ldots, n\}$, there is a partition $J = J_1 \cup J_2$ such that*

$$\left| \sum_{j \in J_1} A_{ij} - \sum_{j \in J_2} A_{ij} \right| \leq 1 \ \text{for } i = 1, \ldots, m \tag{4.70}$$

Proof. First, suppose that A is totally unimodular, and let J be any subset of $\{1, \ldots, n\}$. Let $\delta_j = 1$ when $j \in J$ and $\delta_j = 0$ otherwise, and consider the polyhedron defined by

$$P = \left\{ x \ \middle| \ \lfloor \tfrac{1}{2} A \delta \rfloor \leq Ax \leq \lceil \tfrac{1}{2} A \delta \rceil, \ 0 \leq x \leq \delta \right\}$$

Due to Theorems 4.12 and 4.13, P is integral. Also since $\tfrac{1}{2}\delta \in P$, the polyhedron P is nonempty, and one can choose an arbitrary integral point $y \in P$. Since $\delta_j - 2y_j = \pm 1$, one can partition J by letting

$$J_1 = \{j \in J \mid \delta_j - 2y_j = 1\}$$
$$J_2 = \{j \in J \mid \delta_j - 2y_j = -1\}$$

Now for each i,

$$\sum_{j \in J_1} A_{ij} - \sum_{j \in J_2} A_{ij} = \sum_{j \in J} A_{ij}(\delta_j - 2y_j)$$

$$= \begin{cases} A_i \delta - A_i \delta = 0 & \text{if } A_i \delta \text{ is even} \\ A_i \delta - (A_i \delta \pm 1) = \pm 1 & \text{if } A_i \delta \text{ is odd} \end{cases}$$

For the converse, suppose that for any $J \in \{1, \ldots, n\}$, there is a partition satisfying (4.70). The proof is by induction on the size of J. For $|J| = 1$, (4.70) simply says that the submatrix, and therefore its determinant, is 0 or ± 1. Suppose, then, that the claim is true for any J with $|J| = k - 1 \geq 1$, and let B be an arbitrary $k \times k$ submatrix of A. If one assumes B is nonsingular, it suffices to show $|\det B| = 1$. By Cramer's rule and the induction hypothesis, $B^{-1} = \bar{B}/\det B$ where each $\bar{B}_{ij} \in \{0, \pm 1\}$. Also, $B\bar{B}_1 = |\det B|e^1$, with $\bar{B}_1$ denoting column 1 of $\bar{B}$. Let

$$J_1' = \{i \in J \mid \bar{B}_{i1} = 1\}, \quad J_2' = \{i \in J \mid \bar{B}_{i1} = -1\}$$

with $J = J_1' \cup J_2'$. Note that $J \neq \emptyset$, because otherwise B^{-1} would be singular. Since $B\bar{B}_1 = |\det B|e^1$, for $i = 2, \ldots, n$ one has

$$(B\bar{B}_1)_i = \sum_{j \in J_1'} B_{ij} - \sum_{j \in J_2'} B_{ij} = 0 \tag{4.71}$$

Integer Linear Inequalities 289

By hypothesis, there is a partition J_1, J_2 of J such that

$$\Delta_i = \left| \sum_{j \in J_1} B_{ij} - \sum_{j \in J_2} B_{ij} \right| \leq 1, \quad i = 1, \ldots, k \qquad (4.72)$$

Because partition J_1, J_2 of J can be created by transferring indices from J_1' to J_2' and vice-versa, and because each transfer alters the difference in (4.71) by an even number, (4.71) implies that Δ_i is even for $i = 2, \ldots, k$. Thus, (4.72) implies that $\Delta_i = 0$ for $i = 2, \ldots, k$. Finally, it can be shown as follows that $\Delta_1 = 1$. For if $\Delta_1 = 0$, then due to (4.72) one has $Bz = 0$ when z_j is defined to be 1 for $j \in J_1$, -1 for $j \in J_2$, and 0 otherwise. Since B is nonsingular, this implies $z = 0$, which is impossible since $J \neq \emptyset$. Thus, $\Delta_1 = 1$ and $Bz = \pm e^1$. But since $B\bar{B}_1 = |\det B|e^1$, and since the components of z and $\bar{B}_1$ belong to $\{0, \pm 1\}$, it follows that $\bar{B}_1 = \pm z$ and $|\det B| = 1$. □

As an example, consider the system

$$\begin{aligned} x_1 & \geq b_1 \\ x_1 + x_2 & \geq b_2 \\ & \vdots \\ x_1 + x_2 + \cdots + x_n & \geq b_n \\ x_j \geq 0, \text{ all } j \end{aligned} \qquad (4.73)$$

The coefficient matrix is totally unimodular, as one can see by placing every other column of J in J_1 and the remaining columns of J in J_2 and applying Theorem 4.13. Thus, any extreme point solution of (4.73) is integral if $b_1, \ldots, b_n$ are integral.

A useful consequence of Theorem 4.13 for $0, \pm 1$ matrices is the following.

COROLLARY 4.15 *A matrix A with components in $\{0, \pm 1\}$ is totally unimodular if each column contains no more than two nonzero entries, and any unit column with two nonzeros contains 1 and -1.*

This can be applied to the well-known network flow model. model Let E be the set of arcs (i, j) in a directed network. The net supply s_i of flow is given for each node i, as is the unit cost c_{ij} of flow on arc (i, j). If variable x_{ij} represents the flow from node i to node j, the problem is to find the minimum-cost feasible flow:

$$\min \sum_{(i,j)\in E} c_{ij} x_{ij}$$

$$\sum_{(j,i)\in E} x_{ji} - \sum_{(i,j)\in E} x_{ij} = s_i, \text{ all } i \tag{4.74}$$

$$x_{ij} \geq 0, \text{ all } i,j$$

Since the coefficient matrix satisfies the conditions of Corollary 4.15, it is totally unimodular. This means that the optimal flow is always integral if the net supplies s_i are integral.

4.5.6 Exercises

1 In the proof of Theorem 4.7, show that $T(v_1, \ldots, v_k) \geq 0$ for $k < n$ can be derived from $T(v_1, \ldots, v_k, 0) \geq 0$ and the bound $-x_{k+1} \geq -1$.

2 In the proof of Theorem 4.7, show that $T(v_1, \ldots, v_n) \geq -1$ is the sum of the inequalities (4.45).

3 Consider the integer programming problem

$$\begin{aligned} \min\ & 2x_1 + x_2 \\ & 5x_1 + 4x_2 \geq 10 \\ & x_1 + 3x_2 \geq 3 \\ & x_1, x_2 \geq 0 \text{ and integral} \end{aligned} \tag{4.75}$$

If surplus variables x_3, x_4 are inserted, the optimal solution of the continuous relaxation of (4.75) is $x = (0, \frac{5}{2}, 0, \frac{9}{2})$, with

$$B^{-1} = \begin{bmatrix} \frac{1}{4} & 0 \\ \frac{3}{4} & 1 \end{bmatrix}$$

Write two Gomory cuts in terms of the nonbasic variables. Write the same cuts in terms of a basic variable and the nonbasic variables. Finally, write the two cuts in terms of the original variables, x_1, x_2. Show how to obtain these last two cuts as rank 1 Chvátal-Gomory cuts of the constraints in the continuous relaxation of (4.75). Note that in each case, one of the multipliers corresponds to a bound $x_j \geq 0$.

4 Prove the identity (4.58).

5 Write a mixed integer rounding cut for $\frac{3}{2}y_1 + \frac{4}{3}y_2 - \frac{5}{3}y_3 + 2x_1 - \frac{5}{2}x_2 \geq \frac{5}{2}$, where y_1, y_2, y_3 are integral. What would the cut be if the indices of

integer-valued variables were not partitioned into J_1 and J_2 in the proof of Theorem 4.10?

6 Use Theorem 4.9 to verify that (4.66) is a rounding cut for (4.65).

7 Prove Theorem 4.11 as a corollary of Theorem 4.10.

8 The continuous relaxation of
$$y_1 - 4y_2 + 3x_1 + x_2 = 2$$
$$-3y_1 + y_2 + x_1 - 2x_2 = 1$$
$$x_1, x_2 \geq 0, \ y_1, y_2 \geq 0 \text{ and integral}$$
has a basic solution $y = (0, \frac{1}{7})$, $x = (\frac{6}{7}, 0)$, with
$$B^{-1} = \frac{1}{7} \cdot \begin{bmatrix} -1 & 3 \\ 1 & 4 \end{bmatrix}$$
Write a separating mixed integer rounding cut.

9 Prove Corollary 4.15.

10 A capacitated network flow model has the form (4.74) plus $x_{ij} \leq U_{ij}$ for all $(i, j) \in E$. Show that if each s_i and each U_{ij} is integral, a capacitated network flow problem always has an integral optimal flow.

11 The incidence matrix A for an undirected graph G contains a row for every vertex of G and a column for every edge. $A_{ie} = 1$ when vertex i is incident to edge e, and $A_{ie} = 0$ otherwise. Show that A is totally unimodular if and only if G is bipartite.

12 An *interval matrix* is a 0-1 matrix in which the ones in every row (if any) occur consecutively. Show that any interval matrix is totally unimodular.

4.6   Lagrangean and Surrogate Relaxations

Lagrangean and surrogate relaxations can be used whenever a problem contains inequality constraints. The variables may be continuous or discrete, and the functions involved may be linear or nonlinear. Lagrangean relaxation, which is by far the most popular, "dualizes" some or all of the inequality constraints by moving them into the objective function. This is accomplished by adding terms to the objective function that, roughly speaking, penalize violations of the dualized constraints. Surrogate relaxation replaces some or all of the inequality constraints with a surrogate; that is, a nonnegative linear combination of the constraints. Relaxation duals can be defined for both kinds of relaxation, and one can obtain tighter bounds by solving them.

The Lagrangean and surrogate duals can be viewed as inference duals as well as relaxation duals. They were developed in Section 3.4.2 as inference duals and are derived here as relaxation duals. Section 3.4.2 proved some elementary properties of the Lagrangean dual and applied it to domain reduction. It remains to discuss how to solve the Lagrangean dual, which is done below.

4.6.1 Surrogate Relaxation and Duality

Surrogate relaxation is applied to problems of the form

$$\begin{aligned} &\min f(x) \\ &g(x) \geq 0 \\ &x \in S \end{aligned} \qquad (4.76)$$

where $g(x)$ is a vector of functions $g_1(x), \ldots, g_m(x)$ and $x \in S$ represents an arbitrary constraint set (not necessarily inequalities). As in the case of linear programming (Section 4.2.2), a surrogate relaxation is formed by taking a nonnegative linear combination of inequalities. So, a surrogate relaxation of (4.76) is

$$\begin{aligned} &\min f(x) \\ &ug(x) \geq 0, \quad x \in S \end{aligned} \qquad (4.77)$$

where $u \geq 0$ is fixed. Normally the rationale for formulating a surrogate relaxation is that it may be easier to solve than (4.76) because it has only one inequality constraint. For example, if (4.76) is an integer linear programming problem, then a surrogate relaxation of it is an integer knapsack problem.

Let $\theta(u)$ be the optimal value of (4.77) subject to $u \geq 0$. Then the relaxation dual of (4.76), known in this case as the *surrogate dual*, maximizes $\theta(u)$ subject to $u \geq 0$. An example of a surrogate dual is presented in Section 3.4.1. It was also observed in Section 3.4.1 (Theorem 3.13) that the surrogate dual is equivalent to an inference dual in which inference consists of nonnegative linear combination and implication between inequalities.

4.6.2 Lagrangean Relaxation and Duality

Lagrangean relaxation is applied to problems of the form (4.76), where $x \in S$ again represents an arbitrary constraint set, but in practice is carefully chosen to have special structure. Lagrangean relaxation *dualizes*

Lagrangean and Surrogate Relaxations 293

the constraints $g(x) \geq 0$ by penalizing their violation in the objective function:

$$\begin{aligned} \min_{x} \; & \theta(u, x) \\ & x \in S, \; x \in \mathbb{R}^n \end{aligned} \qquad (4.78)$$

where

$$\theta(u, x) = f(x) - ug(x)$$

is the *Lagrangean* and $u \geq 0$ is a vector of *Lagrange multipliers*. The motivation for using the relaxation is that the special structure of the constraints $x \in S$ makes it easy to solve. The term $ug(x)$ is not, properly speaking, a penalty term, because a penalty term should vanish when x is feasible. Since $g_i(x)$ can be positive when x is feasible, $\theta(u, x)$ is often viewed as a *saddle* function.

Let $\theta(u)$ be the optimal value of (4.78) for $u \geq 0$. The relaxation dual corresponding to Lagrangean relaxation maximizes $\theta(u)$ subject to $u \geq 0$. By Theorem 3.14, this relaxation dual is equivalent to the Lagrangean dual conceived as an inference dual. Section 3.4.2 provides an example of Lagrangean duality and shows that it is equivalent to an inference dual in which implication is nonnegative linear combination and domination (Theorem 3.14). Section 3.4.3 establishes some elementary properties of the dual: weak duality, the concavity of $\theta(u)$, and complementary slackness when there is no duality gap.

4.6.3 Lagrangean Relaxation for Linear Programming

An important application of Lagrangean relaxation is to linear programming problems in which some constraints are hard and some are easy. The hard constraints are dualized, and the resulting relaxation is a linear programming problem with easy constraints.

This type of relaxation is typically used in a branch-and-bound search in which a linear relaxation of the problem is solved at each node of the search tree. At the root node, the optimal Lagrange multipliers can be found by solving the full linear programming relaxation. At subsequent nodes, these same multipliers continue to define Lagrangean relaxations that provide valid bounds, albeit not the tightest possible bounds, since the branching process adds constraints to the problem. Thus, one can obtain a serviceable bound at each node by solving a linear programming problem that is much easier than the full linear relaxation.

Suppose, then, that the linear programming relaxation of a problem to be solved by branching is written

$$\min cx \atop Ax \geq b, \ \ Dx \geq d, \ \ x \geq 0 \qquad (4.79)$$

where the linear system $Dx \geq d$ has some kind of special structure. If $Ax \geq b$ is dualized, then $\theta(u)$ is the optimal value of

$$\min_{\substack{Dx \geq d \\ x \geq 0}} \{\theta(x, u)\} \qquad (4.80)$$

where $\theta(x, u) = cx - u(Ax - b) = (c - uA)x + ub$.

THEOREM 4.16 *If (u^*, w^*) is an optimal dual solution of (4.79) in which u^* corresponds to $Ax \geq b$, then u^* is an optimal solution of the Lagrangean dual problem (4.80).*

Proof. Let x^* be an optimal solution of (4.79). By strong linear programming duality,

$$cx^* = u^*b + w^*d \qquad (4.81)$$

It will be shown below that $\theta(u^*) = cx^*$. This implies that u^* is an optimal dual solution of the Lagrangean dual problem, since $\theta(u)$ is a lower bound on cx^* for any $u \geq 0$.

To see that $\theta(u^*) = cx^*$, note first that x^* is feasible in

$$\min_{x \geq 0} \{(c - u^*A)x \mid Dx \geq d\} \qquad (4.82)$$

and w^* is feasible in its linear programming dual

$$\max_{w \geq 0} \{ud \mid wD \leq c - u^*A\} \qquad (4.83)$$

where the latter is true because (u^*, w^*) is dual feasible for (4.79). But the corresponding objective function value of (4.82) is

$$(c - u^*A)x^* = cx^* + u^*(b - Ax^*) - u^*b = cx^* - u^*b$$

where the second equation is due to complementary slackness. This is equal to the value w^*d of (4.83), due to (4.81). So, $cx^* - u^*b$ is the optimal value of (4.82), which means that cx^* is the optimal value $\theta(u^*)$ of (4.80) when $u = u^*$. □

Due to Theorem 4.16, one can solve the Lagrangean dual of (4.79) at the root node by solving its linear programming dual. Let u^* be the Lagrangean dual solution obtained in this way. At subsequent nodes,

Lagrangean and Surrogate Relaxations 295

one solves the specially structured linear programming problem (4.80), with u set to the value u^* obtained at the root node. Since the linear relaxation (4.79) at non-root nodes contains additional branching constraints, u^* is no longer optimal for the Lagrangean dual of (4.79). Yet due to weak duality, the optimal value of the specially structured linear programming problem (4.80) is a lower bound on the optimal value of (4.79) at that node. This bound can be used to prune the search tree.

4.6.4   Example: Generalized Assignment Problem

The generalized assignment problem is an assignment problem with additional knapsack constraints. By dualizing the knapsack constraints, one can solve a linear relaxation of the generalized assignment problem very rapidly using the Lagrangean technique described in the previous section.

The *generalized assignment problem* has the form

$$
\begin{aligned}
&\min \sum_{ij} c_{ij} x_{ij} &&(a) \\
&\sum_j x_{ij} = \sum_j x_{ji} = 1, \text{ all } i &&(b) \\
&\sum_j a_{ij} x_{ij} \geq \alpha_i, \text{ all } i &&(c) \\
&x_{ij} \in \{0,1\}, \text{ all } i,j
\end{aligned}
\qquad (4.84)
$$

where constraints (b) are the assignment constraints and (c) the knapsack constraints.

Suppose that a linear programming relaxation of the problem is to be solved at each node of the search tree to obtain bounds and perhaps to fix variables using reduced costs. If solving this problem with a general purpose solver is too slow, the complicating constraints (c) can be dualized, resulting in a pure assignment problem

$$
\begin{aligned}
&\min \sum_{ij}(c_{ij} - u_i a_{ij}) x_{ij} + \sum_i u_i \alpha_i \\
&\sum_j x_{ij} = \sum_j x_{ji} = 1, \text{ all } i \\
&x_{ij} \geq 0, \text{ all } i,j
\end{aligned}
\qquad (4.85)
$$

whose optimal value is $\theta(u)$ for $u \geq 0$.

The optimal dual solution u^* at the root node can be obtained by solving the linear relaxation of (4.84) and letting u^* be the optimal dual multipliers associated with constraints (c). At subsequent nodes of the search tree, one can solve the Lagrangean relaxation (4.85) very quickly using, for example, the Hungarian algorithm. The relaxation provides a weak bound, but the dual variables allow useful domain filtering as described in Section 3.3.8. The search branches by setting some x_{ij} to 0 or 1, which in turn can be achieved by giving x_{ij} a very large or very small cost in the objective function of the Lagrangean relaxation, thus preserving the problem structure.

4.6.5 Solving the Lagrangean Dual

If the Lagrangean dual is to provide a bound on the optimal value, it must somehow be solved. Fortunately, the concavity of $\theta(u)$ (Corollary 3.17) implies that any local maximum is a global maximum, which means that $\theta(u)$ can be maximized by a hill-climbing algorithm.

A popular hill-climbing procedure is *subgradient optimization*, which moves from one iterate to another in directions of steepest ascent until a local, and therefore a global, maximum is found. This approach has the advantage that the relevant subgradients are trivial to calculate, but the disadvantage that the step size is difficult to calibrate. In practice, the problem need not be solved to optimality, because only a valid lower bound is required.

Subgradient optimization begins with a starting value $u^0 \geq 0$ and sets each $u^{k+1} = u^k + \alpha_k \xi^k$, where ξ^k is a subgradient of $\theta(u)$ at $u = u^k$. Conveniently, if $\theta(u^k) = \theta(x^k, u^k)$, then $-g(x^k)$ is a subgradient. The stepsize α_k should decrease as k increases, but not so quickly as to cause the iterates to converge before a solution is reached.

A simple option is to set $\alpha_k = \alpha_0/k$, because the resulting step sizes $\alpha_k \to 0$ as $k \to \infty$, while the series $\sum_{i=1}^{k} \alpha_i$ diverges. This can result in a very large number of iterates, however. Another approach is to set

$$\alpha_k = \gamma \frac{\bar{\theta} - \theta(u^k)}{\|\xi^k\|}$$

where $\bar{\theta}$ is a dynamically adjusted upper bound on the maximum value of $\theta(u)$. The step size calibration is highly problem-dependent.

An example from Section 3.4.2 will illustrate the algorithm:

$$\begin{aligned}
&\min 3x_1 + 4x_2 \\
&-x_1 + 3x_2 \geq 0, \quad 2x_1 + x_2 - 5 \geq 0 \\
&x_1, x_2 \in \{0, 1, 2, 3\}
\end{aligned} \quad (4.86)$$

Lagrangean and Surrogate Relaxations

The Lagrangean dual maximizes $\theta(u) = \theta(u_1, u_2)$ subject to $u_1, u_2 \geq 0$, where

$$\theta(u) = \min_{x_j \in \{0,1,2,3\}} \{(3 + u_1 - 2u_2)x_1 + (4 - 3u_1 - u_2)x_2 + 5u_2\}$$

One can easily compute $\theta(u)$ by setting $\theta(u) = \theta(u, x^*)$, where

$$x_1^* = \begin{cases} 0 & \text{if } 3 + u_1 - 2u_2 \geq 0 \\ 3 & \text{otherwise} \end{cases} \qquad x_2^* = \begin{cases} 0 & \text{if } 4 - 3u_1 - u_2 \geq 0 \\ 3 & \text{otherwise} \end{cases}$$
(4.87)

Suppose the first iterate is $u^0 = (0,0)$. Then, by (4.87), $x^0 = (0,0)$ and $\theta(u^0) = 0$. A direction of steepest ascent is given by the subgradient $\xi^0 = g(x^0) = (0,5)$. Using the L_∞ norm, $\|\xi^0\| = \max_j\{|\xi_j^0|\} = 5$. If the upper bound $\bar{\theta}$ is set to 10 and $\gamma = 0.1$, then $\alpha_0 = 0.2$ and $u^1 = u^0 + \alpha_0 \xi^0 = (0, 0.5)$. As the algorithm continues, there is no convergence but the value of $\theta(u^k)$ never exceeds 10. One option at this point is to re-run the algorithm with smaller values of $\bar{\theta}$ until convergence is achieved at a value very close to $\bar{\theta}$. This occurs when $\bar{\theta} \approx 9\frac{2}{7}$, which, as found in Section 3.4.2, is the optimal value of the Lagrangean dual. The iterates u^k converge to $u = (\frac{5}{7}, \frac{13}{7})$, which is the optimal solution. Another option is to reduce $\bar{\theta}$ gradually during the algorithm but increase it somewhat when $\theta(u^k) > \bar{\theta}$.

4.6.6 Exercises

1 The aim is to solve a relaxation of the integer programming problem

$$\min cx$$
$$Ax \geq b$$
$$0 \leq x \leq d, \ x \text{ integral}$$

very quickly at each node of a branching tree. Because solving the full continuous relaxation is too slow, an alternative is to solve a Lagrangean relaxation in which the constraints $Ax \geq b$ are dualized. Describe how to form the Lagrangean relaxation and write a closed-form formula for its optimal solution at each node of the branching tree. Branching is accomplished by splitting the domain of a variable.

2 Given an integer programming problem

$$\min cx$$
$$Ax \geq a$$
$$Bx \geq b$$
$$x \geq 0 \text{ and integral}$$
(4.88)

suppose that $Bx \geq b, x \geq 0$ describes a polyhedron with all integral extreme points. Show that a Lagrangean dual in which $Ax \geq a$ is dualized provides the same bound as the continuous relaxation of (4.88).

3 An optimization problem minimizes $\sum_j f_j(x_j)$ subject to constraint sets $\mathcal{C}_1, \ldots, \mathcal{C}_m$. Each constraint set uses its own variables, except for a few variables that occur in more than one constraint set. Specifically, for each pair k, ℓ ($k < \ell$), constraint sets $\mathcal{C}_k$ and $\mathcal{C}_\ell$ have the variables x_j for $j \in J_{k\ell}$ in common. Show how Lagrangean relaxation might take advantage of this structure to obtain a lower bound on the optimal value.

4 A subgradient of a concave function $\theta(u)$ at $u = u^*$ is a vector ξ such that $\theta(u) - \theta(u^*) \leq \xi(u - u^*)$ for all u. Show that if $\theta(u^*) = f(x^*) - u^* g(x^*)$ for the Lagrangean function $\theta(u)$, then $\xi = -g(x^*)$ is a subgradient of $\theta(u)$ at $u = u^*$.

4.7 Disjunctions of Linear Systems

A broad class of constraints can be written as a disjunction of linear inequality systems. In particular, any constraint that can be expressed by mixed integer linear inequalities can, at least in principle, be written in this form. It is therefore useful to be able to relax disjunctions of linear systems. A variety of relaxation methods are known, including the convex hull relaxation, which is the tightest possible linear relaxation. There are also weaker relaxations that require fewer variables.

A disjunction of linear systems has the form

$$\bigvee_{k \in K} A^k x \geq b^k \qquad (4.89)$$

Each system $A^k x \geq b^k$ represents a polyhedron, and the feasible set of (4.89) is a finite union of polyhedra.

4.7.1 Convex Hull Relaxation

The *convex hull* of a set $S \subset \mathbb{R}^n$ is the set $\text{conv}(S)$ of all convex combinations of points in S. A *convex combination* of points $x^1, \ldots, x^m$ is a point of the form $\sum_{k=1}^m \alpha_k x^k$, where $\sum_k \alpha_k = 1$ and each $\alpha_k \geq 0$.

If the convex hull of a set is a polyhedron, then the inequalities describing that polyhedron provide the tightest possible linear relaxation of the set. The convex hull of a finite union of polyhedra is clearly a

Disjunctions of Linear Systems

polyhedron. Thus, the tightest possible linear relaxation of a disjunction of linear systems can be obtained by describing the convex hull of its feasible set.

Suppose for the moment that each disjunct of (4.89) is feasible. Then, every point in the convex hull of the feasible set of (4.89) can be written as a convex combination $x = \sum_{k \in K} \alpha_k \bar{x}^k$, where each $\bar{x}^k$ lies in the polyhedron described by $A^k x \geq b^k$. The convex hull is therefore described by

$$x = \sum_{k \in K} \alpha_k \bar{x}^k$$
$$A^k \bar{x}^k \geq b^k, \quad k \in K \qquad (4.90)$$
$$\sum_{k \in K} \alpha_k = 1, \quad \alpha_k \geq 0, \quad k \in K$$

This is a nonlinear system, but it can be linearized by the change of variable $x^k = \alpha_k \bar{x}^k$:

$$x = \sum_{k \in K} x^k$$
$$A^k x^k \geq \alpha_k b^k, \quad k \in K \qquad (4.91)$$
$$\sum_{k \in K} \alpha_k = 1, \quad \alpha_k \geq 0, \quad k \in K$$

It will be shown that (4.91) is a *convex hull relaxation* of (4.89), in the sense that the projection of the feasible set of (4.91) onto x is the closure of the convex hull of (4.89).[1] This is true even when not all disjuncts of (4.89) are feasible.

THEOREM 4.17 *System (4.91) is a convex hull relaxation of the disjunction (4.89).*

Proof. Let C be the convex hull of (4.89), P the feasible set of (4.91), and P_x the projection of P onto x. The claim is that $cl(C) = P_x$, where $cl(C)$ is the closure of C.

First show that $cl(C) \subset P_x$. Since P_x is closed, it suffices to show that $C \subset P_x$. It may be assumed that at least one disjunct of (4.89) is feasible, because otherwise C is empty and trivially $C \subset P_x$. Let K_1 be the set of indices $k \in K$ for which $A^k x \geq b^k$ is feasible, and let $K_2 = K \setminus K_1$. Then

---

[1] A *closed* set S is a set that contains all of its limit points. Point x is a limit point of S if, for any $\epsilon > 0$, some point in S is no further than ϵ away from x. The closure of S is the result of adding to S all of its limit points. A polyhedron is clearly closed, as is any projection of a polyhedron.

any $x \in C$ can be written as a convex combination $\sum_{k \in K_1} \alpha_k \bar{x}^k$, where $A^k \bar{x}^k \geq b^k$ for each $k \in K_1$. Introduce the change of variable $x^k = \alpha_k \bar{x}^k$ for $k \in K$. Multiplying $A^k \bar{x}^k \geq b^k$ by the nonnegative number α_k, one obtains $A^k x^k \geq b^k \alpha_k$ for $k \in K_1$. Setting $\alpha_k = 0$ and $x^k = 0$ for $k \in K_2$, one has $x = \sum_{k \in K} x^k$ and $A^k x^k \geq b^k \alpha_k$ for all $k \in K$. Thus x, x^k, and α_k for $k \in K$ satisfy (4.91), and $x \in P_x$.

Now show that $P_x \subset cl(C)$. Define $\hat{P}$ to be the set of points x, x^k, and $\alpha_k > 0$ for $k \in K$ satisfying (4.91). Given any such point in $\hat{P}$, let $\bar{x}^k = x^k/\alpha_k$, and note that $x = \sum_{k \in K} \alpha_k \bar{x}^k$ and $A^k \bar{x}^k \geq b^k$ for $k \in K$. Thus $x \in C$, and $\hat{P}_x \subset C$, which implies $cl(\hat{P}_x) \subset cl(C)$. But since $P = cl(\hat{P})$, it follows that $P_x = cl(\hat{P}_x)$ and $P_x \subset cl(C)$. □

Note that the convex hull relaxation requires that each continuous variable x_j be disaggregated into $|K|$ continuous variables x_j^k. Thus, if $x \in \mathbb{R}^n$, the relaxation requires n new variables x_j^k and one new variable α_k for each disjunct. The convex hull relaxation therefore tends to be more useful when there are only a few disjuncts, or when the relaxation simplifies in a fashion that allows variables to be eliminated.

As an example, consider the disjunction

$$\begin{pmatrix} x_1 - 2x_2 \geq -2 \\ 0 \leq x_j \leq 2, \; j=1,2 \end{pmatrix} \vee \begin{pmatrix} 3x_1 - x_2 \geq 1 \\ 0 \leq x_j \leq 2, \; j=1,2 \end{pmatrix} \quad (4.92)$$

The feasible set of (4.92) is shown in Figure 4.8. The convex hull relaxation (4.91) is

$$\begin{aligned} &x_1 = x_1^1 + x_1^2, \quad x_2 = x_2^1 + x_2^2 \\ &x_1^1 - 2x_2^1 \geq -2\alpha_1, \quad 3x_1^2 - x_2^2 \geq \alpha_2 \\ &0 \leq x_j^1 \leq 2\alpha_1, \quad 0 \leq x_j^2 \leq 2\alpha_2, \quad j=1,2 \\ &\alpha_1 + \alpha_2 = 1, \quad \alpha_1, \alpha_2 \geq 0 \end{aligned}$$

which can be simplified somewhat by eliminating x_j^2 and α_2. If α_1 is renamed α, and x_j^1 is renamed y_j, this yields

$$\begin{aligned} &y_1 - 2y_2 \geq -2\alpha, \quad 3(x_1 - y_1) - (x_2 - y_2) \geq 1 - \alpha \\ &0 \leq y_j \leq 2\alpha, \quad x_j \geq y_j, \quad x_j - y_j \leq 2(1-\alpha), \quad j=1,2 \\ &0 \leq \alpha \leq 1 \end{aligned}$$

4.7.2 Big-M Relaxation

The big-M relaxation for a disjunction of linear systems (4.89) is generally weaker than the convex hull relaxation, but it requires fewer variables—one additional variable for each disjunct. It is obtained by

Disjunctions of Linear Systems

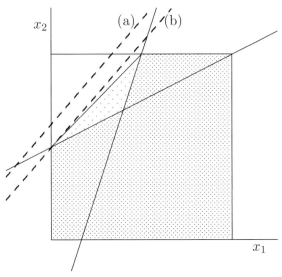

Figure 4.8. Relaxations of the disjunction (4.92). The darker shaded area is the feasible set, and the entire shaded area is its convex hull. Dashed line (a) represents the inequality $\frac{3}{2}x_1 - \frac{4}{3}x_2 \geq -\frac{5}{3}$, which, with the bounds $0 \leq x_j \leq 2$, describes the projection of the big-M relaxation. Dashed line (b) represents the supporting inequality $\frac{3}{2}x_1 - \frac{4}{3}x_2 \geq -\frac{4}{3}$.

first formulating a 0-1 model for the disjunction:

$$\begin{aligned}
A^k x \geq b^k - M^k(1-\alpha_k), \quad k \in K &\quad (a) \\
L \leq x \leq U &\quad (b) \\
\sum_{k \in K} \alpha_k = 1 &\quad (c) \\
\alpha_k \in \{0,1\}, \quad k \in K &
\end{aligned} \quad (4.93)$$

The model assumes that each x_j is bounded ($L_j \leq x_j \leq U_j$). Since one α_k is forced to be 1, the constraints (a) force x to lie in at least one of the polyhedra defined by $A^k x \geq b^k$. The components of the vector M^k should be large enough that $a^k x \geq b^k - M^k(1-\alpha_k)$ does not constrain x when $\alpha_k = 0$. On the other hand, the components of M^k should be as small as possible in order to yield a tighter relaxation. Thus, one can set

$$M_i^k = b_i^k - \min_{L \leq x \leq U}\{A_i^k x\} = b_i^k - \sum_j \min\{0, A_{ij}^k\}U_j - \sum_j \max\{0, A_{ij}^k\}L_j \quad (4.94)$$

where A_i^k is row i of A^k. The integrality condition in (4.93) is now relaxed to obtain the linear relaxation

$$\begin{aligned} A^k x &\geq b^k - M^k(1-\alpha_k), \quad k \in K \\ L &\leq x \leq U \\ \sum_{k \in K} \alpha_k &= 1, \quad \alpha_k \geq 0, \quad k \in K \end{aligned} \quad (4.95)$$

For example, the disjunction (4.92) can be viewed as the disjunction

$$(x_1 - 2x_2 \geq -2) \vee (3x_1 - x_2 \geq 1) \quad (4.96)$$

plus bounds $0 \leq x_j \leq 2$. Since $U_1 = U_2 = 2$, (4.94) sets $M_1^1 = 2$ and $M_1^2 = 3$. So the big-M relaxation (4.93) becomes

$$\begin{aligned} x_1 - 2x_2 &\geq 2\alpha - 4, \quad 3x_1 - x_2 \geq 1 - 3\alpha \\ 0 &\leq x_j \leq 2, \quad j = 1, 2 \\ 0 &\leq \alpha \leq 1 \end{aligned} \quad (4.97)$$

The projection of this relaxation onto (x_1, x_2) is described by

$$\tfrac{3}{2}x_1 - \tfrac{4}{3}x_2 \geq -\tfrac{5}{3} \quad (4.98)$$

and the bounds $0 \leq x_j \leq 2$ (Figure 4.8). It is clearly weaker than the convex hull relaxation.

One way to strengthen the big-M relaxation is to observe that if x does not belong to the polyhedron defined by the kth disjunct, then it must belong to at least one of the other polyhedra. This allows one to set

$$M_i^k = b_i^k - \min_{\ell \neq k} \left\{ \min_{L \leq x \leq U} \left\{ A_i^k x \mid A^\ell x \geq b^\ell \right\} \right\} \quad (4.99)$$

Using these values in (4.95) generally results in a stronger relaxation. In the example (4.96), one obtains $M_1^1 = 1$ and $M_1^2 = 2$, which yields the relaxation

$$\begin{aligned} x_1 - 2x_2 &\geq 2\alpha - 3, \quad 3x_1 - x_2 \geq 1 - 2\alpha \\ 0 &\leq x_j \leq 2, \quad j = 1, 2 \\ 0 &\leq \alpha \leq 1 \end{aligned}$$

The projection of this relaxation onto x_1, x_2 is the single inequality

$$\tfrac{3}{2}x_1 - \tfrac{2}{3}x_2 \geq -\tfrac{4}{3} \quad (4.100)$$

plus the bounds $0 \leq x_j \leq 2$, which, as Figure 4.8 illustrates, is an improvement over (4.97).

Disjunctions of Linear Systems

The drawback of using formula (4.99), in general, is that one must solve $|K|$ linear programming problems to compute each M_i^k. It will be shown in the next section, however, that when relaxing a disjunction of single inequalities plus bounds (as is the case here), there is a closed-form expression that yields (4.100).

4.7.3 Disjunctions of Linear Inequalities

The big-M relaxation simplifies considerably when (4.89) is a system of single linear inequalities:
$$\bigvee_{k \in K} a^k x \geq b_k \tag{4.101}$$

It is again assumed that $L \leq x \leq U$. When projected onto x, the big-M relaxation (4.93) simplifies to a single inequality whose coefficients are trivial to compute.

The big-M relaxation for the disjunction (4.101) is
$$\begin{aligned} a^k x \geq b_k - M_k(1 - \alpha_k), \quad k \in K & \quad (a) \\ L \leq x \leq U & \quad (b) \\ \sum_{k \in K} \alpha_k \geq 1, \quad \alpha_k \geq 0, \quad k \in K & \quad (c) \end{aligned} \tag{4.102}$$

As before, each M_k is chosen so that it is a lower bound on $a^k x - b_k$, for instance, by using (4.94):
$$M_k = b_k - \min_{L \leq x \leq U}\{a^k x\} = b_k - \sum_j \min\{0, a_j^k\} U_j - \sum_j \max\{0, a_j^k\} L_j \tag{4.103}$$

It can be assumed without loss of generality that $M_k > 0$, because otherwise the corresponding inequality is vacuous and can be dropped.

One can now eliminate the variables α_k by taking a linear combination of the inequalities (4.102a), where the kth inequality receives weight $1/M_k$. This yields
$$\left(\sum_{k \in K} \frac{a^k}{M_k}\right) x \geq \sum_{k \in k} \frac{b_k}{M_k} - |K| + 1 \tag{4.104}$$

THEOREM 4.18 *If each M_k is given by (4.103), the inequality (4.104) and the bounds $L \leq x \leq U$ describe the projection of the big-M relaxation (4.93) onto x.*

Proof. Let S be the feasible set of (4.93) and $\bar{S}$ the feasible set of (4.104). The claim is that the projection of S onto x is $\bar{S}$. Clearly, the projection of S onto x is a subset of $\bar{S}$, because as just shown, (4.104) is a nonnegative linear combination of (4.93). It therefore suffices to show that for any $\bar{x} \in \bar{S}$ there are $\alpha_k \in [0,1]$ such that $(\bar{x}, \alpha) \in S$. First "solve" (4.102a) for α_k by setting

$$\bar{\alpha}_k = \frac{1}{M_k}(a^k \bar{x} - b_k + M_k)$$

By construction, $\alpha = (\alpha_1, \ldots, \alpha_{|K|})$ satisfies (4.93a) if $\alpha \leq \bar{\alpha}$. Also,

$$\sum_{k \in K} \bar{\alpha}_k = \left(\sum_{k \in K} \frac{a^k}{M_k}\right) \bar{x} - \sum_{k \in K} \frac{b_k}{M_k} + |K| \geq 1$$

where the inequality follows from (4.104). Further, $\bar{\alpha}_k \geq 0$, because

$$M_k \bar{\alpha}_k = a^k \bar{x} - b_k + M_k = a^k \bar{x} - \sum_j \min\{0, a_j^k\} U_j - \sum_j \max\{0, a_j^k\} L_j \geq 0$$

where the second equality is from (4.103), and the inequality is from the fact that $L \leq x \leq U$. Thus, if one sets

$$\alpha = \frac{\bar{\alpha}}{\sum_{k \in K} \bar{\alpha}_k}$$

then $(\bar{x}, \alpha)$ satisfies (4.93). □

To take an example, the inequality (4.104) for the disjunction (4.96) is precisely (4.98). This inequality (along with bounds $0 \leq x_j \leq 2$) is the projection of the big-M relaxation onto x. The inequality can be strengthened by using the tighter bounds (4.99). As noted earlier, there is a closed-form expression for the relaxation that results from using (4.99) in the special case of a disjunction of single inequalities.

In fact, any valid inequality $cx \geq d$ for the disjunction (4.101) can be strengthened in this way to obtain a *supporting* inequality $cx \geq d^*$ for (4.101). A supporting inequality for a set S is an inequality that is satisfied by every point of S and is satisfied as an equation by at least one point in S. In particular, the inequality (4.104) can be strengthened in this fashion, unless of course it is already a supporting inequality.

The desired right-hand side d^* is the smallest of the minimum values obtained by minimizing cx subject to each of the disjuncts $a^k x \geq b_k$. That is,

$$d^* = \min_{k \in K} d_k^* \tag{4.105}$$

where
$$d_k^* = \min_{L \leq x \leq U} \left\{ cx \mid a^k x \geq b_k \right\}$$

The computation of d_k^* is simplified if $c \geq 0$ and the lower bounds on the variables are zero. To this end, one can introduce the change of variable

$$\hat{x}_j = \begin{cases} x_j - L_j & \text{if } c_j \geq 0 \\ U_j - x_j & \text{otherwise} \end{cases}$$

The strengthened elementary inequality in terms of $\hat{x}$, namely $\hat{c}\hat{x} \geq \hat{d}^*$, can now be computed, where $\hat{c}_j = |c_j|$. The right-hand side of $cx \geq d^*$ can then be recovered from (4.105) by setting

$$d_k^* = \hat{d}_k^* + \sum_{\substack{j \\ c_j > 0}} L_j c_j + \sum_{\substack{j \\ c_j < 0}} U_j c_j \tag{4.106}$$

It remains to compute

$$\hat{d}_k^* = \min_{\hat{x} \geq 0} \left\{ \hat{c}x \mid \hat{a}^k \hat{x} \geq \hat{b}_k \right\}, \tag{4.107}$$

where

$$\hat{a}_j^k = \begin{cases} a_j^k & \text{if } c_j \geq 0 \\ -a_j^k & \text{otherwise} \end{cases}$$

and

$$\hat{b}_k = b_k - \sum_{\substack{j \\ c_j > 0}} L_j a_j^k - \sum_{\substack{j \\ c_j < 0}} U_j a_j^k$$

Because $\hat{c} \geq 0$, linear programming duality applied to (4.107) yields

$$\hat{d}_k^* = \min_{\substack{j \\ \hat{a}_j^k > 0}} \left\{ \frac{\hat{c}_j}{\hat{a}_j^k} \right\} \max\{\hat{b}_k, 0\} \tag{4.108}$$

This proves the next theorem.

THEOREM 4.19 *A valid inequality $cx \geq d$ for the disjunction (4.101) is supporting if and only if $d = d^*$, where d^* is defined by (4.105), (4.106) and (4.108).*

To apply this theorem to the disjunction (4.96), note that the inequalities $\hat{a}^k \hat{x} \geq \hat{b}_k$ for $k = 1, 2$ are $\hat{x}_1 + 2\hat{x}_2 \geq 2$ and $3\hat{x}_1 + \hat{x}_2 \geq 3$. Then, $(\hat{d}_1^*, \hat{d}_2^*) = (\frac{4}{3}, \frac{3}{2})$, and $d^* = \min\{d_1^*, d_2^*\} = \min\{-\frac{4}{3}, -\frac{7}{6}\} = -\frac{4}{3}$. This yields the supporting inequality (4.100), which is illustrated in Figure 4.8.

4.7.4 Disjunctions of Linear Equations

The big-M relaxation for a disjunction of linear equations

$$\bigvee_{k \in K} a^k x = b_k \tag{4.109}$$

has some special structure that will be useful when relaxing the element constraint. Because each $a^k x = b_k$ can be written as a system of two inequalities $a^k x \geq b_k$ and $-a^k x \geq -b_k$, the big-M relaxation of (4.109) is

$$\begin{aligned} a^k x &\geq b_k - M'_k(1 - \alpha_k), \quad k \in K \\ -a^k x &\geq -b_k - M''_k(1 - \alpha_k), \quad k \in K \end{aligned} \tag{4.110}$$

Using (4.99), the big-Ms can be set to

$$\begin{aligned} M'_k &= b_k - \min_{\ell \neq k}\left\{ \min_{L \leq x \leq U}\left\{ a^k x \,\middle|\, a^\ell x = b_\ell \right\} \right\} \\ M''_k &= -b_k - \min_{\ell \neq k}\left\{ \min_{L \leq x \leq U}\left\{ -a^k x \,\middle|\, a^\ell x = b_\ell \right\} \right\} \end{aligned} \tag{4.111}$$

THEOREM 4.20 *If M'_k, M''_k are as given by (4.111), then (4.110) and the bounds $L \leq x \leq U$ provide a relaxation of (4.109).*

For example, if one relaxes the disjunction

$$(x_1 - 2x_2 = -2) \vee (3x_1 - x_2 = 1)$$

with $x_1, x_2 \in [0, 2]$, then $(M'_1, M'_2) = (1, 2)$ and $(M''_1, M''_2) = (\frac{7}{3}, 3)$, and (4.110) becomes

$$\begin{aligned} x_1 - 2x_2 &\geq -2 - \alpha & 3x_1 - x_2 &\geq 1 - 2(1 - \alpha) \\ -x_1 + 2x_2 &\geq 2 - \tfrac{7}{3}\alpha & -3x_1 + x_2 &\geq -1 - 3(1 - \alpha) \\ 0 \leq \alpha &\leq 1 \end{aligned}$$

plus bounds. Projecting out α, this becomes

$$-x_1 + x_2 \leq 1, \quad 6x_1 - 5x_2 \leq 2$$

plus bounds. As it happens, this is a convex hull relaxation, illustrated in Figure 4.9.

Disjunctions of Linear Systems

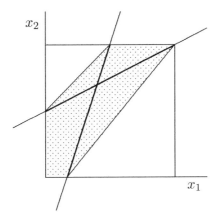

Figure 4.9. Projected big-M relaxation (shaded area) of the feasible set consisting of the two heavy line segments, which is described by bounds $x_1, x_2 \in [0, 2]$ and the disjunction of the equations $x_1 - 2x_2 = -2$, $3x_1 - x_2 = 1$.

4.7.5 Separating Disjunctive Cuts

When the convex hull relaxation or big-M relaxation of a disjunction of linear systems has a large number of variables and constraints, it may be advantageous to generate only a separating cut for the disjunction. Unfortunately, the identification of a separating cut requires solving a linear system that is similar in size to the convex hull relaxation. Thus, if there is only one disjunction in the problem, one may as well put its entire convex hull relaxation into the problem relaxation. But if there are several disjunctions, their combined convex hull relaxations can be quite large. It is much faster to generate only a separating cut for each, because this can be accomplished by solving a separate linear system for each disjunction, and the resulting relaxation contains only the separating cuts.

To find a separating cut for a disjunction (4.89) of linear systems, let $\bar{x}$ be the solution of the current problem relaxation. The goal is to identify a valid cut $dx \geq \delta$ for (4.89) that $\bar{x}$ violates. This can be done by recalling from Corollary 3.6 that $dx \geq \delta$ is valid for the system $A^k x \geq b^k$, $x \geq 0$ if and only if it is dominated by a surrogate of $A^k x \geq b^k$. If the system is feasible, a dominating surrogate can be written $uA^i x \geq ub^i$, where $d \geq uA^i$, $\delta \leq ub$, and $u \geq 0$. But $dx \geq \delta$ is valid for the disjunction as a whole if and only if it is valid for each feasible disjunct.

THEOREM 4.21 *The inequality $dx \geq \delta$ is valid for a disjunction (4.89) of linear systems containing nonnegativity constraints $x \geq 0$ if and only if for each feasible system $A^k x \geq b^k$ there is a $u^k \geq 0$ such that $d \geq u^k A^k$ and $\delta \leq u^k b^k$.*

This allows one to write a linear programming problem that finds a cut $dx \geq \delta$ that $\bar{x}$ violates:

$$\begin{array}{ll}
\max \delta - d\bar{x} & (a) \\
\delta \leq u^k b^k, \ k \in K & (b) \\
d \geq u^k A^k, \ k \in K & (c) \\
-e \leq d \leq e & (d) \\
u^k \geq 0, \ k \in K & \\
\delta, d \text{ unrestricted} &
\end{array} \quad (4.112)$$

The variables in the problem are d, δ, and u^k for $k \in K$. Since a strong cut is desired, it should be designed so that it is in some sense maximally violated by $\bar{x}$. This is the intent of the objective function (a). If the maximum value is less than or equal to zero, there is no separating cut. The constraint (d) is added because, otherwise, the problem would be unbounded; if (d, δ) is feasible, then any scalar multiple $(\alpha d, \alpha \delta)$ is feasible. The problem can, in general, be bounded by placing a bound on some norm of d. One possibility is to bound the L_∞ norm, which is $\max_j\{|d_j|\}$. This is accomplished by constraint (d), in which e is a vector of ones. Another possibility is to bound the L_1 norm, which is $\sum_j |d_j|$. This alternative will be taken up shortly.

As an example, consider the disjunction (4.92), and suppose $\bar{x} = (\frac{1}{2}, 2)$ is the solution of the current problem relaxation (Figure 4.10). To find a separating cut, solve (4.112):

$$\begin{array}{l}
\max \delta - \frac{1}{2}d_1 - 2d_2 \\
\delta \leq -2u_1^1 - 2u_4^1 - 2u_5^1 \\
\delta \leq u_1^2 - 2u_4^2 - 2u_5^2 \\
u_1^1 + u_2^1 - u_4^1 \leq d_1 \\
-2u_1^1 + u_3^1 - u_5^1 \leq d_2 \\
3u_1^2 + u_2^2 - u_4^2 \leq d_1 \\
-u_1^2 + u_3^2 - u_5^2 \leq d_2 \\
-1 \leq d_j \leq 1, \ d = 1, 2 \\
u_i^k \geq 0, \ i = 1, \ldots, 5, \ k = 1, 2
\end{array} \quad (4.113)$$

A solution is

$$d = (1, -1), \ \delta = -1, \ u^1 = (\tfrac{1}{2}, \tfrac{1}{2}, 0, 0, 0), \ u^2 = (\tfrac{1}{3}, 0, 0, 0, \tfrac{2}{3})$$

This yields the separating cut $d_1 - d_2 \geq -1$, which is the facet-defining cut shown in Figure 4.10.

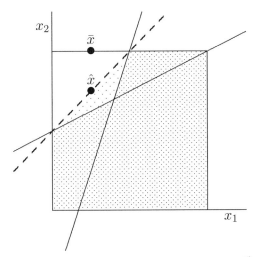

Figure 4.10. Disjunctive cut (dashed line) that separates $\bar{x} = (\frac{1}{2}, 2)$ from the convex hull (shaded area) of the feasible set (dark shading). The point $\hat{x}$ is a closest point in the convex hull set to $\bar{x}$, as measured by the rectilinear distance.

The dual of (4.112) sheds light on the separation problem. Associating dual variables y_k with (b), x^k with (c), and s, s' with (d), the dual is

$$\begin{aligned}
& \min(s + s')e & (a) \\
& \bar{x} - \sum_k x^k = s - s' & (b) \\
& A^k x^k \geq b^k y_k, \quad k \in K & (c) \\
& \sum_k y_k = 1 & (d) \\
& s, s', x^k, y_k \geq 0, \ k \in K &
\end{aligned} \quad (4.114)$$

If $s = s'$, the constraint set of (4.114) is the convex hull relaxation of the disjunction (4.89). Thus, (4.114) finds a point $\sum_k x^k$ in the convex hull that has minimum distance to $\bar{x}$, where the distance is measured by $\sum_j (s_j + s'_j)$. This distance measure is the L_1 norm of $\bar{x} - \sum_k x^k$, also known as the rectilinear distance from $\bar{x}$ to $\sum_k x^k$. Thus, the primal problem (4.112) of finding the strongest separating cut is essentially the same as the dual problem (4.114) of finding the closest point $\hat{x}$ in the convex hull to $\bar{x}$. Furthermore,

THEOREM 4.22 *The separating cut $dx \geq \delta$ found by the primal problem (4.112) contains a closest point $\hat{x}$ in the convex hull of (4.89) to $\bar{x}$. That is, $d\hat{x} = \delta$.*

Proof. Let δ, d, and $\hat{u}^k$ for $k \in K$ be an optimal solution of the primal problem (4.112), and let $\hat{x}^k, \hat{y}_k$ for $k \in K$ be the corresponding optimal solution of the dual problem (4.114). The closest point $\hat{x}$ is $\sum_k \hat{x}^k$, and the claim is that $d \sum_k \hat{x}^k = \delta$. Apply the complementary slackness principle (Corollary 3.9) to (4.112c) to obtain

$$\sum_{k \in K} (\hat{u}^k A^k - d)\hat{x}^k = 0$$

Thus,

$$d \sum_{k \in K} \hat{x}^k = \sum_{k \in K} \hat{u}^k A^k \hat{x}^k = \sum_{k \in K} \hat{u}^k b^k \hat{y}_k = \sum_{k \in K} \delta \hat{y}_k = \delta \quad (4.115)$$

where the second equality is due to complementary slackness applied to (4.114c), the third is due to complementary slackness applied to (4.112b), and the last is due to (4.114d). □

A dual solution for the example problem (4.113) is

$$x^1 = (0, \tfrac{1}{2}), \ x^2 = (\tfrac{1}{2}, 1), \ y = (\tfrac{1}{2}, \tfrac{1}{2}), \ s = (0,0), \ s' = (0, \tfrac{1}{2})$$

So a closest point in the convex hull to $\bar{x}$ is $\hat{x}^1 + \hat{x}^2 = (\tfrac{1}{2}, \tfrac{3}{2})$. As shown in Figure 4.10, the separating cut runs through this point.

Another approach to bounding the separating cut problem (4.112) is to bound the L_1 norm rather than the L_∞ norm of d. For example, one can set $\sum_j |d_j| \leq 1$. This is accomplished by letting $d = d^+ - d^-$, where $d^+, d^- \geq 0$. Then the bound is $e(d^+ + d^-) \leq 1$, where e is a row vector of ones. The problem (4.89) becomes

$$\begin{aligned}
&\max \delta - (d^+ - d^-)\bar{x} \\
&\delta \leq u^k b^k, \ k \in K \\
&d^+ - d^- \geq u^k A^k, \ k \in K \\
&e(d^+ + d^-) \leq 1 \\
&d^+, d^- \geq 0; \ u^k \geq 0, \ k \in K \\
&\delta \text{ unrestricted}
\end{aligned} \quad (4.116)$$

The dual of (4.116) finds a point in the convex hull that is closest to $\bar{x}$, as measured by the L_∞ norm. Again, the separating cut contains this point. Thus, there is a duality of norms. Bounding the L_∞ norm finds a closest point as measured by the L_1 norm, while bounding the L_1 norm finds a closest point as measured by the L_∞ norm.

Disjunctions of Linear Systems 311

4.7.6 Exercises

1 A common modeling situation requires that an operation either observe certain linear constraints or shut down:
$$[Ax \geq b] \vee [x = 0]$$
Write a convex hull relaxation of this disjunction and simplify it.

2 Write a convex hull relaxation of
$$\begin{bmatrix} x_1 = 0 \\ 0 \leq x_2 \leq 1 \end{bmatrix} \vee \begin{bmatrix} x_1 \geq 0 \\ x_2 = 0 \end{bmatrix}$$
Note that the convex hull is not a closed set and that the convex hull relaxation describes its closure.

3 Show by example that a convex hull relaxation of
$$[A^1 x \geq b^1] \vee [A^2 x \geq b^2]$$
combined with $Dx \geq d$ can be weaker than a convex hull relaxation of
$$\begin{bmatrix} A^1 x \geq b^1 \\ Dx \geq d \end{bmatrix} \vee \begin{bmatrix} A^2 x \geq b^2 \\ Dx \geq d \end{bmatrix}$$
even when $Dx \geq d$ is a simple nonnegativity constraint $x \geq 0$.

4 Write a convex hull relaxation for the disjunction
$$[(x_1, x_2) = (0, 1)] \vee [2x_1 - x_2 \geq 2]$$
where each $x_j \in [0, 2]$. Simplify the relaxation so that it contains only the variables x_1, x_2, α where $0 \leq \alpha \leq 1$. Use Fourier-Motzkin elimination to project the relaxation onto x_1, x_2, and verify that it describes the convex hull by drawing a graph of the feasible set. Hint: to obtain the convex hull, the bounds on x_j must be included in the disjuncts (see Exercise 2). The disjunction becomes
$$\begin{bmatrix} x_1 = 0 \\ x_2 = 1 \end{bmatrix} \vee \begin{bmatrix} 2x_1 - x_2 \geq 2 \\ x_1 \leq 2 \\ x_2 \geq 0 \end{bmatrix}$$

5 Write a big-M relaxation of the disjunction in the previous exercise using the big-Ms given by (4.94). Use Fourier-Motzkin elimination to project the relaxation onto x_1, x_2 and draw a graph of the result. Note that the relaxation does not describe the convex hull. Hint:

here the bounds on x_j need not be included in the disjuncts. The following disjunction can therefore be relaxed with bounds $x_j \in [0, 2]$:

$$\begin{bmatrix} -x_1 \geq 0 \\ x_2 \geq 1 \\ -x_2 \geq -1 \end{bmatrix} \vee \begin{bmatrix} 2x_1 - x_2 \geq 2 \end{bmatrix}$$

6 Tighten the relaxation of Exercise 5 by using the big-Ms given by (4.99). Project the relaxation onto x_1, x_2 and draw a graph to verify that the relaxation is in fact tighter.

7 Use Theorem 4.18 to write a big-M relaxation for

$$[-2x_1 + x_2 \geq 0] \vee [-x_1 + 2x_2 \geq 2]$$

subject to the bounds $x_j \in [0, 2]$. Draw a graph of the relaxation along with the feasible set.

8 Use Theorem 4.19 to tighten the relaxation of Exercise 7. Draw a graph and note that the tightened inequality supports the feasible set.

9 Verify that (4.108) is the optimal value of (4.107) by solving the dual of (4.107).

10 Use Theorem 4.20 to write a big-M relaxation of

$$[-x_1 + x_2 = 1] \vee [x_1 - x_2 = 1]$$

subject to the bounds $x_j \in [0, 2]$. Draw a graph of the relaxation along with the feasible set. Note that it is not a convex hull relaxation.

11 Solve a linear programming problem to find an optimal separating cut for the point $\bar{x} = (0, 1)$ and the disjunction

$$\begin{bmatrix} x_1 + x_2 \leq 2 \\ x_1, x_2 \geq 0 \end{bmatrix} \vee \begin{bmatrix} x_1 \leq 1 \\ x_1, x_2 \geq 0 \end{bmatrix}$$

on the assumption that the L_∞ norm of the vector of coefficients in the cut is bounded. Examine the dual solution to identify the point in the convex hull of the feasible set of the disjunction that is closest to $\bar{x}$, as measured by the L_1 norm.

12 Verify the identity (4.115) in the proof of Theorem 4.22.

Disjunctions of Nonlinear Systems 313

13 Write the dual of (4.116) and indicate how the dual solution is used to identify the point in the convex hull of the feasible set of (4.89) that is closest to $\bar{x}$ as measured by the L_∞ norm.

14 Solve the system (4.116) for the problem in Figure 4.10 and identify the optimal separating cut for the L_1 norm. Note that it is the same as the cut obtained for the L_∞ norm. Examine the dual solution and determine the point in the convex hull of the feasible set that is closest to $\bar{x}$ as measured by the L_∞ norm. Hint: the optimal solution of (4.116) has $\delta = -\frac{1}{2}$, $u^1 = (\frac{1}{4}, 0, 0, 0, 0)$, $u^2 = (\frac{1}{6}, 0, 0, 0, \frac{1}{3})$.

4.8 Disjunctions of Nonlinear Systems

When formulating continuous relaxations of nonlinear constraints, the overriding concern is obtaining a convex relaxation. A convex relaxation is one with a convex feasible set (i.e., a feasible set that contains all convex combinations of its points). A convex relaxation is generally much easier to solve, because optimization over a convex feasible set is generally much easier than over more general feasible sets. Most nonlinear programming methods, for instance, are designed to find only a locally optimal solution, which is guaranteed to be optimal only in a small neighborhood surrounding it. Such a solution is globally optimal, however, if one is minimizing a convex function over a convex feasible set (or maximizing a concave function over a convex set).

Fortunately, a disjunction of nonlinear inequality systems can be given a convex relaxation using much the same techniques used earlier to relax disjunctions of linear systems—provided each individual system defines a convex, bounded set. In particular, convex hull relaxations and convex big-M relaxations can be easily formulated for disjunctions of convex nonlinear systems.

4.8.1 Convex Hull Relaxation

The problem is to find a continuous relaxation for the disjunction

$$\bigvee_{k \in K} g^k(x) \leq 0 \qquad (4.117)$$

where each $g^k(x)$ is a vector of functions $g_i^k(x)$ with $x \in \mathbb{R}^n$. It is assumed that $x \in [L, U]$, and $g^k(x)$ is bounded when $x \in [L, U]$. It is further assumed that each $g_i^k(x)$ is a convex function on $[L, U]$, meaning that

$$g_i^k((1-\alpha)x^1 + \alpha x^2) \leq (1-\alpha)g_i^k(x^1) + \alpha g_i^k(x^2)$$

for all $x^1, x^2 \in [L, U]$ and all $\alpha \in [0, 1]$. This implies that the feasible set of each system $g^k(x) \leq 0$ is convex.

To simplify exposition, it is assumed that every disjunct of (4.117) is feasible. The convex hull of (4.117) consists of all points that can be written as a convex combination of points $\bar{x}^k$ that respectively satisfy the disjuncts of (4.117). Thus,

$$x = \sum_{k \in K} \alpha_k \bar{x}^k$$
$$g^k(\bar{x}^k) \leq 0, \quad \text{all } k \in K$$
$$L \leq \bar{x}^k \leq U, \quad \text{all } k \in K$$
$$\sum_{k \in K} \alpha_k = 1, \quad \alpha_k \geq 0, \quad \text{all } k \in K$$

Using the change of variable $x^k = \alpha_k \bar{x}^k$, the following relaxation is obtained:

$$x = \sum_{k \in K} x^k$$
$$g^k\left(\frac{x^k}{\alpha_k}\right) \leq 0, \quad \text{all } k \in K \qquad (4.118)$$
$$\alpha_k L \leq x^k \leq \alpha_k U, \quad \text{all } k \in K$$
$$\sum_{k \in K} \alpha_k = 1, \quad \alpha_k \geq 0, \quad \text{all } k \in K$$

The function $g^k(x^k/\alpha_k)$ is in general nonconvex, but a classical result of convex analysis implies that one can restore convexity by multiplying the second constraint of (4.118) by α_k.

THEOREM 4.23 *Consider the set S consisting of all (x, α) with $\alpha \in [0, 1]$ and $x \in [\alpha L, \alpha U]$. If $g(x)$ is convex and bounded for $x \in [L, U]$, then*

$$h(x, \alpha) = \begin{cases} \alpha g(x/\alpha) & \text{if } \alpha > 0 \\ 0 & \text{if } \alpha = 0 \end{cases}$$

is convex and bounded on S.

Proof. To show convexity of $h(x, \alpha)$, arbitrarily choose points (x^1, α_1), $(x^2, \alpha_2) \in S$. Supposing first that $\alpha_1, \alpha_2 > 0$, convexity can be shown by noting that for any $\beta \in [0,1]$,

$$h\left(\beta x^1 + (1-\beta)x^2, \beta\alpha_1 + (1-\beta)\alpha_2\right)$$
$$= (\beta\alpha_1 + (1-\beta)\alpha_2)\, g\left(\frac{\beta x^1 + (1-\beta)x^2}{\beta\alpha_1 + (1-\beta)\alpha_2}\right)$$
$$= (\beta\alpha_1 + (1-\beta)\alpha_2)\, g\left(\frac{\beta\alpha_1}{\beta\alpha_1 + (1-\beta)\alpha_2}\frac{x^1}{\alpha_1} + \frac{(1-\beta)\alpha_1}{\beta\alpha_1 + (1-\beta)\alpha_2}\frac{x^2}{\alpha_2}\right)$$
$$\leq (\beta\alpha_1 + (1-\beta)\alpha_2)\left[\frac{\beta\alpha_1}{\beta\alpha_1 + (1-\beta)\alpha_2} g\left(\frac{x^1}{\alpha_1}\right) + \frac{(1-\beta)\alpha_1}{\beta\alpha_1 + (1-\beta)\alpha_2} g\left(\frac{x^2}{\alpha_2}\right)\right]$$
$$= \beta h\left(x^1, \alpha_1\right) + (1-\beta)h\left(x^2, \alpha_2\right)$$

where the inequality is due to the convexity of $g(x)$. If $\alpha_1 = \alpha_2 = 0$, then
$$h\left(\beta x^1 + (1-\beta)x^2, \beta\alpha_1 + (1-\beta)\alpha_2\right)$$
$$= h(0,0) = \beta h\left(x^1, \alpha_1\right) + (1-\beta)h\left(x^2, \alpha_2\right)$$

because $\alpha_j L \leq x^k \leq \alpha_j U$ implies $x^k = 0$. If $\alpha_1 = 0$ and $\alpha_2 > 0$,
$$h\left(\beta x^1 + (1-\beta)x^2, \beta\alpha_1 + (1-\beta)\alpha_2\right)$$
$$= h\left((1-\beta)x^2, (1-\beta)\alpha_2\right) = (1-\beta)g\left(\frac{x^2}{\alpha_2}\right)$$
$$= \beta h(0,0) + (1-\beta)h\left(x^2, \alpha_2\right)$$

Finally, $h(x,\alpha) = \alpha g(x/\alpha)$ is bounded because $\alpha \in [0,1]$, $x/\alpha \in [L, U]$, and $g(x)$ is bounded for $x \in [L, U]$. □

Due to Theorem 4.23, multiplying the second constraint of (4.118) by α_k yields a convex hull relaxation of (4.117):

$$\begin{aligned} x &= \sum_{k \in K} x^k \\ \alpha_k g^k\left(\frac{x^k}{\alpha_k}\right) &\leq 0, \quad \text{all } k \in K \\ \alpha_k L &\leq x^k \leq \alpha_k U, \quad \text{all } k \in K \\ \sum_{k \in K} \alpha_k &= 1, \quad \alpha_k \geq 0, \quad \text{all } k \in K \end{aligned} \qquad (4.119)$$

The following can be shown in a manner similar to the proof of Theorem 4.17.

THEOREM 4.24 *Suppose each $g^k(x)$ in (4.117) is convex and bounded for $x \in [L, U]$, and each disjunct contains the constraint $L \leq x \leq U$. Then (4.119) is a convex hull relaxation of (4.117).*

When the boundedness conditions of Theorem 4.24 are violated, the convex hull of a disjunction (4.117) need not be a closed set, even if the disjuncts describe closed sets. For example, the convex hull of

$$\begin{pmatrix} x_1 = 0 \\ x_2 \geq 0 \end{pmatrix} \vee \begin{pmatrix} x_1 \geq 0 \\ x_2 \geq \dfrac{1}{1+x_1} \end{pmatrix}$$

is the nonnegative quadrant except for points $(x_1, 0)$ with $x_1 > 0$. One might say that $x_1, x_2 \geq 0$ is a convex hull relaxation, but only in the sense that it describes the closure of the convex hull.

Since α_k can vanish, it is common in practice to use the constraint

$$(\alpha_k + \epsilon) g^k \left(\frac{x^k}{\alpha_k + \epsilon} \right) \leq 0, \ \ \text{all } k \in K \qquad (4.120)$$

in place of the second constraint of (4.119), for some small $\epsilon > 0$. The introduction of ϵ preserves convexity.

As an example, consider the disjunction

$$\left[x_1^2 + x_2^2 - 1 \leq 0 \right] \vee \left[(x_1 - 2)^2 + x_2^2 - 1 \leq 0 \right] \qquad (4.121)$$

with $x_1 \in [-1, 3]$ and $x_2 \in [-1, 1]$. The feasible set for (4.121) is the union of the discs in Figure 4.11. The convex hull relaxation (4.120) is

$$\begin{bmatrix} x_1 \\ x_2 \end{bmatrix} = \begin{bmatrix} x_{11} \\ x_{21} \end{bmatrix} + \begin{bmatrix} x_{12} \\ x_{22} \end{bmatrix}$$

$$\frac{x_{11}^2 + x_{21}^2}{\alpha + \epsilon} \leq \alpha + \epsilon$$

$$\frac{x_{12}^2 + x_{22}^2}{1 - \alpha + \epsilon} - 4x_{12} + 3(1 - \alpha + \epsilon) \leq 0$$

$$0 \leq \alpha \leq 1$$

(The bounds on x_1, x_2 are redundant and are omitted.) Figure 4.11 shows the projection of the feasible set of this relaxation onto x_1, x_2.

4.8.2 Big-M Relaxation

The big-M relaxation of the disjunction (4.117) introduces a variable α_k for each $k \in K$, where $\alpha_k = 1$ indicates that the kth disjunct is

Disjunctions of Nonlinear Systems

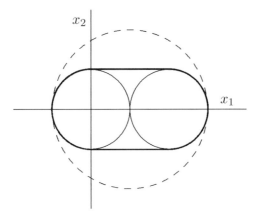

Figure 4.11. Feasible set of a disjunction of two nonlinear systems (two small discs), the convex hull relaxation (area within heavy boundary), and a big-M relaxation (disc with dashed boundary).

enforced. It is assumed that there are bounds $L \leq x \leq U$ on x. The big-M relaxation is

$$\begin{aligned} g^k(x) &\leq M^k(1 - \alpha_k), \text{ all } k \in K \\ L &\leq x \leq U \\ \sum_{k \in K} \alpha_k &= 1, \ \alpha_k \geq 0, \text{ all } k \in K \end{aligned} \quad (4.122)$$

where M^k is a vector of valid upper bounds on the component functions of $g^k(x)$, given that $L \leq x \leq U$. This relaxation is clearly convex, assuming that each $g^k(x)$ is convex.

The bounds M^k can be set to

$$M_i^k = \max_{L \leq x \leq U} \left\{ g_i^k(x) \right\} \quad (4.123)$$

but the tightest bound is

$$M_i^k = \max_{\ell \neq k} \left\{ \max_{L \leq x \leq U} \left\{ g_i^k(x) \mid g^\ell(x) \leq 0 \right\} \right\} \quad (4.124)$$

As an example, again consider the disjunction (4.121) with domains $x_1 \in [-1, 3]$ and $x_2 \in [-1, 1]$. Setting $M^1 = M^2 = 8$ as given by (4.124) yields the big-M relaxation

$$\begin{aligned} x_1^2 + x_2^2 - 1 &\leq 8(1 - \alpha) \\ (x_1 - 2)^2 + x_2^2 - 1 &\leq 8\alpha \\ 0 \leq \alpha &\leq 1 \end{aligned}$$

The large disc in Figure 4.11 depicts the projection of the relaxation onto the x-space. The projection is described by $(x_1 - 1)^2 + x_2^2 \leq 4$.

4.8.3 Exercises

1 Write a convex hull relaxation for
$$\left[\, x_1^2 \leq x_2 \leq 1 \,\right] \vee \left[\, -1 \leq x_2 \leq -x_1^2 \,\right]$$
and simplify it as much as possible.

2 Write a big-M relaxation for the disjunction in the previous exercise, using the big-Ms given by (4.124). Project it onto x_1, x_2, draw a graph, and note that it is a convex hull relaxation.

3 Show that if $g(x)$ satisfies the conditions of Theorem 4.23, then $\bar{h}(x, \alpha) = (\alpha + \epsilon) g(x/(\alpha + \epsilon))$ is convex for any $\epsilon > 0$. Hint: for any convex $f(x)$, it is trivial to show that $\bar{f}(x) = f(x + a)$ is convex.

4.9 MILP Modeling

One general method for obtaining a linear relaxation of a constraint set is to reformulate the constraints as an MILP model, and then drop the integrality condition on the variables.

MILP provides a highly versatile modeling language if one is sufficiently ingenious. There may be several ways, however, to write an MILP model of the same problem, and some models may be more succinct or have tighter relaxations than others. In fact, the more succinct model is often not the tighter one. Formulating a suitable relaxation is more an art than a science.

The theory of MILP modeling tells us that it is essentially tantamount to disjunctive modeling. A problem can be given an MILP model if and only if its feasible set is a finite union of polyhedra that satisfy a certain technical condition. In such cases, the feasible set is described by a disjunction of linear systems that define the polyhedra. The disjunction can then be rewritten as an MILP model using a convex hull formulation or, in many cases, a big-M formulation.

This sometimes provides a practical approach to constructing MILP models, but it often requires an impracticably large number of disjuncts. There are two popular alternatives in such cases: view the problem as several disjunctions to be relaxed separately (discussed below under the heading of disjunctive modeling), or view the problem as several knapsack constraints to be relaxed separately (knapsack modeling). Both strategies may be used within the same problem.

MILP Modeling 319

4.9.1 MILP Representability

It is known precisely what sort of feasible set can be represented by an MILP model. A subset S of $\mathbb{R}^n$ is MILP-representable if and only if S is the union of finitely many polyhedra, all of which have the same recession cone. The *recession cone* of a polyhedron P is the set of directions in which P is unbounded, or more precisely, the set of vectors $r \in \mathbb{R}^n$ such that, for some $u \in P$, $u + \beta r \in P$ for all $\beta \geq 0$.

Representation as an MILP allows the use of auxiliary variables, both continuous and discrete. Thus, the set S is *MILP representable* if there is a constraint set of the following form that represents S, meaning that the projection of its feasible set onto x is S:

$$
\begin{aligned}
& Ax + Bu + Dy \geq b \\
& x \in \mathbb{R}^n, \ u \in \mathbb{R}^m, \ y_k \in \{0,1\}, \ \text{all } k
\end{aligned}
\quad (4.125)
$$

The intuition behind MILP representability is that a finite union of polyhedra is described by a disjunction of linear systems:

$$\bigvee_{k \in K} A^k x \geq b^k \quad (4.126)$$

If the polyhedra have the same recession cone, the disjunction can be reformulated as an MILP model that is identical to the convex hull relaxation (4.91), except that each α_k is replaced by a 0-1 variable y_k:

$$
\begin{aligned}
& x = \sum_{k \in K} x^k \\
& A^k x^k \geq b^k y_k, \ k \in K \\
& \sum_{k \in K} y_k = 1, \ y_k \in \{0,1\}, \ k \in K
\end{aligned}
\quad (4.127)
$$

THEOREM 4.25 *A set $S \subset \mathbb{R}^n$ is MILP representable if and only if S is the union of finitely many polyhedra having the same recession cone. In particular, S is MILP representable if and only if S is the projection onto x of the feasible set of some disjunctive model of the form (4.127).*

Proof. Suppose first that S is the union of polyhedra P_k, $k \in K$, that have the same recession cone. Each P_k is the feasible set of some linear system $A^k x^k \geq b^k$. It can be shown as follows that S is represented by (4.127), and is therefore representable, because (4.127) has the form (4.125). Suppose first that $x \in S$. Then x belongs to some P_{k^*}, which means that x is feasible in (4.127) when $y_{k^*} = 1$, $y_k = 0$ for $k \neq k^*$, $x^{k^*} = x$, and $x^k = 0$ for $k \neq k^*$. The constraint $A^k x^k \geq b^k y_k$ is satisfied by definition when $k = k^*$, and it is satisfied for other k's because $x^k = y_k = 0$.

Now suppose that x, y and x^k for $k \in K$ satisfy (4.127). To show that $x \in S$, note that exactly one y_k, say y_{k^*}, is equal to 1. Then $A^{k^*} x^{k^*} \geq b^{k^*}$ is enforced, which means that $x^{k^*} \in P_{k^*}$. For other k's, $A^k x^k \geq 0$. Thus, $A^k(\beta x^k) \geq 0$ for all $\beta \geq 0$, which implies that x^k is a recession direction for P_k. Because by hypothesis all the P_ks have the same recession cone, each x^k ($k \neq k^*$) is a recession direction for P_{k^*}, which means that $x = x^{k^*} + \sum_{k \neq k^*} x^k$ belongs to P_{k^*} and therefore to the union S of the P_ks.

To prove the converse of the theorem, suppose that S is represented by (4.125). To show that S is a finite union of polyhedra, Let $P(\bar{y})$ be the set of all x that are feasible in (4.125) when $y = \bar{y} \in \{0,1\}^{|K|}$. Obviously $P(\bar{y})$ is a polyhedron, and S is the union of $P(\bar{y})$ over all $\bar{y} \in \{0,1\}^{|K|}$. To show that the $P(\bar{y})$'s have the same recession cone, note that

$$P(\bar{y}) = \left\{ x \;\middle|\; \begin{bmatrix} A & B & D \\ 0 & 0 & 1 \\ 0 & 0 & -1 \end{bmatrix} \begin{bmatrix} x \\ u \\ y \end{bmatrix} \geq \begin{bmatrix} a \\ \bar{y} \\ -\bar{y} \end{bmatrix} \text{ for some } u, y \right\}$$

But x' is a recession direction of $P(\bar{y})$ if and only if (x', u', y') is a recession direction of

$$\left\{ \begin{bmatrix} x \\ u \\ y \end{bmatrix} \;\middle|\; \begin{bmatrix} A & B & D \\ 0 & 0 & 1 \\ 0 & 0 & -1 \end{bmatrix} \begin{bmatrix} x \\ u \\ y \end{bmatrix} \geq \begin{bmatrix} a \\ \bar{y} \\ -\bar{y} \end{bmatrix} \right\}$$

for some u', y'. The latter is true if and only if

$$\begin{bmatrix} A & B & D \\ 0 & 0 & 1 \\ 0 & 0 & -1 \end{bmatrix} \begin{bmatrix} x' \\ u' \\ y' \end{bmatrix} \geq \begin{bmatrix} 0 \\ 0 \\ 0 \end{bmatrix}$$

This means that the recession directions of $P(\bar{y})$ are the same for all $\bar{y}$. □

It is important to note that the continuous relaxation of (4.127) is a valid convex hull relaxation of (4.126), due to Theorem 4.17, even if the disjuncts do not describe polyhedra with the same recession cone. Thus, a finite union of polyhedra can be given a convex hull relaxation even when there is no MILP formulation of the union.

A disjunction (4.126) can often be given a big-M relaxation (4.93), as well as a convex hull relaxation. A big-M relaxation exists if the left-hand side $A^k x$ of each disjunct is bounded, or equivalently, if the big-Ms given by (4.99) are bounded. This occurs when the variables have finite bounds $L \leq x \leq U$, but even if some of the variables are unbounded, the big-Ms given by (4.94) or even (4.99) may be finite. This is illustrated in the next section.

Although a big-M relaxation is generally not as tight as a convex hull relaxation, it may be the better alternative when the convex hull relaxation has a large number of variables. If the big-Ms are large numbers, however, the big-M formulation may provide a very weak relaxation.

In some cases, both the convex hull and big-M relaxations are quite weak, and it is best not to use a relaxation at all. In particular, if one must place large upper bounds on some variables to satisfy the recession cone condition for the convex hull formulation, or to ensure that finite big-Ms exist for a big-M relaxation, then the relaxation that results may be too weak to justify the overhead of solving it. One of the advantages of using a modeling framework more general than MILP is that one is not obliged to incur the overhead of writing an MILP model when it does not provide a useful relaxation.

4.9.2 Example: Fixed-Charge Function

MILP representability is illustrated by the fixed-charge function, which occurs frequently in modeling. Suppose the cost x_2 of manufacturing quantity x_1 of some product is to be minimized. The cost is zero when $x_1 = 0$ and is $f + cx_1$ otherwise, where f is the fixed cost and c the unit variable cost. In general there would be additional constraints in the problem, but the issue here is how to write MILP constraints that represent feasible pairs (x_1, x_2) for the fixed-charge function.

The problem can be viewed as minimizing x_2 subject to $(x_1, x_2) \in S$, where S is the set depicted in Figure 4.12(i). S is the union of two polyhedra P_1 and P_2, and the problem is to minimize x_2 subject to the disjunction

$$\begin{bmatrix} x_1 = 0 \\ x_2 \geq 0 \end{bmatrix} \vee \begin{bmatrix} x_2 \geq cx_1 + f \\ x_1 \geq 0 \end{bmatrix}$$

where the disjuncts correspond respectively to P_1 and P_2.

Note that P_1 and P_2 do not have the same recession cone. The recession cone of P_1 is P_1 itself, and the recession cone of P_2 is the set of all vectors (x_1, x_2) with $x_2 \geq cx_1 \geq 0$. Thus, by Theorem 4.25, S is not MILP representable, and in particular the convex hull representation (4.127) does not correctly represent S:

$$\begin{array}{llll} x_1 = x_1^1 + x_1^2 & x_1^1 \leq 0 & -cx_1^2 + x_2^2 \geq fy_2 & y_1 + y_2 = 1 \\ x_2 = x_2^1 + x_2^2 & x_1^1, x_2^1 \geq 0 & x_1^2 \geq 0 & y_1, y_2 \in \{0, 1\} \end{array}$$
(4.128)

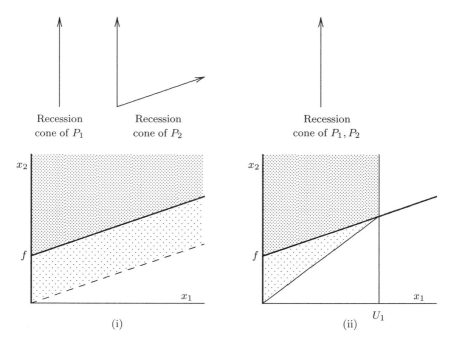

Figure 4.12. (i) *Feasible set of a fixed-charge problem, consisting of the union of polyhedra* P_1 *(heavy vertical line) and* P_2 *(darker shaded area). The convex hull of the feasible set is the entire shaded area, excluding the dashed line.* (ii) *Feasible set of the same problem with the bound* $x_1 \leq L_1$, *where* P_2' *is the darker shaded area. The convex hull of the feasible set is the entire shaded area.*

This can be seen by simplifying the above model. Only one 0-1 variable appears, which can be renamed y. Also, one can set $x_1^2 = x_1$ (since $x_1^1 = 0$) and $x_2^1 = x_2 - x_2^2$, which yields

$$x_1 \geq 0, \quad x_2 - x_2^2 \geq 0, \quad x_2^2 - cx_1 \geq fy, \quad y \in \{0, 1\}$$

Minimizing x_2 subject to this is equivalent to minimizing x_2 subject to

$$x_1 \geq 0, \quad x_2 - cx_1 \geq fy, \quad y \in \{0, 1\}$$

The projection onto (x_1, x_2) is the union of the two polyhedra obtained by setting $y = 0$ and $y = 1$. The projection is therefore the set of all points satisfying $x_2 \geq cx_1$, $x_1 \geq 0$, which is clearly different from $P_1 \cup P_2$. The model is therefore incorrect.

Although (4.128) does not correctly model the problem, its continuous relaxation (formed by replacing $y_j \in \{0, 1\}$ with $y_j \in [0, 1]$) is a valid convex hull relaxation. The projection of its feasible set onto x is the closure of the convex hull. As is evident in Figure 4.12, the convex hull relaxation is quite weak, since x_1 is unbounded.

In practice, one can generally put an upper bound U_1 on x_1 without harm. The problem is now to minimize x_2 subject to

$$\begin{pmatrix} x_1 = 0 \\ x_2 \geq 0 \end{pmatrix} \vee \begin{pmatrix} x_2 \geq cx_1 + f \\ 0 \leq x_1 \leq U_1 \end{pmatrix} \tag{4.129}$$

The recession cone of each of the resulting polyhedra P_1, P_2' (Figure 4.12ii) is the same (namely, P_1), and the feasible set $S' = P_1 \cup P_2'$ is therefore MILP representable. The convex hull formulation (4.127) becomes

$$\begin{array}{llll} x_1^1 \leq 0 & -cx_1^2 + x_2^2 \geq fy_2 & x_1 = x_1^1 + x_1^2 & y_1 + y_2 = 1 \\ x_1^1, x_2^2 \geq 0 & 0 \leq x_1^2 \leq L_1 y_2 & x_2 = x_2^1 + x_2^2 & y_1, y_2 \in \{0, 1\} \end{array}$$

Again the model simplifies. As before, one can set $y = y_2$, $x_1 = x_1^2$, and eliminate x_2^2, resulting in the model

$$x_1 \leq L_1 y, \quad x_2 \geq fy + cx_1, \quad x_1 \geq 0, \quad y \in \{0, 1\} \tag{4.130}$$

Obviously, y encodes whether the quantity produced is zero or positive, in the former case ($y = 0$) forcing $x_1 = 0$, and in the latter case incurring the fixed charge f. The projection onto (x_1, x_2) is $P_1 \cup P_2$, and the model is therefore correct.

Big-M constraints like $x_1 \leq U_1 y$, which are very common in MILP models, can be viewed as originating from scenarios like this one, in which an upper bound is placed on a variable to ensure that the polyhedra concerned have the same recession cone.

The disjunction (4.129) can also be given a big-M model (4.93). Although x_2 is formally unbounded, M_1, M_2 as given by (4.103) are finite. Specifically, $M_1 = U_1$ and $M_2 = f + U_1$, and the big-M model (4.93) becomes

$$\begin{array}{l} x_1 \leq U_1 y, \quad x_2 \geq cx_1 + fy - cU_1(1-y) \\ 0 \leq x_1 \leq U_1, \quad y \in \{0, 1\} \end{array} \tag{4.131}$$

It is interesting to compare the continuous relaxations of the convex hull model (4.130) and the big-M model (4.131). The convex hull model is relaxed by replacing $y \in \{0, 1\}$ with $0 \leq y \leq 1$. Its projection onto x_1, x_2 is the convex hull of the feasible set, which is described by

$$x_2 \geq cx_1 \tag{4.132}$$

and the bounds on x_1, as shown in Figure 4.12(ii).

Due to Theorem 4.18, the continuous relaxation of the big-M model (4.131), when projected onto (x_1, x_2), is given by (4.104) and the bounds

$0 \leq x_1 \leq U_1$, $x_2 \geq 0$. The inequality (4.104), in this case, becomes

$$x_2 \geq \left(\frac{f}{U_1} + 2c\right) x_1 - cU_1 \qquad (4.133)$$

This is weaker than the convex hull relaxation (4.132), as shown in Figure 4.13. Since the inequality (4.133) already supports the feasible set, it cannot be further tightened by using Theorem 4.19.

Due to the succinctness of the convex hull relaxation after it is simplified, it is the clear choice for relaxing this particular fixed-charge problem. Other problems, however, may generate so many axillary variables in the convex hull relaxation that the big-M relaxation is more practical.

4.9.3 Disjunctive Models

MILP modeling devices used in practice are generally based on disjunctions, knapsack-like constraints, or combinations of the two. Disjunctive models are considered in this section, and knapsack models in the next.

Disjunctive modeling devices are useful when one must make several discrete choices from two or more alternatives. For example, each choice y_i might be *no* or *yes*, corresponding to 0 or 1 for $i = 1, \ldots, m$. Choosing $y_i = 0$ enforces the first term, and choosing $y_i = 1$ enforces the second term, of the disjunction

$$\left(A^{i0} x \geq b^{i0}\right) \vee \left(A^{i1} x \geq b^{i1}\right) \qquad (4.134)$$

There may also be constraints on which choices $y = (y_1, \ldots, y_m)$ are mutually possible.

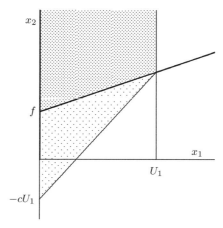

Figure 4.13. Big-M relaxation of the fixed-charge problem in Figure 4.12 (entire shaded area).

MILP Modeling 325

The problem can be formulated as a single disjunction by taking the product of the disjunctions (4.134):

$$\bigvee_{y \in Y} \{A^{iy_i} x \geq b^{iy_i} \mid i = 1, \ldots, m\} \qquad (4.135)$$

where Y is the set of mutually possible choices. This disjunction can, in turn, be given a convex hull or big-M relaxation as described in previous sections. It might be called a *product relaxation* for (4.134). Because (4.135) contains 2^n disjuncts in the worst case, however, the product relaxation may be too large or too complicated for practical use.

Often a more practical approach is to combine individual relaxations of the disjunctions (4.134), along with a relaxation of $y \in Y$, to obtain a relaxation of the entire constraint set. The resulting relaxation, which might be called a *factored relaxation*, is not as tight as a product relaxation in general, but it can be much more succinct.

As a simple illustration of this idea, consider a constraint set consisting of the disjunctions

$$\begin{pmatrix} x_1 = 0 \\ x_2 \in [0,1] \end{pmatrix} \vee \begin{pmatrix} x_2 = 0 \\ x_1 \in [0,1] \end{pmatrix}$$
$$\begin{pmatrix} x_1 = 0 \\ x_2 \in [0,1] \end{pmatrix} \vee \begin{pmatrix} x_2 = 1 \\ x_1 \in [0,1] \end{pmatrix} \qquad (4.136)$$

The two disjunctions, respectively, have the convex hull relaxations (a) and (b) below, which are illustrated in Figure 4.14.

$$\begin{aligned} x_1 + x_2 \leq 1, \quad x_1, x_2 \geq 0 & \quad (a) \\ x_1 \leq x_2, \quad x_1 \geq 0, \quad x_2 \leq 1 & \quad (b) \end{aligned} \qquad (4.137)$$

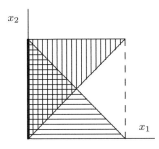

Figure 4.14. Convex hulls of two disjunctions (vertical and horizontal shading, respectively). Their intersection is the feasible set of the factored relaxation. The heavy vertical line segment is the feasible set of the product relaxation.

The feasible set of the factored relaxation (4.137) is the intersection of the two convex hulls.

The product of the disjunctions in (4.136) is

$$\begin{pmatrix} x_1 = 0 \\ x_2 \in [0,1] \end{pmatrix} \vee \begin{pmatrix} x_1 = 0 \\ x_2 = 1 \\ x_1, x_2 \in [0,1] \end{pmatrix} \vee \begin{pmatrix} x_1 = 0 \\ x_2 = 0 \\ x_1, x_2 \in [0,1] \end{pmatrix} \vee \begin{pmatrix} x_2 = 0 \\ x_2 = 1 \\ x_1 \in [0,1] \end{pmatrix}$$

The feasible set of (4.136), and thus of the product, is the heavy line segment in Figure 4.14, which is its own convex hull. Thus, the product relaxation is clearly tighter than the factored relaxation (4.137), although the product relaxation is much more complicated in general—even if it happens to simplify considerably in this example.

A simple *capacitated facility location problem* illustrates how a factored relaxation can be developed in practice. There are m possible locations for facilities, and n customers who obtain products from the facilities. A facility installed at location i incurs fixed cost f_i and has capacity C_i. Each customer j has demand D_j, and the unit cost of shipping from facility i to customer j is c_{ij}. The problem is to decide which facilities to install, and how to supply the customers, so as to minimize total fixed and variable costs.

Let $y_i = 1$ when a facility is installed at location i, and let x_{ij} be the quantity shipped from i to j. Then $y_i = 0$ enforces the first term, and $y_i = 1$ enforces the second term, of the disjunction

$$\begin{pmatrix} x_{ij} = 0, \text{ all } j \\ z_i = 0 \end{pmatrix} \vee \begin{pmatrix} \sum_j x_{ij} \leq C_i \\ x_{ij} \geq 0, \text{ all } j \\ z_i \geq f_i \end{pmatrix} \quad (4.138)$$

Here, z_i represents the fixed cost incurred at location i. In addition, each customer j must receive adequate supply:

$$\sum_i x_{ij} = D_j, \text{ all } j \quad (4.139)$$

This can be viewed as a disjunction with one disjunct. The problem is to minimize $\sum_i z_i + \sum_{ij} c_{ij} x_{ij}$ subject to (4.138) and (4.139).

Rather than relaxing the product of the disjunctions (4.138) and the disjunction (4.139), which is a very complicated matter, one can write a relaxation of each disjunction individually. The convex hull relaxation of (4.138) is

$$x_{ij} \geq 0, \text{ all } j; \quad \sum_j x_{ij} \leq C_i y_i; \quad z_i \geq f_i y_i; \quad 0 \leq y_i \leq 1 \quad (4.140)$$

and (4.139) is its own convex hull relaxation. A factored relaxation can now be obtained by combining (4.139) with (4.140) for all i. This relaxation is succinct enough and tight enough to be useful in practice.

A disjunctive approach to modeling can sometimes lead to tighter relaxations than one would formulate otherwise. A classic beginner's mistake, for example, is to model the *uncapacitated facility location problem* as a special case of the capacitated problem. In the uncapacitated problem, there is no limit on the capacity of each facility, and x_{ij} represents the fraction of customer j's demand supplied by facility i, so that each $D_j = 1$. Although there is no capacity limit, one can observe that each facility will ship at most n units and therefore let $C_i = n$ in the relaxation (4.140) for the capacitated problem. This, along with (4.139), is a valid relaxation for the uncapacitated problem, but there is a much tighter one.

The preferred relaxation can be obtained by starting with a disjunctive conception of the problem. If facility i is not installed, then it supplies nothing. If it is installed, it supplies at most one unit to each customer and incurs cost f_i:

$$\begin{pmatrix} x_{ij} = 0, \text{ all } j \\ z_i = 0 \end{pmatrix} \vee \begin{pmatrix} 0 \leq x_{ij} \leq 1, \text{ all } j \\ z_i \geq f_i \end{pmatrix}$$

The convex hull relaxation of this disjunction is

$$0 \leq x_{ij} \leq y_i, \text{ all } j; \ z_i \geq f_i y_i, \ 0 \leq y_i \leq 1 \quad (4.141)$$

These inequalities for $i = 1, \ldots, m$, along with (4.139), provide a tighter factored relaxation than specializing the capacitated relaxation (4.140). This is also an instance in which the more succinct relaxation is not the tighter one; (4.140), which has $2m$ constraints other than variable bounds, is not as tight as (4.141), which has $m(n+1)$ constraints other than variable bounds.

A *lot sizing problem with setup costs* illustrates logical variables that interrelate. There is a demand D_t for a product in each period t. No more than C_t units of the product can be manufactured in period t, and any excess over demand is stocked to satisfy future demand. If there is no production in the previous period, then a setup cost of f_t is incurred. The unit production cost is p_t, and the unit holding cost per period is h_t. A starting stock level is given. The objective is to choose production levels in each period so as to minimize total cost over all periods.

Let x_t be the production level chosen for period t and s_t the stock level at the end of the period. In each period t, there are three discrete options to choose from: (1) start producing (with a setup cost), (2) continue producing (with no setup cost), and (3) produce nothing. If v_t

is the setup cost incurred in period t, these correspond respectively to the three disjuncts

$$\begin{pmatrix} v_t \geq f_t \\ 0 \leq x_t \leq C_t \end{pmatrix} \vee \begin{pmatrix} v_t \geq 0 \\ 0 \leq x_t \leq C_t \end{pmatrix} \vee \begin{pmatrix} v_t \geq 0 \\ x_t = 0 \end{pmatrix} \quad (4.142)$$

In addition, there are logical connections between the choices in consecutive periods

$$\begin{array}{l} (2) \text{ in period } t \Rightarrow (1) \text{ or } (2) \text{ in period } t-1 \\ (1) \text{ in period } t \Rightarrow \text{ neither } (1) \text{ nor } (2) \text{ in period } t-1 \end{array} \quad (4.143)$$

The inventory balance constraints are

$$s_{t-1} + x_t = D_t + s_t, \quad s_t \geq 0, \quad t = 1, \ldots, n \quad (4.144)$$

where s_t is the stock level in period t and s_0 is given. The problem is to minimize

$$\sum_{t=1}^{n} (p_t x_t + h_t s_t + v_t) \quad (4.145)$$

subject to (4.142)–(4.144).

A convex hull relaxation for (4.142) is

$$\begin{array}{l} v_t = \sum_{k=1}^{3} v_t^k, \quad x_t = \sum_{k=1}^{3} x_t^k, \quad y_t = \sum_{k=1}^{3} y_{tk} \\ v_t^1 \geq f_t y_{t1}, \quad v_t^2 \geq 0, \quad v_t^3 \geq 0 \\ 0 \leq x_t^1 \leq C_t y_{t1}, \quad 0 \leq x_t^2 \leq C_t y_{t2}, \quad x_t^3 = 0 \\ y_{tk} \geq 0, \quad k = 1, 2, 3 \end{array} \quad (4.146)$$

To simply (4.146), define new variables $z_t = y_{t1}$ and $y_t = y_{t2}$ so that $z_t + y_t \leq 1$. Thus, $z_t = 1$ indicates a startup and $y_t = 1$ indicates continued production in period t, while $z_t = y_t = 0$ indicates no production. Since $x_t^3 = 0$, one can set $x_1 = x_t^1 + x_t^2$, which allows the two capacity constraints in (4.146) to be replaced by $0 \leq x_t \leq C_t(z_t + y_t)$. Finally, v_t can replace v_t^1, because v_t is being minimized and v_t^2 and v_t^3 do not appear. The convex hull relaxation (4.146) becomes

$$\begin{array}{l} v_t \geq f_t z_t, \quad z_t \geq 0, \quad y_t \geq 0 \\ 0 \leq x_t \leq C_t(z_t + y_t), \quad z_t + y_t \leq 1 \end{array} \quad (4.147)$$

The logical constraints (4.143) can be formulated

$$y_t \leq z_{t-1} + y_{t-1}, \quad z_t \leq 1 - z_{t-1} - y_{t-1} \quad (4.148)$$

A factored relaxation is now at hand. Minimize (4.145) subject to (4.144) and (4.147)–(4.148) for all t.

A more familiar model for this problem encodes the choices in a slightly different way. The variables z_t are retained, but y_t is replaced by a variable $\bar{y}_t$ that is 1 when something is produced in period t, whether or not there is a startup cost. This model is easily obtained from the above by using the change of variable $\bar{y}_t = z_t + y_t$. (4.147) is equivalent to

$$v_t \geq f_t z_t, \ z_t \geq 0, \ z_t \leq \bar{y}_t \\ 0 \leq x_t \leq C_t \bar{y}_t, \ \bar{y}_t \leq 1 \quad (4.149)$$

and (4.148) is equivalent to

$$z_t \geq \bar{y}_t - \bar{y}_{t-1}, \ z_t \leq 1 - \bar{y}_{t-1} \quad (4.150)$$

Thus, one can relax the problem by minimizing (4.145) subject to (4.144) and (4.149)–(4.150) for all t. This relaxation, however, has no advantage over the one obtained above from disjunctive analysis.

4.9.4 Knapsack Models

A large class of MILP models are based on counting ideas, which can be expressed as knapsack inequalities. Recall that a knapsack covering inequality has the form $ax \geq \alpha$, where each $a_j > 0$ and each x_j must take nonnegative integer values, while a knapsack packing inequality has the form $ax \leq \alpha$. If x_j is interpreted as the number of items of type j chosen for some purpose, the left-hand side of a knapsack inequality counts the total number of items selected (perhaps weighting some more than others), and the right-hand side places a bound on the total number or weight. Knapsack inequalities are therefore useful for formulating problems in which some item count must be bounded.

Perhaps the simplest problem of this sort is a disjunction of boolean variables, such as $\bigvee_{j \in J} x_j$, where each $x_j \in \{0, 1\}$. Since this logical clause states that at least one variable is true, it can be written as an inequality $\sum_{j \in J} x_j \geq 1$. This a special case of a knapsack inequality whose left-hand side counts the number of true variables, and whose right-hand side places a lower bound of 1 on the count. Section 4.10 discusses relaxations for this and several other types of logical propositions.

A similar pattern occurs in set covering, set packing, and set partitioning problems. A *set covering problem* begins with a finite collection $\{S_j \mid j \in J\}$ of sets and a finite set $T \subset \bigcup_{j \in J} S_j$. It seeks the smallest subcollection of sets that covers T; that is, the smallest $J' \subset J$ for which $T \subset \bigcup_{j \in J'} S_j$. For example, if one wants to play a certain set T of songs at a party and S_j is the set of songs on compact disk j, then one might wish to buy the smallest number of compact disks that provide the desired songs. Other applications include crew scheduling, facility location, assembly line balancing, and boolean function simplification.

It is easy to formulate a 0-1 model for the set covering problem. Let i index the elements of T, and let $a_{ij} = 1$ when $i \in S_j$, and $a_{ij} = 0$ otherwise. If binary variable $x_j = 1$ when S_j is selected, the problem is to minimize $\sum_j x_j$ (or, if desired, $\sum_j c_j x_j$) subject to the knapsack covering inequality $\sum_j a_{ij} x_j \geq 1$ for each i.

The *set packing problem* seeks the largest subcollection of sets S_j for which each element in T appears in at most one set. The problem is therefore to maximize $\sum_j c_j x_j$ subject to the knapsack packing inequality $\sum_i a_{ij} x_j \leq 1$ for each i. The *set partitioning problem* seeks to maximize or minimize $\sum_j c_j x_j$ subject to $\sum_i a_{ij} x_j = 1$, which implies both a covering and a packing inequality.

Other knapsack-like constraints include logical constraints that involve a counting element. A *cardinality clause*, for example, requires that at least k of the boolean variables in $\{x_j \mid j \in J\}$ be true. It can, of course, be formulated with the knapsack covering inequality $\sum_{j \in J} x_j \geq k$, which along with $0 \leq x_j \leq 1$ provides a convex hull relaxation. A further generalization is the *cardinality conditional*, which states that if k of the variables in $\{x_j \mid i \in I\}$ are true, then at least ℓ of the variables in $\{x_j \mid j \in J\}$ are true. See Chapter 5 for more information on these constraints.

Two obvious opportunities for knapsack formulations are the *knapsack covering* and *knapsack packing* problems discussed in Section 2.2.3. The *capital budgeting problem* has the same structure as a 0-1 knapsack packing problem. In this problem, there are several possible investments, and each investment j incurs a capital cost of a_j. The return from investment j is c_j, and the objective is to select investments so as to maximize return while staying with the available investment funds α. Thus, the problem is to maximize $\sum_j c_j x_j$ subject to $\sum a_j x_j \leq \alpha$, where $x_j \in \{0, 1\}$ and $x_j = 1$ indicates that investment j is selected.

Section 4.9.3 showed that one can sometimes obtain a better relaxation by starting with a disjunctive conception of a problem. A similar principle applies to knapsack-style models, and a final example illustrates this. The problem is similar to the freight transport problem of Section 2.2.3 but has more detail. A collection of packages are to be shipped by several trucks, and each package j has size a_j. Each available truck i has capacity Q_i and costs c_i to operate. The problem is to decide which trucks to use, and which packages to load on each truck, so as to deliver all the items at minimum cost.

MILP Modeling

The previous section showed that one can sometimes obtain a better relaxation by starting with a disjunctive conception of a problem. A similar principle applies to knapsack-style models, and a final example illustrates this. The problem is similar to the freight transport problem of Section 2.2.3 but has more detail. A collection of packages are to be shipped by several trucks, and each package j has size a_j. Each available truck i has capacity Q_i and costs c_i to operate. The problem is to decide which trucks to use, and which packages to load on each truck, to deliver all the items at minimum cost. The following might be regarded as a standard 0-1 model for the problem:

$$\begin{aligned}
&\min \sum_i c_i y_i & (a) \\
&\sum_j a_j x_{ij} \leq Q_i, \text{ all } i & (b) \\
&\sum_i x_{ij} = 1, \text{ all } j & (c) \\
&x_{ij} \leq y_i, \text{ all } i,j & (d)
\end{aligned} \quad (4.151)$$

where $y_i, x_{ij} \in \{0,1\}$. Variable $y_i = 1$ if truck i is selected, and $x_{ij} = 1$ indicates that item j is loaded on truck i. Constraint (b) observes truck capacity, (c) ensures that every item is shipped, and (d) requires that a truck be selected if it carries cargo.

Formulation (4.151) proves to be hard to solve due to its weak continuous relaxation. The relaxation can be strengthened in two ways. One is to replace (b) with the valid constraint

$$\sum_j a_j x_{ij} \leq Q_i y_i, \text{ all } i$$

This is a fairly standard modeling trick that could have been anticipated without modeling from first principles. A much less obvious improvement, however, is to sum (b) over i and use (c) to obtain constraint (e) in the revised model below.

$$\begin{aligned}
&\min \sum_i c_i y_i & (a) \\
&\sum_j a_j x_{ij} \leq Q_i y_i, \text{ all } i & (b) \\
&\sum_i x_{ij} = 1, \text{ all } j & (c) \\
&x_{ij} \leq y_i, \text{ all } i,j & (d) \\
&\sum_i Q_i y_i \geq \sum_j a_j & (e)
\end{aligned} \quad (4.152)$$

The new constraint (e) says that the trucks chosen should have enough total capacity to carry all the items. It turns out that this model is dramatically easier to solve than (4.151), because knapsack cuts can be generated for (e), and these make the continuous relaxation much tighter. That fact that constraint (e) is helpful, however, violates the conventional wisdom in MILP modeling, since it is a linear combination of the other constraints.

If one conceives the problem from an overall knapsack point of view, however, the better model (4.152) falls out immediately. Like the freight transport problem of Section 2.2.3, this problem is primarily a knapsack covering problem—choose a minimum-cost collection of trucks that can carry the load. This is precisely constraint (e). Once the trucks are chosen, there is the secondary problem of selecting which items each truck will carry, and the selected items must satisfy a knapsack packing constraint. The covering and packing knapsack constraints are connected by disjunctive reasoning: if truck i is selected, then the items loaded onto it must fit into space Q_i, and cost c_i is incurred; while if it is not selected, there is no cost and no cargo. This poses the disjunction

$$\begin{pmatrix} z_i \geq c_i \\ \sum_j a_j x_{ij} \leq Q_i \\ 0 \leq x_{ij} \leq 1, \text{ all } j \end{pmatrix} \vee \begin{pmatrix} z_i \geq 0 \\ x_{ij} = 0 \end{pmatrix} \qquad (4.153)$$

where z_i is the cost incurred by truck i. Note that the first disjunct contains the continuous relaxation of the knapsack packing constraint, since the goal is to obtain a continuous relaxation of the problem. The convex hull relaxation of (4.153) is

$$z_i \geq c_i y_i, \quad \sum_j a_j x_{ij} \leq Q_i y_i, \quad 0 \leq x_{ij} \leq y_i$$

This yields (a), (d), and the tricky constraint (b). Finally, each item j must be loaded onto some truck, which is constraint (c). One therefore obtains the continuous relaxation of (4.152) by modeling from first principles.

An *assignment problem* can be formulated with knapsack inequalities, as discussed in Section 4.12.2. The *traveling salesman problem* contains knapsack-like constraints, which are presented as a relaxation for the *circuit* constraint in Section 4.14.2.

Sections 2.2.3 and 4.4.2 describe cutting planes that can be used to tighten the continuous relaxation of knapsack formulations. Cutting planes have also been developed specifically for set covering, set packing and set partitioning problems.

4.9.5 Exercises

1. Show that the disjunction representation (4.127) does not correctly model the union of the line segment $\{(0, x_2) \mid 0 \leq x_2 \leq 1\}$ with the ray $\{(x_1, 0) \mid x_1 \geq 0\}$, and explain why. Show that the continuous relaxation of (4.127) nonetheless describes the closure of the convex hull. Truncate the ray to $\{(x_1, 0) \mid 0 \leq x_1 \leq M\}$ in order to make the union MILP representable, and write an MILP model for it.

2. Use the disjunctive representation (4.127) to write an MILP model of the union of the cone $\{(x_1, x_2) \mid 0 \leq x_1 \leq x_2\}$ with the cone $\{(x_1, x_2) \mid 1 \leq x_1 \leq x_2 + 1\}$. Note that the model can be simplified so that it contains only the original variables x_1, x_2 and $\alpha \in \{0, 1\}$.

3. Use the disjunctive representation (4.127) to write an MILP model of the union of the cone $\{(x_1, x_2) \mid 0 \leq x_1 \leq x_2\}$ with the cone $\{(x_1, x_2) \mid 1 \leq x_1 \leq 2x_2 + 1\}$. Show that the model does not correctly represent the union. Why?

4. In a scheduling problem, Jobs 1 and 2 both have duration 1. Let x_j be the start time of job j. Since the jobs cannot overlap,
$$[x_2 \geq x_1 + 1] \vee [x_1 \geq x_2 + 1]$$
Show that this disjunction has no MILP model. Place an upper bound U on x_1, x_2 and write an MILP representation (4.127) of the resulting disjunction. Draw a graph of the feasible set and its convex hull. Note that because U is typically much larger than 1, the convex hull relaxation is quite weak. The overhead of an MILP formulation may therefore be unjustified.

5. Write the tightest possible big-M model of the scheduling disjunction of Exercise 4 by using the big-Ms in (4.99). Project its continuous relaxation onto x_1, x_2 and draw a graph. Note that the big-M model is useless for obtaining a relaxation because its continuous relaxation is redundant of the bounds $x_j \in [0, U]$. However, the relaxation is somewhat better if there are time windows. Write the tightest possible big-M relaxation if job j has release time r_j and deadline d_j.

6. Let $x \geq 0$ be a continuous variable and y a 0-1 variable. Identify which of the following conditions has an MILP formulation, and explain why. Hint: consider the feasible set in x-y space.

   (a) $x = 0$ if $y = 0$.
   (b) $x = 0$ if $y = 0$, and $x \leq M$ otherwise.
   (c) $y = 0$ if $x = 0$.

7 There are n manufacturing plants that must produce a total of at least R widgets in a given period. If plant j operates at level A, it produces a_j widgets at cost f_j. If it operates at level B, it produces b_j widgets at cost g_j. The plant can also be shut down, in which case it incurs no cost. Write a factored disjunctive model that minimizes cost subject to the output requirement, and simplify it as much as possible.

8 Suppose in Exercise 7 that when a plant operates at level A, its output must be in the interval $[a_j, A_j]$, and it incurs cost $f_j + c_j x_j$. When it operates at level B, $x_j \in [b_j, B_j]$, and the cost is $g_j + d_j x_j$. Write a factored disjunctive model.

9 The classical assignment problem minimizes $\sum_i c_{i x_i}$ subject to the constraint alldiff$(x_1, \ldots, x_m)$, where each $x_i \in \{1, \ldots, n\}$ and $m \leq n$. Show that the standard 0-1 model

$$\min_{y_{ij} \in \{0,1\}} \left\{ \sum_{ij} c_{ij} y_{ij} \,\middle|\, \sum_j y_{ij} = 1, \text{ all } i; \; \sum_i y_{ij} \leq 1, \text{ all } j \right\}$$

can be obtained by disjunctive modeling. Since the coefficient matrix is totally unimodular, $y_{ij} \in \{0, 1\}$ can be replaced by $0 \leq y_{ij} \leq 1$. Hint: there are two sets of decisions. For each i, write a disjunction expressing the choice of value for x_i, using 0-1 variables y_{ij}. For each j, write a disjunction expressing the choice of which x_i will have value j (if any), using 0-1 variables $\bar{y}_{ji}$. Add the logical constraint that $y_{ij} = \bar{y}_{ji}$ and simplify.

10 The quadratic assignment problem minimizes $\sum_{ij} c_{ij} d_{x_i x_j}$ subject to alldiff$(x_1, \ldots, x_m)$. It can be viewed as assigning facilities i to locations x_i to minimize the cost of traffic between all pairs of facilities, where c_{ij} is the cost per unit distance of traffic between i and j, and $d_{k\ell}$ is the distance from location k to location ℓ. A quadratic 0-1 model can be written

$$\min_{y_{ik} \in \{0,1\}} \left\{ \sum_{ijk\ell} c_{ij} d_{k\ell} y_{ik} y_{j\ell} \,\middle|\, \sum_k y_{ik} = 1, \text{ all } i; \; \sum_i y_{ik} \leq 1, \text{ all } k \right\}$$

Use disjunctive reasoning similar to that of the previous exercise to obtain a linear 0-1 model. Hint: for each pair i, j ($i \neq j$), formulate a disjunction, and similarly for each pair k, ℓ ($k \neq \ell$).

11 In a disjunctive scheduling problem, each of n jobs j has a duration p_j, release time r_j, and deadline d_j. Only one job can run at a time. The schedule can be viewed as a sequence of events k, each of which

Propositional Logic 335

is the start of some job. Thus, one can write a disjunction of n alternatives for each job j, corresponding to the n events that might be the start of j. Let t_k be the time at which event k occurs, and write a factored disjunctive model for the problem. Use assignment constraints to make sure that no two jobs are assigned to the same event.

12 In a cumulative scheduling problem, each of n jobs j has a duration p_j, release time r_j, deadline d_j, and rate of resource consumption c_j. Jobs can run simultaneously as long as the total rate of resource consumption at any time is at most C. The schedule can be viewed as a sequence of $2n$ events, each of which is the start or end of some job. So, for each job j, one can write a disjunction of alternatives corresponding to the pairs k, k' of events $(k < k')$ that might represent the start and end of job j. Let t_k be the time at which event k occurs, and let z_k be the total rate of resource consumption immediately after event k occurs. Thus, when event k is the start of job j, $z_k = z_{k-1} + c_j$, and similarly when event k is the end of a job. Write a factored disjunctive model, using assignment constraints to make sure that no two jobs involve the same event.

4.10   Propositional Logic

The boolean values *true* and *false* have no structure, aside from the fact that there are two of them. However, if they are interpreted as the numbers 1 and 0, the rich structure of the unit hypercube becomes available for building continuous relaxations of logical propositions.

The most straightforward way to relax a logical formula is to convert it to a set of logical clauses (Section 3.5.1), since each logical clause can be written as a 0-1 linear inequality. However, relaxations of the individual clauses may provide a weak relaxation for the clause set as a whole. Three strategies can address this problem:

- develop tight relaxations for elementary logical formulas other than clauses
- process the clause set with some version of the resolution algorithm to obtain a tighter relaxation (this is effect generates cutting planes)
- derive separating cuts for the clause set.

4.10.1   Common Logical Formulas

Certain logical expressions tend to occur frequently, and it is useful to build a catalog of convex hull relaxations for them.

Logical clauses are perhaps the most basic logical expression, because any logical formula can be converted to clausal form, and a logical clause has an obvious 0-1 representation. For example, the clause $x_1 \vee x_2 \vee \neg x_3$ can be represented as the 0-1 linear inequality $x_1 + x_2 + (1-x_3) \geq 1$, which is relaxed by replacing $x_j \in \{0,1\}$ with $0 \leq x_j \leq 1$. It is convenient to call this the *clausal inequality* that corresponds to the clause $x_1 \vee x_2 \vee \neg x_3$.

More generally, a logical clause C may be written

$$\bigvee_{j \in P} x_j \vee \bigvee_{j \in N} \neg x_j \tag{4.154}$$

It is relaxed by the corresponding clausal inequality and bounds on x_j:

$$\sum_{j \in P} x_j + \sum_{j \in N} (1 - x_j) \geq 1, \quad 0 \leq x_j \leq 1 \text{ for } j \in P \cup N \tag{4.155}$$

It is convenient to denote the clausal inequality for C by linear(C).

THEOREM 4.26 *The system (4.155) is a convex hull relaxation of the logical clause (4.154).*

Proof. Suppose, without loss of generality, that all the literals in (4.154) are positive, so that $P = \{1, \ldots, n\}$ and $N = \emptyset$ (negated variables can be complemented). Let S be the polyhedron described by (4.155). Since the integral solutions of (4.155) are the feasible solutions of (4.154), it suffices to show that S is an integral polyhedron. Since the right-hand sides of (4.155) are integral, for this it suffices by Theorem 4.13 to show that the coefficient matrix for (4.155) is totally unimodular. The matrix has the form

$$\begin{bmatrix} e \\ -I \end{bmatrix}$$

where e is a row of ones and I is the identity matrix. This matrix satisfies the condition of Corollary 4.15 and is therefore totally unimodular. $\square$

Table 4.1 displays convex hull relaxations of several common logical formulas. Formula 1 is covered by Theorem 4.26. Formula 2 is equivalent to clause $\neg x_1 \vee x_2$ and therefore has the convex hull relaxation $(1-x_1) + x_2 \geq 1$ (plus bounds). Formula 3 is also a clause. The relaxation given for Formula 4 describes the convex hull because it is totally unimodular (left as an exercise). Formula 5 is the result of complementing every variable in Formula 4, and so its convex hull relaxation can be derived from that of Formula 4. The relaxation of Formula 7, however, is not totally unimodular, and a different sort of proof is required.

Propositional Logic

Table 4.1. Convex hull relaxations of some common propositional formulas. The index sets J_1, J_2 are disjoint. The bounds $0 \leq x_j \leq 1$ are understood to be part of the relaxation.

	Formula	Convex Hull Relaxation
1.	$\bigvee_{j \in J} x_j$	$\sum_{j \in J} x_j \geq 1$
2.	$x_1 \to x_2$	$x_1 \leq x_2$
3.	$\bigwedge_{j \in J_1} x_j \to \bigvee_{j \in J_2} x_j$	$\sum_{j \in J_1} x_j \leq \sum_{j \in J_2} x_j + |J_1| - 1$
4.	$\bigvee_{j \in J_1} x_j \to \bigvee_{j \in J_2} x_j$	$x_i \leq \sum_{j \in J_2} x_j$, all $i \in J_1$
5.	$\bigwedge_{j \in J_1} x_j \to \bigwedge_{j \in J_2} x_j$	$\sum_{j \in J_1} x_j \leq x_i + |J_1| - 1$, all $i \in J_2$
6.	$x_1 \equiv x_2$	$x_1 = x_2$
7.	$\bigvee_{j \in J_1} x_j \equiv \bigvee_{j \in J_2} x_j$	$x_i \leq \sum_{j \in J_2} x_j$, all $i \in J_1$ $x_i \leq \sum_{j \in J_1} x_j$, all $i \in J_2$
8.	$\bigwedge_{j \in J_1} x_j \equiv \bigwedge_{j \in J_2} x_j$	$\sum_{j \in J_2} x_j \leq x_i + |J_2| - 1$, all $i \in J_1$ $\sum_{j \in J_1} x_j \leq x_i + |J_1| - 1$, all $i \in J_2$
9.	$\bigwedge_{j \in J_1} x_j \equiv \bigvee_{j \in J_2} x_j$	$\sum_{j \in J_1} x_j \leq \sum_{j \in J_2} x_j + |J_1| - 1$ $x_i \leq \sum_{j \in J_2} x_j$, all $i \in J_1$
10.	$(x_1 \equiv x_2) \equiv x_3$	$x_1 + x_2 + x_3 \geq 1$ $x_2 + x_3 \leq x_1 - 1$ $x_1 + x_3 \leq x_2 - 1$ $x_1 + x_2 \leq x_3 - 1$
11.	$(x_1 \vee x_2) \wedge$ $(x_1 \vee x_3) \wedge$ $(x_2 \vee x_3)$	$x_1 + x_2 + x_3 \geq 2$
12.	$(x_1 \vee x_2 \vee x_3) \wedge$ $(x_1 \vee x_4) \wedge$ $(x_2 \vee x_4) \wedge$ $(x_3 \vee x_4)$	$x_1 + x_2 + x_3 + 2x_4 \geq 3$

THEOREM 4.27 *A convex hull relaxation of the formula*

$$\bigvee_{j \in J_1} x_j \equiv \bigvee_{j \in J_2} x_j \qquad (4.156)$$

is given by

$$\begin{aligned}
x_i &\leq \sum_{j \in J_2} x_j, \quad \text{all } i \in J_1 & (a) \\
x_i &\leq \sum_{j \in J_1} x_j, \quad \text{all } i \in J_2 & (b) \\
0 &\leq x_j \leq 1, \quad \text{all } j \in J_1 \cup J_2 & (c)
\end{aligned} \qquad (4.157)$$

Proof. It suffices to show that all vertices of the polyhedron described by (4.157) are integral. The vertices can be enumerated by solving all linearly independent subsets S of $|J_1| + |J_2|$ of the following equations:

$$\begin{aligned}
x_i &= \sum_{j \in J_2} x_j, \quad \text{all } i \in J_1 & (i) \\
x_i &= \sum_{j \in J_1} x_j, \quad \text{all } i \in J_2 & (ii) \\
x_j &= 0, \quad \text{all } j \in J_1 \cup J_2 & (iii) \\
x_j &= 1, \quad \text{all } j \in J_1 & (iv) \\
x_j &= 1, \quad \text{all } j \in J_2 & (v)
\end{aligned} \qquad (4.158)$$

Case I. S consists of the equations (i) and (ii). The coefficient matrix for these equations has the form

$$\begin{bmatrix} -I & E \\ E & -I \end{bmatrix} \qquad (4.159)$$

where E is a matrix of all ones, and it is clearly nonsingular. Since the right-hand side is zero, the solution is $x = 0$, which is integral.

Case II. S contains at least one of the equations (iv) but none of the equations (v). Since $x_j = 1$ for some $j \in J_1$, the constraints (4.157b) are redundant, and the equations (ii) need not be considered for S. The coefficient matrix of S is therefore some subset of rows of

$$A = \begin{bmatrix} -I & E \\ I & 0 \\ 0 & I \\ I & 0 \end{bmatrix}$$

Propositional Logic 339

But A is totally unimodular, due to Theorem 4.14. Given any subset $\bar{J}$ of columns of A, the desired partition of $\bar{J}$ can be obtained by placing the last $\lceil |J_2 \cap \bar{J}|/2 \rceil$ columns in one set and the remaining columns in the other. Thus, any solution obtained is integral.

Case III. S contains at least one of the equations (v) but none of the equations (iv). This is analogous to Case II.

Case IV. S contains at least one of the equations (iv) and at least one of the equations (v). Since $x_j = 1$ for some $j \in J_1$ and for some $j \in J_2$, constraints (4.157a) and (4.157b) are redundant, and equations (i) and (ii) need not be considered for S. Any independent set of the remaining equations has an integral solution.

Case V. S contains at least one of the equations (iii) but none of the equations (iv) and (v). In this case, one can eliminate the variables set to zero, and the resulting system of equations again has a coefficient matrix of the form (4.159). Since the right-hand side is zero, the solution is $x = 0$ and therefore integral. $\square$

Because Formula 8 is the result of complementing all variables in Formula 7, its convex hull relaxation can be obtained from that of Formula 7. A special case of Formula 8, namely $(x_1 \wedge x_2) \equiv x_3$, occurs frequently in MILP modeling because it expresses such relations as $x_1 x_2 = x_3$. Another valid MILP model for this relation is

$$x_3 \geq x_1 + x_2 - 1$$
$$x_1 + x_2 \geq 2x_3$$

This is not a convex hull formulation, however, because the nonintegral point $x = (1, 0, \frac{1}{2})$, for instance, is a vertex of the polyhedron it defines but does not lie in the convex hull. As for Formula 9, the proof of its convex hull formulation is similar to that for Formula 7.

Every inequality listed in Table 4.1 defines a facet of the associated convex hull; that is, an $(n-1)$-dimensional face of the convex hull. The inequality $\sum_j x_j \geq 1$, for example, defines a facet of the convex hull of $\bigvee_j x_j$ because n affinely independent[2] points in the convex hull lie on the hyperplane defined by $\sum_j x_j = 1$. In particular, the affinely independent points $e^1, \ldots, e^n$ lie on the hyperplane, where e^j is the jth unit vector.

The facet-defining inequality for Formula 11 is the smallest example (in terms of the number of variables) of a nonclausal facet-defining inequality for a set of 0-1 points. The inequality for Formula 12 is the smallest example of a facet-defining inequality with a coefficient not in $\{0, 1, -1\}$.

---

[2]Points $a^1, \ldots, a^n$ are affinely independent when $a^2 - a^1, \ldots, a^n - a^1$ are linearly independent.

4.10.2 Resolution as a Tightening Technique

Any formula of propositional logic can be written as a clause set S, which can, in turn, be given a factored relaxation. This relaxation, denoted linear(S), contains the inequality linear(C) corresponding to each clause C in S, along with bounds $0 \leq x_j \leq 1$. Although this relaxation tends to be weak, it can generally be tightened by first applying some form of resolution (Section 3.5.2) to S. If S' is the result of applying resolution to S, then linear(S') is generally tighter than linear(S).

Resolution can clearly tighten a relaxation. The clause set S consisting of $x_1 \vee x_2$ and $x_1 \vee \neg x_2$, for example, has the feasible set $\{(1,0),(1,1)\}$. Linear(S) consists of

$$x_1 + x_2 \geq 1, \quad x_1 + (1-x_2) \geq 1, \ 0 \leq x_j \leq 1, \ j=1,2 \qquad (4.160)$$

This is not a convex hull relaxation, because $(x_1, x_2) = (\frac{1}{2}, \frac{1}{2})$ satisfies linear(S) but does not belong to the convex hull of the feasible set, which is the line segment connecting the two feasible points. However, applying the resolution algorithm to S yields the single clause x_1, and linear$(\{x_1\})$ is

$$x_1 \geq 1, \ 0 \leq x_j \leq 1, \ j=1,2$$

This is a convex hull relaxation and therefore tighter than linear(S).

A resolvent always tightens the factored relaxation of its parents because it is a rank 1 Chvátal-Gomory cut (Section 4.4.1). More precisely,

THEOREM 4.28 *Given two clauses C, D with resolvent R, linear(R) is a rank 1 Chvátal-Gomory cut for linear$(\{C, D\})$.*

For example, clause (c) below is the resolvent R of (a) and (b):

$$\begin{aligned} x_1 \vee x_2 \vee x_3 & \qquad (a) \\ \neg x_1 \vee x_2 \quad \vee \neg x_4 & \qquad (b) \\ x_2 \vee x_3 \vee \neg x_4 & \qquad (c) \end{aligned}$$

Linear(R) is $x_2 + x_3 + (1-x_4) \geq 1$, which can be obtained by computing the weighted sum

$$\begin{array}{rl} x_1 + x_2 + x_3 \geq 1 & (\tfrac{1}{2}) \\ -x_1 + x_2 \quad - x_4 \geq -1 & (\tfrac{1}{2}) \\ x_3 \quad \geq 0 & (\tfrac{1}{2}) \\ -x_4 \geq -1 & (\tfrac{1}{2}) \\ \hline x_2 + x_3 - x_4 \geq -\tfrac{1}{2} & \end{array}$$

Propositional Logic   341

using the weights indicated on the right (note that the constants on the left-hand side of the clausal inequalities are moved to the right). By rounding up the $-\frac{1}{2}$ on the right-hand side of the weighted sum, one obtains the rank 1 Chvátal-Gomory cut $x_2 + x_3 - x_4 \geq 0$, which is linear(R).

In general, one can obtain linear(R) for the resolvent R of clauses C and D by taking a nonnegative linear combination of linear(C), linear(D), bounds $x_j \geq 0$ for all literals x_j that occur in C or D but not both, and bounds $-x_j \geq -1$ for all $\neg x_j$ that occur in C or D but not both. Weight $\frac{1}{2}$ is assigned to each of these, and the right-hand side of the result is rounded up.

Each iteration of the resolution algorithm generates additional cuts. The algorithm does not in general result in a convex hull relaxation, however. For example, the clause set S consisting of $x_1 \vee x_2$, $x_1 \vee x_3$, $x_2 \vee x_3$ is unchanged by the resolution algorithm. Yet the point $(x_1, x_2, x_3) = (\frac{1}{2}, \frac{1}{2}, \frac{1}{2})$ satisfies linear(S):

$$x_1 + x_2 \geq 1, \quad x_1 + x_3 \geq 1, \quad x_2 + x_3 \geq 1, \quad 0 \leq x_j \leq 1, \; j = 1, 2, 3$$

and does not lie within the convex hull of the feasible set

$$\{(0, 1, 1), (1, 0, 1), (1, 1, 0), (1, 1, 1)\}$$

Although resolution does not generally yield a convex hull relaxation, it can nonetheless tighten the relaxation considerably. When the clause set is large, the full resolution algorithm may consume too much time and it may be advantageous to use an incomplete form of resolution.

One incomplete form, unit resolution (Section 3.5.3), is very fast but unhelpful. It is easy to see that if R is the resolvent of a unit clause U with another clause C, then linear(R) is simply the sum of linear(U) and linear(C). Linear$(\{U, C, R\})$ therefore describes the same polyhedron as linear$(\{U, C\})$. This means that applying the unit resolution algorithm to a clause set S has no effect on the tightness of linear(S).

Input resolution, which lies between full resolution and unit resolution in strength, is a more attractive alternative. The input resolution algorithm is similar to the full resolution algorithm, except that at least one parent of each resolvent must belong to the original clause set. Input resolution therefore requires less computation than full resolution, but tightens the factored relaxation somewhat.

Some indication of how much input resolution tightens the relaxation is provided by an interesting fact to be established in Section 4.10.4: input resolution generates all rank 1 Chvátal-Gomory cuts that have clausal form. That is, applying the input resolution algorithm to a clause set S generates precisely the clauses C such that linear(C) is a rank 1 cut for linear(S). So applying input resolution to a clause set before relaxing it is tantamount is relaxing it first and then generating all rank 1 clausal cuts.

The input resolution algorithm goes as follows. Let S_0 be the original clause set S. In each iteration $k \geq 1$, let S_k contain every clause R such that (a) R is the resolvent of a clause in S_{k-1} with a clause in S_0, and (b) R is absorbed by no clause in $\bigcup_{i=0}^{k-1} S_i$. The algorithm terminates at step m if S_m is empty, at which point all the clauses in $S' = \bigcup_{i=0}^{m-1} S_i$ are implied by S. One may then delete from S' all clauses that are absorbed by other clauses in S'.

Consider, for example, the clause set $S = S_0$:

$$\begin{array}{l} \neg x_1 \vee x_2 \vee x_3 \vee x_4 \\ x_2 \vee \neg x_3 \\ x_1 \quad\ \vee x_3 \vee x_4 \\ \neg x_2 \quad\quad\quad \vee x_5 \end{array} \quad (4.161)$$

S_1 contains all resolvents of clauses in S_0:

$$\begin{array}{l} \neg x_1 \vee x_2 \quad\ \vee x_4 \\ x_2 \vee x_3 \vee x_4 \\ \neg x_1 \quad\ \vee x_3 \vee x_4 \vee x_5 \\ x_1 \vee x_2 \quad\ \vee x_4 \\ \neg x_3 \vee \quad\ \vee x_5 \end{array} \quad (4.162)$$

S_2 contains all resolvents of clauses in S_1 with those in S_0:

$$\begin{array}{l} \neg x_1 \quad\quad\ \vee x_4 \vee x_5 \\ x_2 \quad\ \vee x_4 \\ x_3 \vee x_4 \vee x_5 \\ \neg x_1 \vee x_2 \quad\ \vee x_4 \vee x_5 \\ x_1 \quad\quad\ \vee x_4 \vee x_5 \end{array} \quad (4.163)$$

S_3 contains the single clause

$$x_4 \vee x_5 \quad (4.164)$$

which is the only resolvent of a clause in S_2 with one in S_0 that is not absorbed by a clause previously generated. Now S' consists of all the clauses (4.161)–(4.164). Most of these clauses are absorbed by others,

Propositional Logic

reducing S' to the set of clauses on the left:

$$\begin{array}{ll} x_1 \vee x_3 \vee x_4 & x_1 + x_2 + x_4 \geq 1 \\ x_2 \vee \neg x_3 & x_2 + (1 - x_3) \geq 1 \\ x_2 \vee x_4 & x_2 + x_4 \geq 1 \\ \neg x_3 \vee x_5 & (1 - x_3) + x_5 \geq 1 \\ x_4 \vee x_5 & x_4 + x_5 \geq 1 \\ & 0 \leq x_j \leq 1, \ j = 1, \ldots, 5 \end{array} \quad (4.165)$$

The inequalities on the right comprise linear(S'), which is a tighter relaxation of S than linear(S).

4.10.3 Refutation by Linear Relaxation

Before establishing the connection between input resolution and rank 1 cuts, it is useful to investigate the ability of input and unit resolution to prove infeasibility. This inquiry is also interesting in its own right. It is easy to show that input resolution proves infeasibility precisely when the linear relaxation is infeasible, and similarly for unit resolution. It follows that input resolution and unit resolution have the same power to detect infeasibility.

Let a *unit refutation* of clause set S be a unit resolution proof that S is infeasible, and similarly for an *input refutation*.

THEOREM 4.29 *Clause set S has a unit refutation if and only if linear(S) is infeasible.*

Proof. First suppose that S has a unit refutation. It was observed in Section 4.10.2 that adding unit resolvents to S has no effect on the feasible set of linear(S). Because one of the unit resolvents added during the unit refutation is the empty clause, which corresponds to the inequality $0 \geq 1$, linear(S) is infeasible.

For the converse, suppose linear(S) has no unit refutation. Apply the unit resolution algorithm to S to obtain the clause set S'. Make every unit clause in S' true by setting its variable to 0 or 1 as required. Because the remaining variables x_j occur only in clauses C with two or more literals, setting each $x_j = \frac{1}{2}$ satisfies linear(C). Linear(S') is therefore feasible. Since linear(S) has the same feasible set, it is likewise feasible. □

The converse of the following theorem is also true, but its proof must await the next section.

THEOREM 4.30 *Clause set S has an input refutation if linear(S) is infeasible.*

Proof. It suffices to show that if S has no input refutation, then linear(S) is feasible. Let S' be the result of applying the input resolution algorithm to S. If the unit clause x_j occurs in S' but the unit clause $\neg x_j$ does not, then set $x_j = 1$. Similarly, if the unit clause $\neg x_j$ occurs in S' but x_j does not, set $x_j = 0$. We know that no unit clause $\neg x_j$ for which x_j is set to 1 occurs in S, because otherwise it would have resolved with x_j to produce an input refutation. Similarly, no unit clause x_j for which x_j is set to 0 occurs in S. Now setting all remaining variables in S to $\frac{1}{2}$ satisfies linear(S). $\square$

4.10.4 Input Resolution and Rank 1 Cuts

Recall that a rank 1 Chvátal-Gomory cut for a 0-1 system $Ax \geq b$ is the result of rounding up the coefficients and right-hand side of some surrogate inequality of $Ax \geq b$. Section 4.10.2 relied on the following theorem:

THEOREM 4.31 *The input resolution algorithm applied to a clause set S generates precisely the clauses C for which linear(C) is a rank 1 Chvátal-Gomory cut for linear(S).*

It is convenient to prove the theorem in two parts, one showing that input resolution generates nothing but rank 1 clausal cuts, and the other showing that it generates all of them.

THEOREM 4.32 *If the input resolution algorithm applied to S generates clause C, then linear(C) is a rank l cut for linear(S).*

The idea of the proof can be seen in an example. The clause $x_4 \vee x_5$ was derived from the clause set S in (4.161) by the following series of input resolutions:

$$\begin{array}{cc} & \neg x_1 \vee x_2 \vee x_3 \vee x_4 \\ x_1 \vee x_3 \vee x_4 & x_2 \vee x_3 \vee x_4 \\ x_2 \vee \neg x_3 & x_2 \vee x_4 \\ \neg x_2 \vee x_5 & x_4 \vee x_5 \end{array} \quad (4.166)$$

Each clause on the right (after the first) is the resolvent of the clause above and the clause to the left. Since the clause to the left is always a premise in S, these are input resolutions (the first clause on the right also belongs to S). To show that the inequality $x_4 + x_5 \geq 1$ corresponding to $x_4 \vee x_5$ is a rank 1 cut, construct a surrogate inequality as follows. First sum the inequalities that correspond to the parents of the first resolvent

Propositional Logic 345

to obtain $x_2 + 2x_3 + 2x_4 \geq 1$ as shown below, where the inequalities on the left correspond to the premises on the left in (4.166):

$$
\begin{array}{ll}
& -x_1 + x_2 + x_3 + x_4 \geq 0 \\
(1) \quad x_1 + x_3 + x_4 \geq 1 & x_2 + 2x_3 + 2x_4 \geq 0 \\
(2) \quad x_2 - x_3 \geq 0 & 3x_2 + 2x_4 \geq 1 \\
(3) \quad -x_2 + x_5 \geq 0 & 2x_4 + 3x_5 \geq 1
\end{array}
\qquad (4.167)
$$

The next step is to take a weighted sum of $x_2 + 2x_3 + 2x_4 \geq 1$ with the premise $x_2 - x_3 \geq 0$ by giving the latter a weight of 2 (shown in parentheses), to cancel out x_3. Do the same for the rest of the premises as shown. The resulting weights provide a linear combination of premises that eliminates variables x_1, x_2, and x_3 and yields the weighted sum $2x_4 + 3x_5 \geq 1$:

$$
\begin{array}{rl}
-x_1 + x_2 + x_3 + x_4 & \geq 0 \quad (1) \\
x_1 + x_3 + x_4 & \geq 1 \quad (1) \\
x_2 - x_3 & \geq 0 \quad (2) \\
-x_2 + x_5 & \geq 0 \quad (3) \\ \hline
2x_4 + 3x_5 & \geq 1
\end{array}
\qquad (4.168)
$$

$$
\begin{array}{rl}
x_4 & \geq 0 \quad (5) \\
x_5 & \geq 0 \quad (4) \\ \hline
7x_4 + 7x_5 & \geq 1
\end{array}
$$

By adding to $2x_4 + 3x_5 \geq 1$ multiples of the bounds $x_4 \geq 0$ and $x_5 \geq 0$ as shown, dividing the sum by 7 (the sum of the premise weights), and rounding up the $\frac{1}{7}$ on the right, one obtains the desired $x_4 + x_5 \geq 1$, which is therefore a rank 1 cut for linear(S).

For the proof of Theorem 4.32, it is convenient to write a clausal inequality in the form $ax \geq 1 + n(a)$, where $n(a)$ is the sum of the negative components of a.

Proof of Theorem 4.32. Let $P_0, \ldots, P_m$ be the premises used in the input resolution proof of C (repeated use of a premise is possible). Let linear(C) be $bx \geq 1 + n(b)$, and let linear(P_i) be $a^i x \geq 1 + n(a^i)$. The goal is to show that $bx \geq 1 + n(b)$ is a rank 1 cut. The ith step of the input resolution proof is

$$
a^i x \geq 1 + n(a^i) \qquad \begin{array}{l} b^{i-1} x \geq 1 + n(b^{i-1}) \\ b^i x \geq 1 + n(b^i) \end{array}
$$

Now let $u^i x \geq v_i$ be the weighted sum of the inequalities above and to the left, as shown below, using weights 1 and w_i, respectively:

$$u^{i-1} x \geq v_{i-1}$$
$$a^i x \geq 1 + n(a^i) \qquad u^i x \geq v_i$$

where $u^0 x \geq v_0$ is $a^0 x \geq 1+n(a^0)$. Let $\text{sgn}(\alpha) = 1$ if $\alpha > 0$, -1 if $\alpha < 0$, and 0 if $\alpha = 0$. First show inductively that

$$\begin{aligned} b_j^i &= \text{sgn}(u_j^i) \text{ for all } j \quad (a) \\ v_i &= 1 + n(u^i) \qquad\qquad (b) \end{aligned} \qquad (4.169)$$

which is trivially true for $i = 1$. Assume then that (4.169) holds for $i - 1$. Let x_k be the variable on which resolution takes place, so that $a_k^i = -b_k^{i-1}$, and a_j^i and b_j^{i-1} do not have opposite signs for $j \neq k$. The weight w_i needed to make $a_k x_k$ and $u_k^{i-1} x_k$ cancel is a positive number (namely, $|u_k^{i-1}|$) because the induction hypothesis and $a_k^i = -b_k^{i-1}$ imply that a_k^i and u_k^{i-1} have opposite signs. Also, since a_j^i and b_j^{i-1} do not have opposite signs for $j \neq k$, the induction hypothesis implies that a_j^i and u_j^{i-1} do not have opposite signs for $j \neq k$, and (a) follows. To show (b), note that

$$\begin{aligned} v_i &= w_i(1+n(a^i)) + v_{i-1} = w_i(1+n(a^i)) + 1 + n(u^{i-1}) \\ &= w_i + n(w_i a^i + u^{i-1}) - w_i + 1 = 1 + n(u^i) \end{aligned}$$

where the first equality is by definition of v_i, the second is by the induction hypothesis, and the third is due to the fact that one negative component cancels in the sum $w_i a^i + u^{i-1}$.

To complete the proof, it suffices to show that $b^m x \geq 1 + n(b^m)$ can be obtained by rounding a nonnegative linear combination of the inequality $u^m x \geq 1 + n(u^m)$ and bounds. Let $W = w_1 + \cdots + w_m$ and add to $u^m x \geq 1 + n(u^m)$ the inequalities $(W - u_j^m) x_j \geq 0$ for each j for which $u_j^m > 0$ and $(-W - u_j^m) x_j \geq -W - u_j^m$ for each j for which $u_j^m < 0$. Due to (4.169), this yields the inequality

$$W b^m x \geq 1 + n(u^m) - \sum_{u_j^m < 0} (W + u_j^m) = 1 + W n(b^m) \qquad (4.170)$$

Dividing by W and rounding up on the right yields the desired inequality $b^m x \geq 1 + n(b^m)$. □

At this point, the work of the previous section can be completed.

Propositional Logic 347

COROLLARY 4.33 *Clause set S has an input refutation if and only if linear(S) is infeasible.*

Proof. From Theorem 4.30, if S has no input refutation, then linear(S) is feasible. Conversely, if S has an input refutation, then by the proof of Theorem 4.32, $0 \geq \alpha$ is a surrogate of linear(S) for some positive α. This implies linear(S) is infeasible. □

This and Theorem 4.29 imply

COROLLARY 4.34 *A clause set has a unit refutation if and only if it has an input refutation.*

This does not say that the unit resolution and input resolution algorithms derive the same clauses. It says that they detect infeasibility in the same instances.

Before moving to the converse of Theorem 4.32, it is necessary to prove a lemma. A variable x_j is *monotone* for C in a clause set S if it always occurs with the same sign as in C.

LEMMA 4.35 *If linear(C) is a rank 1 cut for linear(S), then it is a rank 1 cut for linear(S') for some subset S' of S in which every variable of C is monotone for C.*

Proof. Suppose, without loss of generality, that C contains only positive literals, and let linear(C) be $bx \geq 1 + n(b)$. Since $bx \geq 1 + n(b)$ is a rank 1 cut, it is the result of rounding up some surrogate $ux \geq v$ of linear(S). Suppose that some variable x_k in C is negated in an inequality $ax \geq 1 + n(a)$ that has weight, say, w in the nonnegative linear combination that yields $ux \geq v$. Then one can remove $ax \geq 1 + n(a)$ from the linear combination and compensate by adding inequalities, as follows:

$$\text{For } x_j \text{ in } C, \text{ add} \begin{cases} 2wx_j \geq 0 & \text{if } a_j = 1 \\ wx_j & \text{if } a_j = 0 \\ \text{nothing} & \text{if } a_j = -1 \end{cases}$$

$$\text{For } x_j \text{ not in } C, \text{ add} \begin{cases} wx_j \geq 0 & \text{if } a_j = 1 \\ \text{nothing} & \text{if } a_j = 0 \\ -wx_j \geq -w & \text{if } a_j = -1 \end{cases}$$

Let $u'x \geq v'$ be the resulting linear combination. If x_j is not in C, then $u'_j = u_j = 0$, and otherwise $u'_j = u_j + w$. Also, if s is the number of variables in C that are negated in $ax \geq 1 + n(a)$, one can check that $v' = v + (s-1)w$. Thus, $u'x \geq v'$ can be written

$$\frac{1+w}{w} u \geq v + (s-1)w$$

Multiplying this inequality by $w/(1+w)$, one obtains

$$ux \geq \frac{w}{1+w}v + \frac{s-1}{w} \geq \frac{w}{1+w}v$$

where the second inequality is due to $s \geq 0$. This, after rounding, yields $bx \geq 1 + n(b)$. □The converse of Theorem 4.32 can now be proved.

THEOREM 4.36 *If linear(C) is a rank 1 cut for linear(S), then the input resolution algorithm applied to S generates clause C.*

Proof. Suppose, without loss of generality, that C contains only positive literals, and let linear(C) be $bx \geq 1 + n(b)$. By Lemma 4.35, linear(C) is a rank 1 cut for linear(S'), where S' is some subset of C in which all the variables of C are monotone for C. Then there is a surrogate $ux \geq v$ of linear(S') that rounds up to $bx \geq 1 + n(b)$, which means, in particular, that $u_j \leq b_j$ for all j and $v > 0$. For each x_j not in C, add $(b_j - u_j)x_j \geq 0$ to $ux \geq v$, and let the result be $u'x \geq v$. Thus, $u'x \geq v'$ is a surrogate of linear(S) such that $u'_j = 0$ for all j not in C. Let clause set S' be the result of removing the variables in C from all the clauses in S. Then the same set of multipliers that produced $u'x \geq v$ yields a surrogate $0 \geq v'$ of linear(S'). Because all the literals removed from S are positive, due to the monotonicity of the variables in C, the right-hand sides in linear(S') are the same as in linear(S). Thus, $v' \geq v > 0$, which means that linear(S') is infeasible. By Theorem 4.30, there is an input refutation of S'. Since all the variables in C occur positively in S, one can restore these variables to the premises in the input refutation and obtain an input resolution proof of C. □

4.10.5 Separating Resolvents

An alternative to generating all resolvents, or all input resolvents, is to generate only separating resolvents. There are efficient algorithms for doing so. Generating separating resolvents is usually much faster than generating all resolvents, but one may be obliged to solve several linear relaxations before accumulating enough resolvents to obtain a tight relaxation.

Given a clause set S and a solution $\bar{x}$ of linear(S), a *separating resolvent* of S is a resolvent R such that $\bar{x}$ violates linear(R). One way to identify separating resolvents is to identify clauses of S that are potential parents of a separating resolvent. Then, when searching for separating resolvents, it is necessary only to examine pairs of the identified clauses.

There is a simple method for identifying potential parents of a separating resolvent. For a given clause C in S and a solution $\bar{x}$ of linear(S),

Propositional Logic

let
$$\bar{x}_j^C = \begin{cases} \bar{x}_j & \text{if literal } x_j \text{ is in } C \\ 1 - \bar{x}_j & \text{if } \neg x_j \text{ is in } C \\ 0 & \text{otherwise} \end{cases}$$

Then a resolvent R of clauses in S is separating if and only if $\sum_j \bar{x}_j^R < 1$. Now consider either parent C of R. All the literals of C occur in R except the variable x_k on which the resolution takes place. Thus, R can be separating only if

$$\sum_{j \neq k} \bar{x}_j^C < 1 \tag{4.171}$$

which implies $\sum_j \bar{x}_j^C < 2$. Furthermore, $\sum_j \bar{x}_j^C \geq 1$ because $\bar{x}$ satisfies linear(C). Supposing that x_k occurs positively in C, this inequality and (4.171) imply

$$\bar{x}_k = \sum_j \bar{x}_j^C - \sum_{j \neq k} \bar{x}_j^C > 0$$

Since $\neg x_k$ occurs in R's other parent, similar reasoning shows $1 - \bar{x}_k > 0$. Thus, one need only examine pairs of clauses that respectively contain literals $x_k, \neg x_k$ with $0 < \bar{x}_k < 1$. The following has been shown:

THEOREM 4.37 *If $\bar{x}$ is a solution of linear(S) for clause set S, then a clause C in S can be a parent of a separating resolvent of S only if $\sum_j \bar{x}_j^C < 2$. Furthermore, a separating resolvent can be obtained from C only by resolving on a variable x_k for which $\bar{x}_k$ is fractional.*

Consider for example the clause set S:

$x_1 \vee x_2 \vee x_3$				C_1
$\neg x_1 \vee x_3 \vee \neg x_4$				C_2
$\neg x_2 \vee x_3 \vee x_4$				C_3
$x_1 \vee x_2 \vee \neg x_3$				C_4
$x_2 \vee x_3 \vee \neg x_4$				C_5
0	$\frac{1}{2}$	$\frac{1}{2}$	0	$\bar{x}$

Suppose for the sake of the example that the objective is to minimize the number of true variables while satisfying S. A solution of linear(S) that minimizes $\sum_j x_j$ is $(\bar{x}_1, \bar{x}_2, \bar{x}_3, \bar{x}_4) = (0, \frac{1}{2}, \frac{1}{2}, 0)$, shown along the bottom. The values of $\sum_j \bar{x}_j^{C_i}$ for $i = 1, \ldots, 5$ are given in Table 4.2. These values indicate that only clauses C_1, C_3, and C_4 need be considered as potential parents of a separating resolvent. Furthermore, since only $\bar{x}_2$ and $\bar{x}_3$ are nonintegral, it suffices to consider pairs of clauses that resolve on x_2 and x_3. Two pairs (C_1, C_3 and C_1, C_4) satisfy these conditions. They

Table 4.2. Values used to screen out clauses that cannot be parents of a separating resolvent.

Clause C_i	$\sum_j \bar{x}_j^{C_i}$
C_1	1
C_2	$2\frac{1}{2}$
C_3	1
C_4	1
C_5	2

respectively yield the resolvents $x_1 \vee x_3 \vee x_4$ and $x_1 \vee x_2$. As it happens, both are separating. If these two clauses are added to S, then a solution of linear(S) that minimizes $\sum_j x_j$ is $(\bar{x}_1, \bar{x}_2, \bar{x}_3, \bar{x}_4) = (1, 0, 0, 0)$, which is integral and therefore an optimal solution of the original problem.

4.10.6 Exercises

1. Show that the relaxation given for Formula 4 in Table 4.1 is a convex hull relaxation by showing that its coefficient matrix is totally unimodular.

2. Show that the coefficient matrix for the relaxation of Formula 7 in Table 4.1 is not totally unimodular.

3. Use the information in Table 4.1 to write a convex hull relaxation for $(x_1 \to x_2) \to (x_3 \to x_4)$.

4. Write a convex hull relaxation for $\bigwedge_{j \in J_1} x_j \vee \bigwedge_{j \in J_2} x_j$, and prove it is a convex hull relaxation.

5. Consider the clause set

$$x_1 \vee x_2$$
$$x_1 \vee \neg x_2 \vee x_3$$
$$x_1 \vee \neg x_2 \vee \neg x_3$$

Show that input resolution yields the same result as full resolution but requires more steps. Also show, using this example, that input resolution can infer fewer clauses if absorbed clauses are deleted from the original clause set before the algorithm is completed. Show that this is not true for full resolution.

Propositional Logic 351

6 Consider the clause set S, consisting of

$$x_1 \lor x_2 \lor x_3$$
$$x_1 \lor x_2 \lor \neg x_3 x_3$$
$$x_1 \lor \neg x_2 \lor x_3$$
$$x_1 \lor \neg x_2 \lor \neg x_3$$

Let S' be the result of applying input resolution. Show that S' does not contain all prime implications of S (which means input resolution is weaker than full resolution), but linear(S') is a tighter relaxation than linear(S).

7 Show that although input resolution has the same refutation power as unit resolution, it can infer clauses that unit resolution cannot.

8 The proof of Theorem 4.30 contains a method for constructing a solution of linear(S) when input resolution does not infer a contradiction from S. Use this method when S is

$$x_1 \lor x_2$$
$$\neg x_1 \lor x_3$$
$$\neg x_2 \lor x_3$$
$$x_1 \lor x_2 \lor \neg x_3$$

after applying input resolution to S. One can also find an input refutation for S when linear(S) is infeasible, but this is more complicated.

9 A unit clause C was derived in the input resolution proof of Exercise 8. Use the construction in the proof of Theorem 4.31 to show that linear(C) is a rank 1 Chvátal-Gomory cut for linear(S).

10 In the proof of Theorem 4.31, verify that (4.170) is a nonnegative linear combination of $u^m x \geq 1 + n(u^m)$ and bounds.

11 Input resolution derives the clause $x_1 \lor x_3$ from the clause set S, consisting of

$$x_1 \lor x_2 \quad (a)$$
$$\neg x_1 \lor x_2 \lor x_3 \quad (b)$$
$$x_1 \lor \neg x_2 \lor x_3 \quad (c)$$

by resolving (a) and (b), and resolving the resulting clause with (c). Using the mechanism of Theorem 4.31, construct a set of multipliers to show that linear($x_1 \lor x_3$) is a rank 1 cut for linear(S). Now use the procedure in the proof of Lemma 4.35 to show that linear($x_1 \lor x_3$) is a rank 1 cut for a monotone subset of clauses, namely (a) and (c).

12 If S is the clause set

$$\begin{array}{cccccc} x_1 \vee & x_2 \vee & x_3 & & \vee & x_5 \vee \neg x_6 \\ x_1 \vee & x_2 & & \vee & x_4 & \vee \\ x_1 & & \vee & x_3 \vee x_4 & & \vee \neg x_6 \\ & x_2 \vee & x_3 \vee x_4 & & \vee \neg x_6 \\ \neg x_1 \vee & x_2 \vee & x_3 & & \vee \neg x_5 \\ & \neg x_2 \vee & x_3 \vee x_4 & \vee & x_5 \\ & & \neg x_3 \vee x_4 & & \vee & x_6 \end{array}$$

then linear(S) has a basic solution $x = (\frac{1}{3}, \frac{1}{3}, \frac{1}{3}, \frac{1}{3}, 0, 1)$. Identify which clauses can be parents of separating resolvents. Derive their resolvents, and identify which one(s) are separating.

4.11 The Element Constraint

There are several continuous relaxations for the element constraint, most of them based on the relaxations for disjunctive constraints presented in Section 4.7. These relaxations are particularly useful when the variables have fairly tight bounds. There are also effective relaxations for the indexed linear element constraint. Section 3.8 discusses modeling applications for the various forms of the element constraint.

The goal is to derive a continuous relaxation for element$(y, z \,|\, a)$ in terms of z, for element(y, x, z) in terms of z and $x = (x_1, \ldots, x_m)$, and for the indexed linear constraint element$(y, x, z \,|\, a)$ in terms of x and z. Thus, if S is the feasible set of element$(y, z \,|\, a)$, then a relaxation should describe a set that contains the projection of S onto z, and onto x and z for the other element constraints. A convex hull relaxation in each case describes the convex hull of the corresponding projection.

A *vector-valued* generalization of the element constraint is important for some modeling applications. It typically has the form

$$\text{element}(y, z \,|\, (a^1, \ldots, a^m))$$

where z and each a^i are tuples. There is also a vector-valued indexed linear constraint

$$\text{element}(y, x, z \,|\, (a^1, \ldots, a^m))$$

where x is a scalar variable. It sets z equal to the yth tuple in the list $xa^1, \ldots, xa^m$. Both of these have useful continuous relaxations. Section 4.12.4 shows how a vector-valued element constraint can help formulate and solve the quadratic assignment problem.

The Element Constraint 353

4.11.1 Convex Hull Relaxations

The simplest element constraint, element$(y, z \,|\, a)$, has an obvious convex hull relaxation:
$$\min_{k \in D_y}\{a_k\} \leq z \leq \max_{k \in D_y}\{a_k\} \tag{4.172}$$

Thus, a convex hull relaxation for element$(y, (2, 4, 5), z)$ is $2 \leq z \leq 5$, if $D_y = \{1, 2, 3\}$.

A useful convex hull relaxation can be derived for element(y, x, z) if there are lower and/or upper bounds $L \leq x \leq U$ on the variables. The first step is to observe that element(y, x, z) and these bounds imply the disjunction
$$\bigvee_{k \in D_y} \begin{pmatrix} z = x_k \\ L \leq x \leq U \end{pmatrix} \tag{4.173}$$

where $x = (x_1, \ldots, x_m)$. Theorem 4.17 can now be applied.

THEOREM 4.38 *If $L_i \leq x_i \leq U_i$ for all $i \in D_y$, then the following is a convex hull relaxation of element$(y, (x_1, \ldots, x_m), z)$:*

$$z = \sum_{k \in D_y} x_k^k, \quad \sum_{k \in D_y} y_k = 1, \quad x_i = \sum_{k \in D_y} x_i^k, \; \text{all } i \in D_y \tag{4.174}$$
$$Ly_k \leq x^k \leq Uy_k, \quad y_k \geq 0, \; \text{all } k \in D_y$$

For example, a convex hull relaxation of
$$\text{element}(y, (x_1, x_2), z), \quad 1 \leq x_1 \leq 4, \quad 3 \leq x_2 \leq 5 \tag{4.175}$$

is
$$z = x_1^1 + x_2^2, \quad x_1 = x_1^1 + x_1^2, \quad x_2 = x_2^1 + x_2^2$$
$$y \leq x_1^1 \leq 4y, \quad y \leq x_2^1 \leq 4y$$
$$3(1-y) \leq x_1^2 \leq 5(1-y), \quad 3(1-y) \leq x_2^2 \leq 5(1-y)$$
$$0 \leq y \leq 1$$

The projection onto the (z, x) space is
$$\tfrac{2}{3}x_1 + \tfrac{1}{3} \leq z \leq 5, \quad \tfrac{4}{3}x_1 + x_2 - \tfrac{16}{3} \leq z \leq \tfrac{2}{3}x_1 + x_2 + \tfrac{2}{3}$$
$$1 \leq x_1 \leq 4, \quad 3 \leq x_2 \leq 5$$

In general, the projection is quite complex and is not computed explicitly in this fashion.

When every variable x_i has the same bounds, the convex hull relaxation simplifies considerably.

THEOREM 4.39 *Suppose that $L_0 \leq x_i \leq U_0$ for all $i \in D_y$. Then a convex hull relaxation of $\text{element}(y, (x_1, \ldots, x_m), z)$ is given by*

$$\sum_{i \in D_y} x_i - (|D_y| - 1)U_0 \leq z \leq \sum_{i \in D_y} x_i - (|D_y| - 1)L_0 \qquad (4.176)$$

and the bounds $L_0 \leq z \leq U_0$, $L_0 \leq x_i \leq U_0$ for $i \in D_y$.

Proof. Without loss of generality, the origin can be moved to $x = (L_0, \ldots, L_0)$ and $z = L_0$, and (4.176) becomes

$$\sum_{i \in D_y} x_i - (|D_y| - 1)U_0 \leq z \leq \sum_{i \in D_y} x_i \qquad (4.177)$$

It suffices to show that any point $(\bar{z}, \bar{x})$ that satisfies (4.177) and bounds $0 \leq x_i \leq U_0$, $0 \leq z \leq U_0$ is a convex combination of points that satisfy the disjunction (4.173). Due to the second inequality in (4.177), one can write $\bar{z} = \alpha \sum_i \bar{x}_i$ for $\alpha \in [0, 1]$. For convenience, suppose $D_y = \{1, \ldots, m\}$. It will be shown that $(\bar{z}, \bar{x})$ is the following convex combination:

$$\begin{bmatrix} \bar{z} \\ \bar{x}_1 \\ \bar{x}_2 \\ \bar{x}_3 \\ \vdots \\ \bar{x}_m \end{bmatrix} = \frac{\alpha \bar{x}_1}{U_0} \begin{bmatrix} U_0 \\ U_0 \\ b_2 \\ b_3 \\ \vdots \\ b_m \end{bmatrix} + \frac{\alpha \bar{x}_2}{U_0} \begin{bmatrix} U_0 \\ b_1 \\ U_0 \\ b_3 \\ \vdots \\ b_m \end{bmatrix} + \cdots + \frac{\alpha \bar{x}_m}{U_0} \begin{bmatrix} U_0 \\ b_1 \\ b_2 \\ b_3 \\ \vdots \\ U_0 \end{bmatrix}$$

$$+ \frac{1}{m}\left(1 - \frac{\alpha}{U_0}\sum_i \bar{x}_i\right) \left(\begin{bmatrix} 0 \\ 0 \\ c_2 \\ c_3 \\ \vdots \\ c_m \end{bmatrix} + \begin{bmatrix} 0 \\ c_1 \\ 0 \\ c_3 \\ \vdots \\ c_m \end{bmatrix} + \cdots + \begin{bmatrix} 0 \\ c_1 \\ c_2 \\ c_3 \\ \vdots \\ 0 \end{bmatrix} \right)$$

where b_i, c_i will be defined shortly. Note that each vector

$$(U_0, b_1, \ldots, b_{i-1}, U_0, b_{i+1}, \ldots, b_m)$$

and each vector

$$(0, c_1, \ldots, c_{i-1}, 0, c_{i+1}, \ldots, c_m)$$

satisfy the disjunct $z = x_i$. It must also be shown that

(i) $0 \leq b_i \leq U_0$ for all i
(ii) $0 \leq c_i \leq U_0$ for all i
(iii) $(\bar{z}, \bar{x})$ is equal to the linear combination shown
(iv) the linear combination is a convex combination

For (iv), it is enough to show that the coefficients are nonnegative, because they obviously sum to one. But this follows from

$$0 \leq \frac{\alpha}{U_0} \sum_i \bar{x}_i \leq 1$$

where the first inequality is due to $\bar{x}_i \geq 0$ and the second to the fact that

$$\alpha \sum_i \bar{x}_i = \bar{z} \leq U_0 \tag{4.178}$$

Now, for each i, there are two cases.

Case I. $\bar{x}_i \leq \bar{z}$. In this case, set

$$b_i = U_0 \left(\frac{1-\alpha}{\alpha} \right) \left(\frac{\bar{x}_i}{\sum_i \bar{x}_i - \bar{x}_i} \right)$$

Obviously, $b_i \geq 0$. Also, $b_i \leq U_0$, due to the fact that

$$b_i = U_0 \frac{\bar{x}_i - \alpha \bar{x}_i}{\alpha \sum_i \bar{x}_i - \alpha \bar{x}_i} \leq U_0 \frac{\bar{z} - \alpha \bar{x}_i}{\bar{z} - \alpha \bar{x}_i} = U_0$$

where the inequality is due to the case hypothesis and $\bar{z} = \alpha \sum_i \bar{x}_i$. This establishes (i). One can now set $c_i = 0$, so that (ii) holds, and direct calculation verifies (iii).

Case II. $\bar{x}_i > \bar{z}$. In this case, set $b_i = U_0$ so that (i) obviously holds. Also set

$$c_i = \frac{\bar{x}_i - \alpha \sum_j \bar{x}_j}{1 - \frac{\alpha}{U_0} \sum_j \bar{x}_j}$$

Here, (iii) is easily verified. It remains to show (ii). The fact that $c_i \geq 0$ follows from the case hypothesis and (4.178). To see that $c_i \leq U_0$, write

$$c_i \leq \frac{U_0 - \sum_j \bar{x}_j + (m-1)U_0}{1 - \frac{1}{U_0}\sum_j \bar{x}_j + (m-1)} = U_0$$

where the inequality is due to $\bar{x}_i \leq U_0$, the first inequality in (4.177), and (4.178). This completes the proof. □

When the variables x_i have different bounds L_i, U_i, the inequality (4.176) remains valid if L_0 and U_0 are set to the most extreme bounds.

COROLLARY 4.40 *If $L_i \leq x_i \leq U_i$ for all $i \in D_y$, the following is a relaxation of* element$(y, (x_1, \ldots, x_m), z)$:

$$\sum_{i \in D_y} x_i - (|D_y| - 1)U_0 \leq z \leq \sum_{i \in D_y} x_i - (|D_y| - 1)L_0 \quad (4.179)$$
$$L_0 \leq z \leq U_0, \quad L_0 \leq x_i \leq U_0, \ i \in D_y$$

where $L_0 = \min_{i \in D_y}\{L_i\}$ and $U_0 = \max_{i \in D_y}\{U_i\}$.

For example, a valid relaxation can be obtained for (4.175) by setting $L_0 = 1$, $U_0 = 5$:

$$x_1 + x_2 - 5 \leq z \leq x_1 + x_2 - 1, \quad 1 \leq z \leq 5, \quad 1 \leq x_i \leq 5, \ i = 1, 2$$

A convex hull relaxation is easy to write for the indexed linear element constraint, because the relaxation contains only two variables.

COROLLARY 4.41 *If $L \leq x \leq U$, then the following is a convex hull relaxation of* element$(y, x, z \mid (a_1, \ldots, a_m))$:

$$z \geq \frac{U_{\min} - L_{\min}}{U - L}x + \frac{UL_{\min} - LU_{\min}}{U - L}$$
$$z \leq \frac{U_{\max} - L_{\max}}{U - L}x + \frac{UL_{\max} - LU_{\max}}{U - L}$$
$$L \leq x \leq U$$

where

$$L_{\min} = \min_{i \in D_y}\{a_i L\}, \quad U_{\min} = \min_{i \in D_y}\{a_i U\}$$
$$L_{\max} = \max_{i \in D_y}\{a_i L\}, \quad U_{\max} = \max_{i \in D_y}\{a_i U\}$$

The Element Constraint

If $L = 0$, the relaxation simplifies to

$$\min_{i \in D_y}\{a_i\} x \leq z \leq \max_{i \in D_y}\{a_i\} x, \quad 0 \leq x \leq U$$

For the constraint $\text{element}(y, x, z \,|\, (-1, 2))$ with $x \in [-3, 2]$,

$$(L_{\min}, U_{\min}) = (-6, -2) \quad \text{and} \quad (L_{\max}, U_{\max}) = (3, 4)$$

So, the convex hull relaxation is

$$\tfrac{4}{5} x - \tfrac{18}{5} \leq z \leq \tfrac{1}{5} x + \tfrac{18}{5}, \quad -3 \leq x \leq 2$$

as illustrated in Figure 4.15.

4.11.2 Big-M Relaxations

Big-M relaxations of the element constraint are generally not as tight as a convex hull relaxation, but involve fewer variables and constraints. A relaxation can be formed by applying Theorem 4.20 to the disjunction

$$\bigvee_{k \in D_y} (-z + x_k = 0) \tag{4.180}$$

Using (4.111),

$$M'_k = U_k^{\max} - L_k, \quad M''_k = U_k - L_k^{\min}$$

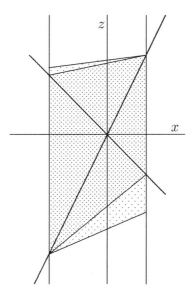

Figure 4.15. Convex hull relaxation (dark shading) and big-M relaxation (entire shaded area) for an indexed linear element constraint.

where
$$U_k^{\max} = \max_{\ell \neq k}\{U_k\}, \quad L_k^{\min} = \min_{\ell \neq k}\{L_\ell\}$$

A direct application of Theorem 4.20 yields the following corollary.

COROLLARY 4.42 *The big-M relaxation of element(y, x, z) with bounds $L \leq x \leq U$ is equivalent to the following:*

$$\begin{aligned} & z \geq x_k - \left(U_k - L_k^{\min}\right)(1 - \alpha_k), \text{ all } k \in D_y \\ & z \leq x_k + \left(U_k^{\max} - L_k\right)(1 - \alpha_k), \text{ all } k \in D_y \\ & \sum_{k \in D_y} \alpha_k = 1, \quad \alpha_k \geq 0, \text{ all } k \in D_y \end{aligned} \quad (4.181)$$

When all the lower bounds L_j are equal to L_0, and all upper bounds U_i are equal to U_0, the relaxation (4.181) is dominated by the convex hull relaxation (4.179). When the lower bounds or upper bounds differ, however, it may be advantageous to use both relaxations.

The relaxation (4.179) for example (4.175) was found in the previous section to be

$$x_1 + x_2 - 5 \leq z \leq x_1 + x_2 - 1 \quad (4.182)$$

plus bounds. The big-M relaxation (4.181) for this example, when projected onto x_1 and x_2, is

$$\begin{aligned} 4x_1 + x_2 - 4 &\leq z \leq -x_1 + x_2 + 1 \\ x_1 - 1 &\leq z \leq x_1 + 4 \\ x_2 - 4 &\leq z \leq x_2 + 1 \end{aligned}$$

plus bounds. Neither relaxation is redundant of the other.

There is no point in using a big-M relaxation for the indexed linear constraint element$(y, (a_1x, \ldots, a_mx)z)$, because the convex hull relaxation is already quite simple. A big-M relaxation for the vector form of the constraint element$(y, (A_1x, \ldots, A_mx), z)$ could be desirable, however, and is represented in the next section.

4.11.3 Vector-Valued Element

The vector-valued element constraint

$$\text{element}(y, z \mid (a^1, \ldots, a^m)) \quad (4.183)$$

can be very useful because its continuous relaxation is much tighter than the result of relaxing each of the component constraints individually.

This can be seen in an example. Suppose that the production level of a shop is a_1^y, and the corresponding cost is a_2^y, where $y \in \{1, 2, 3, 4\}$. The

The Element Constraint   359

possible production levels are 10, 20, 40, and 90 and the corresponding costs are 100, 150, 200, and 250. Perhaps there are various constraints on y and other problem variables, but the constraints of interest here are

$$\begin{aligned}\text{element}(y, z_1 \,|\, (10, 20, 40, 90)) \\ \text{element}(y, z_2 \,|\, (100, 150, 200, 250))\end{aligned} \quad (4.184)$$

which define z_1 to be the production level and z_2 to be the cost. The individual convex hull relaxations of the two element constraints are $10 \leq z_1 \leq 90$ and $100 \leq z_2 \leq 150$. However, a vector-valued element constraint that combines the two

$$\text{element}\left(y, \begin{bmatrix} z_1 \\ z_2 \end{bmatrix} \,\Big|\, \left(\begin{bmatrix} 10 \\ 100 \end{bmatrix}, \begin{bmatrix} 20 \\ 150 \end{bmatrix}, \begin{bmatrix} 40 \\ 200 \end{bmatrix}, \begin{bmatrix} 90 \\ 250 \end{bmatrix}\right)\right)$$

has a much tighter convex hull relaxation (Figure 4.16).

In general, one can obtain a convex hull relaxation for (4.183) by computing a convex hull description of the points $a^1, \ldots, a^m$ in z-space. This is practical when the dimensionality of z is 2 or 3, using various algorithms developed for computational geometry. One may also be able to exploit special structure in the values a^i, as illustrated by the quadratic assignment problem in Section 4.12.4.

For higher dimensionality, it may be useful to find a separating cut $dz \geq \delta$ that is maximally violated by the solution $\bar{z}$ of the current relaxation. That is, d and δ are chosen so that $\bar{z}$ maximally violates $dx \geq \delta$, and $a^1, \ldots, a^m$ satisfy it. This may be done by solving the linear opti-

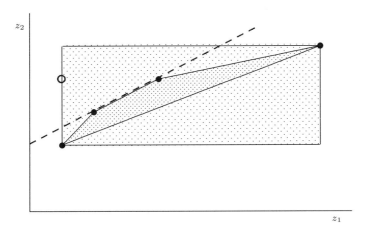

Figure 4.16. Convex hull relaxation of a vector-valued element constraint (dark shading) and of two element constraints considered separately (entire shaded area). The dashed line is an optimal separating cut for the point represented by the small circle.

mization problem

$$\begin{aligned}\max\ & \delta - d\bar{z} \\ & \|d\| \leq 1, \quad da^i \geq \delta,\ i = 1, \ldots, m\end{aligned} \quad (4.185)$$

where d, δ are unrestricted in sign and $\|\cdot\|$ is a convenient norm. If the L_∞ norm ($\max_j\{|d_j|\}$) is used, the constraint $\|d\| \leq 1$ becomes $-1 \leq d_j \leq 1$ for all j.

If $\bar{z} = (10, 200)$ in the example of Figure 4.16, the optimal separating cut using the L_∞ norm is $z_1 - 0.4z_2 \geq -40$, as shown in the figure.

There is also a vector-valued form of the indexed linear element constraint:

$$\text{element}(y, x, z \,|\, (a^1, \ldots, a^m)) \quad (4.186)$$

It can be used to implement indexed linear expressions of the form xa^y, where each a^k is a vector and x a scalar variable, by replacing xa^y with z and adding the constraint (4.186). This type of expression occurs, for instance, in the product configuration problem of Section 2.2.7.

Because (4.186) implies the disjunction

$$\bigvee_{k \in D_y} \begin{pmatrix} z = xa^k \\ L \leq x \leq U \end{pmatrix}$$

Theorem 4.17 can be applied to obtain a convex hull relaxation for (4.186).

COROLLARY 4.43 *If* $L \leq x \leq U$, $\text{element}(y, x \,|\, (a^1, \ldots, a^m))$ *has the convex hull relaxation*

$$z = \sum_{k \in D_y} a^k x_k, \quad x = \sum_{k \in D_y} x_k, \quad \sum_{k \in D_y} y_k = 1 \quad (4.187)$$

$$L y_k \leq x_k \leq U y_k, \quad y_k \geq 0,\ \text{all}\ k \in D_y$$

A big-M relaxation can be written for (4.186), as follows. Suppose, as before, that $L \leq x \leq U$. Since z will be equated with xa^k for some k, it can be assumed that $\bar{L} \leq z \leq \bar{U}$, where

$$\begin{aligned}\bar{L}_j &= \min_{k \in D_y} \left\{\min\{a_j^k L, a_j^k U\}\right\},\ \text{all}\ j \\ \bar{U}_j &= \max_{k \in D_y} \left\{\max\{a_j^k L, a_j^k U\}\right\},\ \text{all}\ j\end{aligned} \quad (4.188)$$

Using (4.94), the big-Ms are

$$M^k = \begin{bmatrix} \bar{U} - L^k \\ U^k - \bar{L} \end{bmatrix}$$

where
$$L^k = \min\{0, a^k\}U + \max\{0, a^k\}L$$
$$U^k = \min\{0, a^k\}L + \max\{0, a^k\}U$$
and the minimum and maximum are taken componentwise. So, the big-M relaxation (4.95) becomes

$$xa^k - (U^k - \bar{L})(1 - y_k) \leq z \leq xa^k + (\bar{U} - L^k)(1 - y_k)$$
$$L \leq x \leq U, \quad \sum_{k \in D_y} y_k = 1, \quad y_k \geq 0, \text{ all } k \in D_y$$

Section 4.11.1 derived a convex hull relaxation for the constraint element$(y, x \mid (-1, 2))$ with $x \in [-3, 2]$. Although a big-M relaxation would not in practice be used for this constraint, it is interesting to contrast it with the convex hull relaxation. The projection of the big-M relaxation onto z, x is

$$\tfrac{8}{19}x - \tfrac{90}{19} \leq z \leq \tfrac{1}{8}x + \tfrac{15}{4}, \quad -3 \leq x \leq 2$$

This is illustrated along with the convex hull relaxation in Figure 4.15.

4.11.4 Exercises

1 Use Theorem 4.38 to write a convex hull relaxation for the constraint element$(y, (x_1, x_2), z)$, where $y \in \{1, 2\}$, $x_1 \in [-1, 2]$, and $x_2 \in [0, 3]$.

2 Prove Theorem 4.38 as a corollary of Theorem 4.17.

3 Use Theorem 4.39 to write a convex hull relaxation for the constraint element$(y, (x_1, x_2), z)$ where $y \in \{1, 2\}$ and $x_1, x_2 \in [-1, 3]$.

4 The point $(z, x_1, x_2) = (3, 2, 2)$ is not feasible for element$(y, (x_1, x_2), z)$ in Exercise 3 but belongs to the convex hull of the feasible set. Use the mechanism in the proof of Theorem 4.39 to construct a convex combination of feasible points that yields $(3, 2, 2)$. Note that the origin must be shifted before applying the proof.

5 Use Corollary 4.40 to write an alternate relaxation for the element constraint in Exercise 1.

6 Use Corollary 4.41 to write a convex hull relaxation for the indexed linear constraint element$(y, x, z, \mid (-2, 3))$ with $x \in [-5, 6]$.

7 Prove Corollary 4.41.

8 Use Corollary 4.42 to write a big-M relaxation for the element constraint in Exercise 1. Compare it with the convex hull relaxation, as well as the relaxation obtained for the same constraint in Exercise 5.

9 Prove Corollary 4.42.

10 Write the linear programming problem whose solution obtains an optimal separating cut for the point $(10, 200)$ in Figure 4.16, using the L_∞ norm. Verify that the cut $z_1 - 0.4z_2 \geq -40$ is optimal.

11 Prove Corollary 4.43.

4.12 The All-Different Constraint

As noted in Section 3.9, the constraint

$$\text{alldiff}(x_1, \ldots, x_n)$$

frequently occurs in constraint programming models, not only because many modeling situations call for it, but also because there are fast and effective domain filtering algorithms for the constraint. When the variables x_i take numerical values, continuous relaxations can be created for the constraint as well. One possible relaxation uses only the original variables $x_1, \ldots, x_n$, while another is based on an MILP model of the constraint and introduces 0-1 variables.

While both relaxations assume that the variables x_i take numerical values, in practice the variables may refer to tasks or workers that are labeled with nonnumeric names or symbols. The relaxations can nonetheless be useful. The MILP relaxation can dispense with the x_i variables if cost and other constraints involving these variables can be expressed in terms of the 0-1 variables instead. The relaxation in the original variables requires numeric values, but one may be able to take advantage of the fact that the numeric labels have no particular meaning by selecting labels that result in a good relaxation. This idea inspires a model of the well-known quadratic assignment problem, conceived as a location problem, that may be able to exploit the structure of the distance metric.

4.12.1 Convex Hull Relaxation

The constraint $\text{alldiff}(x_1, \ldots, x_n)$ can be given a continuous relaxation in the original variables $x_1, \ldots, x_n$, if they take numerical values. If all the variable domains are the same, it is a convex hull relaxation. Unfortunately, the relaxation contains exponentially many constraints, but there is a simple way to generate separating cuts.

Suppose first that every variable has the same domain. A continuous relaxation of the alldiff constraint can be based on the fact that the sum

The All-Different Constraint 363

of any k variables must be at least as large as the k smallest domain elements.

THEOREM 4.44 *If the domain D_{x_j} is $\{v_1, \ldots, v_m\}$ for $j = 1, \ldots, n$, where $v_1 < \cdots < v_m$ and $m \geq n$, then the following is a convex hull relaxation of alldiff$(x_1, \ldots, x_n)$:*

$$\sum_{j=1}^{|J|} v_j \leq \sum_{j \in J} x_j \leq \sum_{j=m-|J|+1}^{m} v_j, \quad \text{all } J \subset \{1, \ldots, n\} \tag{4.189}$$

Proof. Let S be the set of feasible points for the alldiff constraint. Clearly all points in S satisfy (4.189). It remains to show that every point satisfying (4.189) is a convex combination of points in S.

Rather than show this directly, it is convenient to use a dual approach. Since the convex hull of S is the intersection of all half planes containing S, it suffices to show that any half plane containing S contains all points satisfying (4.189). That is, it suffices to show that any valid inequality $ax \geq b$ for the alldiff is implied by (4.189), or equivalently, is dominated by a surrogate (nonnegative linear combination) of (4.189).

The first step is to prove the theorem when $a \geq 0$ or $a \leq 0$. Index the variables so that $a_1 \geq \cdots \geq a_n$, and consider a linear combination of the following inequalities from (4.189):

$$\sum_{j=1}^{i} x_j \geq \sum_{j=1}^{i} v_j, \quad i = 1, \ldots, n-1 \quad (a_i - a_{i+1} \text{ if } a \geq 0, \ 0 \text{ otherwise})$$

$$\sum_{j=1}^{n} x_j \geq \sum_{j=1}^{n} v_j \qquad (a_n \text{ if } a \geq 0, \ 0 \text{ otherwise})$$

$$-\sum_{j=i}^{n} x_j \geq -\sum_{j=m-n+i}^{m} v_j, \quad i = 2, \ldots, n \quad (a_{i-1} - a_i \text{ if } a \leq 0, \ 0 \text{ otherwise})$$

$$-\sum_{j=1}^{n} x_j \geq -\sum_{j=m-n+1}^{m} v_j \qquad (-a_1 \text{ if } a \leq 0, \ 0 \text{ otherwise})$$

where each inequality has the nonnegative multiplier shown on the right. The result of the linear combination is $ax \geq a\bar{v}$ where

$$\bar{v} = \begin{cases} (v_1, \ldots, v_n) & \text{if } a \geq 0 \\ (v_{m-n+1}, \ldots, v_m) & \text{if } a \leq 0 \end{cases}$$

But since $x = \bar{v}$ is a feasible solution of the alldiff, the validity of $ax \geq b$ implies $a\bar{v} \geq b$, which means that $ax \geq b$ is dominated by the surrogate $ax \geq a\bar{v}$ of (4.189).

Now, take an arbitrary valid inequality $ax \geq b$ in which the components of a may have any sign. Index the variables so that $a_1 \geq \cdots \geq a_n$, and suppose that $a_j \geq 0$ for $j = 1, \ldots, k$ and $a_j < 0$ for $j = k+1, \ldots, n$. Since $v_1 < \cdots < v_m$, it is clear that $ax \geq b$ cannot be valid unless $b \leq av^*$, where $v^* = (v_1, \ldots, v_k, v_{m-n+k+1}, \ldots, v_m)$. Also,

$$\sum_{j=1}^{k} a_j x_j \geq \sum_{j=1}^{k} a_j v_j \qquad (4.190)$$

is valid because $a_1 \geq \cdots \geq a_k \geq 0$ and $v_1 < \cdots < v_m$, and

$$\sum_{j=k+1}^{n} (-a_j) x_j \leq \sum_{j=m-n+k+1}^{m} (-a_j) v_j$$

is valid because $0 \leq -a_{k+1} \leq \cdots \leq -a_n$ and $v_1 < \cdots < v_m$. The last inequality can be written

$$\sum_{j=k+1}^{n} a_j x_j \geq \sum_{j=m-n+k+1}^{m} a_j v_j \qquad (4.191)$$

The sum of (4.190) and (4.191) is $ax \geq av^*$, which dominates $ax \geq b$ because $b \leq av^*$. But since (4.190) and (4.191) are valid and have coefficients that are all nonnegative or all nonpositive, they are surrogates of (4.189). The same is therefore true of $ax \geq b$, and the theorem follows. □

For example, consider the constraint

$$\text{alldiff}(x_1, x_2, x_3), \quad x_j \in \{1, 5, 8\}, \ j = 1, 2, 3$$

The convex hull relaxation is

$$\begin{aligned}
&x_1, x_2, x_3 \geq 1 &&(a)\\
&x_1 + x_2 \geq 6, \quad x_1 + x_3 \geq 6, \quad x_2 + x_3 \geq 6 &&(b)\\
&x_1 + x_2 + x_3 \geq 14 &&(c)\\
&x_1 + x_2 + x_3 \leq 14 &&(d)\\
&x_1 + x_2 \leq 13, \quad x_1 + x_3 \leq 13, \quad x_2 + x_3 \leq 13 &&(e)\\
&x_1, x_2, x_3 \leq 8 &&(f)
\end{aligned}$$

Since in this case each domain contains exactly n elements, constraints (e) and (f) are redundant, and the relaxation simplifies to

$$\begin{aligned}
&x_1, x_2, x_3 \geq 1 &&(a)\\
&x_1 + x_2 \geq 6, \quad x_1 + x_3 \geq 6, \quad x_2 + x_3 \geq 6 &&(b)\\
&x_1 + x_2 + x_3 = 14 &&(c)
\end{aligned} \qquad (4.192)$$

The All-Different Constraint

COROLLARY 4.45 *If $m = n$ in Theorem 4.44, then the following is a convex hull relaxation of alldiff$(x_1, \ldots, x_n)$:*

$$\sum_{j \in J} x_j \geq \sum_{j=1}^{|J|} v_j, \text{ all } J \subset \{v_1, \ldots, v_n\} \text{ with } |J| < n \quad (a)$$
$$\sum_{j=1}^{n} x_j = \sum_{j=1}^{n} v_j \quad (b)$$
(4.193)

The relaxation in Theorem 4.44 contains exponentially many constraints because there are exponentially subsets J. However, one can begin by using only the constraints

$$\sum_{j=1}^{n} v_j \leq \sum_{j=1}^{n} x_j \leq \sum_{j=m-n+1}^{m} v_j \quad (4.194)$$

and bounds on the variables, and then generate separating cuts as needed. Let $\bar{x}$ be the solution of the current relaxation of the problem, and renumber the variables so that $\bar{x}_1 \leq \cdots \leq \bar{x}_n$. Then, for $i = 2, \ldots, n-1$, one can generate the cut

$$\sum_{j=1}^{i} x_j \geq \sum_{j=1}^{i} v_j \quad (4.195)$$

whenever

$$\sum_{j=1}^{i} \bar{x}_j < \sum_{j=1}^{i} v_j$$

Also, for each $i = n-1, \ldots, 2$, generate the cut

$$\sum_{j=i}^{n} x_j \leq \sum_{j=m-n+i}^{m} v_j$$

whenever

$$\sum_{j=i}^{n} \bar{x}_j > \sum_{j=m-n+i}^{m} v_j$$

There is no separating cut if $\bar{x}$ lies within the convex hull of the alldiff feasible set. If $m = n$, one can start with (4.193b) and bounds, and generate the cut (4.195) for $i = 1, \ldots, n-1$ whenever

$$\sum_{j=1}^{i} \bar{x}_j < \sum_{j=1}^{i} v_j$$

For example, suppose one wishes to solve the problem

$$\begin{array}{l} \min\ 2x_1 + 3x_2 + 4x_3 \\ x_1 + 2x_2 + 3x_3 \geq 32,\ \text{alldiff}(x_1, x_2, x_3) \\ x_j \in \{1, 5, 8\},\ j = 1, 2, 3 \end{array} \quad (4.196)$$

If one replaces the alldiff and domains in (4.196) with (4.192), the solution of the resulting relaxation is $\bar{x} = (1, 8, 5)$. Since this is feasible for the alldiff and domains, no branching is necessary and the problem is solved at the root node. Alternatively, once can use only (4.192c) and bounds $1 \leq x_j \leq 8$ in the relaxation, and add separating cuts as needed. In this case, the solution is again $\bar{x} = (1, 8, 5)$ and there is no need for a separating cut.[3] One can check that no separating cut exists by noting that $\bar{x}_1 \geq 1$ and $\bar{x}_1 + \bar{x}_3 \geq 6$.

A similar relaxation can be written when the variable domains are arbitrary finite sets, although it is not in general a convex hull relaxation. One can write valid inequalities of the form

$$L(J) \leq \sum_{j \in J} x_j \leq U(J),\ \text{all}\ J \subset \{1, \ldots, n\} \quad (4.197)$$

where

$$L(J) = \min \left\{ \sum_{j \in J} x_j\ \bigg|\ \text{alldiff}(x_j \mid j \in J),\ x_j \in D_{x_j}\ \text{for all}\ j \in J \right\}$$

$$U(J) = \max \left\{ \sum_{j \in J} x_j\ \bigg|\ \text{alldiff}(x_j \mid j \in J),\ x_j \in D_{x_j}\ \text{for all}\ j \in J \right\}$$

THEOREM 4.46 *If $L(J), U(J)$ are defined as above, then (4.197) is a relaxation of alldiff$(x_1, \ldots, x_n)$.*

One can compute $L(J)$ and $U(J)$ for a given J by solving a minimum and a maximum cost network flow problem (4.74), for which fast specialized algorithms are known. The network contains a node i for each $i \in J$, representing variable x_i, and a node v for each $v \in D_J = \bigcup_{j \in J} D_{x_j}$. An arc with cost $c_{iv} = v$ runs from each i to each $v \in D_{x_i}$, and an arc with capacity $U_{vt} = 1$ runs from each $v \in D_J$ to a sink node t. All other costs are zero, and all other capacities infinite. Also, $s_i = 1$ for $i \in J$, $s_v = 0$

---

[3] One could have predicted that the same solution would result, because in this small example the constraints (b) are implied by (c) and the bounds.

The All-Different Constraint

for $v \in D_J$, and $s_t = -|J|$. A flow of $y_{iv} = 1$ represents assigning value v to x_i. $L(J)$ is the cost of a minimum-cost flow, and $U(J)$ the cost of a maximum cost flow.

For example, consider the problem (4.196) where the domains are

$$x_1 \in \{1, 4, 6\}, \quad x_2, x_3 \in \{5, 6\} \tag{4.198}$$

The relaxation (4.197) is

$$\begin{array}{ll}
x_1 + x_2 + x_3 \geq 12 & x_1 + x_2 + x_3 \leq 15 \\
x_1 + x_2 \geq 6 & x_1 + x_2 \leq 11 \\
x_1 + x_3 \geq 6 & x_1 + x_3 \leq 11 \\
x_2 + x_3 \geq 11 & x_2 + x_3 \leq 11 \\
& 1 \leq x_1 \leq 6 \\
& 5 \leq x_2 \leq 6 \\
& 5 \leq x_3 \leq 6
\end{array} \tag{4.199}$$

The minimum-cost network flow problems for obtaining the two constraints on the top line are illustrated in Figure 4.17. The optimal solution of (4.196) with relaxation (4.199) replacing the alldiff and domains is $\bar{x} = (4, 5, 6)$. Since this is feasible for the alldiff and domains, the problem is solved at the root node.

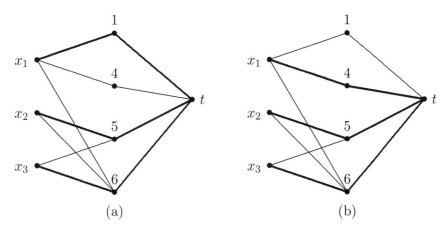

Figure 4.17. Network flow model for a relaxation of alldiff with unequal variable domains. Heavy lines show a minimum-cost flow in diagram (a) and a maximum cost flow in (b).

4.12.2 Convex Hull MILP Formulation

The constraint alldiff$(x_1,\ldots,x_n)$ can be given an MILP formulation whose continuous relaxation provides a convex hull relaxation for arbitrary variable domains. Again, it is assumed that the x_is take numerical values. This approach requires the introduction of mn binary variables, however, where m is the number of distinct domain elements.

Let $\{v_1,\ldots,v_m\}$ be the union of the variable domains D_{x_i}. Define binary variables y_{ij} that take the value 1 when $x_i = v_j$. The alldiff constraint can be formulated

$$
\begin{aligned}
& x_i = \sum_{j=1}^{m} v_j y_{ij}, \quad i=1,\ldots,n \quad &(a) \\
& \sum_{j=1}^{m} y_{ij} = 1, \quad i=1,\ldots,n \quad &(b) \\
& \sum_{i=1}^{n} y_{ij} \leq 1, \quad j=1,\ldots,m \quad &(c) \\
& y_{ij} = 0 \ \text{all}\ i,j\ \text{with}\ j \notin D_{x_i} \quad &(d) \\
& y_{ij} \in \{0,1\}, \ \text{all}\ i,j &
\end{aligned}
\qquad (4.200)
$$

The continuous relaxation of (4.200) is formed by replacing $y_{ij} \in \{0,1\}$ with $y_{ij} \geq 0$.

THEOREM 4.47 *The continuous relaxation of (4.200) projected onto $x_1,\ldots,x_n$ is a convex hull relaxation of alldiff$(x_1,\ldots,x_n)$ and $x_i \in D_{x_i}$ for all i.*

Proof. The constraints of (4.200) are clearly valid for alldiff$(x_1,\ldots,x_n)$ and $x_i \in D_{x_i}$. It remains to show that the projection onto x of every feasible solution of (4.200) is a convex combination of feasible solutions of the alldiff and $x_i \in D_{x_i}$. So let $(\bar{x},\bar{y})$ be a feasible solution of (4.200). Note first that the coefficient matrix of constraints (b)–(d) is totally unimodular, using Theorem 4.14. So all vertices of the polyhedron defined by (b)–(d) and $y_{ij} \geq 0$ are 0-1 points, which means that $\bar{y}$ is a convex combination $\sum_k \alpha_k y^k$ of 0-1 points y^k satisfying (b)–(d). Now since $\bar{x}_i = \sum_j v_j \bar{y}_{ij}$,

$$\bar{x}_i = \sum_j v_i \left(\sum_k \alpha_k y^k_{ij}\right) = \sum_k \alpha_k \left(\sum_j v_j y^k_{ij}\right) = \sum_k \alpha_k x^k_i$$

where $x^k_i = \sum_j v_j y^k_{ij}$. But because $\sum_j y^k_{ij} = 1$, $y^k_{ik} \in \{0,1\}$, and $y^k_{ij} = 0$ when $v_j \notin D_{x_i}$, it follows that $x^k_i \in D_{x_i}$. Also since $\sum_i y^k_{ij} \leq 1$ for all j

The All-Different Constraint 369

and k, one can infer alldiff$(x_1^k, \ldots, x_n^k)$ for each k. Thus, $\bar{x}$ is a convex combination of points x^k that are feasible for the alldiff and $x_i \in D_{x_i}$. □

For example, alldiff(x_1, x_2, x_3) with domains (4.198) has the convex hull relaxation

$$
\begin{aligned}
x_1 &= y_{11} + 4y_{12} + 6y_{14} & y_{11} + y_{12} + y_{14} &= 1 \\
x_2 &= 5y_{23} + 6y_{24} & y_{23} + y_{24} &= 1 \\
x_3 &= 5y_{33} + 6y_{34} & y_{33} + y_{34} &= 1 \\
& & y_{23} + y_{33} &\leq 1 \\
& & y_{14} + y_{24} + y_{34} &\leq 1 \\
& & y_{ij} &\geq 0, \text{ all } i, j
\end{aligned}
\quad (4.201)
$$

If this relaxation replaces the alldiff and domains in (4.196), the optimal solution is $\bar{x} = (4, 5, 6)$, with $\bar{y}_{12} = \bar{y}_{23} = \bar{y}_{34} = 1$ and all other $\bar{y}_{ij} = 0$. Since this is feasible for the alldiff and domains, the problem is solved without branching. Incidentally, the projection of (4.201) onto x simplifies in this case to $x_2 + x_3 = 11$ with bounds $x_1 \in [1, 4]$ and $x_2, x_3 \in [5, 6]$. In general, the projection is quite complex and is not computed.

4.12.3 Modeling Costs with Alldiff

Both of the relaxations presented above for alldiff assume that the constraint variables take numerical values. This may seem to limit the applicability of the relaxations. The variables often denote tasks or workers with nonnumeric labels, or with numeric labels that have no particular meaning as numbers. However, the relaxations are often useful despite this fact. Their primary function is to provide a bound on cost, and they can play this role even when the tasks and workers do not have meaningful numerical labels.

This requires some explanation. Consider first the MILP relaxation (4.200). As noted in Section 3.11.1, the alldiff constraint is frequently associated with the cost function $\sum_i c_{ix_i}$, where c_{ix_i} can be viewed as the cost of assigning worker x_i to task i. The total cost w can be written in terms of the 0-1 variables of the relaxation:

$$w = \sum_{ij} c_{ij} y_{ij} \quad (4.202)$$

Thus, when the MILP relaxation is used with nonnumeric variables x_i, the constraint (4.200a) that connects the x_is to the relaxation can be dropped, and (4.202) added.

Because the variables y_{ij} are generated specifically for the alldiff relaxation and are not part of the original model, they are not available for the user to express cost in terms of them. It is therefore convenient to provide a version of the alldiff constraint that does this automatically:

$$\text{alldiff}(x, w \mid c) \tag{4.203}$$

Here, $x = (x_1, \ldots, x_n)$, variable w represents total cost, and c is a matrix of costs coefficients c_{ij}. The constraint enforces alldiff(x) and $w = \sum_i c_{ix_i}$.

When an MILP relaxation (4.200) is generated for the alldiff constraint (4.203), the definition (4.202) of w becomes part of the relaxation. The original variables x_i remain in the original model, and their domains can be filtered, but they do not appear in the relaxation. The cost variable w can now be constrained or minimized in the original model as desired. For example, the classical assignment problem is written by minimizing w subject to (4.203). The solver can recognize that (4.203) is the only constraint and solve the relaxation by a specialized algorithm, such as the Hungarian algorithm.

If the alldiff constraint (4.203) is given a relaxation (4.197) in the original variables, then there is no escaping the fact that those variables must take numeric values, which may have no meaning as numbers. One can perhaps take advantage of this fact, however, if there is some kind of structure in the cost matrix, by selecting numeric values that result in a good relaxation. Since the cost expression $w = \sum_i c_{ix_i}$ contains a variable index, it can be unpacked by replacing it with $w = \sum_i w_i$ and

$$\text{element}(x_i, w_i \mid (c_{i1}, \ldots, c_{in})), \quad i = 1, \ldots, n \tag{4.204}$$

A relaxation for the element constraints in (4.204) can now be combined with $w = \sum_i w_i$ and the convex hull relaxation (4.197) to obtain a relaxation for the alldiff constraint (4.203). The point here is that it may be possible to obtain a good relaxation for the element constraints by assigning the right numerical indices to the tasks and workers.

This is best seen in an example. Suppose that the costs c_{ij} of assigning worker j to task i are as given in Table 4.3(a), where the task and worker indices are the usual 1,2,3. The table can be viewed as defining a cost function $c(i, j) = c_{ij}$. The cost function shows some structure, since $c(i, j)$ increases monotonically with both i and j. (It may be necessary to permute the indices to see such a pattern.) Clearly, some workers receive higher pay across all machines, and some tasks are more expensive to perform regardless of who performs them. This suggests that the cost function $c(i, j)$ can be approximated by an affine function. In fact, the approximation $3 + i + 2j$ is almost exact if the task and worker indices are adjusted as in Table 4.3(b). The entry 7 in the middle of the table lies slightly below the plane defined by the other entries.

The All-Different Constraint

Table 4.3. (a) Cost data c_{ij} for an assignment problem. (b) The same cost data with indices adjusted to provide a better continuous relaxation.

(a)

	j		
i	1	2	3
1	5	7	11
2	6	7	12
3	8	10	14

(b)

	j		
i	0.5	1.5	3.5
1	5	7	11
2	6	7	12
4	8	10	14

Since $c(i,j)$ is closely approximated by an affine function, the convex hull of the points (i, j, c_{ij}) is quite thin and gives a close approximation to $z = c(i,j)$. The convex hull can be rapidly computed using techniques from computational geometry and in this case is

$$
\begin{aligned}
z &\geq 3.5 + i + j \\
z &\geq 1 + 1.5i + 2j \\
z &\geq 1.25 + i + 2.5j \\
z &\geq 4 + 2j \\
z &\leq 1 + i + 2j \\
1 &\leq i \leq 4, \ 0.5 \leq j \leq 3.5
\end{aligned}
\quad (4.205)
$$

Let $S(\sigma_1, \sigma_2, \sigma_3)$ denote the inequality set (4.205) in which the symbols $\sigma_1, \sigma_2, \sigma_3$ are respectively substituted for i, j, z. Now the element constraints (4.204) can be collectively relaxed by writing the inequalities in the sets $S(1, x_1, w_1)$, $S(2, x_2, w_2)$, and $S(4, x_4, w_4)$. These, plus the following convex hull relaxation of alldiff(x_1, x_2, x_4) (derived from Corollary 4.45)

$$
\begin{aligned}
&0.5 \leq x_i \leq 3.5, \ i = 1, 2, 4 \\
&x_1 + x_2 \geq 2, \ x_1 + x_4 \geq 2, \ x_2 + x_4 \geq 2 \\
&x_1 + x_2 + x_4 = 5.5
\end{aligned}
$$

comprise a relaxation of the alldiff constraint (4.203).

If one wishes to solve the classical assignment problem, for example, one can minimize $w = w_1 + w_2 + w_4$ subject to this relaxation and obtain $x = (0.5, 1.5, 3.5)$ with $w = 26$. Since alldiff(x) is satisfied and each x_i is in its domain $\{0.5, 1.5, 3.5\}$, there is no need to branch, and the problem is solved. Assign tasks 1, 2, and 4 to workers 0.5, 1.5, and 3.5, respectively. If some x_i were to receive an intermediate value, such as

$x_1 = 2.5$, one could branch by forcing x_1 to be in the domain $\{0.5, 1.5\}$ and then in the domain $\{3.5\}$.

A similar idea is used in the next section to relax the quadratic assignment problem.

4.12.4 Example: Quadratic Assignment Problem

The *quadratic assignment problem* illustrates how continuous relaxations of alldiff and element can help exploit problem structure. This problem commonly arises when facilities are located to minimize the amount of travel between them. For instance, one may wish to locate hospital wards in a building when there is a known volume of traffic between each pair of wards, and the objective is to minimize the total amount of travel.

Let a_{ij} be the volume of traffic from facility i to facility j, and let $b_{k\ell}$ be the distance from location k to location ℓ. Then, if facilities i and j are respectively located at k and ℓ, the total amount of travel from i to j is $a_{ij}b_{k\ell}$. If y_i is the location assigned to facility i, the problem can be succinctly written

$$\min \sum_{ij} a_{ij} b_{y_i y_j} \qquad (4.206)$$
$$\text{alldiff}(y_1, \ldots, y_n)$$

After unpacking the variable indices, (4.206) becomes

$$\min \sum_{ij} a_{ij} z_{ij}$$
$$\text{element}((y_i, y_j), z_{ij} \,|\, B), \text{ all } i,j \qquad (4.207)$$
$$\text{alldiff}(y_1, \ldots, y_n)$$

where B is a matrix of distances $b_{k\ell}$.

The quadratic assignment problem is notoriously hard to solve, and one reason for its difficulty is that the general problem exhibits little structure. In a particular instance, however, the distance matrix B may have a good deal of structure that most solution methods do not exploit. For instance, the distances may reflect a rectilinear or Euclidian metric. One can take advantage of this kind of structure by introducing a vector-valued element constraint and combining its relaxation with a relaxation of the alldiff.

To simplify exposition, suppose for the moment that each location k can be represented by a single number p_k. For instance, all the locations might lie on a long hallway, in which case p_k would be the distance of location k from the end of the hallway. Then, if $x_1, \ldots, x_n$ are the positions assigned to facilities $1, \ldots, n$, alldiff$(x_1, \ldots, x_n)$ is a valid constraint. The model (4.207) can therefore be written

$$\min \sum_{ij} a_{ij} z_{ij} \quad (a)$$

$$\text{element}\left((y_i, y_j), \begin{bmatrix} x_i \\ x_j \\ z_{ij} \end{bmatrix} \middle\| H\right), \text{ all } i, j \quad (b) \quad (4.208)$$

$$\text{alldiff}(x_1, \ldots, x_n) \quad (c)$$

where each y_i has domain $\{1, \ldots, n\}$ and each component of matrix H is

$$h_{ij} = \begin{bmatrix} p_i \\ p_j \\ b_{ij} \end{bmatrix}$$

Constraint (b) sets x_i equal to position p_{y_i}, x_j equal to position p_{y_j}, and z_{ij} equal to distance $b_{y_i y_j}$.

A continuous relaxation of constraint (b) can now capture the relationship between pairs of positions (p_k, p_ℓ) and the distance between them. For example, if the p_ks are positions along a hallway, then $b_{k\ell} = b_{\ell k} = |p_k - p_\ell|$, and constraint (b) has the relaxation

$$z_{ij} \geq x_i - x_j, \quad z_{ij} \geq x_j - x_i, \quad \text{all } i, j$$
$$\min D_{x_i} \leq x_i \leq \max D_{x_i}, \quad \text{all } i \quad (4.209)$$

One can now solve the quadratic assignment problem by branching on the x_is and solving a relaxation at each search node that minimizes (4.208a) subject to (4.209) and a relaxation of the alldiff constraint (4.208c). If the relaxation values of all x_is belong to $\{p_1, \ldots, p_n\}$ and satisfy the alldiff, but the relaxation value of some $z_{y_i y_j}$ is less than $b_{y_i y_j}$, then one can branch on a y_i.

A more realistic model represents each location k with a pair of coordinates $p_k = (p_{k1}, p_{k2})$. In this case, x_i is likewise a pair (x_{i1}, x_{i2}), and the model (4.208) can again be used. The relaxation methods for alldiff discussed above do not apply, however, because x_i is a vector of two variables. To overcome this problem, one can replace (4.208c) with

$$\text{alldiff}(x_{11}, \ldots, x_{n1}), \quad \text{alldiff}(x_{12}, \ldots, x_{n2}) \quad (4.210)$$

To make sure (4.210) is valid, it may be necessary to perturb some of the positions p_i slightly. For example, if there are two points $p_1 = (\alpha, \beta)$ and $p_2 = (\alpha, \gamma)$, one can perturb the latter to $(\alpha + \epsilon, \gamma)$ to satisfy alldiff(p_{11}, p_{21}). Once this is done, either one of the alldiffs in (4.210) is sufficient to define the feasible set if (4.208b) is correctly propagated. This is because reducing the set $D_{x_{i1}}$ of possible values of x_{i1} reduces the set of possible values of x_{i2} to $\{p_{k2} \mid p_{k1} \in D_{x_{i1}}\}$, and vice-versa.

If B reflects a rectilinear metric, then

$$b_{k\ell} = b_{\ell k} = |p_{k1} - p_{\ell 1}| + |p_{k2} - p_{\ell 2}|$$

In this case, the vector-valued element constraint (4.208b) has the relaxation

$$\begin{array}{l} z_{ij} = z_{ij1} + z_{ij2}, \text{ all } i,j \\ z_{ijt} \geq x_{it} - x_{jt}, \quad z_{ijt} \geq x_{jt} - x_{it}, \quad t=1,2, \text{ all } i,j \\ \min_{p_k \in D_{x_i}} \{p_{kt}\} \leq x_{it} \leq \max_{p_k \in D_{x_i}} \{p_{kt}\}, \quad t=1,2, \text{ all } i \end{array} \quad (4.211)$$

4.12.5 Exercises

1. Write a convex hull relaxation for alldiff(x_1, x_2, x_3) when each $x_j \in \{1, 3, 5, 6\}$.

2. Minimize $2x_1 + 3x_2 + 4x_3$ subject to $x_1 + 2x_2 + 3x_3 \geq 14$ and the alldiff constraint of Exercise 1 by replacing the alldiff with its convex hull relaxation. Branching should be unnecessary. In many cases, however, a good deal of branching is required, because the convex hull relaxation of alldiff is quite weak. The relaxation is least effective when the objective function and other constraints tend to push all the x_js near the middle of their possible range.

3. Solve the problem of Exercise 2 but with a partial relaxation of the alldiff that contains (4.189) for $J = \{1, 2, 3\}$ only, along with bounds $1 \leq x_j \leq 6$ for $j = 1, 2, 3$. Generate a separating alldiff cut for the resulting solution and re-solve the relaxation with this cut. This should yield a feasible solution.

4. The proof of Theorem 4.44 simplifies considerably when $m = n$ (as in Corollary 4.45). Write this simpler proof.

5. Use Theorem 4.46 to write a relaxation for alldiff(x_1, x_2, x_3) with $x_1 \in \{1, 3, 5\}$, $x_2 \in \{1, 6\}$, and $x_3 \in \{3, 5, 6\}$. Solve the appropriate minimum and maximum cost flow problems to obtain the right-hand sides.

6. Minimize $2x_1 + 3x_2 + 4x_3$ subject to $x_1 + 2x_2 + 3x_3 \geq 14$ and the domains of Exercise 5 by replacing the alldiff with the relaxation obtained in that exercise. Is the resulting solution feasible?

The Cardinality Constraint 375

7 Formulate the convex hull MILP relaxation of alldiff(x_1, x_2, x_3) using the domains of Exercise 5.

8 Minimize $2x_1 + 3x_2 + 4x_3$ subject to $x_1 + 2x_2 + 3x_3 \geq 14$ and the domains of Exercise 5 by replacing the alldiff with the MILP-based relaxation of Exercise 7. Is the resulting solution feasible?

9 Verify that the constraint matrix of (b)–(c) in (4.200) is totally unimodular as claimed in the proof of Theorem 4.47.

10 Show that either one of the alldiffs in (4.210) is sufficient to satisfy alldiff($x_1, \ldots, x_n$) if domain completeness is achieved for each element constraint in (4.208).

11 Show that (4.211) relaxes the vector-valued element constraints in (4.208) if the distances in B represent a rectilinear metric.

12 Formulate a relaxation of the element constraints in (4.208) if the distances in B represent an L_∞ metric. That is, if location k has coordinates $p_k = (p_{k1}, p_{k2})$,

$$b_{k\ell} = b_{\ell k} = \max\{|p_{k1} - p_{\ell 1}|, |p_{k2} - p_{\ell 2}|\}$$

This metric can occur, for example, when moving items on warehouse shelves. If p_{k1} and p_{k2} represent horizontal and vertical coordinates, respectively, the time required for a forklift to move from p_k to p_ℓ is governed by the maximum of the horizontal and vertical transit time. The L_∞ distance is also a lower bound on the Euclidean distance and can be used as a relaxation when B reflects Euclidean distances.

4.13 The Cardinality Constraint

The cardinality constraint generalizes the alldiff constraint and has convex hull relaxations analogous to those given for alldiff in Section 4.12.

Recall that the constraint

$$\text{cardinality}(x \mid v, \ell, u)$$

requires each value v_j to occur at least ℓ_j times, and at most u_j times, among the variables $x = (x_1, \ldots, x_n)$. For purposes of relaxation, it is assumed that the v_js are numbers. It is also assumed that the domain of each x_i is a subset of $\{v_1, \ldots, v_m\}$. This incurs no loss of generality because any domain element that does not occur in v can be placed in v with upper and lower bounds of 0 and n.

4.13.1 Convex Hull Relaxation

The cardinality constraint has a continuous relaxation in the original variables $x_1, \ldots, x_n$ that is similar to that given for alldiff in Theorem 4.44. Again, it is a convex hull relaxation if all variables have the same domain.

Any sum $\sum_{j \in J} x_j$ must be at least as large the sum of $|J|$ values from $\{v_1, \ldots, v_m\}$, selecting the smallest value as many times as possible, then selecting the second smallest value as many times as possible, and so forth. Similarly, $\sum_{j \in J} x_j$ must be at least as small as the largest feasible sum of $|J|$ values from $\{v_1, \ldots, v_m\}$. Thus, if $v_1 < \cdots < v_m$, one can state

$$\sum_{i=1}^{m} p(|J|, i) v_i \leq \sum_{j \in J} x_j \leq \sum_{i=1}^{m} q(|J|, i) v_i, \text{ all } J \subset \{1, \ldots, n\} \quad (4.212)$$

where $p(k, i)$ is the largest number of times one can select v_i when minimizing a sum of k x_is, and $q(k, i)$ the largest number of times one can select v_i when maximizing a sum of k x_is. Thus,

$$p(k, i) = \min\left\{ p_i, k - \sum_{j=1}^{i-1} p(k, j) \right\}, \quad q(k, i) = \min\left\{ q_i, k - \sum_{j=i+1}^{m} q(k, j) \right\}$$

where

$$p_i = \min\left\{ u_i, n - \sum_{j=1}^{i-1} p_j - \sum_{j=i+1}^{m} \ell_j \right\}, \quad i = 1, \ldots, m$$

$$q_i = \min\left\{ u_i, n - \sum_{j=i+1}^{m} q_j - \sum_{j=1}^{i-1} \ell_j \right\}, \quad i = m, \ldots, 1$$

For example, consider the constraint

$$\text{cardinality}(\{x_1, \ldots, x_5\} \mid (20, 30, 60), (1, 2, 1), (3, 3, 1))$$

Here, $(p_1, p_2, p_3) = (2, 2, 1)$ and $(q_1, q_2, q_3) = (1, 3, 1)$. The inequalities (4.212) are

$$1 \cdot 20 \leq x_j \leq 1 \cdot 60, \text{ all } j$$

$$40 = 2 \cdot 20 \leq \sum_{j \in J} x_j \leq 1 \cdot 60 + 1 \cdot 30 = 90, \text{ all } J \text{ with } |J| = 2$$

$$70 = 2 \cdot 20 + 1 \cdot 30 \leq \sum_{j \in J} x_j \leq 1 \cdot 60 + 2 \cdot 30 = 120, \text{ all } J \text{ with } |J| = 3$$

The Cardinality Constraint 377

$$100 = 2 \cdot 20 + 2 \cdot 30 \leq \sum_{j \in J} x_j \leq 1 \cdot 60 + 3 \cdot 30 = 150, \text{ all } J \text{ with } |J| = 4$$

$$160 = 2 \cdot 20 + 2 \cdot 30 + 1 \cdot 60 \leq \sum_{j=1}^{5} x_j \leq 1 \cdot 60 + 3 \cdot 30 + 1 \cdot 20 = 170$$

where $J \subset \{1, \ldots, 5\}$. This is, in fact, a convex hull relaxation.

THEOREM 4.48 *If $x = (x_1, \ldots, x_n)$, $v = (v_1, \ldots, v_m)$ with $v_1 < \cdots < v_m$, and $D_{x_i} = \{v_1, \ldots, v_m\}$ for $i = 1, \ldots, n$, then (4.212) is a convex hull relaxation of cardinality$(X \mid v, \ell, u)$.*

The proof is very similar to the proof of Theorem 4.44, and is left as an exercise.

The separation algorithm is similar to that given for alldiff in Section 4.12.1. Constraints

$$\sum_{i=1}^{m} p_i v_i \leq \sum_{j=1}^{n} x_j \leq \sum_{i=1}^{m} q_i v_i$$

and variable bounds can be included in the initial relaxation. Let $\bar{x}$ be the solution of the relaxation, and renumber the variables so that $\bar{x}_1 \leq \cdots \leq \bar{x}_n$. Then, for each $i = 2, \ldots, n-1$, one can generate the cut

$$\sum_{j=1}^{i} x_j \geq \sum_{j=1}^{m} p(i,j) v_j$$

whenever

$$\sum_{j=1}^{i} \bar{x}_j < \sum_{j=1}^{m} p(i,j) v_j$$

Also for each $i = n-1, \ldots, 2$ one can generate the cut

$$\sum_{j=i}^{n} x_j \leq \sum_{j=1}^{m} q(i,j) v_j$$

whenever

$$\sum_{j=i}^{n} \bar{x}_j > \sum_{j=1}^{m} q(i,j) v_j$$

In the above example, suppose $(\bar{x}_1, \ldots, \bar{x}_5) = (20, 20, 20, 30, 70)$. Then one can generate the separating cuts

$$x_1 + x_2 + x_3 \geq 70, \quad x_1 + x_2 + x_3 + x_4 \geq 100, \quad x_4 + x_5 \leq 90$$

4.13.2 Convex Hull MILP Formulation

The constraint cardinality$(x \mid v, \ell, u)$ can be given an MILP formulation very similar to that given in Section 4.12.2 for alldiff. Its continuous relaxation is a convex hull relaxation for arbitrary variable domains, provided the domain elements are numbers.

Let $\{v_1, \ldots, v_m\}$ be the union of the variable domains D_{x_i}. Define binary variables y_{ij} that take the value 1 when $x_i = v_j$. The cardinality constraint can be formulated

$$\begin{aligned}
x_i &= \sum_{j=1}^{m} v_j y_{ij}, \quad i = 1, \ldots, n \\
\sum_{j=1}^{m} y_{ij} &= 1, \quad i = 1, \ldots, n \\
\ell_j &\leq \sum_{i=1}^{n} y_{ij} \leq u_j, \quad j = 1, \ldots, m \\
y_{ij} &= 0 \text{ all } i, j \text{ with } j \notin D_{x_i} \\
y_{ij} &\in \{0, 1\}, \text{ all } i, j
\end{aligned} \quad (4.213)$$

The continuous relaxation of (4.213) is formed by replacing $y_{ij} \in \{0, 1\}$ with $y_{ij} \geq 0$.

THEOREM 4.49 *Let $x = (x_1, \ldots, x_n)$ and $v = (v_1, \ldots, v_m)$. The continuous relaxation of (4.213) projected onto $x_1, \ldots, x_n$ is a convex hull relaxation of cardinality$(X \mid v, \ell, u)$ and $x_i \in D_{x_i}$ for all i.*

The proof is almost the same as the proof of Theorem 4.47.

4.13.3 Exercises

1 Write a convex hull relaxation for

$$\text{cardinality}((x_1, \ldots, x_5) \mid (2, 5, 6, 7), (2, 1, 1, 1), (3, 3, 2, 2))$$

where each $x_j \in \{2, 5, 6, 7\}$. Note that the lower and upper bounds on $\sum_{j=1}^{5} x_j$ are equal.

2 Suppose that $(x_1, \ldots, x_5) = (2, 2, 4, 7, 7)$, which is infeasible for the cardinality constraint in Exercise 1. Identify the separating cuts. One can also branch on x_3 by setting $x_3 \leq 2$ and $x_3 \geq 5$.

3 Prove Theorem 4.48.

4 Identify a family of valid cuts for the cardinality constraint when the domains differ, following the pattern of Theorem 4.46. Formulate

minimum and maximum flow problems that can be used to obtain the right-hand sides.

5 Write a convex hull MILP relaxation for the cardinality constraint in Exercise 1.

6 Prove Theorem 4.49.

4.14 The Circuit Constraint

Because the circuit constraint is the sole constraint in one formulation of the traveling salesman problem (Section 3.11.1), relaxations developed for the traveling salesman problem can serve as relaxations for circuit. One particular 0-1 programming formulation of the problem has been studied in depth, with the aim of discovering strong cutting planes and separation algorithms. These cutting planes are not sufficient to describe the convex hull of the feasible set, which is extremely complex, but some of them have proved quite useful, even essential, for solving large instances of the traveling salesman problem. They can help provide tight relaxations of the circuit constraint.

4.14.1 0-1 Programming Model

The traveling salesman problem (TSP) minimizes $\sum_i c_{ix_i}$ subject to

$$\text{circuit}(x_1, \ldots, x_n) \qquad (4.214)$$

where each D_{x_i} is a subset of $\{1, \ldots, n\}$. The problem can be viewed as defined on a directed graph G with a vertex for each x_i and an edge (i, j) when $j \in D_{x_i}$. Every feasible solution corresponds to a *tour*, or a hamiltonian cycle in G. The objective is to find a minimum-length tour, where c_{ij} is the length of edge (i, j).

The TSP is said to be *symmetric* when $c_{ij} = c_{ji}$ and $j \in D_{y_i}$ if and only if $i \in D_{x_j}$ for all i, j. Otherwise, it is *asymmetric*. The symmetric problem is normally associated with an undirected graph, on which the objective is to find a minimum-cost undirected tour. The symmetric and asymmetric problems receive somewhat different, albeit related, analyzes in the literature. The focus here is on the asymmetric problem, because it subsumes the symmetric problem as a special case.

The most widely studied integer programming model of the asymmetric TSP is the following. Let the 0-1 variable y_{ij} (for $i \neq j$) take the value 1 when $x_i = j$; that is, when edge (i, j) is part of the selected tour.

Then, the objective is to minimize $\sum_{ij} c_{ij} y_{ij}$ subject to

$$\sum_j y_{ij} = \sum_j y_{ji} = 1, \text{ all } i \qquad (a)$$

$$\sum_{(i,j) \in \delta(S)} y_{(i,j)} \geq 1, \text{ all } S \subset \{1, \ldots, n\} \text{ with } 2 \leq |S| \leq n-2 \quad (b)$$

$$y_{ij} \in \{0, 1\}, \text{ all } i, j$$

$$(4.215)$$

Here, $\delta(S)$ is the set of edges (i, j) of G for which $i \in S$ and $j \notin S$. If $j \notin D_{x_i}$, the variable y_{ij} is omitted from (4.215). Constraints (a) are vertex-degree constraints that require every vertex to have one incoming edge and one outgoing edge in the tour. They are identical to assignment problem constraints (Section 3.11.1) but have a different interpretation. The *subtour-elimination constraints* (b) exclude hamiltonian cycles on all proper subsets S of $\{1, \ldots, n\}$ by requiring that at least one edge connect a vertex in S to one outside S. The constraints (4.215), together with

$$x_i = \sum_j j y_{ij}, \text{ all } i \qquad (4.216)$$

provide a 0-1 model of circuit$(x_1, \ldots, x_n)$.

4.14.2 Continuous Relaxations

A relaxation of circuit can be obtained by combining (4.216) with a continuous relaxation of (4.215). The simplest relaxation of (4.215) is the *assignment relaxation*, which consists of the assignment constraints (a) and $y_{ij} \geq 0$.

The assignment relaxation can be strengthened by adding subtour-elimination inequalities (b). It is impractical to add all of these, however, because there are exponentially many. The usual practice is to add separating inequalities as needed. One can also exclude subtours on two vertices *a priori* by adding all constraints of the form $y_{ij} + y_{ji} \leq 1$.

Separating inequalities can be found as follows. If $\bar{y}$ is a solution of the current continuous relaxation, let the *capacity* of edge (i, j) be $\bar{y}_{ij}$. Select a proper subset S of the vertices for which the total capacity of edges leaving S (i.e., the *outflow capacity* of S) is a minimum. The subtour-elimination constraint (b) corresponding to S is a separating cut if this minimum outflow capacity is less than 1.

Consider for example the graph in Figure 4.18, which corresponds to the constraint circuit$(x_1, \ldots, x_6)$ with domains $D_{x_1} = \{2, 3\}$, $D_{x_2} = \{3, 5\}$, $D_{x_3} = \{1, 4\}$, $D_{x_4} = \{5, 6\}$, $D_{x_5} = \{2, 6\}$, and $D_{x_6} = \{1\}$. The

The Circuit Constraint

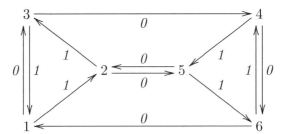

Figure 4.18. *Graph for a circuit constraint. An optimal solution of the assignment relaxation is indicated by the numbers on the edges.*

assignment relaxation is

$$\begin{aligned}
&y_{12} + y_{13} = 1, \quad y_{23} + y_{25} = 1, \quad y_{31} + y_{34} = 1 \\
&y_{45} + y_{46} = 1, \quad y_{52} + y_{56} = 1, \quad y_{64} + y_{61} = 1 \\
&y_{31} = 1, \quad y_{12} + y_{32} + y_{62} = 1, \quad y_{13} + y_{23} + y_{63} = 1 \\
&y_{54} = 1, \quad y_{25} = y_{45} + y_{65} = 1, \quad y_{26} + y_{46} + y_{56} = 1 \\
&y_{13} + y_{31} \leq 1, \quad y_{25} + y_{52} \leq 1, \quad y_{46} + y_{64} \leq 1 \\
&x_{ij} \geq 0, \quad \text{all } i, j
\end{aligned} \quad (4.217)$$

If the costs are $c_{34} = c_{25} = c_{52} = c_{61} = 1$ and $c_{ij} = 0$ for other i, j, then the optimal solution of the relaxation, shown in Figure 4.18, defines two subtours. There are two minimum-capacity vertex sets, namely $S = \{1, 2, 3\}$ and $S = \{4, 5, 6\}$, both of which have outflow capacity of zero. They correspond to the violated subtour-elimination inequalities

$$y_{34} + y_{25} \geq 1, \quad y_{52} + y_{61} \geq 1 \quad (4.218)$$

When these are added to the relaxation (4.217), the resulting solution is $y_{ij} = \frac{1}{2}$ for all i, j. This satisfies all subtour-elimination inequalities. Yet, it is infeasible because it is nonintegral, and one must either generate more cutting planes or branch.

There are fast algorithms for finding an S with a minimum-outflow capacity (i.e., for finding a *minimum-capacity cut*) [123, 263], but in practice a simple heuristic is often used. Known in this context as the *max back* heuristic (Figure 4.19), it adds vertices one at a time to an initial set S (perhaps a single vertex) and keeps track of the resulting outflow capacity. The process continues until S contains all the vertices. The set with the smallest outflow capacity is selected to generate a subtour-elimination inequality, provided its outflow capacity is strictly less than 1. The vertex i added to S in each iteration is one with the largest *max back value* $b_i = \sum_{j \in S} \bar{y}_{ji}$, on the theory that S will have smaller outflow capacity if large-capacity edges are brought within S.

Let $S = S_0$, $S_{\min} = S_0$, $C_{\min} = \sum_{(i,j) \in \delta(S_0)} \bar{y}_{ij}$, $C = C_{\min}$.
For all $i \notin S$ let $b_i = \sum_{j \in S} \bar{y}_{ji}$.
While $S \neq \{1, \ldots, n\}$ repeat:
  Select a vertex $i \notin S$ that maximizes b_i.
  Let $S = S \cup \{i\}$, $C = C + 2 - 2b_i$.
  For all $j \notin S$, let $b_j = b_j + \bar{y}_{ij}$.
  If $C < C_{\min}$ then let $C_{\min} = C$ and $S_{\min} = S$.

Figure 4.19. Max back heuristic for finding a minimum capacity vertex set containing S_0. The quantity $C_{\min}$ records the minimum outflow capacity found so far. At the termination of the heuristic, a subtour-elimination inequality is generated for $S_{\min}$ if $C_{\min} < 1$.

The procedure can be restarted at different vertices until one or perhaps several separating inequalities have been found.

In the example of Figure 4.18, one might start the max back heuristic with $S = \{1\}$, which has outflow capacity 1. Vertex 2 is added next, and then vertex 3, at which point the outflow capacity is 0. Since subsequent sets cannot have smaller outflow capacity, $S_{\min} = \{1, 2, 3\}$ is selected and the subtour-elimination inequality $y_{25} + y_{34} \geq 1$ is generated. If further separating inequalities are desired, one might restart the max back heuristic at a vertex outside $S_{\min}$.

4.14.3 Comb Inequalities

Several classes of cutting planes have been developed to strengthen a continuous relaxation of (4.215). By far the most widely used are *comb inequalities*. Suppose H is a subset of vertices of G, and $T_1, \ldots, T_m$ are pairwise disjoint sets of vertices (where m is odd) such that $H \cap T_k$ and $T_k \setminus H$ are nonempty for each k. The subgraph of G induced by the vertices of H and the T_ks is a *comb* with *handle* H and *teeth* T_k. The associated comb inequality is

$$\sum_{(i,j) \in \delta(H)} y_{ij} + \sum_{k=1}^{m} \sum_{(i,j) \in \delta(T_k)} y_{ij} \geq \tfrac{1}{2}(3m+1) \qquad (4.219)$$

For example, the comb shown in Figure 4.20 corresponds to the inequality

$$(y_{42} + y_{59} + y_{6,11} + y_{78}) + (y_{10,6} + y_{11,12} + y_{74} + y_{78}) \geq 5$$

THEOREM 4.50 *Any comb with handle H and teeth T_k for $k = 1, \ldots, m$ gives rise to a valid inequality (4.219).*

The Circuit Constraint

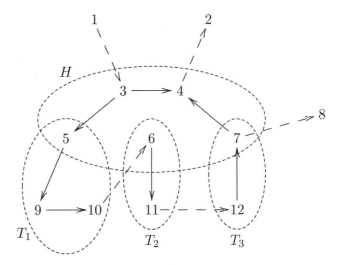

Figure 4.20. Example of a comb with handle H and teeth T_1, T_2, T_3. The solid arrows indicate edges of the comb, and the dashed arrows show other edges of the graph.

Proof. Define $T_k^o = T_k \setminus H$ and $T_k' = T_k \cap H$. Also, for a vertex set S let $y(S)$ abbreviate $\sum_{(i,j) \in \delta(S)} y_{ij}$. The following will be shown:

$$y(H) + \sum_k y(T_k) \geq \tfrac{1}{2}\left[\sum_k y(T_k^o) + \sum_k y(T_k') + \sum_k y(T_k)\right] \geq \tfrac{3}{2}m \tag{4.220}$$

Because $y(H) + \sum_k y(T_k)$ is integral, and m is odd, the right-hand side of (4.220) can be rounded up to $\tfrac{1}{2}(3m+1)$ and the theorem follows. To show (4.220), note that the second inequality follows from the fact that each term of the middle expression must be at least one, because any tour must contain at least one edge leaving a given vertex set. The first inequality of (4.220) can be established by a bookkeeping argument. Let O be the set of vertices outside the comb, let $H^o = H \setminus \bigcup_k T_k$, and let $y(S_1, S_2)$ abbreviate $\sum_{i \in S_1, j \in S_2} y_{ij}$. Then the first inequality of (4.220) can be verified by writing each expression in terms of $y(H^o, O)$, $y(T_k', T_\ell^o)$, $y(T_k', O)$, $y(H^o, T_k')$, $y(T_k', T_\ell')$, $y(T_k', O)$, and $y(T_k^o, O)$. □

Several separation heuristics have been developed for comb inequalities. One goes as follows. Define a graph to be *2-connected* if the removal of any one vertex does not disconnect it. Given a solution $\bar{y}$ of the current relaxation, let the *support graph* $\bar{G}$ be the subgraph of G whose edges (i,j) correspond to fractional $\bar{y}_{ij}$s. Select a 2-connected component of $\bar{G}$ and let it be the handle H. Let $\{i,j\}$ be a tooth T_k if $\bar{y}_{ij} = 1$ and exactly one of i, j is in H. If $\bar{y}$ violates the corresponding comb inequality (4.220), then the comb inequality is separating. Otherwise a variation of the max back heuristic can be used. Start adding more teeth to the comb in the following way, keeping track of how the comb inequality (4.220) changes each time a tooth is added. Select a vertex of H that belongs to another 2-connected component of $\bar{G}$, and let this connected component be a tooth. Note that all teeth added in this way will be pairwise disjoint. Continue until the comb inequality becomes violated, or until no further teeth can be added.

As an example, consider again the graph of Figure 4.18, only this time with different costs: let $c_{34} = c_{61} = 0$ and $c_{ij} = 1$ for all other i, j. The optimal solution of the relaxation puts $y_{34} = y_{61} = 1$, $y_{31} = y_{64} = 0$, and $y_{ij} = \frac{1}{2}$ for all other i, j. The resulting support graph $\bar{G}$ is shown in Figure 4.21. There are three 2-connected components of $\bar{G}$, one on vertices $\{1, 2, 3\}$, one on $\{4, 5, 6\}$, and one on $\{2, 5\}$. Using the heuristic just described, one can let the first component be the handle $H = \{1, 2, 3\}$ and include two teeth $T_1 = \{3, 4\}$ and $T_2 = \{1, 6\}$. The corresponding comb inequality is not violated by $\bar{y}$ (in fact it is not actually a comb inequality, because there are an even number of teeth). Applying the max back heuristic, the 2-connected component containing vertex 2 can be added as a third tooth $T_3 = \{2, 5\}$. The corresponding comb inequality

$$(y_{34} + y_{25}) + (y_{31} + y_{45} + y_{46} + y_{23} + y_{56} + y_{12} + y_{13} + y_{64}) \geq 5 \quad (4.221)$$

is violated by $\bar{y}$, and there is no need to add more teeth. The separating inequality (4.221) is added to the relaxation (4.217), which now has solution $\bar{y}_{12} = \bar{y}_{23} = \bar{y}_{34} = \bar{y}_{45} = \bar{y}_{56} = \bar{y}_{61} = 1$ and $\bar{y}_{ij} = 0$ for other i, j. This defines a tour and is therefore an optimal solution.

Since comb and subtour-elimination inequalities incompletely describe the convex hull, they may not furnish a separating cut. This is illustrated by the relaxation (4.217)–(4.218) obtained in Section 4.14.2, whose solution is $\bar{y} = \frac{1}{2}$ for all i, j. This solution satisfies all comb and subtour-

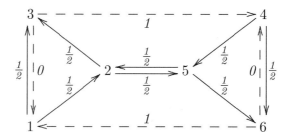

Figure 4.21. A solution of the assignment relaxation that satisfies all subtour-elimination constraints. The edges of the support graph are shown with solid arrows.

Disjunctive Scheduling 385

elimination inequalities. To find an integral solution, one must branch or identify another family of cutting planes.

4.14.4 Exercises

1. Write the assignment relaxation for the problem of minimizing $\sum_i c_{ix_i}$ subject to circuit$(x_1, \ldots, x_6)$ and draw the associated graph, where $c_{24} = c_{61} = 1$ and all other $c_{ij} = 0$. The domains are $x_1 \in \{2\}$, $x_2 \in \{3,4\}$, $x_3 \in \{1,4\}$, $x_4 \in \{5\}$, and $x_6 \in \{1,3,4\}$. Solve the relaxation and note that the solution is infeasible.

2. Use the max back heuristic to identify one or more separating subtour-elimination inequalities for the solution obtained in the previous exercise. Re-solve the relaxation with these cuts added and obtain an optimal tour.

3. Given a graph with directed edges $(1,2)$, $(1,3)$, $(2,3)$, $(3,1)$, $(3,4)$, $(3,6)$, $(4,5)$, $(5,6)$, $(6,7)$, $(7,1)$, and $(7,4)$, write one or more comb inequalities and verify that they are satisfied by the only tour in the graph.

4. Verify the first inequality in (4.220).

4.15 Disjunctive Scheduling

The constraint disjunctive$(s \mid p)$ introduced in Section 3.13 requires that jobs be scheduled sequentially so as not to overlap in time. Here $s = (s_1, \ldots, s_n)$ is a vector of start time variables for jobs $1, \ldots, n$, and the parameter $p = (p_1, \ldots, p_n)$ is a vector of processing times. Each job j also has a release time r_j and deadline d_j. The time windows are implicit in the initial domain $[r_j, d_j - p_j]$ of each variable s_j.

The disjunctive constraint can be relaxed in at least three ways. One is to create convex hull or big-M relaxations for individual disjunctions implied by the constraint. The convex hull relaxations add a large number of auxiliary variables, while the big-M relaxations use only the original variables but are considerably weaker. Another approach is to use the continuous relaxation of a mixed integer model of the constraint. Such models can be formulated for both discrete and continuous time domains. A third approach is to write a family of valid inequalities that are obtained by solving a certain optimization problem in advance. The second and third relaxation methods will also prove useful for cumulative scheduling.

None of the relaxations presented here for disjunctive and cumulative scheduling are particularly tight. This is probably because not even the convex hull relaxations, were they known, would be tight. It is therefore important to use relaxations in conjunction with the sophisticated filtering methods that have been developed for these constraints, some of which are described in Sections 3.13 and 3.14.

Relaxations of disjunctive scheduling problems are important in logic-based Benders methods, as indicated in Section 2.3.7. These relaxations are expressed in terms of master problem variables rather than the start time variables of the scheduling subproblems. They will be discussed in connection with cumulative scheduling in Section 4.16. Relaxations for the disjunctive scheduling problem are a special case of those for the cumulative scheduling problem, and there is no advantage in examining this case separately.

4.15.1 Disjunctive Relaxations

The most straightforward way to relax disjunctive$(s\,|\,p)$ is to relax each disjunction

$$(s_i + p_i \leq s_j) \vee (s_j + p_j \leq s_i) \tag{4.222}$$

individually—thus creating a factored relaxation (Section 4.9.3). The disjunctions can be given either a convex hull or a big-M relaxation. Unfortunately, both tend to result in a weak relaxation for the scheduling problem, except perhaps when the time windows are small.

The convex hull relaxation introduces a large number of auxiliary variables. Applying Theorem 4.17, the convex hull relaxation of each disjunction (4.222) is

$$\begin{array}{ll} -s'_i + s'_j \geq p_i y & s''_i - s''_j \geq p_i(1-y) \\ r_i y \leq s'_i \leq \ell_i y & r_i(1-y) \leq s''_i \leq \ell_i(1-y) \\ r_j y \leq s'_j \leq \ell_j y & r_j(1-y) \leq s''_j \leq \ell_j(1-y) \\ s_i = s'_i + s''_i, \ s_j = s'_j + s''_j, \ 0 \leq y \leq 1 \end{array} \tag{4.223}$$

where $\ell_i = d_i - p_i$ and where s'_i, s''_i, s'_j, s''_j and y are new variables. Because a separate copy of the new variables must be made for each disjunction, the relaxation of disjunctive$(s\,|\,p)$ becomes

$$\left.\begin{array}{ll} -s'_{ij} + s'_{ji} \geq p_i y_{ij} & s''_{ij} - s''_{ji} \geq p_i(1-y_{ij}) \\ r_i y_{ij} \leq s'_{ij} \leq \ell_i y_{ij} & r_i(1-y_{ij}) \leq s''_{ij} \leq \ell_i(1-y_{ij}) \\ r_j y_{ij} \leq s'_{ij} \leq \ell_j y_{ij} & r_j(1-y_{ij}) \leq s''_{ji} \leq \ell_j(1-y_{ij}) \end{array}\right\} \begin{array}{l} \text{all } i,j \\ \text{with} \\ i<j \end{array}$$

$$s_i = s'_{ij} + s''_{ij}, \ \text{all } i,j \text{ with } i \neq j$$

$$0 \leq y_{ij} \leq 1, \ \text{all } i,j \text{ with } i < j \tag{4.224}$$

As an example, consider the scheduling problem illustrated in Figure 4.22. The problem is to minimize the sum of the completion times

$$(s_1 + p_1) + (s_2 + p_2) + (s_3 + p_3) = s_1 + s_2 + s_3 + 6 \tag{4.225}$$

subject to disjunctive$(s \mid p)$ and the time windows shown in the figure. The optimal solution, which has value 13, is also shown in the figure.

In this instance, there are three disjunctions

$$\begin{aligned} (s_1 + p_1 \leq s_2) &\vee (s_2 + p_2 \leq s_1) \\ (s_1 + p_1 \leq s_3) &\vee (s_3 + p_3 \leq s_1) \\ (s_2 + p_2 \leq s_3) &\vee (s_3 + p_3 \leq s_2) \end{aligned} \tag{4.226}$$

Figure 4.23 shows the feasible set of the relaxation (4.223) of each disjunction, projected onto the variables s_j. Although each disjunction receives a convex hull relaxation, the relaxation (4.224) as a whole does not describe the convex hull of the feasible set. This is evident in the fact that the relaxed problem of minimizing (4.225) subject to (4.224) has a nonintegral optimal solution $(y_{12}, y_{13}, y_{23}) = (0, \frac{1}{2}, 1)$. Also, the optimal value 11 of the relaxation is less than the optimal value 13 of the original problem, although it provides a reasonably good lower bound in this small example.

One can simplify matters considerably by using a big-M relaxation of each disjunction rather than a convex hull relaxation. Due to Theorem 4.18, each disjunction (4.222) has the relaxation

$$\left(\frac{1}{d_j - r_i} - \frac{1}{d_i - r_j}\right) s_i + \left(\frac{1}{d_i - r_j} - \frac{1}{d_j - r_i}\right) s_j \\ \geq \frac{p_i}{d_i - r_j} + \frac{p_j}{d_j - r_i} - 1 \tag{4.227}$$

Figure 4.22. Time windows (horizontal lines) for a 3-job disjunctive scheduling problem. The heavy lines show the solution that minimizes the sum of completion times. Note that the processing times are $(p_1, p_2, p_3) = (3, 1, 2)$.

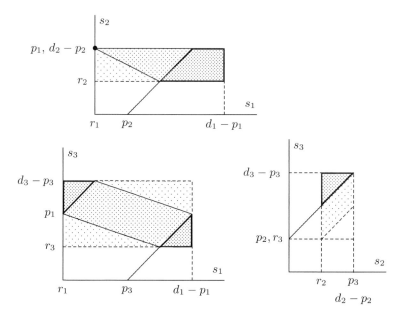

Figure 4.23. Feasible sets (dark shading) of three disjunctions associated with the scheduling problem of Figure 4.22. The dark and medium shading represent the feasible set of the convex hull relaxation of each disjunction, and the entire shaded area represents the feasible set of the big-M relaxation.

along with the time-window bounds. By writing this inequality for each pair i, j with $i < j$, and the bounds $r_j \leq s_j \leq d_j - p_j$ for each j, one can obtain a continuous relaxation without adding any variables.

Unfortunately, this relaxation tends to be weak. In the example of Figure 4.22, it simplifies to

$$s_1 - s_2 \geq 2, \quad s_3 \geq s_2 \\ 0 \leq s_1 \leq 4, \quad 1 \leq s_2 \leq 2, \quad 1 \leq s_3 \leq 3 \tag{4.228}$$

As the figure illustrates, the relaxation is only slightly tighter than the time-window bounds alone. In fact, when the relaxation is solved by minimizing (4.225) subject to (4.228), the solution is the same as would be obtained by using only the bounds; namely, $(s_1, s_2, s_3) = (0, 1, 1)$, with an optimal value of 8.

4.15.2 MILP Relaxations

An alternative to using factored disjunctive relaxations is to use continuous relaxations of time-indexed or event-indexed MILP models. The time-indexed models may be practical when time can be discretized into a reasonably small number of equal units. The event-indexed models use continuous time variables but are more complicated and tend to be weaker. Yet, they may be more practical when the application requires a time horizon and granularity that result in many time increments.

The time-indexed model is the easiest to formulate. Let the 0-1 variable x_{jt} be 1 when job i starts at discrete time t. The variable appears for a particular pair i, t only when job j can start at time t without violating its time window; that is, when $r_j \leq t < d_j - p_j$. Note that, due to the discreteness of time, the job must finish strictly before its deadline. An MILP model can be written

$$\begin{aligned} \sum_j \sum_{t' \in T_{jt}} x_{jt'} \leq 1, \quad \text{all } t \quad (a) \\ \sum_t x_{jt} = 1, \quad \text{all } j \quad (b) \end{aligned} \quad (4.229)$$

where each $x_{jt} \in \{0, 1\}$ and $T_{jt} = \{t' \mid t - p_j < t' \leq t\}$ is the set of times job j could have started if it is running at time t. Constraint (a) ensures that at most one job is running at any one time. Constraint (b) requires each job to be assigned exactly one start time. As usual, the continuous relaxation is formed by replacing (c) with $0 \leq x_{jt} \leq 1$. A variation on this model introduces an inventory variable z_t that is equal to 1 when a job is running at time t:

$$\begin{aligned} z_t = z_{t-1} + \sum_j x_{jt} - \sum_j x_{j,t-p_j} \quad \text{all } t \\ \sum_t x_{jt} = 1, \quad \text{all } j \\ z_t \leq 1, \quad \text{all } t \end{aligned} \quad (4.230)$$

where each $x_{jt} \in \{0, 1\}$. If the first time is $t = 0$, the initial inventory at time $t = -1$ is $z_{-1} = 0$. The models (4.229) and (4.230) have equivalent relaxations, because (4.229) can be obtained from (4.230) by substitution of variables.

The example of Figure 4.22 has the time-indexed relaxation

$$\begin{aligned} x_{10} + x_{11} + x_{21} + x_{31} \leq 1 \\ x_{10} + x_{11} + x_{12} + x_{22} + x_{31} + x_{32} \leq 1 \\ x_{11} + x_{12} + x_{13} + x_{32} + x_{33} \leq 1 \\ x_{12} + x_{13} + x_{14} + x_{33} \leq 1 \\ \sum_{t=0}^{4} x_{1t} = \sum_{t=1}^{2} x_{2t} = \sum_{t=1}^{3} x_{3t} = 1, \\ 0 \leq x_{jt} \leq 1, \quad \text{all } j, t \end{aligned}$$

Three redundant constraints are omitted. The objective function can be written $\sum_{jt}(t+p_j)x_{jt} = \sum_{jt} tx_{jt} + \sum_j p_j$. This relaxation is stronger than either of the disjunctive relaxations solved in the previous section, because its optimal value is 13, the same as the optimal value of the original problem. In fact, the optimal solution of the relaxation is the solution shown in Figure 4.22.

When there are a large number of discrete times, it may be advantageous to use an event-indexed model in which time is continuous. In such a model, there are n events, each of which is the start of a job, and the 0-1 variable x_{jk} is 1 if event k is the start of job j. The continuous variable t_k is the time at which event k occurs. The model for disjunctive$(s\,|\,p)$ is

$$\sum_k x_{jk} = 1, \quad \text{all } j \tag{a}$$

$$\sum_j x_{jk} = 1, \quad \text{all } k \tag{b}$$

$$t_{k+1} - t_k \geq \sum_j p_j x_{jk}, \quad \text{all } k < n \tag{c}$$

$$t_k \geq \sum_j x_{jk} r_j, \quad \text{all } k \tag{d}$$

$$t_k \leq d_{\max}\left(1 - \sum_j x_{jk}\right) + \sum_j (d_i - p_j) x_{jk} \quad \text{all } k \tag{e}$$

$$x_{jk} \in \{0,1\}, \quad \text{all } j, k$$

(4.231)

where $d_{\max} = \max_j\{d_j\}$. Symmetry-breaking constraints $t_k \leq t_{k+1}$ can also be added. Constraints (a) and (b) ensure that each job is assigned to one event and vice-versa. Constraint (c) prevents jobs from overlapping, and (d) enforces the release times. Constraint (e) is the model's weak point, because it must use a big-M construction (where $d_{\max}$ is the big-M) to enforce the deadlines. In the small example of Figure 4.22, however, the continuous relaxation of (4.231) yields the optimal solution.

Another MILP model for the disjunctive scheduling problem is suggested in Exercise 11 of Section 4.9.

4.15.3 A Class of Valid Inequalities

One way to relax a constraint is to generate valid inequalities of the form $f(x) \geq \alpha$, where $f(x)$ is some function of the constraint variables x. Here, α is the minimum of $f(x)$ subject to a relaxation R of the constraint that is chosen to make α easy to compute in advance. One can also maximize $f(x)$ subject to R and obtain a valid inequality $f(x) \leq \alpha$.

Valid inequalities of this sort can then be used to form or enrich a relaxation of the constraint—perhaps a continuous relaxation. This can be advantageous when the resulting relaxation R' is tighter than R, or when the objective function of interest can be easily optimized subject to R' but not subject to R.

A simple way to apply this technique to disjunctive scheduling is to let $f(s)$ be a weighted sum $\sum_{j \in J} a_j s_j$ of some subset of start times. Let the relaxation R be the original disjunctive constraint, but with time windows relaxed so that every release time is $r_J = \min_{j \in J}\{r_j\}$ and every deadline is $d_J = \max_{j \in J}\{d_j\}$. Let $J = \{j_1, \ldots, j_k\}$ where the jobs are indexed so that $p_{j_1} \leq \cdots \leq p_{j_k}$. Now, if $a_{j_1} \geq \cdots \geq a_{j_k} \geq 0$, the problem of minimizing $f(s)$ subject to R can be solved by a greedy method: let job j_1 start at r_J, job j_2 immediately after it at $r_J + p_{j_1}$, and so forth. Then job j_i starts at

$$s_i^{\min} = r_J + \sum_{\ell=1}^{i-1} p_{j_\ell} \qquad (4.232)$$

This clearly minimizes the weighted sum of finish times $\sum_{j \in J} a_j(s_j + p_j)$, and therefore minimizes the weighted sum of start times $\sum_{j \in J} a_j s_j$. One can also maximize $f(s)$ by scheduling the jobs in nonincreasing order of processing time, with the last job finishing at time d_J and job j_i starting at

$$s_i^{\max} = d_J - \sum_{\ell=1}^{i} p_{j_\ell} \qquad (4.233)$$

So, one obtains the following result:

THEOREM 4.51 *If $J = \{j_1, \ldots, j_k\} \subset \{1, \ldots, n\}$, $p_{j_1} \leq \cdots \leq p_{j_k}$, and $a_{j_1} \geq \cdots \geq a_{j_k} \geq 0$, then*

$$\sum_{j \in J} a_j s_j \geq \sum_{i=1}^{k} a_{j_i} s_i^{\min} \qquad \sum_{j \in J} a_j s_j \leq \sum_{i=1}^{k} a_{j_i} s_i^{\max} \qquad (4.234)$$

are valid inequalities for disjunctive$(s \mid p)$, where $s_i^{\min}$ and $s_i^{\max}$ are given by (4.235) and (4.233). In particular (letting each $a_j = 1$),

$$\sum_{j \in J} s_j \geq k r_J + \sum_{i=1}^{k-1} (k-i) p_{j_i}$$
$$\sum_{j \in J} s_j \leq k d_J - \sum_{i=1}^{k} (k-i+1) p_{j_i} \qquad (4.235)$$

are valid inequalities.

Let $J(t_1, t_2)$ be the set of jobs with time windows in the interval $[t_1, t_2]$; that is,
$$J(t_1, t_2) = \{j \mid t_1 \leq r_j, d_j \leq t_2\}$$
It is clear that all valid inequalities (4.234)–(4.235) are dominated by those corresponding to sets $J = J(r_j, d_k)$ for which $r_j < d_k$. In the example of Figure 4.22, these sets are $J(0,3) = J(1,3) = \{2\}$, $J(0,5) = J(1,5) = J(1,7) = \{2,3\}$, and $J(0,7) = \{1,2,3\}$. The corresponding valid inequalities (4.235), omitting the singleton $\{2\}$, are

$$\begin{aligned} 3 \leq s_2 + s_3 \leq 6 \\ 4 \leq s_1 + s_2 + s_3 \leq 11 \end{aligned} \quad (4.236)$$

The two upper bounds are redundant of the time-window bounds $r_j \leq s_j \leq d_j - p_j$, but the two lower bounds are not.

One way to use the inequalities (4.236) is to add them to the time-window bounds and big-M relaxation (4.228) to obtain a continuous relaxation R' in the original variables s. In this particular case, one can easily minimize the desired objective function (4.225) subject to R, by using the greedy method, but it is still advantageous to use R', because it is tighter than R and easy to solve as a linear programming problem. The minimum of (4.225) subject to R is 10, but its minimum subject to R' is 12.

Since the strength of the inequalities (4.234) tends to rise rapidly with the size of the set J, in larger problems it is reasonable to select a few large sets $J(r_j, d_k)$ to generate inequalities. It is not hard to design selection heuristics that yield strong inequalities. Such heuristics should recognize, however, that an inequality corresponding to $[r_k, d_k]$ does not necessarily dominate one corresponding to a proper subset of $[r_j, d_k]$.

4.15.4 Exercises

1 Verify that (4.223) is a convex hull relaxation of (4.222).

2 Consider the problem of minimizing $s_1 + s_2$ subject to the constraint disjunctive$((s_1, s_2) \mid (2, 2))$, with domains $s_1 \in [0, 2]$ and $s_2 \in [1, 3]$. Draw a graph of the problem and identify the feasible set. Write a disjunctive relaxation using the convex hull relaxation of each disjunction in the model. Since there is only one disjunction in this problem, this yields a convex hull relaxation for the entire problem. Given that the optimal solution is obvious upon inspection, indicate what must be the optimal value of each variable in the relaxation without actually computing a solution of the relaxation.

3 Show that (4.227) is a valid big-M relaxation of the disjunction (4.222).

4. Write the relaxation (4.227) for the problem of Exercise 2. Note that it and the bounds define a convex hull relaxation, although this is not true in general.

5. Write a discrete-time MILP model for the problem of Exercise 2.

6. Write a continuous-time MILP model for the problem of Exercise 2. In this case, solution of its continuous relaxation is feasible in the original problem.

7. Show that if $a_1 \geq \cdots \geq a_n$, $p_1 \leq \cdots \leq p_n$, and each $x_j \geq 0$, the minimum of $\sum_j a_j x_j$ subject to disjunctive$(x \mid p)$ is the greedy solution $x_j = \sum_{i=1}^{j-1} p_i$.

8. Verify the validity of cuts (4.235) in Theorem 4.51.

9. What are the cuts (4.235) for the example of Exercise 2?

10. It is possible to obtain cuts by applying Theorem 4.51 for a given J with $a \neq (1, \ldots, 1)$ that are not redundant of the cuts obtained with $a = (1, \ldots, 1)$?

11. Valid cuts are obtained above by defining a relaxation of the constraint disjunctive$(s \mid p)$ that can by solved by a greedy algorithm. In particular, the constraint is relaxed by replacing each time window $[r_j, d_j]$ with $[r_J, d_J]$. Other relaxations can be solved in a greedy fashion—for example a relaxation obtained by changing each time window to $[r_j, d_J]$ and each p_j to $\min_{j \in J}\{p_j\}$. Show how to obtain cuts from this relaxation. The resulting cuts could be useful when the processing times are about the same, but the release times differ substantially. Symmetric cuts can, of course, be obtained by setting each time window to $[r_J, d_j]$.

4.16 Cumulative Scheduling

Cumulative scheduling requires that jobs be scheduled so that their total rate of resource consumption never exceeds a given limit. Section 3.14 presents several filtering methods for the most popular cumulative scheduling constraint. The task of the present section is to derive some continuous relaxations for it.

Two of the three relaxation methods presented for disjunctive scheduling in Section 4.15 can be extended to cumulative scheduling. Continuous relaxations of MILP models can be written for the constraint and valid inequalities can be derived—in this case inequalities based on energetic reasoning.

Recall that in cumulative$(s \mid p, c, C)$, variables $s = (s_1, \ldots, s_n)$ represent the start time of the jobs. The parameter $p = (p_1, \ldots, p_n)$ contains the processing time p_j of each job j, and $c = (c_1, \ldots, c_n)$ contains the rate c_j of resource consumption for each job j. The constraint requires that the total rate of resource consumption of the jobs running at any time t never exceed C:

$$\sum_{\substack{j \\ s_j \leq t \leq s_j + p_j}} c_j \leq C, \quad \text{for all times } t$$

There is also a time window $[r_j, d_j]$ for each job j, which is reflected in the initial domain $[r_j, d_j - p_j]$ of s_j.

4.16.1 MILP Models

As in the case of disjunctive scheduling, one can formulate time-indexed and event-indexed models for cumulative scheduling. The time-indexed model is a straightforward generalization of the disjunctive scheduling model (4.229). Again the 0-1 variable x_{jt} is 1 when job i starts at discrete time t. The variable appears for a particular pair i, t only when $r_j \leq t < d_j - p_j$. The model is

$$\begin{aligned}
\sum_j \sum_{t' \in T_{jt}} c_j x_{jt'} \leq C, \quad &\text{all } t \quad (a) \\
\sum_t x_{jt} = 1, \quad &\text{all } j \quad (b) \\
x_{jt} \in \{0, 1\}, \quad &\text{all } j, t \quad (c)
\end{aligned} \quad (4.237)$$

where $T_{jt} = \{t' \mid t - p_j < t' \leq t\}$. Constraint (a) ensures that the total rate of resource consumption is at most C at any one time. Constraint (b) requires each job to be assigned exactly one start time. The continuous relaxation is formed by replacing (c) with $0 \leq x_{jt} \leq 1$.

A variation on the model uses an inventory variable z_t that keeps track of the resource consumption at each time t:

$$\begin{aligned}
z_t = z_{t-1} + \sum_j c_j x_{jt} - \sum_j c_j x_{j,t-p_j} \quad &\text{all } t \\
\sum_t x_{jt} = 1, \quad &\text{all } j \\
z_t \leq C, \quad &\text{all } t \\
x_{jt} \in \{0, 1\}, \quad &\text{all } j, t
\end{aligned} \quad (4.238)$$

Cumulative Scheduling

The initial inventory at time $t = -1$ is $z_{-1} = 0$. The models (4.229) and (4.230) have equivalent continuous relaxations.

A small cumulative scheduling problem appears in Figure 4.24. If the objective is to minimize the sum of the finish times, the optimal value is 11. The time-indexed relaxation (4.237) for this problem instance is

$$
\begin{aligned}
& x_{10} + x_{11} + 3x_{21} + 2x_{31} \leq 3 \\
& x_{10} + x_{11} + x_{12} + 3x_{22} + 2(x_{31} + x_{32}) \leq 3 \\
& x_{11} + x_{12} + 2(x_{32} + x_{33}) \leq 3 \\
& x_{10} + x_{11} + x_{12} = x_{21} + x_{22} = x_{31} + x_{32} + x_{33} = 1 \\
& x_{jt} \geq 0, \quad \text{all } j, t
\end{aligned}
\quad (4.239)
$$

where two redundant capacity constraints are omitted. The minimum value of the objective function $\sum_{jt} tx_{jt} + \sum_j p_j$ subject to (4.239) is $9\frac{5}{6}$, which is a lower bound on the optimal value of the original problem.

When there are a large number of discrete times, it may be advantageous to use one of two event-indexed models that employ continuous time variables. In these models, there are $2n$ events, each of which can be the start of a job or the finish of a job. In one model, the 0-1 variable $x_{jkk'}$ is 1 if event k is the start of job j and event k' is the finish of job j. The continuous variable t_k is the time at which event k occurs. The inventory variable z_k keeps track of how much resource is being consumed when event k occurs; obviously one wants $z_k \leq C$. The model

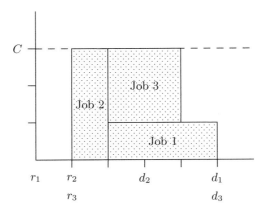

Figure 4.24. Optimal solution of a small cumulative scheduling problem in which the objective is to minimize the sum of completion times. The horizontal axis represents time and the vertical axis represents resource consumption.

for cumulative$(s \mid p, c, C)$ is

$$z_k = z_{k-1} + \sum_j \sum_{k'>k} c_j x_{jkk'} - \sum_j \sum_{k'<k} c_j x_{jk'k}, \text{ all } k \quad (a)$$

$$z_0 = 0, \ 0 \leq z_k \leq C, \text{ all } k \quad (b)$$

$$\sum_k \sum_{k'>k} x_{jkk'} = 1, \text{ all } j \quad (c)$$

$$\sum_j \sum_{k'>k} x_{jkk'} = \sum_j \sum_{k'<k} x_{jk'k} = 1, \text{ all } k \quad (d)$$

$$t_{k'} - t_k \geq \sum_j p_j x_{jkk'}, \text{ all } k, k' \text{ with } k < k' \quad (e)$$

$$t_k \geq \sum_j \sum_{k'>k} x_{jkk'} r_j, \text{ all } k \quad (f)$$

$$t_k \leq d_{\max}\left(1 - \sum_j \sum_{k'<k} x_{jk'k}\right) + \sum_j \sum_{k'<k} x_{jk'k} d_j, \text{ all } k \quad (g)$$

$$x_{jkk'} \in \{0, 1\}, \text{ all } j, k, k'$$

$$(4.240)$$

where $d_{\max} = \max_j\{d_j\}$ and M is a large number (e.g., the total length of the time horizon). Constraint (a) keeps track of how much resource is being consumed, and (b) imposes the upper limit. Constraint (c) makes sure that each job starts once and ends once. Constraint (d) requires each event to be associated with exactly one job. Constraint (e) presents jobs from overlapping, and (f) enforces the release times. Constraint (g) uses a big-M construction (where $d_{\max}$ is the big-M) to enforce the deadlines.

If one wishes to relate the event times t_k to the start times s_j, the following constraints can be added

$$t_k - d_{\max}\left(1 - \sum_j \sum_{k'>k} x_{jkk'}\right) \leq s_j \leq$$

$$t_k + d_{\max}\left(1 - \sum_j \sum_{k'>k} x_{jkk'}\right), \text{ all } j, k$$

However, this constraint may be unnecessary in a particular context. For instance, if the objective is to minimize latest finish time, one can minimize y and add the constraint $y \geq t_k$ for all k, and the start time variables s_j are unnecessary.

The objective for the example of Figure 4.24, which is to minimize the sum of finish times, can also be written in terms of the event times t_k without using the start time variables s_k. If $\sum_S$ is the sum of start times and $\sum_F$ the sum of finish times, then $\sum_F = \sum_S + \sum_j p_j$. So, $2\sum_F = (\sum_S + \sum_F) + \sum_j p_j = \sum_j t_j + \sum_j p_j$, which implies $\sum_F = \frac{1}{2}\sum_j t_j + \frac{1}{2}\sum_j p_j$. Thus, one can minimize this quantity subject to the relaxation (4.240). The resulting minimum value is 5.316, which is a rather weak lower bound on the optimal value of the original problem.

Model (4.231) is quite large due to the triply indexed variables $x_{jkk'}$. An alternative is to use separate variables for start-events and finish-events, which requires that the deadlines be enforced in a different way. This reduces the triple index to a double index at the cost of producing a weaker relaxation. Let the 0-1 variable x_{jk} be 1 when event k is the start of job j, and $y_{jk} = 1$ when event k is the finish of job j. The new continuous variable f_j is the finish time of job j:

$$z_k = z_{k-1} + \sum_j c_j x_{jk} - \sum_j c_j y_{jk}, \text{ all } k \quad (a)$$

$$z_0 = 0, \quad z_k \leq C, \text{ all } k \quad (b)$$

$$\sum_k x_{jk} = 1, \quad \sum_k y_{jk} = 1, \text{ all } j \quad (c)$$

$$\sum_j x_{jk} + y_{jk} = 1, \text{ all } k \quad (d)$$

$$s_{k-1} \leq s_k, \quad x_{jk} \leq \sum_{k'<k} y_{jk'}, \text{ all } k > 1 \quad (e) \quad (4.241)$$

$$s_k \geq \sum_j r_j x_{jk}, \text{ all } k \quad (f)$$

$$s_k + p_j x_{jk} - d_{\max}(1 - x_{jk}) \leq f_j$$
$$\leq s_k + d_{\max}(1 - y_{jk}), \text{ all } j, k \quad (g)$$

$$f_j \leq d_j, \text{ all } j \quad (h)$$

$$x_{jk}, y_{jk} \in \{0, 1\}$$

Constraints (a) and (b) perform the same function as before. Constraints (c) and (d) require that each job start once and end once, but not as the same event. Constraints (e) are redundant but may tighten the relaxation. One constraint requires the events to occur in chronological order, and one requires a job's start-event to have a smaller index than its finish-event. Constraint (f) observes the release times. The new element is constraint (g). The first inequality defines the finish time f_j of each job by forcing it to occur no earlier than p_j time units after the start time. The second inequality forces the time associated with the finish-event to be no earlier than the finish time. Finally, constraint (h) enforces the deadlines.

Again, the objective function that measures the sum of finish times in the example of Figure 4.24 is $\frac{1}{2}\sum_j s_j + \frac{1}{2}\sum_j p_j$. The minimum value of this function subject to the relaxation (4.241) is 4.2, a weaker lower bound than obtained from the triply indexed model (4.240).

Another MILP model for the cumulative scheduling problem is suggested in Exercise 12 of Section 4.9.

4.16.2 A Class of Valid Inequalities

As in the case of disjunctive scheduling, one can develop a class of valid inequalities $\sum_{j \in J} a_j s_j \geq \alpha$ for cumulative scheduling by letting α be the minimum of $\sum_{j \in J} a_j s_j$ subject to a relaxation R of the constraint.

In this case, the relaxation R not only relaxes the time windows to $[r_J, d_J]$ for every job, but it also allows any job j to be replaced by any job with the same energy consumption $p_j c_j$ and with resource consumption rate at most C. Let the jobs be indexed so that $J = \{j_1, \ldots, j_k\}$ and $p_{j_1} c_{j_1} \leq \cdots \leq p_{j_k} c_{j_k}$. Then, one feasible solution of R is a *compressed* schedule that replaces each job j with a job having duration $p_j c_j / C$ and resource consumption rate C, and schedules jobs $j_1, \ldots, j_k$ consecutively with job j_1 starting at r_J. Then each job j_i in the compressed schedule begins at

$$s_i^{\min} = r_J + \frac{1}{C} \sum_{\ell=1}^{i-1} p_{j_\ell} c_{j_\ell} \qquad (4.242)$$

and finishes at $s_i^{\min} + p_{j_i} c_{j_i}/C$. The compressed schedule for the problem of Figure 4.24 is illustrated in Figure 4.25.

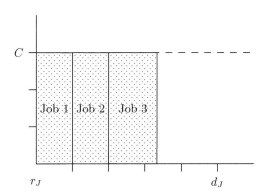

Figure 4.25. Compressed schedule for the cumulative scheduling problem of Figure 4.24. The order of jobs 1 and 2 could be reversed.

Cumulative Scheduling 399

It will be shown that if $a_{j_1} \geq \cdots \geq a_{j_k} \geq 0$ and f_j is the finish time of job j, the compressed schedule minimizes $\sum_{j \in J} a_j f_j = g(f)$ subject to R. One can also maximize $g(f)$ by scheduling each job j_i at

$$s_i^{\max} = d_J - \frac{1}{C} \sum_{\ell=1}^{i} p_{j_\ell} c_{j_\ell} \qquad (4.243)$$

This leads to the following theorem.

THEOREM 4.52 *If* $J = \{j_1, \ldots, j_k\} \subset \{1, \ldots, n\}$, $p_{j_1} c_{j_1} \leq \cdots \leq p_{j_k} c_{j_k}$, *and* $a_{j_1} \geq \cdots \geq a_{j_k} \geq 0$, *then*

$$\sum_{j \in J} a_j s_j \geq \sum_{i=1}^{k} a_{j_i} s_i^{\min} - \sum_{j \in J} a_j p_j \left(1 - \frac{c_j}{C}\right) \qquad (4.244)$$

and

$$\sum_{j \in J} a_j s_j \leq \sum_{i=1}^{k} a_{j_i} s_i^{\max} \qquad (4.245)$$

are valid inequalities for cumulative$(s \mid p, c, C)$, *where* $s_i^{\min}$ *and* $s_i^{\max}$ *are given by (4.242) and (4.243). In particular (letting each $a_j = 1$),*

$$\begin{aligned} \sum_{j \in J} s_j &\geq k r_J + \frac{1}{C} \sum_{i=1}^{k} (k-i+1) p_{j_i} c_{j_i} - \sum_{i=1}^{k} p_{j_i} \\ \sum_{j \in J} s_j &\leq k d_J - \frac{1}{C} \sum_{i=1}^{k} (k-i+1) p_{j_i} c_{j_i} \end{aligned} \qquad (4.246)$$

are valid inequalities.

Proof. Let $\hat{R}$ be a restriction of R in which each job j has duration $p_j c_j / C$ and resource consumption rate C. Then it is clear that the compressed schedule $\hat{f}$ minimizes $g(f)$ subject to $\hat{R}$. It will be shown that $g(\hat{f})$ is less than or equal to the minimum of $g(f)$ subject to R. Since R is a relaxation of $\hat{R}$, it follows that $\hat{f}$ minimizes $g(f)$ subject to R, which implies that

$$\sum_{j \in J} a_j f_j \geq \sum_{i=1}^{k} a_{j_i} \left(s_i^{\min} + \frac{p_{j_i} c_{j_i}}{C} \right)$$

is a valid inequality. Since $f_j = s_j - p_j$ in the original problem, inequality (4.244) follows. Inequality (4.245) is derived in a similar fashion.

It remains to show that $g(\hat{f}) \leq g(\bar{f})$ for any feasible solution $\bar{f}$ of R. This will be shown by induction on the number k of jobs. For $k = 1$, it suffices to note that if a job j finishes at $\bar{f}_j$, then because the job starts no earlier than r_J and has duration of at least $p_j c_j/C$, the compressed schedule $\hat{f}_j = r_J + p_j c_j/C$ results in $\hat{f}_j \leq \bar{f}_j$. This implies $g(\hat{f}_j) \leq g(\bar{f}_j)$.

Now suppose that the claim is true for $k-1$ jobs and show it is true for k jobs. Take any feasible schedule $\bar{f}$ of R, and index the jobs so that job k is the last job to finish. By the induction hypothesis, there is a compressed schedule $(\hat{f}_1, \ldots, \hat{f}_{k-1})$ on jobs $1, \ldots, k-1$ for which

$$g(\hat{f}_1, \ldots, \hat{f}_{k-1}) \leq g(\bar{f}_1, \ldots, \bar{f}_{k-1}) \tag{4.247}$$

It must be shown that the compressed schedule $\hat{f}$ for jobs $1, \ldots, k$ is feasible and satisfies $g(\hat{f}) \leq g(\bar{f})$. Note that $\hat{f}_k \leq \bar{f}_k$, because no schedule for k jobs can finish all the jobs sooner than the compressed schedule does. Thus $\hat{f}_k \leq d_J$, since $\bar{f}_k \leq d_J$, and $\hat{f}$ is therefore feasible. Also, $\hat{f}_k \leq \bar{f}_k$ and (4.247) imply $g(\hat{f}) \leq g(\bar{f})$. $\square$

As in the case of disjunctive scheduling, one need only consider inequalities corresponding to sets $J = J(r_j, d_k)$ with $r_j < d_k$. As noted in Section 4.15.3, these sets for the example of Figure 4.24 are $\{1, 2, 3\}$ and $\{2, 3\}$ (ignoring the singleton $\{2\}$). The corresponding inequalities (4.246) are

$$\tfrac{1}{3} \leq s_1 + s_2 + s_3 \leq 8\tfrac{2}{3}$$
$$2\tfrac{1}{3} \leq s_2 + s_3 \leq 6\tfrac{2}{3}$$

Two of the bounds ($8\tfrac{2}{3}$ and $2\tfrac{1}{3}$) are nonredundant of the time-window bounds.

The inequalities (4.244)–(4.245) grow rapidly in strength as the size of J increases. It is therefore advisable to design a heuristic that selects a few large sets J to generate inequalities.

4.16.3 Relaxation of Benders Subproblems

The success of Benders methods for planning and scheduling can depend critically on the inclusion of a subproblem relaxation in the master problem. Since the subproblem is a disjunctive or cumulative scheduling problem, it is essential to develop relaxations for this type of problem. This type of relaxation differs from those discussed above, because it must be expressed in terms of the master problem variables, which assign jobs to facilities, rather than variables that specify start times. The relaxation should also take the form of linear inequalities, because the master problem is frequently modeled as an integer linear programming problem. The relaxations presented here are designed for cumulative scheduling, but are readily specialized to disjunctive scheduling.

The Benders approach to planning and scheduling is described in Section 2.3.7, which presents a problem in which several jobs are assigned to facilities and scheduled on them. Sections 3.13.3 and 3.14.5 discuss in more detail how Benders cuts may be generated for the master problem.

Time-Window Relaxation

The simplest relaxation is based on the fact that the total running time of any subset of jobs assigned to the same facility must fit within the earliest release time and latest deadline for those jobs. As illustrated in the example of Section 2.3.6, a relaxation of this sort can contain many redundant inequalities. The following analysis enables one to identify the redundant inequalities and delete them.

To review the notation, p_{ij} is the processing time of job j on facility i, and c_{ij} is the rate of resource consumption. The maximum rate of resource consumption on facility i is C_i. The master problem contains 0-1 variables x_{ij} that are equal to 1 when job j is assigned to facility i. The subproblem contains variables s_j that represent the start time of job j. The task is to write a linear relaxation of the time-window constraints on s_j in terms of the variables x_{ij}. Such a relaxation is useful whenever there are hard constraints on the release time and finish time of the jobs, such as in the minimum-cost problem (3.130).

Recall that the energy consumed by job j on machine i is $p_{ij}c_{ij}$, and $p_{ij}c_{ij}/C_i$ is the task interval of the job. Let $J(t_1, t_2)$ be the set of jobs j whose time windows lie in the interval $[t_1, t_2]$, so that

$$J(t_1, t_2) = \{j \mid [r_j, d_j] \subset [t_1, t_2]\}$$

The total task interval of jobs in $J(t_1, t_2)$ assigned to a given machine i must fit into the time interval $[t_1, t_2]$:

$$\frac{1}{C_i} \sum_{j \in J_i(t_1, t_2)} p_{ij} c_{ij} x_{ij} \leq t_2 - t_1 \qquad (4.248)$$

It is convenient to refer to the inequality (4.248) as $R_i(t_1, t_2)$. Let $r_1 < \cdots < r_p$ be the distinct release times and $d_1 < \cdots < d_q$ the distinct deadlines. Then, the following is a valid linear relaxation for facility i:

$$R_i(r_j, d_k), \text{ all } j \in \{1, \ldots, p\} \text{ and } k \in \{1, \ldots, q\} \text{ with } r_j < d_k$$

This relaxation can be added to the master problem for each facility i. A relaxation for disjunctive scheduling is obtained by setting $C_i = c_{ij} = 1$ for all i, j.

Redundant inequalities in the relaxation are identified as follows. Let

$$T_i(t_1, t_2) = \frac{1}{C_i} \sum_{j \in J(t_1, t_2)} p_{ij} c_{ij} - t_2 + t_1$$

measure the *tightness* of $R_i(t_1, t_2)$. The following is easily verified.

LEMMA 4.53 $R_i(t_1, t_2)$ *dominates* $R_i(u_1, u_2)$ *when* $[t_1, t_2] \subset [u_1, u_2]$ *and* $T_i(t_1, t_2) \geq T_i(u_1, u_2)$.

The algorithm of Figure 4.26 now allows one to delete redundant inequalities. It has $O(n^3)$ complexity in the worst case, where n is the number of jobs, because it is possible that none of the inequalities are eliminated. This occurs, for example, when each job j has release time $j-1$ and deadline j, and $p_{ij} = 2$.

Relaxation for Minimizing Makespan

The minimum makespan problem is formulated in (3.135). If M is the makespan of a schedule on facility i, then for any time t the total task interval of the jobs in $J(t, \infty)$ must fit in the interval $[t, M]$. This leads to a lower bound on the makespan:

$$M \geq t + \frac{1}{C_i} \sum_{j \in J(t, \infty)} p_{ij} c_{ij} x_{ij}$$

Let this be inequality $R_i(t)$. Then if $r_1 < \cdots < r_p$ are the distinct release times, the following relaxation can be added to 0-1 formulation of the master problem (3.136) for each facility i:

$$R_i(r_j), \text{ all } j \in \{1, \ldots, p\}$$

A relaxation of this sort is illustrated in Section 2.3.7.

Let $\mathcal{R}_i = \emptyset$.
For $j = 1, \ldots, p$:
  Set $k' = 0$.
  For $k = 1, \ldots, q$:
    If $d_k > r_j$ and $T_i(r_j, d_k) > T_i(r_j, d_{k'})$ then
      Remove from $\mathcal{R}_i$ all $R_i(r_{j'}, d_k)$ for which $j' < j$ and $T_i(r_j, d_k) \geq T_i(r_{j'}, d_k)$.
      Add $R_i(r_j, d_k)$ to $\mathcal{R}_i$ and set $k' = k$.

Figure 4.26. Algorithm for generating an inequality set $\mathcal{R}_i$ that relaxes the time-window constraints for facility i, where $r_1 < \cdots < r_p$ are the distinct release times and $d_1 < \cdots < d_q$ the distinct deadlines. By convention, $d_0 = -\infty$ and $T_i(r_j, -\infty) = 0$.

Let the tightness of $R_i(t)$ be

$$T_i(t) = t + \frac{1}{C_i} \sum_{j \in J(t,\infty)} p_{ij} c_{ij}$$

Then, $R_i(t)$ dominates $R_i(u)$ if $t \leq u$ and $T_i(t) \geq T_i(u)$. An $O(n)$ algorithm for eliminating redundant inequalities appears in Figure 4.26.

Relaxation for Minimizing Late Jobs

The problem of minimizing the number of late jobs is formulated in (3.141). A linear lower bound for the number of late jobs can be derived much as for the makespan and included in the 0-1 formulation of the master problem.

As before, the total running time of the jobs in $J(t_1, t_2)$ assigned to facility i is at least their total task interval

$$\frac{1}{C_i} \sum_{\ell \in J(t_1,t_2)} c_{i\ell} p_{i\ell} x_{i\ell}$$

If this quantity is greater than $t_2 - t_1$, then at least one job is late. In fact the number L_i of late jobs on facility i can be bounded as follows:

$$L_i \geq \frac{\frac{1}{C_i} \sum_{\ell \in J(t_1,t_2)} c_{i\ell} p_{i\ell} x_{i\ell} - t_2 + t_1}{\max_{\ell \in J(t_1,t_2)} \{p_{i\ell}\}}$$

If this inequality is denoted $\bar{R}_i(t_1, t_2)$, the one can add to the master problem the following bound on the total number L of late jobs. Again, let $r_1, \ldots, r_p$ be the distinct release times and $d_1, \ldots, d_q$ the distinct deadlines.

$$L \geq \sum_i L_i$$

$\bar{R}_i(r_j, d_k)$, all $j \in \{1, \ldots, p\}$ and $k \in \{1, \ldots, q\}$ with $r_j < d_k$, all i
$L_i \geq 0$, all i

(4.249)

Let $\mathcal{R}_i = \emptyset$ and set $k = q + 1$.
For $j = p, \ldots, 1$:
  If $T_i(r_j) > T_i(r_k)$ then add $R_i(r_j)$ to $\mathcal{R}_i$ and set $k = j$.

Figure 4.27. Algorithm for generating a relaxation $\mathcal{R}_i$ for minimizing makespan, where $r_1 < \cdots < r_p$ are the distinct release times. By convention, $r_{q+1} = \infty$ and $T_i(\infty) = 0$.

As an example, suppose that four jobs are to be assigned to, and scheduled on, two facilities. The problem data appear in Table 4.4. There is one distinct release time $r_1 = 0$ and four distinct deadlines $(d_1, d_2, d_3, d_4) = (2, 3, 4, 5)$. The relaxation (4.249) is

$$
\begin{aligned}
&L \geq L_1 + L_2 \\
&L_1 \geq x_{11} - 1, \quad L_1 \geq \tfrac{1}{2}x_{11} + \tfrac{2}{3}x_{12} - \tfrac{3}{4}, \\
&L_1 \geq \tfrac{2}{5}x_{11} + \tfrac{8}{15}x_{12} + \tfrac{1}{3}x_{13} - \tfrac{4}{5}, \\
&L_1 \geq \tfrac{1}{3}x_{11} + \tfrac{4}{9}x_{12} + \tfrac{5}{18}x_{13} + \tfrac{1}{3}x_{14} - \tfrac{5}{6} \\
&L_2 \geq x_{21} - \tfrac{1}{2}, \quad L_2 \geq \tfrac{4}{5}x_{11} + \tfrac{2}{3}x_{22} - \tfrac{3}{5}, \\
&L_2 \geq \tfrac{2}{3}x_{21} + \tfrac{5}{9}x_{22} + \tfrac{1}{3}x_{23} - \tfrac{2}{3}, \\
&L_2 \geq \tfrac{2}{3}x_{21} + \tfrac{5}{9}x_{22} + \tfrac{1}{3}x_{23} + \tfrac{5}{6}x_{24} - \tfrac{5}{6} \\
&L_1, L_2 \geq 0
\end{aligned}
\tag{4.250}
$$

For instance, if Jobs 1, 2 and 3 are assigned to Facility 1 (this occurs when $x_{11} = x_{12} = x_{13} = 1$ and $x_{14} = 0$), the bounds in (4.250) for Facility 1 corresponding to d_1, d_2, d_3 become $L_1 \geq 0$, $L_1 \geq \tfrac{5}{12}$, and $L_1 \geq \tfrac{7}{15}$, respectively (the bound for d_4 is redundant). This implies that at least $\lceil \tfrac{7}{15} \rceil = 1$ job is late on Facility 1. In fact, two jobs must be late, as in the optimal schedule of Figure 4.28.

Relaxation for Minimizing Total Tardiness

The problem of minimizing total tardiness has almost the same formulation (3.141) as that for minimizing the number of late jobs. The differences are that the objective function is $\sum_j T_j$ (where T_j is the tardiness of job j) and the conditional constraint is dropped.

Table 4.4. Data for a planning and scheduling problem instance with four jobs and two facilities. The release times r_j are all zero, and each facility i has capacity $C_i = 3$.

		Facility 1			Facility 2		
j	d_j	p_{1j}	c_{1j}	$p_{1j}c_{1j}$	p_{2j}	c_{2j}	$p_{2j}c_{2j}$
1	2	2	3	6	4	3	12
2	3	4	2	8	5	2	10
3	4	5	1	5	6	1	6
4	5	6	3	18	5	3	15

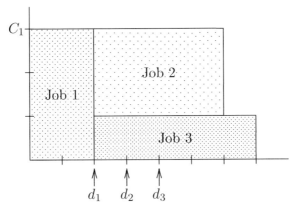

Figure 4.28. Schedule for Jobs 1–3 on Facility 1 that minimizes the number of late jobs (2) and minimizes total tardiness (6).

Two types of bounds can be developed for the total tardiness. The first and simpler bound is similar to that obtained for minimizing the number of late jobs. It is based on the following lemma. Let α^+ denote $\max\{0, \alpha\}$.

LEMMA 4.54 *Consider a minimum total tardiness problem in which jobs $j = 1, \ldots, n$ with time windows $[r_j, d_j]$ are scheduled on a single facility i, where $\min_j\{r_j\} = 0$. The total tardiness incurred by any feasible solution is bounded below by*

$$\left(\frac{1}{C_i} \sum_{j \in J(0, d_k)} p_{ij} c_{ij} - d_k\right)^+$$

for each $k = 1, \ldots, n$.

Proof. For any k, the last scheduled job in the set $J(0, d_k)$ can finish no earlier than time $t = \frac{1}{C_i} \sum_{j \in J(0, d_k)} p_{ij} c_{ij}$. Since the last job has due date no later than d_k, its tardiness is no less than $(t - d_k)^+$. Thus, total tardiness is no less than $(t - d_k)^+$. □

In the example of Table 4.4, suppose again that Jobs 1–3 are assigned to Facility 1. The bounds of Lemma 4.54 are

for $J(0, d_1) = \{1\}$: $\left(\frac{1}{3}(6) - 2\right)^+ = 0$
for $J(0, d_2) = \{1, 2\}$: $\left(\frac{1}{3}(6 + 8) - 3\right)^+ = 1\frac{2}{3}$
for $J(0, d_3) = \{1, 2, 3\}$: $\left(\frac{1}{3}(6 + 8 + 5) - 4\right)^+ = 2\frac{1}{3}$

Since the data are integral, the minimum tardiness is at least $\lceil 2\frac{1}{3} \rceil = 3$; it is 6 in the optimal schedule of Figure 4.28.

Lemma 4.54 gives rise to the following bounds, which can be added to a master problem in which T represents the total tardiness:

$$T \geq \sum_i T_i^L$$
$$T_i^L \geq \frac{1}{C_i} \sum_{j \in J(0,d_k)} p_{ij} c_{ij} x_{ij} - d_k, \quad \text{all } i, k \qquad (4.251)$$
$$T_i^L \geq 0, \quad \text{all } i$$

In the example, the relaxation (4.251) becomes

$$\begin{aligned}
& T \geq T_1^L + T_2^L \\
& T_1^L \geq 2x_{11} - 2, \quad T_1^L \geq 2x_{11} + \tfrac{8}{3}x_{12} - 3, \\
& T_1^L \geq 2x_{11} + \tfrac{8}{3}x_{12} + \tfrac{5}{3}x_{13} - 4, \\
& T_1^L \geq 2x_{11} + \tfrac{8}{3}x_{12} + \tfrac{5}{3}x_{13} + 2x_{14} - 5 \\
& T_2^L \geq 4x_{21} - 2, \quad L_2 \geq 4x_{11} + \tfrac{10}{3}x_{22} - 3, \\
& T_2^L \geq 4x_{21} + \tfrac{10}{3}x_{22} + 2x_{23} - 4, \\
& L_2 \geq 4x_{21} + \tfrac{10}{3}x_{22} + 2x_{23} + 5x_{24} - 5 \\
& T_1^L, T_2^L \geq 0
\end{aligned} \qquad (4.252)$$

A second type of bound can be developed on the basis of the following lemma. For each facility i, let π_i be a permutation of $\{1, \ldots, n\}$ that orders the jobs by increasing energy on facility i; that is, $p_{i\pi_i(1)} c_{i\pi_i(1)} \leq \cdots \leq p_{\pi_i(n)} c_{\pi_i(n)}$. One can obtain a lower bound on the tardiness incurred by the job with the kth latest deadline by supposing that it finishes no sooner than the task interval of the k jobs with the smallest energies. More precisely:

LEMMA 4.55 *Consider a minimum total tardiness problem in which jobs $1, \ldots, n$ with time windows $[r_j, d_j]$ are scheduled on a single facility i. Assume $\min_j\{r_j\} = 0$ and index the jobs so that $d_1 \leq \cdots \leq d_n$. Then the total tardiness T of any feasible solution is bounded below by $\underline{T} = \sum_{k=1}^n \underline{T}_k$, where*

$$\underline{T}_k = \left(\frac{1}{C_i} \sum_{j=1}^k p_{i\pi_i(j)} c_{i\pi_i(j)} - d_k \right)^+, \quad k = 1, \ldots, n$$

Proof. Consider any feasible solution of the one-facility minimum tardiness problem, in which jobs $1, \ldots, n$ are respectively scheduled at

Cumulative Scheduling 407

times $t_1, \ldots, t_n$. The minimum tardiness is

$$T^* = \sum_{k=1}^n (t_k + p_{ik} - d_k)^+ \qquad (4.253)$$

Let $\sigma_0(1), \ldots, \sigma_0(n)$ be the order in which jobs are scheduled in this solution, so that $t_{\sigma_0(1)} \leq \cdots \leq t_{\sigma_0(n)}$. For an arbitrary permutation σ of $\{1, \ldots, n\}$, let

$$\underline{T}_k(\sigma) = \left(\frac{1}{C_i} \sum_{j=1}^k p_{i\pi_i(j)} c_{i\pi_i(j)} - d_{\sigma(k)} \right)^+ \qquad (4.254)$$

and $\underline{T}(\sigma) = \sum_{k=1}^n \underline{T}_k(\sigma)$.

It will first be shown that $T^* \geq \underline{T}(\sigma_0)$. Since σ_0 is a permutation, one can write (4.253) as

$$T^* = \sum_{k=1}^n \left(t_{\sigma_0(k)} + p_{i\sigma_0(k)} - d_{\sigma_0(k)} \right)^+$$

Observe that

$$T^* \geq \sum_{k=1}^n \left(\frac{1}{C_i} \sum_{i=1}^k p_{i\sigma_0(j)} c_{i\sigma_0(j)} - d_{\sigma_0(k)} \right)^+$$

$$\geq \sum_{k=1}^n \left(\frac{1}{C_i} \sum_{j=1}^k p_{i\pi_0(j)} c_{\pi_0(j)} - d_{\sigma_0(k)} \right)^+ = \underline{T}(\sigma_0)$$

where the first inequality is based on the energy required by jobs, and the second inequality is due to the definition of π_i.

Now, perform a bubble sort on the integers $\sigma_0(1), \ldots, \sigma_0(n)$ to put them in increasing order, and let $\sigma_0, \ldots, \sigma_P$ be the resulting series of permutations. Thus $(\sigma_P(1), \ldots, \sigma_P(n)) = (1, \ldots, n)$, and σ_{p+1} is obtained from σ_p by swapping two adjacent terms $\sigma_p(k)$ and $\sigma_p(k+1)$, where $\sigma_p(k) > \sigma_p(k+1)$. This means σ_p and σ_{p+1} are the same, except that $\sigma_{p+1}(k) = \sigma_p(k+1)$ and $\sigma_{p+1}(k+1) = \sigma_p(k)$. Because $T^* \geq \underline{T}(\sigma_0)$ and $\underline{T}(\sigma_P) = \underline{T}$, to prove the theorem it suffices to show $\underline{T}(\sigma_0) \geq \cdots \geq \underline{T}(\sigma_P)$.

For any two adjacent permutations σ_p, σ_{p+1}, it can be shown as follows that $\underline{T}(\sigma_p) \geq \underline{T}(\sigma_{p+1})$. One can observe that

$$\underline{T}(\sigma_p) = \sum_{j=1}^{k-1} \underline{T}_j(\sigma_p) + \underline{T}_k(\sigma_p) + \underline{T}_{k+1}(\sigma_p) + \sum_{j=k+2}^{n} \underline{T}_j(\sigma_p)$$

$$\underline{T}(\sigma_{p+1}) = \sum_{j=1}^{k-1} \underline{T}_j(\sigma_p) + \underline{T}_k(\sigma_{p+1}) + \underline{T}_{k+1}(\sigma_{p+1}) + \sum_{j=k+2}^{n} \underline{T}_j(\sigma_p) \tag{4.255}$$

Using (4.254), note that $\underline{T}_k(\sigma_p) = (a - B)^+$, $\underline{T}_{k+1}(\sigma_p) = (A - b)^+$, $\underline{T}_k(\sigma_{p+1}) = (a - b)^+$, and $\underline{T}_{k+1}(\sigma_{p+1}) = (A - B)^+$ if one sets

$$a = \frac{1}{C_i} \sum_{j=1}^{k} p_{i\pi_i(j)} c_{i\pi_i(j)}, \quad A = \frac{1}{C_i} \sum_{j=1}^{k+1} p_{i\pi_i(j)} c_{i\pi_i(j)}$$

$$b = d_{\sigma_p(k+1)}, \quad B = d_{\sigma_p(k)}$$

Clearly, $a \leq A$. Also, $b \leq B$ because $\sigma_p(k) > \sigma_p(k+1)$ and $d_1 \leq \cdots \leq d_n$. From (4.255), one has

$$\underline{T}(\sigma_p) - \underline{T}(\sigma_{p+1}) = (a - B)^+ + (A - b)^+ - (a - b)^+ - (A - B)^+ \tag{4.256}$$

It is straightforward to check that this quantity is always nonnegative when $a \leq A$ and $b \leq B$. The theorem follows. □

In the example, suppose again that Jobs 1–3 are assigned to Facility 1. The permutation π_1 is $(\pi_1(1), \pi_1(2), \pi_1(3)) = (3, 1, 2)$. The lower bound $\underline{T}$ of Lemma 4.55 is $\underline{T}_1 + \underline{T}_2 + \underline{T}_3$, where

$$\begin{aligned} \underline{T}_1 &= \left(\tfrac{1}{3}(5) - 2\right)^+ = 0 \\ \underline{T}_2 &= \left(\tfrac{1}{3}(5+6) - 3\right)^+ = \tfrac{2}{3} \\ \underline{T}_3 &= \left(\tfrac{1}{3}(5+6+8) - 4\right)^+ = 2\tfrac{1}{3} \end{aligned} \tag{4.257}$$

The bound is $\underline{T} = 3$, which in this case is slightly stronger than the bound of $2\tfrac{1}{3}$ obtained from Lemma 4.54.

The bound of Lemma 4.55 can be written in terms of the variables x_{ik}:

$$\sum_{k=1}^{n} \underline{T}'_{ik} x_{ik}$$

where

$$\underline{T}'_{ik} \geq \frac{1}{C_i} \sum_{j=1}^{k} p_{i\pi_i(j)} c_{i\pi_i(j)} x_{i\pi_i(j)} - d_k, \quad k = 1, \ldots, n$$

and $\underline{T}'_{ik} \geq 0$. One can linearize the bound by writing it as

$$\sum_{k=1}^{n} \underline{T}_{ik} \tag{4.258}$$

where

$$\underline{T}_{ik} \geq \frac{1}{C_i}\sum_{j=1}^{k} p_{i\pi_i(j)} c_{i\pi_i(j)} x_{i\pi_i(j)} - d_k - (1-x_{ik})U_{ik}, \quad k=1,\ldots,n \tag{4.259}$$

and $\underline{T}_{ik} \geq 0$. The big-M term U_{ik} is given by

$$U_{ik} = \frac{1}{C_i}\sum_{j=1}^{k} p_{i\pi_i(j)} c_{i\pi_i(j)} - d_k$$

Note that, although U_{ik} can be negative, the right-hand side of (4.259) is never positive when $x_{ik} = 0$. Finally, to obtain a bound on total tardiness, sum (4.258) over all facilities and write

$$T \geq \sum_{i=1}^{m}\sum_{k=1}^{n} \underline{T}_{ik} \tag{4.260}$$

One can now add to the master problem a bound consisting of (4.251), (4.260), and (4.259) for $i = 1,\ldots,m$. The bound is valid only when jobs are indexed so that $d_1 \leq \cdots \leq d_n$. In the example, the bound consists of

$$\begin{aligned}
&T \geq \underline{T}_{11} + \underline{T}_{12} + \underline{T}_{13} + \underline{T}_{21} + \underline{T}_{22} + \underline{T}_{23} \\
&\underline{T}_{11} \geq \tfrac{5}{3}x_{11} - 2 + \tfrac{1}{3}(1 - x_{11}) \\
&\underline{T}_{12} \geq \tfrac{5}{3}x_{11} + 2x_{12} - 3 - \tfrac{2}{3}(1 - x_{12}) \\
&\underline{T}_{13} \geq \tfrac{5}{3}x_{11} + 2x_{12} + \tfrac{8}{3}x_{13} - 4 - \tfrac{7}{3}(1 - x_{13}) \\
&\underline{T}_{14} \geq \tfrac{5}{3}x_{11} + 2x_{12} + \tfrac{8}{3}x_{13} + 6x_{14} - 5 - \tfrac{22}{3}(1 - x_{14}) \\
&\underline{T}_{21} \geq 2x_{21} - 2 \\
&\underline{T}_{22} \geq 2x_{21} + \tfrac{10}{3}x_{12} - 3 - \tfrac{7}{3}(1 - x_{22}) \\
&\underline{T}_{23} \geq 2x_{21} + \tfrac{10}{3}x_{12} + 4x_{13} - 4 - \tfrac{16}{3}(1 - x_{23}) \\
&\underline{T}_{24} \geq 2x_{21} + \tfrac{10}{3}x_{12} + 4x_{13} + 5x_{24} - 5 - \tfrac{28}{3}(1 - x_{24}) \\
&\underline{T}_{ik} \geq 0
\end{aligned} \tag{4.261}$$

When the master problem contains the bounds (4.257) and (4.261), its solution assigns Jobs 1–3 to Facility 1 and Job 4 to Facility 2.

4.16.4 Exercises

1. Consider the constraint cumulative$(s\,|\,p,c,C)$ with $p = (2,1,1,2)$, $c = (1,2,3,2)$, and $C = 3$. The release times are $r = (0,0,0,1)$ and the deadlines $d = (5,5,5,5)$. Write all the valid, nonredundant inequalities given by Theorem 4.52 with $a = (1,1,1,1)$.

2. The problem is to minimize $\sum_j s_j$ subject to the cumulative constraint and time windows in Exercise 1. What is the optimal solution and its optimal value? What is the minimum value of $\sum_j s_j$ subject to the relaxation obtained in Exercise 1? The relaxation should include the bounds on s_j.

3. Show how the valid inequalities (4.244) and (4.245) in Theorem 4.52 are derived.

4. Verify that (4.246) follows from (4.244)–(4.245).

5. Theorem 4.52 obtains valid inequalities from a relaxation of the cumulative constraint that can be solved by a greedy algorithm. In particular, the relaxation replaces each time window $[r_j, d_j]$ with $[r_J, d_J]$ and allows p_j, c_j for each job to be replaced by any values p'_j, c'_j such that $p'_j c'_j = p_j c_j$. Another relaxation that can be solved in a greedy fashion replaces each $[r_j, d_j]$ with $[r_j, d_J]$ and allows p_j, c_j to be replaced by any values p'_j, c'_j for which $p'_j c'_j \geq \min_i \{p_i c_i\}$. Derive valid cuts from this relaxation. These cuts could be useful when the energies $p_j c_j$ of the jobs are about the same but the release times differ considerably. Symmetrical cuts can be derived by replacing $[r_j, d_j]$ with $[r_J, d_j]$.

6. Use Lemma 4.53 to identify the nonredundant inequalities (2.42) for Machine B in the example problem of Section 2.3.7.

7. Prove Lemma 4.53.

8. Suppose that four jobs are to be assigned to machines and scheduled so as to minimize the number of late jobs. The jobs have release times $r = (0,0,0,1)$ and deadlines $d = (2,2,2,2)$. On Machine 1, the job durations are $(p_{11}, \ldots, p_{14}) = (2,1,1,2)$, and the rates of resource consumption are $(c_{11}, \ldots, c_{14}) = (1,2,3,2)$. The resource capacity of Machine 1 is $C_1 = 3$. Write two nonredundant valid bounds on the minimum number of late jobs on Machine 1 in terms of 0-1 variables x_{1j}, where $x_{1j} = 1$ when job j is assigned to machine 1. If all four jobs are assigned to Machine 1, what is the lower bound on the number of late jobs? What is the actual minimum number of late jobs?

9 Suppose that all four jobs are assigned to Machine 1 in Exercise 8. Use Lemmas 4.54 and 4.55 to obtain lower bounds on the minimum total tardiness. What is the actual minimum?

10 Show that inequality (4.256) in the proof of Lemma 4.55 is valid.

11 Show that the right-hand side of (4.259) is nonpositive when $x_{ik} = 0$.

12 Exhibit an example in which Lemma 4.54 provides a tighter bound than Lemma 4.55.

4.17 Bibliographic Notes

Section 4.1. Relaxation duals, particularly the Lagrangean dual, have long been studied. The abstract concept of a relaxation dual is discussed in [185].

Section 4.2. The simplex algorithm for linear programming is due to George Dantzig [97, 98]. Good expositions may be found in [85] and [325], the latter of which also presents interior point methods for linear programming. An early discussion of surrogate relaxations is [148]. Linear programming duality is based on the Farkas Lemma and but is usually credited to John von Neumann, who formulated it in connection with the theory of noncooperative two-person games [99].

The convex hull of a semicontinuous piecewise linear function is used by [259, 260, 279] in an integrated problem-solving context. Fast methods for computing the convex hull in two dimensions can be found in [308] (pages 351–352) and [270].

Section 4.4. Theorem 4.3, which states that all valid inequalities for a 0-1 system are Chvátal-Gomory cuts, is due to Chvátal [82]. The cuts are named for Gomory as well due to his pioneering work in cutting plane theory [152, 154] and the fact that Gomory cuts are a special case of Chvátal-Gomory cuts (Theorem 4.8).

Cover inequalities for knapsack constraints originate in [15, 166, 262, 335]. The sequential lifting procedure presented here for cover inequalities is developed in [15, 166, 335] and summarized in [230]. The dual dynamic programming recursion for computing $h^*(t)$ is described in Section II.6.1 of [251] (Proposition 1.6). The approach to sequence-independent lifting described here (Theorem 4.4) is based on [164, 165, 335].

The odd cycle inequalities (Theorem 4.5) are shown in [261] to be facet-defining when the odd cycle is an odd hole. A polynomial-time separation algorithm can be based on Lemma 9.1.11 of [162]. The clique inequalities [138] are shown in [261] to be facet-defining when the clique is maximal. Separation algorithms are discussed in [63]. See the entry for the set packing constraint in Chapter 5.

Section 4.5. The completeness proof of Chvátal-Gomory cuts for general integer inequality systems (Theorem 4.7) appears in [82]. Gomory cuts originate in [152, 154]. Mixed integer rounding cuts (Theorems 4.9–4.11) are due to [231].

The equivalence of integrality and total unimodularity (Theorem 4.13) is proved in [178], and a simpler proof (essentially the one presented here) is provided in [327]. The necessary and sufficient condition for total unimodularity (Theorem 4.14) is proved in [143].

Lagrangean relaxation was first applied to integer programming in [170, 171], where the Lagrangean dual is solved by subgradient optimization. A good exposition of the application of Lagrangean duality to integer programming can be found in [125]. Algorithms for subgradient optimization are presented in [25, 250].

Section 4.7. The convex hull relaxation of a disjunction of linear systems (Theorem 4.17) is due to [16]. The tightened big-M model using (4.99) is described in [198], which gives a necessary and sufficient condition for its integrality. This article also points out that the solution of the big-M relaxation of a disjunction may be fractional, even when it satisfies the disjunction. The simplified big-M relaxation for a disjunction of individual linear inequalities (Theorem 4.18) originates with [33]. The procedure for tightening a big-M cut to a supporting cut (Theorem 4.19) is given in [198]. Disjunctive cuts, characterized by Theorem 4.21 [16], play a role in several areas of cutting plane theory. The perspective given here (Theorem 4.22) appears in [198].

Section 4.8. Theorem 4.23, which provides the basis for the convex hull relaxation of a disjunction of nonlinear systems, can be found in the convex analysis literature, e.g. [176]. The convex hull formulation (Theorem 4.24) is due to [223, 310].

Section 4.9. The necessary and sufficient condition for MILP representability (Theorem 4.25) is proved in [205], which proves a similar result for convex mixed nonlinear/integer programming. General discussions of MILP modeling can be found in [331, 332]. The package shipment problem of Section 4.9.4 is taken from [320].

Section 4.10. The connection between resolution and cutting planes (e.g., Theorem 4.28) is observed in [90, 179, 330]. The equivalence of unit refutation and linear programming on clause sets (Theorem 4.29) is remarked in [54]. The fact that input resolution generates the clauses that are rank 1 cuts (Theorems 4.31, 4.32, 4.36) is proved in [180]. The equivalence of input and unit refutation (Corollary 4.34), which follows from

Bibliographic Notes 413

Theorem 4.31, is given a purely logic-based proof in [78]. The characterization of separating resolvents (Theorem 4.37) appears in [197], which states an algorithm that identifies all separating resolvents.

Section 4.11. The convex hull relaxation of the element constraint (Theorem 4.38) is described in [185]. The simplified relaxation that results when all variables have the same bounds (Theorem 4.39) is also proved in [185].

Section 4.12. Theorem 4.44 describes the convex hull of an alldiff constraint in which the n variables have the same domain. The special case in which the domain size is n (i.e., Corollary 4.45) is proved in [185, 333].

Section 4.14. The popular 0-1 formulation of the traveling salesman problem given here is due to [100]. Comb inequalities were first described in a restricted form by [83]. The larger family of inequalities described here appears in [163]. A number of heuristics have been proposed for solving the separation problem for the subtour-elimination and comb inequalities. The max back heuristic described here is adapted from a similar heuristic for the symmetric problem described in [247, 248, 249]. Comprehensive discussions of polyhedral analysis of the traveling salesman problem include [21, 247]. Solution techniques for the problem have been intensively studied. One survey of this voluminous literature may be found in [208].

Section 4.16. The MILP model (4.238) for cumulative scheduling is similar to a model for the scheduling of chemical plants proposed in [74], and model (4.241) is suggested by a model for a related problem in [323].

The time-window relaxation for the Benders scheduling subproblem and Lemma 4.53 appear in [189]. The relaxations for minimizing the number of late jobs and for minimizing total tardiness (Lemmas 4.54 and 4.55) are from [191].

Chapter 5

DICTIONARY OF CONSTRAINTS

This book advocates modeling with metaconstraints, which are constraints that represent structured groups of more elementary constraints. As explained in Chapter 1, this allows modelers to write succinct formulations and to communicate the structure of the problem to the solver. Ideally, modelers would have at hand a menu of constraints, organized by problem domain, or in other ways that guide them to the right choice of constraints. This chapter provides a sampling of some metaconstraints that might appear on that menu.

There is no pretense of comprehensiveness. The list contains most of the better-known global constraints from the constraint programming world, and some of the major problem classes in the mathematical programming tradition. Generally, a global constraint is included only if filtering algorithms are discussed in the open literature. Some of the mathematical programming structures appear in the guise of global constraints, such as *flow* (capacitated network flows), *path* (shortest path), and *network design* (fixed-charge network flows). Others appear under global constraints because they are formulated in order to relax the constraint; thus, the assignment problem appears under *alldiff*, the traveling salesman problem under *circuit*, and the vehicle routing problem under *cycle*. Several of the constraints are known under multiple names. There is no attempt here to dictate a standard name but only to choose a suggestive name that is reasonably well known. Some constraints have variations and extensions that are mentioned in passing.

For brevity, the dictionary refers to a filter that achieves domain completeness as a *complete* filter. *Boolean* variables are variables whose two values can be identified with *true* and *false* in some natural way. To simplify notation, a list of variables may be shown as a tuple even when the order is not important. For example, the order of the variables $x = (x_1, \ldots, x_n)$ is important in element(y, x, z) but not in alldiff(x). When a constraint is written in the form *constraint-name*$(x, y \,|\, a, b)$, the symbols before the vertical bar are variables (or tuples of variables), and those after the bar are parameters (or tuples of parameters). This distinction is important, because achieving domain completeness becomes harder when parameters are treated as variables whose domains must be filtered.

When two or more instances of the same metaconstraint appear in a model, they may for convenience be grouped together under that name. For example, the two constraints alldiff(x) and alldiff(y) may be written

$$\text{Alldiff:} \begin{cases} (x) \\ (y) \end{cases}$$

The two constraints are nonetheless filtered and relaxed separately. To filter or relax them jointly, one must use a constraint designed for this purpose (the k-alldiff constraint). If there are two or more instances of a metaconstraint that is specified as a set of elementary constraints, they should be combined into a single metaconstraint. Thus if there are two *linear integer* metaconstraints, the inequalities they list should be merged to form a single integer linear metaconstraint (or if they are not, the modeling system should merge them automatically).

Examples of modeling with many of these constraints appear in Chapter 2 and elsewhere in the book. These passages are referenced below, as are numerous sources from the literature. In particular, see [37] for a catalog of global constraints used in the constraint programming community.

0-1 linear

A *0-1 linear* constraint consists of a system of linear inequalities or equations in 0-1 variables.

Notation. A 0-1 linear system can be written

$$Ax \geq b$$

where A is an $m \times n$ matrix and x is a tuple $(x_1, \ldots, x_n)$ of variables with domain $\{0, 1\}$. A system of equations $Ax = b$ is a special case because it can be written $Ax \geq b$, $-Ax \geq -b$.

Usage. 0-1 linear inequalities are a subset of mixed integer linear inequalities, which provide a versatile modeling language. Modeling principles and examples are presented in Section 4.9 as well as [331, 332].

Inference. Inferences may be drawn from each individual 0-1 inequality, or from the entire set considered jointly. Section 3.6.1 shows how to check for domination between individual knapsack inequalities, Section 3.6.2 shows how to derive implied logical clauses from a single inequality, and Section 3.6.3 shows how to derive implied cardinality clauses [29, 179]. 0-1 resolution [179, 181] is a complete inference method for sets of 0-1 linear inequalities considered jointly (Section 3.6.4). A weaker form of 0-1 resolution achieves k-completeness (Section 3.6.5), and a still weaker form achieves strong k-consistency (Section 3.6.6). Inference duality provides the basis for deducing Benders cuts [185, 199] (Section 3.7.3). One can use bounds propagation (Section 3.3.2) and dual multipliers for the continuous relaxation (Section 3.3.2) to fix variables.

Relaxation. A continuous relaxation can be obtained by replacing each $x_j \in \{0, 1\}$ with $0 \leq x_j \leq 1$. The relaxation of a single knapsack inequality can be strengthened by the addition of 0-1 knapsack cuts, also known as *cover inequalities* when generated for knapsack packing constraints (Section 4.4.2). They can often be strengthened by lifting techniques [15, 96] (Sections 4.4.3–4.4.4). Knapsack cuts and lifting are discussed in [230, 251, 338] and generalized in [122, 337]. Cutting planes can also be inferred from the system of inequalities considered jointly. They can all, in principle, be generated by Chvátal's procedure [82] (Section 4.4.1). General separating cuts, known as Gomory cuts [152, 154], can be derived as well (Section 4.5.2). Other general cuts include lift-and-project cuts [19]. There are a number of specialized cuts that presuppose special structure in the problem, some of which are discussed in [230, 251, 338].

Related constraints. There is also an *integer linear* constraint, in which the variables take general integer values.

All-different

The *all-different* (*alldiff*) constraint requires a set of variables to take distinct values.

Notation. The constraint is written

$$\text{alldiff}(x)$$

where x is a tuple $(x_1, \ldots, x_n)$ of variables with finite domains. It requires that each x_i take a distinct value. The constraint can also be formulated with a cost variable w:

$$\text{alldiff}(x, w \,|\, c) \tag{5.1}$$

where c is a matrix whose elements c_{ij} indicate the cost of assigning value j to x_i. Constraint (5.1) enforces alldiff(x) and $w = \sum_i c_{ix_i}$. This has been called the minimum weight alldiff [298].

Usage. The constraint is used in a wide variety of discrete models, particularly where assignment, scheduling, and allocation are involved. The classical assignment problem minimizes w subject to (5.1), as discussed in Section 4.12.3. The alldiff constraint is illustrated in Sections 2.2.5 (employee scheduling), 3.9 (job assignment), 3.11.1 (contrast with *circuit*, traveling salesman), 4.12 (use with *element* and variable indices), and 4.12.4 (quadratic assignment problem).

Inference. A complete filter for alldiff(x) based on maximum cardinality matching [93, 281] is described in Section 3.9.2. A complete filter for bounds completeness is based on finding Hall intervals [272, 226] or on the convexity of the graph [237], and the latter is presented in Section 3.9.3.

Relaxation. A convex hull relaxation in the original variables for alldiff with n-element numerical domains appears in [185, 333]. It is generalized to numerical domains of arbitrary but equal cardinality in Section 4.12.1. A convex hull relaxation for arbitrary domains of any cardinality can be written by adding $O(n^2)$ new variables (Section 4.12.2).

Related constraints. A set-valued alldiff constraint, alldiff($X_1, \ldots, X_n$), requires the sets $X_1, \ldots, X_n$ to be distinct and has been offered by commercial solvers. One can define a k-alldiff constraint that contains k alldiffs with some variables in common, which would be jointly filtered. Relaxations of k-alldiff are discussed in [7, 8]. Other generalizations of alldiff include *nvalues* and *cardinality*.

Among

The *among* constraint bounds the number of variables that take one of a given set of values.

Notation. Several forms of the constraint are described in [39], one of which is
$$\text{among}(x \mid v, \ell, u) \tag{5.2}$$
where x is a tuple $(x_1, \ldots, x_n)$ of variables, v a set of values, and ℓ and u are nonnegative integers. It requires that at least ℓ and at most u of the variables take values in the set v. Another form is
$$\text{among}(x \mid v, \ell, u, s) \tag{5.3}$$

Dictionary of Constraints

in which s is a positive integer. It requires that in any sequence of s consecutive variables, at least ℓ and at most u variables take a value in v.

Usage. Constraint (5.2) can be used in recourse-constrained sequencing models. Several product types are to be sequenced on an assembly line, and the product types in v require a certain resource. One can specify that at most u products using the resource will be processed at a certain station during a given time period. Let x_i be the product type assigned to slot i of the sequence, and let $x_1, \ldots, x_n$ represent the slots that are to be processed at the station during the time period. Such constraints may be imposed for several resources and stations. Constraint (5.3) can prevent uneven usage of a resource over time.

Related constraints. The *cardinality* constraint is similar to (5.2).

Bin packing

The (one-dimensional) *bin packing* constraint require that a given set of items be packed into bins so that the weight in no bin exceeds a given limit.

Notation. The constraint is

$$\text{bin-packing}(x \mid w, u, k) \tag{5.4}$$

where x is an n-tuple of variables x_i with domain $\{1, \ldots, k\}$ indicating which bin holds item i, w an n-tuple of item weights, u the bin capacity (same for every bin), and k the number of bins. The constraint requires that

$$\sum_{\substack{i \\ x_i = j}} w_i \leq u, \quad j = 1, \ldots, k$$

The number k of bins is treated as a variable in

$$\text{bin-packing}(x, k \mid w, u) \tag{5.5}$$

The classical bin packing problem minimizes k subject to (5.5). One can also treat u as a variable

$$\text{bin-packing}(x, u \mid w, k) \tag{5.6}$$

and minimize the maximum load u subject to (5.6).

Usage. Many resource-allocation problems have a bin packing component. The problem has received attention in the operations research community at least since 1974 [206].

Inference. Achieving domain completeness is NP-hard. Incomplete filtering algorithms for (5.4) are discussed in [246, 243, 244, 245, 304].

Relaxation. MILP-based and other relaxations are discussed in [232, 233, 104]. There are many approximation algorithms for the problem [87].

Related constraints. Constraint (5.4) is a special case of the *cumulative scheduling* constraint with all processing times equal to one. See also the *0-1 linear* constraint.

Cardinality

The *cardinality* (*distribute, gcc, generalized cardinality*) constraint bounds the number of variables that take each of a given set of values.

Notation. The constraint is written

$$\text{cardinality}(x \mid v, \ell, u) \tag{5.7}$$

where x is a tuple $(x_1, \ldots, x_n)$ of variables, v an m-tuple of values, and ℓ, u are m-tuples of nonnegative integers. The constraint requires, for $j = 1, \ldots, m$, that at least ℓ_j, and at most u_j, of the variables take the value v_j. Another form of the constraint treats the number of occurrences as a variable:

$$\text{cardinality}(x, y \mid v) \tag{5.8}$$

where $y = (y_1, \ldots, y_m)$. This constraint says that exactly y_j variables must take the value v_j. The number of occurrences of v_j can be bounded by bounding y_j.

Usage. This is a highly generic constraint that often finds application when there are bounds on cardinalities. It is illustrated in Section 2.2.5 (employee scheduling).

Inference. The simplest complete filtering algorithm for (5.7) [282] is based on a network flow model and is presented in Sections 3.10.2–3.10.3. Some improved algorithms appear in [273], which shows that achieving domain completeness for (5.8) is NP-hard. Bounds consistency algorithms for (5.7) generalize those for alldiff. One [274] finds Hall intervals, and another [210] exploits convexity of the graph. The algorithm in [210] also computes bounds completeness for (5.8).

Relaxation. A convex hull relaxation in the original variables for cardinality with numerical domains of equal cardinality is presented in

Section 4.13.1. A convex hull relaxation for arbitrary domains can be written by adding $O(n^2)$ new variables (Section 4.13.2).

Related constraints. The very similar *among* constraint can be used in resource-constrained sequencing models. At-least and at-most constraints have also been defined to bound the number of occurrences of a given value in a set of variables. Cardinality-atleast and cardinality-atmost constraints bound the number of times the values in a given set of value occur in a set of variables.

Cardinality clause

A *cardinality clause* specifies that at least a certain number of boolean variables must be true.

Notation. The constraint can be written

$$\sum_{j=1}^{n} x_j \geq k$$

where $x_1, \ldots, x_n$ are boolean variables and k is a nonnegative integer. It requires that at least k of the variables be true. If desired, the constraint can be generalized to

$$\sum_{j=1}^{n} L_j \geq k$$

where each L_j is a literal x_j or $\neg x_j$, to require that at least k literals be true. This constraint can also be written as a 0-1 inequality

$$ax \geq k + (a) \tag{5.9}$$

where each $a_j \in \{0, 1, -1\}$ and $n(a)$ is the sum of the negative components of a. Setting x_j to 1 or 0 corresponds to making x_j true or false, respectively.

Usage. Cardinality clauses allow cardinality conditions to be expressed in a quasi-logical language for which inference is much easier than general 0-1 inequalities.

Inference. It is easy to check when a 0-1 inequality implies a cardinality clause, or when one cardinality clause implies another [179] (Section 3.6.3). There is a method for generating all cardinality clauses implied by a 0-1 linear system [29].

Relaxation. The inequality (5.9), together with $x_j \in [0, 1]$ for all j, is a convex hull representation of the cardinality clause it represents.

Related constraints. The cardinality clause becomes a logical clause when $k=1$ and is generalized by a *cardinality conditional*.

Cardinality conditional

A *cardinality conditional* specifies that, if at least k variables in one set of boolean variables are true, then at least ℓ of another set must be true.

Notation. The constraint can be written

$$\left(\sum_{i=1}^{m} x_i \geq k\right) \rightarrow \left(\sum_{j=1}^{n} y_j \geq \ell\right)$$

where $x_1, \ldots, x_n$ and $y_1, \ldots, y_m$ are boolean variables and k, ℓ are integers. It states that if at least k of the variables x_i are true, then at least ℓ of the variables y_j must be true. If desired, one can permit negated variables as in a cardinality clause.

Usage. A cardinality conditional allows one to express a conditional of this kind conveniently without using the general *conditional* constraint and thereby losing the special structure. A large number of 0-1 inequalities may be required to capture the meaning.

Relaxation. A convex hull relaxation is given in [340]. This result is generalized in [18].

Related constraints. The cardinality conditional becomes a *cardinality clause* when $k=0$.

Change

The *change* constraint counts the number of times a given type of change occurs in a sequence of variables.

Notation. The constraint is

$$\text{change}(x, k \,|\, \text{rel})$$

where $x = (x_1, \ldots, x_n)$, k is an integer-valued variable, and *rel* is a binary relation such as $=, \neq, \leq, \geq, <,$ or $>$. The constraint requires that k be the number of times two consecutive variables x_i, x_{i+1} satisfy x_i rel x_{i+1}.

Usage. The constraint can be used in employee scheduling and other timetabling problems to constrain the number of times a change in shift or work assignment occurs.

Inference. Filtering algorithms are discussed in [35].

Related constraints. The *stretch* constraint can limit the type of changes that occur.

Circuit

The *circuit* constraint describes a hamiltonian cycle on a directed graph.

Notation. The constraint is
$$\text{circuit}(x)$$
where $x = (x_1, \ldots, x_n)$ is a tuple of variables whose domains are subsets of $\{1, \ldots, n\}$. The constraint requires that $y_1, \ldots, y_n$ be a permutation of $1, \ldots, n$, where each $y_{i+1} = x_{y_i}$ and y_{n+1} is identified with y_1. The constraint can be viewed as describing a hamiltonian cycle on a directed graph G that contains an edge (i, j) if and only if j belongs to the domain of x_i. An edge (i, j) of G is selected when $x_i = j$, and *circuit* requires that the selected edges form a hamiltonian cycle. One can add a cost variable w to obtain the constraint

$$\text{circuit}(x, w \,|\, c) \tag{5.10}$$

where c is a matrix of cost coefficients c_{ij} that indicate the cost incurred when j immediately follows i in a hamiltonian cycle. Constraint (5.10) enforces circuit(x) and $w = \sum_i c_{ix_i}$.

Usage. Section 3.11.1 discusses modeling with *circuit*. The traveling salesman problem minimizes w subject to (5.10) (Section 3.1.2).

Inference. Achieving domain completeness for circuit(x) is NP-hard. Section 3.11.2 describes some elementary incomplete filtering methods, which are related to those in [73, 306]. Sections 3.11.3–3.11.4 strengthen the filter by analyzing a separator graph [211].

Relaxation. The circuit constraint can be relaxed by adding 0-1 variables and writing valid inequalities for the traveling salesman problem, as described in Sections 4.14.1–4.14.3. Valid inequalities, facet-defining inequalities, and separating cuts for this problem have been intensively studied; see [21, 247] for surveys.

Related constraints. The *cycle* constraint generalizes *circuit* by specifying the number of subtours allowed.

Clique

A *clique* constraint requires that a given graph contain a clique of a specified size.

Notation. One way to write a clique constraint is

$$\text{clique}(x, k \,|\, G) \qquad (5.11)$$

where x is an n-tuple of boolean variables, k is the number of true x_js, and G is an undirected graph. The constraint requires that the set V of vertices i for which $x_i = true$ induce a clique of G (i.e, every pair of vertices in V is connected by an edge of G). The *maximum clique* problem maximizes k subject to (5.11).

Usage. The maximum clique problem has received much attention and has seen applications in coding theory, fault diagnosis, pattern recognition, cutting plane generation, and other areas; see [61] for a survey.

Inference. Achieving domain completeness is NP-hard. Incomplete filtering algorithms are presented in [119, 284].

Relaxation. MILP-based and other relaxations are described in [20, 22].

Related constraints. See the *set packing* constraint.

Conditional

A *conditional* constraint states conditions under which a given constraint set is enforced.

Notation. A conditional constraint has the form

$$\mathcal{D} \Rightarrow \mathcal{C}$$

where $\mathcal{D}$ and $\mathcal{C}$ are constraint sets. It states that the constraints in $\mathcal{C}$ must hold if the constraints in $\mathcal{D}$ hold. If the conditional is to be implemented in practice, the set $\mathcal{D}$ must be limited to constraints for which the solver can easily check whether *all* solutions belonging to the current domains satisfy $\mathcal{D}$. One option is to restrict $\mathcal{D}$ to domain constraints (constraints of the form $x_j \in D$) or logical propositions in which the atoms are domain constraints, such as $(x_1 \in D_1 \wedge x_2 \in D_2) \to x_3 \notin D_3$. A conditional constraint should be distinguished from a logical conditional $F \to G$, which is a formula of propositional logic (Section 3.5) and can be used in a *logic* constraint.

Usage. Conditional constraints are a very convenient device but can be overused (see below).

Inference. The solver posts the constraints in $\mathcal{C}$ whenever the search process reduces domains to the point that $\mathcal{D}$ is satisfied by any set of values belonging to the current domains. Logical conditionals should not be written as conditional constraints, because this prevents the application of inference algorithms designed especially for propositional logic.

Relaxation. Conditionals should not be used when a more specialized metaconstraint is available. For example, a set of conditionals in which the consequent $\mathcal{C}$ is a linear system can often be re-expressed as one or more linear disjunctions that exploit the problem structure for relaxation purposes. This is illustrated by the production planning problem of Section 2.2.4.

Related constraints. See the discussion of *linear disjunctions*.

Cumulative scheduling

A *cumulative scheduling* constraint requires that jobs be scheduled so that the total rate of resource consumption at no time exceeds a given limit.

Notation. The constraint is written

$$\text{cumulative}(s \mid p, c, C)$$

where $s = (s_1, \ldots, s_n)$ is a tuple of real-valued[1] variables s_j representing the start time of job j. The parameter $p = (p_1, \ldots, p_n)$ is a tuple of processing times for each job, $c = (c_1, \ldots, c_n)$ resource consumption rates, and C a limit on total resource consumption rate at any one time. The constraint requires that the total rate of resource consumption of the jobs underway at any time t be at most C:

$$\sum_{\substack{j \\ s_j \leq t \leq s_j + p_j}} c_j \leq C, \quad \text{for all times } t$$

The current domains $[L_j, U_j]$ of the start time variables s_j impose release times L_j and deadlines $U_j + p_j$.

Variations of cumulative scheduling allow *preemptive* scheduling (in which one job can interrupt another), multiple resources, and resource limits that are variable over time.

---

[1] These variables are traditionally integer-valued in constraint programming systems, but this is not necessary in an integrated solver.

Usage. The cumulative scheduling constraint is widely used for scheduling tasks subject to one or more resource constraints. It is illustrated in Sections 3.14 and 4.16.

Inference. Achieving bounds completeness is NP-hard. Incomplete filtering methods include timetabling methods [133, 264, 309], edge finding [72, 255, 256] (Section 3.14.1), extended edge finding [254, 24] (Section 3.14.2), not-first/not-last rules [255, 256] (Section 3.14.3), and energetic reasoning [116, 117] (Section 3.14.4). These and other methods are described in [24]. Generation of logic-based Benders cuts [185, 190, 191] is described in Section 3.14.5.

Relaxation. The *cumulative* constraint can be given MILP relaxations (Section 4.16.1) or relaxations in the original variables (Section 4.16.2). When it appears as the subproblem of a Benders formulation, there are relaxations for minimizing cost and makespan [190, 196], as well as number of late jobs and total tardiness [191, 196] (Section 4.16.3).

Related constraints. When c is a tuple of ones and $C = 1$, the cumulative scheduling constraint becomes a *disjunctive scheduling* constraint.

Cutset

A *cutset* constraint requires that a set of vertices cut all directed cycles in a graph.

Notation. The cutset constraint is

$$\text{cutset}(x, k \mid G) \tag{5.12}$$

where x is an n-tuple of boolean variables, k is the number of true x_js, and G is a directed graph. The constraint requires that the set V' of vertices i for which $x_i = true$ be a cutset; that is, if V is the set of vertices in G, then $V \setminus V'$ induces a subgraph of G that contains no cycles. The *minimum cutset* problem minimizes k subject to (5.12).

Usage. The minimum cutset problem has applications in deadlock breaking, program verification, and Bayesian inference [289, 303, 341].

Inference. Filtering algorithms are presented in [118].

Cycle

The *cycle* constraint [39] specifies the number of subtours that must cover a directed graph.

Notation. The constraint is

$$\text{cycle}(y, x)$$

Dictionary of Constraints 427

where y is an integer-valued variable and x a tuple $(x_1, \ldots, x_n)$ with $D_{x_i} \subset \{1, \ldots, n\}$ for each i. Let G be a graph with vertices $1, \ldots, n$, in which (i, j) is an edge if and only if $j \in D_{x_i}$. The variables x select an edge (i, j) when $x_i = j$. The constraint requires that x select edges that form exactly y directed cycles, such that every vertex belongs to exactly one cycle.

Usage. The cycle constraint can be used in vehicle routing problems where x_i is the city a vehicle visits immediately after city i. The constraint requires that exactly y vehicles cover all the cities.

Inference. The elementary filtering methods for *circuit* in Section 3.11.2 can be adapted to *cycle*.

Relaxation. The cycle constraint can be given an integer programming model for vehicle routing, which can in turn be relaxed by dropping integrality. Formulations include the two-index, three-index, set partitioning, and multicommodity flow models; several are surveyed in [91, 124, 318].

Related constraints. The cycle constraint becomes *circuit* when y is fixed to 1.

Diffn

The *diffn* constraint arranges a given set of multidimensional boxes in n-space so that they do not overlap.

Notation. The basic diffn constraint [3, 39] is

$$\text{diffn}((x^1, \Delta x^1), \ldots, (x^m, \Delta x^m))$$

where each x^i is an n-tuple of variables indicating the coordinates of one corner of the ith box to be placed, and Δx^i is an n-tuple of variables indicating the size of the box along each dimension. The constraint requires that for every pair of boxes i, i', there is at least one dimension j such that $x_j^{i'} \geq x_j^i + \Delta x_j^i$ or $x_j^i \geq x_j^{i'} + \Delta x^{i'}$. Bounds on the corner-positioning and size of the boxes are implicit in the initial variable domains. More elaborate versions of the constraint allow one to bound the volume of each box, to bound the area within which each box may lie (which is slightly different than bounding x^i and Δx^i), and to bound the distance between boxes. The constraint has been generalized to convex polytopes [40, 286].

Usage. The constraint is used for space or space-time packing. The two-dimensional version is used for resource-constrained scheduling, in which each box represents a job whose horizontal dimension is the processing time and vertical dimension is the rate of resource consumption.

Inference. A sweep algorithm for filtering for the two-dimensional case appears in [36]. Propagation for the general case is discussed in [40] and generalized to convex polytopes in [40, 286].

Related constraints. The two-dimensional ndiff constraint is closely related to *cumulative scheduling*.

Disjunctive scheduling

A *disjunctive scheduling* constraint requires that jobs be scheduled sequentially without overlapping.

Notation. A disjunctive scheduling constraint has the form

$$\text{disjunctive}(s \mid p)$$

where $s = (s_1, \ldots, s_n)$ is a tuple of real-valued[2] variables s_j indicating the start time of job j. The parameter $p = (p_1, \ldots, p_n)$ consists of processing times for each job. The constraint enforces

$$(s_i + p_i \leq s_j) \vee (s_j + p_j \leq s_i)$$

for all i, j with $i \neq j$. The current domains $[L_j, U_j]$ of the start time variables s_j impose release times L_j and deadlines $U_j + p_j$. A variation of the constraint allows for *preemptive scheduling*, in which one job may interrupt another.

Usage. The disjunctive scheduling constraint is widely used for scheduling tasks sequentially (i.e., under a *unary resource* constraint). It is illustrated in Section 2.3.7 (machine scheduling).

Inference. Achieving bounds completeness is NP-hard. Incomplete filtering methods are based on timetabling, edge finding, and not-first/not-last rules. Timetabling filters appear in [133, 264, 309]. Edge finding originated in [69]. An $O(n^2)$ edge finding algorithm (where n is the number of jobs) based on the Jackson preemptive schedule (Section 3.13.1) is given in [70]. Another $O(n^2)$ algorithm appears in [24, 254, 257]. An algorithm that achieves $O(n \log n)$ complexity with complex data structures is given in [71], and an $O(n^3)$

---

[2]These variables are traditionally integer-valued in constraint programming systems, but this is not necessary in an integrated solver.

Dictionary of Constraints 429

algorithm that allows incremental updates in [72]. Extensions that take setup times into account are presented in [66, 130]. Propagation algorithms for not-first/not-last rules appear in [23, 110, 317] (Section 3.13.2). A comprehensive treatment of scheduling is [24]. Generation of logic-based Benders cuts [185, 203, 190, 191] is described in Section 3.13.3.

Relaxation. The disjunctive scheduling constraint can be given disjunctive relaxations (Section 4.15.1) or MILP relaxations (Section 4.15.2). When it appears as the subproblem of a Benders formulation, there are relaxations for minimizing cost and makespan [190, 196] as well as number of late jobs and total tardiness [191, 196] (Section 4.16.3).

Related constraints. The *cumulative scheduling* constraint allows jobs to be scheduled in parallel, subject to a resource limit.

Element

The *element* constraint selects a specified value from a list and assigns it to a variable.

Notation. The simplest form of the constraint is written

$$\text{element}(y, z \mid a) \tag{5.13}$$

where y is a variable representing a positive integer, z is a variable, and a is a tuple $(a_1, \ldots, a_m)$ of values. The constraint requires z to take the yth value in the tuple. Another form is

$$\text{element}(y, x, z) \tag{5.14}$$

where x is a tuple $(x_1, \ldots, x_m)$ of variables. The constraint requires z to take the same value as the yth variable in the tuple. The constraint becomes multidimensional when index variable y is a p-tuple, in which case a and x become p-dimensional arrays.

A *vector-valued element* constraint

$$\text{element}(y, z \mid (a^1, \ldots, a^m)) \tag{5.15}$$

requires the tuple $z = (z_1, \ldots, z_m)$ of variables to take the values in the tuple a^y.

Usage. The element constraint is very useful for implementing *variable indices* (*variable subscripts*). An expression of the form a_y, in which y is a variable index, is processed by replacing it with z and adding constraint (5.13). An expression of the form x_y, where x is a variable, can be similarly processed with constraint (5.14). When y is a tuple representing multiple indices, a multidimensional element is used.

Illustrations of *element* appear in Sections 2.2.5 (employee scheduling), 2.2.7 (product configuration), and 4.12 (job assignment). Examples of the vector-valued constraint (5.15) are given in Section 4.11.3 (production planning) and 4.12.4 (quadratic assignment problem).

Inference. Filtering for domain and bounds completeness for (5.13) and (5.14) is straightforward. Domain completeness is discussed in Section 3.8.1 and bounds completeness in Section 3.8.2. Filtering for the multidimensional constraint is the same as for the one-dimensional constraint.

Relaxation. Relaxation of (5.13) is trivial (Section 4.11.1). Relaxation of (5.14) is based on relaxations for a disjunction of linear systems. Section 4.11.1 describes the convex hull relaxation and Section 4.11.2 a big-M relaxation. Relaxation of the multidimensional constraint is the same as for the one-dimensional constraint. Relaxation of the vector-valued constraint is discussed in Sections 2.2.7 and 4.11.3.

Related constraints. The indexed linear element constraint

$$\text{element}(y, (A_1 x, \ldots, A_m x), z)$$

is used in the *indexed linear* metaconstraint and is discussed in connection with it. The constraint also has a vector-valued form. The *sum* constraint implements index sets with variable indices.

Flow

The *flow* constraint requires flow conservation on a capacitated network and computes variable costs.

Notation. One way to notate the constraint is

$$\text{flow}\,(x, z \mid N, s, A, \ell, u, c) \qquad (5.16)$$

where x is a tuple of real-valued flow variables x_{ij}, and z is a real-valued variable representing cost. N is a tuple of nodes, and s is a tuple of net supply values corresponding to the nodes in N. A is a tuple of directed arcs, and ℓ, u, c are tuples of lower flow bounds, upper flow bounds, and unit costs corresponding to the arcs in A.

Dictionary of Constraints 431

The constraint requires that

$$z = \sum_{(i,j) \in A} c_{ij} x_{ij}$$

$$\sum_{(i,j) \in A} x_{ij} - \sum_{(j,i) \in A} x_{ji} = s_i, \text{ all } i \in N$$

$$\ell_{ij} \leq x_{ij} \leq u_{ij}, \text{ all } (i,j) \in A$$

The classical minimum-cost network flow problem minimizes z subject to (5.16). A slight generalization of the model allows *gains*, which specify that the flow leaving (i, j) is α_{ij} times the flow entering it. There is a multicommodity version of the problem, but it does not enjoy a very fast specialized solution algorithm as does the single-commodity problem.

Usage. The capacitated network flow model has countless applications [4, 5, 30]. It is also used to filter several constraints, such as *cardinality* (Section 3.10.2) and *circuit* (Section 3.11.4). For other constraint programming applications, see [59].

Related constraints. The *network design* constraint selects the arcs that will belong to the network and charges a fixed cost for each.

Indexed linear

The *indexed linear* constraint is used to implement linear inequalities in which the coefficients are indexed by variables.

Notation. The constraint has the form

$$\sum_{i \in I} z^i + \sum_{i \in I'} A_i x_i \geq b \qquad (a)$$

$$\text{element}\left(y_i, x_i, z^i \mid A_i\right), \quad i \in I \qquad (b)$$

(5.17)

where each x_i is a real-valued variable, each z^i is a tuple of real-valued variables, each y_i is a variable representing a positive integer, and each A_i is a tuple of the same length as z^i. Each element constraint i requires that z^i be equal to $A_{iy_i} x_i$. This form of the element constraint is an *indexed linear element* constraint.

Usage. The indexed linear constraint is used to implement linear inequality systems of the form

$$\sum_{i \in I} A_{iy_i} x_i + \sum_{i \in I'} A_i x_i \geq b \qquad (5.18)$$

It is more efficient to define a single constraint of this structure than simply to replace each term $A_{iy_i}x_i$ in (5.18) with z^i and write the element constraint (5.17b) for each. A single constraint allows the solver to generate cover inequalities for the knapsack constraint (5.17a) on the basis of information about the coefficients in the element constraints, as explained in Section 2.2.7. Normally, an indexed linear constraint would not occur explicitly in a model but would be generated automatically by the modeling system when a constraint of the form (5.18) is present. A model that uses the constraint is presented in Section 2.2.7 (product configuration).

Inference. Filtering is described in Sections 3.8.1–3.8.2.

Relaxation. Relaxation methods appear in [315] and Sections 4.11.1–4.11.2.

Related constraints. Other forms of the *element* constraint are listed under that entry.

Integer linear

An *integer linear* constraint consists of a system of linear inequalities or equations in integer-valued variables.

Notation. An integer linear system can be written

$$Ax \geq b$$

where A is an $m \times n$ matrix and x is a tuple $(x_1, \ldots, x_n)$ of variables whose domain is the set of integers. A system of equations $Ax = b$ is a special case because it can be written $Ax \geq b, -Ax \geq -b$.

Usage. Integer linear inequalities are a subset of mixed integer linear inequalities, which provide a versatile modeling language. Modeling principles and examples are presented in Section 4.9 as well as [331, 332].

Inference. Bounds propagation for integer linear inequalities is the same as for linear inequalities (Section 3.3.2), except that bounds can be rounded to the appropriate integer value. Dual multipliers for the continuous relaxation (Section 3.3.8) can be used to fix variables, again with rounding. Inference duality provides the basis for deducing Benders cuts [185, 199] (Section 3.7.3).

Relaxation. A continuous relaxation can be obtained by dropping the integrality requirement. The relaxation of an individual inequality can be strengthened by the addition of integer knapsack cuts [14]

(Section 2.2.3). They, in turn, can often be strengthened by lifting techniques [13, 165]. Cutting planes can also be inferred from the system of inequalities considered jointly. All valid cuts can, in principle, be generated by Chvátal's procedure [82] (Section 4.5.1). General separating cuts, known as Gomory cuts [152, 154], can be derived as well (Section 4.5.2). For a general discussion of cutting planes, see [137, 230, 251, 338]. Another approach to relaxation is based on group theory and Gröbner bases [155, 156, 311], surveyed in [1, 313].

Related constraints. There is also a *0-1 linear* constraint, in which the variables take 0-1 values.

Lex-greater

The *lex-greater* constraint requires that one tuple of variables be lexicographically greater than another.

Notation. The constraint is
$$x >_{\text{lex}} y$$
where x, y are n-tuples of variables whose domains are subsets of a totally ordered set. The constraint requires that x be lexicographically greater than y; that is, $x \geq_{\text{lex}} y$, and $x_n > y_n$ if $x_i = y_i$ for $i = 1, \ldots, n-1$. The relation $x \geq_{\text{lex}} y$ holds if and only if $x_1 \geq y_1$ and, for $i = 2, \ldots, n$, $x_i \geq y_i$ whenever $x_j = y_j$ for all $j < i$.

Usage. The constraint is used for symmetry breaking in matrices of variables [126] as, for example, in sports scheduling [175], and for multicriteria optimization.

Inference. Complete filtering algorithms are given in [136, 214].

Related constraints. Lex-greater is one of several lex ordering constraints [136].

Linear disjunction

A *linear disjunction* is a disjunction of linear systems.

Notation. The constraint is written
$$\bigvee_{k \in K} \begin{bmatrix} y_k \\ A^k x \geq b^k \end{bmatrix} \tag{5.19}$$

where x is a tuple $(x_1, \ldots, x_n)$ of real-valued variables and each y_k is a boolean variable. The constraint enforces

$$\bigvee_{k \in K} y_k \quad (a)$$
$$y_k \to A^k x \geq b^k, \text{ all } k \in K \quad (b) \tag{5.20}$$

Usage. It is common for linear systems to be enforced under certain conditions y_k, as expressed by conditionals like those in (5.20b). When a disjunction like (5.20a) is known to hold, one can obtain a much stronger relaxation by writing a single linear disjunction (5.19), rather than the individual constraints in (5.20). Disjunctions such as (5.20a) can be derived by computing the prime implications of the logical constraints in the problem. Section 2.2.4 illustrates the idea with a production planning example.

Relaxation. The constraint (5.19) implies

$$\bigvee_{\substack{k \\ 1 \in D_{y_k}}} A^k x \geq b^k$$

which can be given a convex hull relaxation [16] (Section 4.7.1) or a big-M relaxation (Section 4.7.2). There are specialized relaxations for the case where each disjunct is a single inequality [33] (Section 4.7.3) or a single equation (Section 4.7.4). Separating cuts can also be generated [198] (Section 4.7.5).

Related constraints. Certain *nonlinear disjunctions* can be given convex hull relaxations.

Logic

A *logic* constraint consists of logical clauses. The conjunction of the clauses is a propositional formula in conjunctive normal form. The constraint can be extended to allow other types of logical propositions as well.

Notation. A logic constraint has the form

$$\bigvee_{j \in J_i} x_j \vee \bigvee_{j \in \bar{J}_i} \neg x_j, \quad i \in I$$

where each x_j is a boolean variable in the set $\{x_1, \ldots, x_n\}$. Each $i \in I$ corresponds to a logical *clause* that contains the literals x_j for $j \in J_i$ and $\neg x_j$ (not x_j) for $j \in \bar{J}_i$. The clause requires that at least one of the literals be true.

Dictionary of Constraints   435

It may be convenient to allow all formulas of propositional logic, and not just logical clauses, to appear in a logic constraint, because they can be automatically converted to conjunctive normal form, as described in Section 3.5.1.

Usage. The logic constraint expresses logical relations between variables, generally boolean variables that take the values F, T or $0, 1$. The constraint is illustrated in Section 2.3.2 (employee staffing).

Inference. The resolution method is a complete inference method for general clause sets and therefore achieves domain completeness (Sections 3.5.2 and 3.5.4). Unit resolution does the same for Horn sets (Section 3.5.3). k-resolution achieves strong k-consistency on general clause sets (Section 3.5.5). Inference duality permits the deduction of nogoods (conflict clauses) from a clause set (Section 2.3.2). Parallel resolution [185] makes the nogood set easy to satisfy (Sections 2.3.4 and 3.5.6).

Relaxation. Convex hull relaxations of common logical formulas [195] appear in Section 4.10.1. General clause sets can be relaxed by dropping the integrality constraints from a 0-1 formulation that is strengthened by resolvents (Section 4.10.2). In particular, input resolution generates all clausal rank 1 cuts [180] (Section 4.10.4). Section 4.10.5 describes how to generate separating resolvents [197].

Related constraints. Variables that appear in a logic constraint are often related to other constraints by means of *conditional, disjunctive linear,* or *disjunctive nonlinear* constraints. For example, to enforce at least one of the constraints C_1, C_2, C_3, one can write $x_1 \vee x_2 \vee x_3$ and the conditionals $x_i \rightarrow C_i$ for $i = 1, 2, 3$. If each C_i is a linear system, the conditionals should be replaced by the disjunctive linear constraint
$$\begin{bmatrix} x_1 \\ C_1 \end{bmatrix} \vee \begin{bmatrix} x_2 \\ C_2 \end{bmatrix} \vee \begin{bmatrix} x_3 \\ C_3 \end{bmatrix}$$
and similarly if the C_is are nonlinear systems.

Logical clauses can be generalized to include disjunctions of the form $\bigvee_{j \in J}(x_j \in D_j)$, where x_j is any finite domain variable. A complete multivalent resolution method for such clauses is described in [185, 198].

MILP

An *MILP* or *mixed integer linear programming* constraint consists of a system of linear inequalities or equations in which at least some of the variables must take integer values.

Notation. An MILP constraint can be written

$$Ax + By \geq b$$

where A and B are matrices, x is a tuple of integer-valued variables, and y a tuple of real-valued variables.

Usage. Mixed integer linear inequalities provide a versatile modeling language. Modeling principles and examples are presented in Section 4.9 as well as [331, 332]. Many constraints can be relaxed by formulating them as an MILP constraint and taking its continuous relaxation.

Inference. Bounds propagation (Section 3.3.2) and dual multipliers for the continuous relaxation (Section 3.3.8) can be used to reduce domains.

Relaxation. A continuous relaxation can be obtained by dropping the integrality requirement. The relaxation can be strengthened by the addition of Gomory's mixed integer cuts [153], mixed integer rounding cuts [231] (Sections 4.5.3–4.5.4), and lift-and-project cuts [19], among others. For a general discussion of cutting planes, see [137, 230, 251, 338].

Related constraints. See the *integer linear* and *0-1 linear* constraints for additional inference and relaxation techniques.

Min-n

The *min-n* constraint selects the rth smallest value taken by a set of variables.

Notation. The constraint

$$\text{min-}n(x, v \mid r)$$

requires that v be the rth smallest distinct value taken by the finite domain variables $x = (x_1, \ldots, x_n)$, or the largest value if there are fewer than r distinct values.

Usage. The constraint has been used to require that a task not start until at least r other tasks have started.

Inference. A complete filtering algorithm is given in [34].

Related constraints. There is also a max-n constraint. Min-n is closely related to *nvalues*.

Dictionary of Constraints 437

Network design

The *network design (fixed-charge network flow)* constraint requires that capacitated arcs be selected and flows placed on them to satisfy flow conservation constraints. It also computes variable and fixed costs.

Notation. The constraint can be written

$$\text{network-design}\,(x, y, z \mid N, s, A, u, c, f) \qquad (5.21)$$

where x is a tuple of real-valued flow variables x_{ij}, y a tuple of corresponding integer variables y_{ij}, and z a real-valued variable representing cost. N a tuple of nodes, and s a tuple of net supply values corresponding to the nodes in N. A is a tuple of directed arcs, and u, c, and f are tuples of flow capacities, unit variable costs, and fixed costs corresponding to the arcs in A. The variables y_{ij} can take integral values greater than one to allow for the possibility that multiple arcs of a given capacity may connect nodes i and j, as when one lays two or more telecommunication cables. Constraint (5.21) requires that

$$\begin{aligned} z &= \sum_{(i,j) \in A} c_{ij} x_{ij} + \sum_{(i,j) \in A} f_{ij} y_{ij} \\ &\sum_{(i,j) \in A} x_{ij} - \sum_{(j,i) \in A} x_{ji} = s_i, \text{ all } i \in N \\ &0 \leq x_{ij} \leq u_{ij} y_{ij}, \text{ all } (i,j) \in A \\ &y_{ij} \geq 0 \text{ and } y_{ij} \text{ integral, all } i, j \end{aligned} \qquad (5.22)$$

The fixed-charge network flow problem minimizes z subject to (5.21). There is a multicommodity version of the problem that allows one to require that messages from one node travel to a certain other node.

Usage. The model (5.22) and variations of it are widely used for the design of telecommunication and other networks. Special cases of the problem include certain lot sizing problems [26] and the capacitated facility location problem [112].

Relaxation. The model (5.22) can be relaxed by dropping the integrality requirement. Several types of cutting planes, known as *flow cuts*, have been developed for the model and its variations [230].

Related constraints. The *flow* constraint is a special case in which the choice of arcs is fixed.

Nonlinear disjunction

A *nonlinear disjunction* is a disjunction of convex, bounded nonlinear systems.

Notation. The constraint is written

$$\bigvee_{k \in K} g^k(x) \leq 0 \tag{5.23}$$

where each $g^k(x)$ is a tuple of functions $g_i^k(x)$, and x a tuple of real-valued variables. It is assumed that $\ell \leq x \leq u$, and $g^k(x)$ is bounded when $\ell \leq x \leq u$. It is further assumed that each $g_i^k(x)$ is a convex function on $[\ell, u]$. The constraint enforces

$$\bigvee_{k \in K} y_k$$

$$y_k \rightarrow \left(g^k(x) \leq 0\right), \text{ all } k \in K$$

Usage. Nonlinear disjunctions are used in a manner analogous to *linear disjunctions*.

Relaxation. The constraint (5.23) implies

$$\bigvee_{\substack{k \\ 1 \in D_{y_k}}} g^k(x) \leq 0$$

which can be given a convex hull relaxation [223, 310] (Section 4.8.1) or a big-M relaxation (Section 4.8.2).

Related constraints. There is a special constraint for *linear disjunctions*.

Nvalues

The *nvalues* constraint bounds the number of distinct values taken by a set of variables.

Notation The simplest form of the constraint is

$$\text{nvalues}(x \mid \ell, u) \tag{5.24}$$

where x is a tuple $(x_1, \ldots, x_n)$ of variables with finite domains, and ℓ and u are nonnegative integers. The constraint requires that the variables take at least ℓ and at most u different values. In other versions of the constraint, the lower bound is a variable:

$$\text{nvalues}(x, \ell \mid u) \tag{5.25}$$

or the upper bound is a variable:

$$\text{nvalues}(x, u \mid \ell) \tag{5.26}$$

Usage. The constraint is illustrated in Section 2.2.5 (employee scheduling).

Inference. As Section 3.10.4 notes, a network flow model provides complete filtering for (5.24). A more elaborate model [267] provides the same for (5.25). Achieving domain completeness for (5.26) is NP-hard [51]. An incomplete filter is presented in [50], and a less thorough one in [34].

Related constraints. Alldiff is a special case of *nvalues* in which $\ell = u = n$. A *weighted nvalues* constraint is presented in [38], which also describes a filter. There are a number of other variations, such as ninterval (counts values as distinct only when they lie in different intervals) and nclass (counts values as distinct only when they lie in different sets) [34].

Path

The *path* constraint finds a path in a graph having at most a given length.

Notation. The constraint can be written

$$\text{path}(x, w \mid G, c, s, t) \tag{5.27}$$

where x is a tuple of boolean variables corresponding to the vertices of a directed or undirected graph G, w is an integer, G is a directed or undirected graph on n vertices, c contains an integral edge length c_{ij} for every edge (i, j) of G, and s, t are vertices of G. It is assumed there are no cycles in G of negative length. The constraint requires that G contain a simple path (no repeated vertices) of length at most w from s to t, where x_i is true if and only if vertex i lies on the path. The *shortest path problem* on G minimizes w subject to (5.27) when the domain of every x_j is $\{true, false\}$.

Usage. Shortest path models have countless applications. An application of a specially structured path constraint is presented in Section 2.2.9.

Inference. When the domain of every x_j contains *false*, the minimum value of w subject to (5.27) can be computed in polynomial time by a shortest path algorithm on an induced subgraph of G. Shortest-path algorithms, some of which compute shortest paths between all pairs of vertices, are explained in [4, 30, 221] and exhaustively surveyed in [109]. Finding a shortest path that contains certain vertices is NP-hard, however, as is achieving domain completeness for (5.27). Incomplete filtering methods are presented in [299]. If the graph is acyclic, complete filtering can be accomplished in polynomial time [120] (Section 2.2.9).

Related constraints. One variation of the path constraint partitions the vertex set of G to define a given number of disjoint paths.

Piecewise linear

The *piecewise linear* constraint sets a variable equal to a semicontinuous piecewise linear function of one variable.

Notation. The constraint can be written

$$\text{piecewise}(x, z \mid L, U, c, d)$$

where x and z are real-valued variables. $L = (L_1, \ldots, L_m)$ and $U = (U_1, \ldots, U_m)$ define the intervals $[L_k, U_k]$ on which the semicontinuous piecewise linear function is defined. Adjacent intervals should intersect in at most a point. On each interval, the graph of the function is a line segment connecting (L_k, c_k) with (U_k, d_k). The constraint sets z equal to

$$\frac{U_k - x}{U_k - L_k} c_k + \frac{x - L_k}{U_k - L_k} d_k \text{ for } x \in [L_k, U_k], \ k = 1, \ldots, m$$

Usage. A continuous or semicontinuous piecewise linear constraint is a versatile modeling device. Piecewise linear functions commonly occur in practice, and moreover they can approximate many nonlinear functions that would be hard to relax directly. A separable nonlinear function $\sum_j h_j(x_j)$ can be approximated to a high degree of accuracy by replacing each nonlinear function $h_j(x)$ with a variable z_j that is set equal to a piecewise linear function having sufficiently many linear pieces. This is illustrated in Section 4.3, and applications are described in [259, 260, 279].

Inference. The piecewise linear function can interact with variable bounds to tighten the latter, as described in Section 4.3 and [279].

Dictionary of Constraints 441

Relaxation. A convex hull relaxation can be generated in linear time without using additional variables. Fast methods for computing the convex hull in two dimensions can be found in [308] (pages 351–352) and [270].

Same

The *same* constraint requires that two sets of equally many variables take the same multiset of values.

Notation. The constraint is

$$\text{same}(x, y)$$

where x and y are n-tuples of variables having finite domains. The constraint requires that the multiset of values taken by $x_1, \ldots, x_n$ equal the multiset of values taken by $y_1, \ldots, y_n$.

Usage. The following example appears in [42]. The organization Doctors without Borders wishes to pair doctors and nurses for emergency missions. Each x_i is the date on which doctor i departs, and the domain of x_i consists of the dates the doctor is available to depart. Each y_i has the same meaning for nurses. In another example [37], x_i is the shift one person works on day i, and y_i is the same for another person. The same constraint is enforced for all pairs of persons to ensure fairness, in that all work the same multiset of shifts.

Inference. A complete filter based on network flows (but a different model than used for *alldiff* and *cardinality*) appears in [42].

Related constraints. The constraint usedby(x, y) allows x to contain more variables than y and requires that the multiset of values used by y be contained in the multiset used by x [41]. The sort(x, y) constraint [55, 237] is slightly stronger than *same*—it requires that variables y be the result of sorting variables x.

Set covering

Given a collection of sets, the *set covering* constraint selects at most k sets that have the same union as the collection.

Notation. The constraint can be written

$$\text{set-covering}(x, k \mid S_1, \ldots, S_n) \tag{5.28}$$

where x is an n-tuple of boolean variables x_j that are true when set S_j is selected, and k is a positive integer. The constraint requires that at most k sets be selected whose union is $\bigcup_{j=1}^n S_j$. A 0-1 representation of the constraint is also natural:

$$Ax \geq e, \quad \sum_{j=1}^n x_j \leq k, \quad x_j \in \{0,1\}, \; j=1,\ldots,n \qquad (5.29)$$

Here, e is an m-tuple of ones, and A is an $m \times n$ 0-1 matrix in which $A_{ij} = 1$ when $i \in S_j$. The *set covering problem* is to minimize k subject to (5.28), or more generally to minimize $\sum_j c_j x_j$.

Usage. Set covering models arise in a wide variety of contexts [328]. For example, one may wish to buy the fewest possible CDs that contain all of one's favorite songs.

Inference. Achieving domain completeness is NP-hard, because the set covering problem is NP-hard. The upper bound on k can be tightened by solving the set covering problem approximately. The problem has a long history of algorithmic development. One survey of algorithms is [265].

Relaxation. A continuous relaxation can be obtained by dropping the integrality constraints from the 0-1 model (5.29) and adding valid cuts [253, 293], which are surveyed in [63].

Related constraints. The *set packing* constraint is the analogous constraint for $\leq$ inequalities.

Set packing

Given a collection of sets, the *set packing* constraint selects at least k sets that are pairwise disjoint.

Notation. The constraint can be written

$$\text{set-packing}(x, k \mid S_1, \ldots, S_n) \qquad (5.30)$$

where x is an n-tuple of boolean variables x_j that are true when set S_j is selected, and k is a positive integer. The constraint requires that at least k pairwise disjoint sets be selected. A 0-1 representation of the constraint is also natural:

$$Ax \leq e, \quad \sum_{j=1}^n x_j \geq k, \quad x_j \in \{0,1\}, \; j=1,\ldots,n \qquad (5.31)$$

Here, e is an m-tuple of ones, and A is an $m \times n$ 0-1 matrix in which $A_{ij} = 1$ when $i \in S_j$. The *set packing problem* is to maximize k subject to (5.30), or more generally to maximize $\sum_j c_j x_j$.

Dictionary of Constraints 443

Usage. Set packing models arise in wide variety of contexts [328]. For example, in a combinatorial auction, each bid is for a bundle of items. The auctioneer may wish to select bids to maximize revenue, but with no two bids claiming the same item. Set packing inequalities are generated as knapsack cuts in mixed integer solvers and strengthened with cutting planes (Section 4.4.5).

Inference. Achieving domain completeness is NP-hard, because the set packing problem is NP-hard. The lower bound on k can be tightened by solving the set packing problem approximately. The problem has a long history of algorithmic development. One survey of algorithms is [265].

Relaxation. A continuous relaxation can be obtained by dropping the integrality constraints from the 0-1 model (5.31) and adding valid cuts. These include odd cycle inequalities [261] (Section 4.4.5), for which there is a polynomial-time separation algorithm (Lemma 9.1.11 of [162]); clique inequalities [138, 261] (Section 4.4.5), for which separation algorithms are surveyed in [63]; orthonormal representation inequalities, which include clique inequalities and can be separated in polynomial time [162]; and other families surveyed in [63]. General discussions of these cutting planes may be found in [137, 230]

Related constraints. The *set covering* constraint is the analogous constraint for $\geq$ inequalities.

Soft alldiff

The *soft alldiff* constraint limits the degree to which a set of variables fail to take all different values.

Notation. Two varieties of the *soft* alldiff constraint have been studied, both written
$$\text{alldiff}(x, k)$$
One requires k to be an upper bound on the number of pairs of variables that have the same value. The other requires k to be an upper bound on the minimum number of variables that must change values to satisfy alldiff(x).

Usage. The constraints are designed for overconstrained problems in which the object is to obtain a solution that is close to feasibility.

Inference. An incomplete filter for the first soft alldiff constraint appears in [267] and a complete one in [324]. A complete filter for the second constraint is given in [267].

Related constraints. Soft alldiff is an extension of the *all-different* constraint. Soft constraints in general have received much recent attention in constraint programming. An introduction to the area can be found in [28].

Stretch

The *stretch* constraint was designed for scheduling workers in shifts. It specifies limits on how many consecutive days a worker can be assigned to each shift, and which shifts can follow another.

Notation. The constraint is

$$\text{stretch}(x \mid v, \ell, u, P)$$

where $x = (x_1, \ldots, x_n)$ is a tuple of variables with finite domains. Perhaps x_i denotes the shift that a given employee will work on day i. Also, v is an m-tuple of possible values of the variables, ℓ an m-tuple of lower bounds, and u an m-tuple of upper bounds. A stretch is a maximal sequence of consecutive variables that take the same value. Thus, $x_j, \ldots, x_k$ is a stretch if for some value v, $x_j, \ldots, x_k = v$, $x_{j-1} \neq v$ (or $j = 1$), and $x_{k+1} \neq v$ (or $k = n$). The stretch constraint requires that for each $j \in \{1, \ldots, m\}$, any stretch of value v_j in x have length at least ℓ_j and at most u_j. In addition, P is a set of *patterns*, which are pairs of values $(v_j, v_{j'})$. The constraint requires that when a stretch of value v_j immediately precedes a stretch of value $v_{j'}$, the pair $(v_j, v_{j'})$ must be in P.

There is also a cyclic version of the stretch constraint, *stretch-cycle*, that recognizes stretches that continue from x_n to x_1. It can be used when every week must have the same schedule.

Usage. Examples of scheduling with *stretch* appear in Section 3.12 and [266].

Inference. A polynomial-time dynamic programming algorithm achieves domain completeness [172] (Section 3.12).

Related constraints. Stretch is similar to the pattern constraint [64]. The *change* constraint limits the number of times a given type of shift change occurs.

Sum

The *sum* constraint computes a sum over an index set that depends on the value of a variable.

Dictionary of Constraints 445

Notation. The simplest form of the constraint is written

$$\text{sum}(y, z \mid a, S)$$

where y is a variable representing a positive integer, z is a variable, a is a tuple $(a_1, \ldots, a_m)$ of values, and S a tuple $(S_1, \ldots, S_n)$ of index sets, with each $S_i \subset \{1, \ldots, m\}$. The constraint requires z to be equal to $\sum_{j \in S_y} a_j$. Another form is

$$\text{sum}(y, x, z \mid S)$$

where x is a tuple $(x_1, \ldots, x_m)$ of variables. The constraint enforces $z = \sum_{j \in S_y} x_j$.

Usage. The constraint implements sums over variable index sets, which frequently occur in modeling. An application to a production planning problem with sequence-dependent cumulative costs is described in [342].

Inference. A complete filter is described in [185].

Relaxation. A convex hull relaxation appears in [342].

Related constraints. The sum constraint can be viewed as an extension of *element*.

Symmetric alldiff

Given a set of items, the *symmetric alldiff* constraint pairs each with another compatible item in the set.

Notation. One way to write the constraint is simply

$$\text{symalldiff}(x)$$

where x is a tuple $(x_1, \ldots, x_n)$ of variables with domains that are subsets of $\{1, \ldots, n\}$. The constraint requires $\text{alldiff}(x)$ and $x_{x_i} = i$ for $i = 1, \ldots, n$. In [283], the constraint is written with a bijection that associates the domain elements with the variables.

Usage. The constraint can be used to pair items with compatible items. For example, x_i can denote a sports team that is to play team i. If team i plays team j ($x_i = j$), then team j plays team i ($x_j = i$), and no two teams play the same team—$\text{alldiff}(x)$. The domain of x_i contains the teams that team i is allowed to play. Other applications include staffing problems in which people are assigned to work in pairs.

Inference. A complete filter and a faster, incomplete filter are presented in [283].

Related constraints. This is a restriction of *alldiff*.

Symmetric cardinality

The *symmetric cardinality* constraint bounds the number of values assigned to each variable, as well as the number of variables to which each value is assigned.

Notation. The constraint can be written

$$\text{symcardinality}(X \mid \bar{\ell}, \bar{u}, v, \ell, u)$$

where X is a tuple $(X_1, \ldots, X_n)$ of set-valued variables, $\bar{\ell}, \bar{u}$ are n-tuples of nonnegative integers, v an m-tuple of values, and ℓ, u are m-tuples of nonnegative integers. The constraint requires that (a) each x_i be assigned a set with cardinality at least $\bar{\ell}_i$ and at most $\bar{u}_i$, and (b) for $j = 1, \ldots, m$, at least ℓ_j and at most u_j of the variables be assigned a set containing v_j.

Usage. The constraint is used to limit how many workers are assigned to each task, and how many tasks are assigned to each worker. The workers that are appropriate for each task i are indicated by the initial domain of x_i.

Inference. A complete filter based on network flows is given in [215].

Related constraints. This is an extension of *cardinality*.

Value precedence

Given two values s and t, and a set of variables, the (integer) *value precedence* constraint requires that if a variable takes value t then a variable with a lower index takes value s.

Notation. The constraint is

$$\text{integer-value-precedence}(x \mid s, t)$$

where x is a tuple of variables with integer domains and s, t are integers. The constraint requires that whenever a variable x_j takes value t, a variable x_i with $i < j$ takes value s.

Usage. The constraint is used for symmetry breaking, in particular when two values are interchangeable in a set of variables [220].

Inference. Filtering algorithms are presented in [220].

Related constraints. There is an analogous constraint for set-valued variables [220].

References

[1] K. Aardal, R. Weismantel, and L. Wolsey. Non-standard approaches to integer programming. *Discrete Applied Mathematics*, 123:5–74, 2002.

[2] T. Achterberg. SCIP: A framework to integrate constraint and mixed integer programming. ZIB-report 04-19, Konrad-Zuse-Zentrum für Informationstechnik Berlin, 2004.

[3] A. Aggoun and N. Beldiceanu. Extending CHIP in order to solve complex scheduling and placement problems. *Mathematical and Computer Modelling*, 17:57–73, 1993.

[4] R. K. Ahuja, T. L. Magnanti, and J. B. Orlin. *Network Flows: Theory, Algorithms, and Applications, 3rd ed.* Prentice-Hall, Upper Saddle River, NJ, 1993.

[5] R. K. Ahuja, T. L. Magnanti, and J. B. Orlin. Applications of network optimization. In M. O. Ball, T. L. Magnanti, C. L. Monma, and G. L. Nemhauser, editors, *Network Models*, Handbooks in Operations Research and Management Science, pages 1–84. Elsevier, Amsterdam, 1995.

[6] H. Alt, N. Blum, K. Mehlhorn, and M. Paul. Computing maximum cardinality matching in time $O(n^{1.5}\sqrt{m/\log n})$. *Information Processing Letters*, 37:237–240, 1991.

[7] G. Appa, D. Magos, and I. Mourtos. Linear programming relaxations of multiple all-different predicates. In J. C. Régin and M. Rueher, editors, *Integration of AI and OR Techniques in Constraint Programming for Combinatorial Optimization Problems (CPAIOR 2004)*, volume 3011 of *Lecture Notes in Computer Science*, pages 364–369. Springer, 2004.

[8] G. Appa, D. Magos, and I. Mourtos. On the system of two all-different predicates. *Information Processing Letters*, 94:99–105, 2004.

[9] G. Appa, I. Mourtos, and D. Magos. Integrating constraint and integer programming for the orthogonal Latin squares problem. In P. Van Hentenryck, editor, *Principles and Practice of Constraint Programming (CP 2002)*, volume 2470 of *Lecture Notes in Computer Science*, pages 17–32. Springer, 2002.

[10] S. Arnborg, D. G. Corneil, and A. Proskurowski. Complexity of finding embeddings in a k-tree. *SIAM Journal on Algebraic and Discrete Mathematics*, 8:277–284, 1987.

[11] S. Arnborg and A. Proskurowski. Characterization and recognition of partial k-trees. *SIAM Jorunal on Algebraic and Discrete Mathematics*, 7:305–314, 1986.

[12] I. Aron, J. N. Hooker, and T. H. Yunes. SIMPL: A system for integrating optimization techniques. In J. C. Régin and M. Rueher, editors, *Integration of AI and OR Techniques in Constraint Programming for Combinatorial Optimization Problems (CPAIOR 2004)*, volume 3011 of *Lecture Notes in Computer Science*, pages 21–36. Springer, 2004.

[13] A. Atamtürk. Sequence independent lifting for mixed-integer programming. *Operations Research*, 52:487–490, 2004.

[14] A. Atamtürk. Cover and pack inequalities for (mixed) integer programming. *Annals of Operations Research*, 139:21–38, 2005.

[15] E. Balas. Facets of the knapsack polytope. *Mathematical Programming*, 8:146–164, 1975.

[16] E. Balas. Disjunctive programming. *Annals of Discrete Mathematics*, 5:3–51, 1979.

[17] E. Balas. Disjunctive programming and a hierarchy of relaxations for discrete optimization problems. *SIAM Journal on Algebraic and Discrete Methods*, 6:466–485, 1985.

[18] E. Balas, A. Bockmayr, N. Pisaruk, and L. Wolsey. On unions and dominants of polytopes. *Mathematical Programming*, 99:223–239, 2004.

[19] E. Balas, S. Ceria, and G. Cornuéjols. A lift-and-project cutting plane algorithm for mixed 0-1 programs. *Mathematical Programming*, 58:295–324, 1993.

[20] E. Balas, S. Ceria, G. Cornuéjols, and G. Pataki. Polyhedral methods for the maximum clique problem. In D. S. Johnson and M. A. Trick, editors, *Cliques, Colorings and Satisfiability: 2nd DIMACS Implementation Challenge, 1993*, pages 11–28. American Mathematical Society, 1996.

[21] E. Balas and M. Fischetti. Polyhedral theory for the asymmetric traveling salesman problem. In G. Gutin and A. P. Punnen, editors, *The Traveling Salesman Problem and its Variations*, pages 117–168. Kluwer, Dordrecht, 2002.

[22] E. Balas and C. S. Yu. Finding a maximum clique in an arbitrary graph. *SIAM Journal on Computing*, 14:1054–1068, 1986.

[23] P. Baptiste and C. Le Pape. Edge-finding constraint propagation algorithms for disjunctive and cumulative scheduling. In *Proceedings of the Fifteenth Workshop of the U.K. Planning Special Interest Group*, Liverpool, U.K., 1996.

[24] P. Baptiste, C. Le Pape, and W. Nuijten. *Constraint-Based Scheduling: Applying Constraint Programming to Scheduling Problems*. Kluwer, Dordrecht, 2001.

REFERENCES

[25] F. Barahona and R. Anbil. The volume algorithm: Producing primal solutions with a subgradient algorithm. *Mathematical Programming*, 87:385–399, 2000.

[26] I. Barany, T. J. van Roy, and L. A. Wolsey. Strong formulations for multi-item capacitated lot-sizing. *Management Science*, 30:1255–1261, 1984.

[27] C. Barnhart, E. L. Johnson, G. L. Nemhauser, M. W. P. Savelsbergh, and P. H. Vance. Branch-and-price: Column generation for solving huge integer programs. *Operations Research*, 46:316–329, 1998.

[28] R. Barták. Modelling soft constraints: A survey. *Neural Network World*, 12:421–431, 2002.

[29] P. Barth. *Logic-based 0-1 Constraint Solving in Constraint Logic Programming*. Kluwer, Dordrecht, 1995.

[30] M. S. Bazaraa, J. J. Jarvis, and H. D. Sherali. *Linear Programming and Network Flows, 3rd ed.* Wiley, New York, 2004.

[31] M. S. Bazaraa, H. D. Sherali, and C. M. Shetty. *Nonlinear Programming: Theory and Algorithms*. Wiley, New York, 1993.

[32] P. Beame, H. Kautz, and A. Sabharwal. Understanding the power of clause learning. In *International Joint Conference on Artificial Intelligence (IJCAI 2003)*, 2003.

[33] N. Beaumont. An algorithm for disjunctive programs. *European Journal of Operational Research*, 48:362–371, 1990.

[34] N. Beldiceanu. Pruning for the *minimum* constraint family and for the *number of distinct values* constraint family. In T. Walsh, editor, *Principles and Practice of Constraint Programming (CP 2001)*, volume 2239 of *Lecture Notes in Computer Science*, pages 211–224. Springer, 2001.

[35] N. Beldiceanu and M. Carlsson. Revisiting the cardinality operator and introducing the cardinality-path constraint family. In P. Codognet, editor, *International Conference on Logic Programming (ICLP 2001)*, volume 2237 of *Lecture Notes in Computer Science*, pages 59–73. Springer, 2001.

[36] N. Beldiceanu and M. Carlsson. Sweep as a generic pruning technique applied to the non-overlapping rectangles constraints. In T. Walsh, editor, *Principles and Practice of Constraint Programming (CP 2001)*, volume 2239 of *Lecture Notes in Computer Science*, pages 377–391. Springer, 2001.

[37] N. Beldiceanu, M. Carlsson, and J.-X. Rampon. Global constraint catalog. SICS technical report T2005:08, Swedish Institute of Computer Science, 2005.

[38] N. Beldiceanu, M. Carlsson, and S. Thiel. Cost-filtering algorithms for the two sides of the *sum of weights of distinct values* constraint. SICS technical report T2002:14, Swedish Institute of Computer Science, 2002.

[39] N. Beldiceanu and E. Contejean. Introducing global constraints in CHIP. *Mathematical and Computer Modelling*, 12:97–123, 1994.

[40] N. Beldiceanu, Q. Guo, and S. Thiel. Non-overlapping constraints between convex polytopes. In T. Walsh, editor, *Principles and Practice of Constraint Programming (CP 2001)*, volume 2239 of *Lecture Notes in Computer Science*, pages 392–407. Springer, 2001.

[41] N. Beldiceanu, I. Katriel, and S. Thiel. Filtering algorithms for the Same and UsedBy constraints. Research report MPI-I-2004-1-001, Max-Planck-Institut für Informatik, Saarbrücken, 2004.

[42] N. Beldiceanu, I. Katriel, and S. Thiel. Filtering algorithms for the same constraint. In J. C. Régin and M. Rueher, editors, *Integration of AI and OR Techniques in Constraint Programming for Combinatorial Optimization Problems (CPAIOR 2004)*, volume 3011 of *Lecture Notes in Computer Science*, pages 65–79. Springer, 2004.

[43] J. F. Benders. Partitioning procedures for solving mixed-variables programming problems. *Numerische Mathematik*, 4:238–252, 1962.

[44] T. Benoist, E. Gaudin, and B. Rottembourg. Constraint programming contribution to Benders decomposition: A case study. In P. Van Hentenryck, editor, *Principles and Practice of Constraint Programming (CP 2002)*, volume 2470 of *Lecture Notes in Computer Science*, pages 603–617. Springer, 2002.

[45] T. Benoist, F. Laburthe, and B. Rottembourg. Lagrange relaxation and constraint programming collaborative schemes for traveling tournament problems. In C. Gervet and M. Wallace, editors, *Proceedings of the International Workshop on Integration of Artificial Intelligence and Operations Research Techniques in Constraint Programming for Combintaorial Optimization Problems (CPAIOR 2001)*, Ashford, U.K., 2001.

[46] C. Berge. Two theorems in graph theory. *Proceedings of the National Academy of Sciences*, 43:842–844, 1957.

[47] C. Berge. *Graphes et Hypergraphes*. Dunod, Paris, 1970.

[48] U. Bertele and F. Brioschi. *Nonserial Dynamic Programming*. Academic Press, New York, 1972.

[49] D. P. Bertsekas. *Dynamic Programming and Optimal Control*, volume 1 and 2. Athena Scientific, Nashua, NH, 2001.

[50] C. Bessiere, E. Hebrard, B. Hnich, Z. Kiziltan, and T. Walsh. Filtering algorithms for the nvalue constraint. In R. Barták and M. Milano, editors, *Integration of AI and OR Techniques in Constraint Programming for Combinatorial Optimization Problems (CPAIOR 2005)*, volume 3524 of *Lecture Notes in Computer Science*, pages 79–93. Springer, 2005.

[51] C. Bessiere, E. Hebrard, B. Hnich, and T. Walsh. The complexity of global constraints. In *National Conference on Artificial Intelligence (AAAI 2004)*, pages 112–117, 2004.

[52] C. E. Blair and R. G. Jeroslow. The value function of a mixed integer program: I. *Discrete Applied Mathematics*, 19:121–138, 1977.

REFERENCES

[53] C. E. Blair and R. G. Jeroslow. The value function of a mixed integer program. *Mathematical Programming*, 23:237–273, 1982.

[54] C. E. Blair, R. G. Jeroslow, and J. K. Lowe. Some results and experiments in programming techniques for propositional logic. *Computers and Operations Research*, 13:633–645, 1988.

[55] N. Bleuzen-Guernalec and A. Colmerauer. Narrowing a block of sortings in quadratic time. In G. Smolka, editor, *Principles and Practice of Constraint Programming (CP 1999)*, volume 1330 of *Lecture Notes in Computer Science*, pages 2–16. Springer, 1997.

[56] C. Bliek. Generalizing dynamic and partial order backtracking. In *National Conference on Artificial Intelligence (AAAI 1998)*, pages 319–325, Madison, WI, 1998.

[57] A. Bockmayr and T. Kasper. Branch-and-infer: A unifying framework for integer and finite domain constraint programming. *INFORMS Journal on Computing*, 10:287–300, 1998.

[58] A. Bockmayr and N. Pisaruk. Detecting infeasibility and generating cuts for mixed integer programming using constraint progrmaming. In M. Gendreau, G. Pesant, and L.-M. Rousseau, editors, *Proceedings of the International Workshop on Integration of Artificial Intelligence and Operations Research Techniques in Constraint Programming for Combintaorial Optimization Problems (CPAIOR 2003)*, Montréal, 2003.

[59] A. Bockmayr, N. Pisaruk, and A. Aggoun. Network flow problems in constraint programming. In T. Walsh, editor, *Principles and Practice of Constraint Programming (CP 2001)*, volume 2239 of *Lecture Notes in Computer Science*, pages 196–210. Springer, 2001.

[60] S. Bollapragada, O. Ghattas, and J. N. Hooker. Optimal design of truss structures by mixed logical and linear programming. *Operations Research*, 49:42–51, 2001.

[61] I. M. Bomze, M. Budinich, P. M. Pardalos, and M. Pelillo. The maximum clique problem. In D.-Z. Du and P. M. Pardalos, editors, *Handbook of Combinatorial Optimization, Supplement Volume A*, pages 1–74. Kluwer, Dordrecht, 1999.

[62] G. Boole. *Studies in Logic and Probability* (ed. by R. Rhees). Open Court Publishing Company, La Salle, IL, 1952.

[63] R. Borndörfer. *Aspects of set packing, partitioning, and covering*. Shaker Verlag, Aachen, Germany, 1998.

[64] S. Bourdais, P. Galinier, and G. Pesant. Hibiscus: A constraint programming application to staff scheduling in health care. In F. Rossi, editor, *Principles and Practice of Constraint Programming (CP 2003)*, volume 2833 of *Lecture Notes in Computer Science*, pages 153–167. Springer, 2003.

[65] R. G. Brown, J. W. Chinneck, and G. M. Karam. Optimization with constraint programming systems. In R. Sharda et al., editor, *Impact of Recent Computer*

Advances on Operations Research, volume 9 of *Publications in Operations Research Series*, pages 463–473, Williamsburg, VA, 1989. Elsevier.

[66] P. Brucker and O. Thiele. A branch and bound method for the general-shop problem with sequence-dependent setup times. *OR Spektrum*, 18:145–161, 1996.

[67] H. Cambazard, P.-E. Hladik, A.-M. Déplanche, N. Jussien, and Y. Trinquet. Decomposition and learning for a hard real time task allocation problem. In M. Wallace, editor, *Principles and Practice of Constraint Programming (CP 2004)*, volume 3258 of *Lecture Notes in Computer Science*, pages 153–167. Springer, 2004.

[68] J. Carlier. One machine problem. *European Journal of Operational Research*, 11:42–47, 1982.

[69] J. Carlier and E. Pinson. An algorithm for solving the job-shop problem. *Management Science*, 35:164–176, 1989.

[70] J. Carlier and E. Pinson. A practical use of Jackson's preemptive schedule for solving the job shop problem. *Annals of Operations Research*, 26:269–287, 1990.

[71] J. Carlier and E. Pinson. Adjustment of heads and tails for the job-shop problem. *European Journal of Operational Research*, 78:146–161, 1994.

[72] Y. Caseau and F. Laburthe. Improved CLP scheduling with task intervals. In *Proceedings of the Eleventh International Conference on Logic Programming (ICLP 1994)*, pages 369–383. MIT Press, 1994.

[73] Y. Caseau and F. Laburthe. Solving small TSPs with constraints. In L. Naish, editor, *Proceedings, Fourteenth International Conference on Logic Programming (ICLP 1997)*, volume 2833, pages 316–330. The MIT Press, 1997.

[74] P. M. Castro and I. E. Grossmann. An efficient MILP model for the short-term scheduling of single stage batch plants. technical report, Departamento de Modelação e Simulação de Processos, INETI, Lisbon, 2006.

[75] S. Ceria, C. Cordier, H. Marchand, and L. A. Wolsey. Cutting planes for integer programs with general integer variables. *Mathematical Programming*, 81:201214, 1998.

[76] A. Chabrier. A cooperative CP and LP optimizer approach for the pairing generation problem. In *Proceedings of the International Workshop on Integration of Artificial Intelligence and Operations Research Techniques in Constraint Programming for Combintaorial Optimization Problems (CPAIOR 1999)*, Ferrara, Italy, 2000.

[77] A. Chabrier. Heuristic branch-and-price-and-cut to solve a network design problem. In M. Gendreau, G. Pesant, and L.-M. Rousseau, editors, *Proceedings of the International Workshop on Integration of Artificial Intelligence and Operations Research Techniques in Constraint Programming for Combintaorial Optimization Problems (CPAIOR 2003)*, Montréal, 2003.

[78] C. L. Chang. The unit proof and the input proof in theorem proving. *Journal of the ACM*, 14:698–707, 1970.

[79] D. Chhajed and T. J. Lowe. Solving structured multifacility location problems efficiently. *Transportation Science*, 28:104–115, 1994.

[80] Y. Chu and Q. Xia. Generating Benders cuts for a class of integer programming problems. In J. C. Régin and M. Rueher, editors, *Integration of AI and OR Techniques in Constraint Programming for Combinatorial Optimization Problems (CPAIOR 2004)*, volume 3011 of *Lecture Notes in Computer Science*, pages 127–141. Springer, 2004.

[81] Y. Chu and Q. Xia. A hybrid algorithm for a class of resource-constrained scheduling problems. In R. Barták and M. Milano, editors, *Integration of AI and OR Techniques in Constraint Programming for Combinatorial Optimization Problems (CPAIOR 2005)*, volume 3524 of *Lecture Notes in Computer Science*, pages 110–124. Springer, 2005.

[82] V. Chvátal. Edmonds polytopes and a hierarchy of combinatorial problems. *Discrete Mathematics*, 4:305–337, 1973.

[83] V. Chvátal. Edmonds polytopes and weakly hamiltonian graphs. *Mathematical Programming*, 5:29–40, 1973.

[84] V. Chvátal. Tough graphs and hamiltonian circuits. *Discrete Mathematics*, 5:215–228, 1973.

[85] V. Chvátal. *Linear Programming*. W. H. Freeman, New York, 1983.

[86] V. Chvátal. Hamiltonian cycles. In E. L. Lawler, J. K. Lenstra, A. H. G. Rinooy Kan, and D. B. Shmoys, editors, *The Traveling Salesman Problem: A Guided Tour of Combinatorial Optimization*, pages 403–430. Wiley, New York, 1985.

[87] E. G. Coffman, M. R. Garey, and D. S. Johnson. Approximation algorithms for bin-packing: A survey. In D. Hochbaum, editor, *Approximation Algorithms for NP-hard Problems*, pages 46–93. PWS Publishing Company, Boston, 1997.

[88] Y. Colombani and S. Heipcke. Mosel: An extensible environment for modeling and programming solutions. In N. Jussien and F. Laburthe, editors, *Proceedings of the International Workshop on Integration of Artificial Intelligence and Operations Research Techniques in Constraint Programming for Combinatorial Optimization Problems (CPAIOR 2002)*, pages 277–290, Le Croisic, France, 2002.

[89] Y. Colombani and S. Heipcke. Mosel: An overview. White paper, DASH Optimization, 2004.

[90] W. Cook, C. R. Coullard, and G. Turán. On the complexity of cutting plane proofs. *Discrete Applied Mathematics*, 18:25–38, 1987.

[91] J.-F. Cordeau and G. Laporte. Modeling and optimization of vehicle routing and arc routing problems. In G. Appa, L. Pitsoulis, and H. P. Williams, editors, *Handbook on Modelling for Discrete Optimization*, pages 151–191. Springer, 2006.

[92] A. I. Corréa, A. Langevin, and L. M. Rousseau. Dispatching and conflict-free routing of automated guided vehicles: A hybrid approach combining constraint programming and mixed integer programming. In J. C. Régin and M. Rueher, editors, *Integration of AI and OR Techniques in Constraint Programming for Combinatorial Optimization Problems (CPAIOR 2004)*, volume 3011 of *Lecture Notes in Computer Science*, pages 370–378. Springer, 2004.

[93] M.-C. Costa. Persistency in maximum cardinality bipartite matchings. *Operations Research Letters*, 15:143–149, 1994.

[94] Y. Crama, P. Hansen, and B. Jaumard. The basic algorithm for pseudoboolean programming revisited. *Discrete Applied Mathematics*, 29:171–185, 1990.

[95] W. Cronholm and Farid Ajili. Strong cost-based filtering for Lagrange decomposition applied to network design. In M. Wallace, editor, *Principles and Practice of Constraint Programming (CP 2004)*, volume 3258 of *Lecture Notes in Computer Science*, pages 726–730. Springer, 2004.

[96] H. Crowder, E. Johnson, and M. W. Padberg. Solving large-scale zero-one linear programming problems. *Operations Research*, 31:803–834, 1983.

[97] G. B. Dantzig. Maximization of a linear function of variables subject to linear inequalities. In T. C. Koopmans, editor, *Activity Ananlysis of Production and Allocation*, pages 339–347. Wiley, New York, 1951.

[98] G. B. Dantzig. *Linear Programming and Extensions*. Princeton University Press, 1963.

[99] G. B. Dantzig. Linear programming. In J. K. Lenstra, A. H. G. Rinnooy Kan, and A. Schrijver, editors, *History of Mathematical Programming: A Collection of Personal Reminiscences*, Handbooks in Operations Research and Management Science, pages 19–31. CWI, North-Holland, Amsterdam, 1991.

[100] G. B. Dantzig, D. R. Fulkerson, and S. M. Johnson. Solution of a large scale traveling salesman problem. *Operations Research*, 2:393–410, 1954.

[101] E. Davis. Constraint propagation with intervals labels. *Artificial Intelligence*, 32:281–331, 1987.

[102] M. Davis and H. Putnam. A computing procedure for quantification theory. *Journal of the ACM*, 7:201–215, 1960.

[103] M. Dawande and J. N. Hooker. Inference-based sensitivity analysis for mixed integer/linear programming. *Operations Research*, 48:623–634, 2000.

[104] J. Valerio de Carvalho. Exact solution of bin-packing problems using column generation and branch-and-bound. *Annals of Operations Research*, 86:629–659, 1999.

[105] R. Dechter. Bucket elimination: A unifying framework for several probabilistic inference algorithms. In *Proceedings of the Twelfth Annual Conference on Uncertainty in Artificial Intelligence (UAI 96)*, pages 211–219, Portland, OR, 1996.

REFERENCES

[106] S. Demassey, C. Artiques, and P. Michelon. A hybrid constraint propagation-cutting plane procedure for the RCPSP. In N. Jussien and F. Laburthe, editors, *Proceedings of the International Workshop on Integration of Artificial Intelligence and Operations Research Techniques in Constraint Programming for Combinatorial Optimization Problems (CPAIOR 2002)*, Le Croisic, France, 2002.

[107] S. Demassey, G. Pesant, and L.-M. Rousseau. Constraint-programming based column generation for employee timetabling. In R. Barták and M. Milano, editors, *Integration of AI and OR Techniques in Constraint Programming for Combinatorial Optimization Problems (CPAIOR 2005)*, volume 3524 of *Lecture Notes in Computer Science*, pages 140–154. Springer, 2005.

[108] E. V. Denardo. *Dynamic Programming: Models and Applications*. Dover Publications, Mineola, NY, 2003.

[109] N. Deo and C.-Y. Pang. Shortest-path algorithms: Taxonomy and annotation. *Networks*, 14:275–323, 1984.

[110] U. Dorndorf, E. Pesch, and T. Phan-Huy. Solving the open shop scheduling problem. *Journal of Scheduling*, 4:157–174, 2001.

[111] W. F. Dowling and J. H. Gallier. Linear-time algorithms for testing the satisfiability of propositional Horn formulae. *Journal of Logic Programming*, 1:267–284, 1984.

[112] Z. Drezner. *Facility Location: A Survey of Applications and Methods*. Springer, New York, 1995.

[113] K. Easton, G. Nemhauser, and M. Trick. The traveling tournament problem description and benchmarks. In T. Walsh, editor, *Principles and Practice of Constraint Programming (CP 2001)*, volume 2239 of *Lecture Notes in Computer Science*, pages 580–584. Springer, 2001.

[114] K. Easton, G. Nemhauser, and M. Trick. Solving the traveling tournament problem: A combined integer programming and constraint programming approach. In *Proceedings of the International Conference on the Practice and Theory of Automated Timetabling (PATAT 2002)*, 2002.

[115] K. Easton, G. Nemhauser, and M. Trick. CP based branch and price. In M. Milano, editor, *Constraint and Integer Programming: Toward a Unified Methodology*, pages 207–231. Kluwer, Dordrecht, 2004.

[116] J. Erschler, P. Lopez, and P. Esquirol. Ordonnancement de tâches sous contraintes: Une approche énergétique. *RAIRO Automatique, Productique, Informatique Industrielle*, 26:453–481, 1992.

[117] J. Erschler, P. Lopez, and C. Thuriot. Raisonnement temporel sous contraintes de ressource et problèmes d'ordonnancement. *Revue d'Intelligence Artificielle*, 5:7–32, 1991.

[118] F. Fages and A. Lal. A global constraint for cutset problems. In M. Gendreau, G. Pesant, and L.-M. Rousseau, editors, *Proceedings of the International Workshop on Integration of Artificial Intelligence and Operations Research Techniques in Constraint Programming for Combintaorial Optimization Problems (CPAIOR 2003)*, Montréal, 2003.

[119] T. Fahle. Cost based filtering vs. upper bounds for maximum clique. In N. Jussien and F. Laburthe, editors, *Proceedings of the International Workshop on Integration of Artificial Intelligence and Operations Research Techniques in Constraint Programming for Combinatorial Optimization Problems (CPAIOR 2002)*, pages 93–108, Le Croisic, France, 2002.

[120] T. Fahle, U. Junker, S. E. Karish, N. Kohn, M. Sellmann, and B. Vaaben. Constraint programming based column generation for crew assignment. *Journal of Heuristics*, 8:59–81, 2002.

[121] G. Farkas. A Fourier-féle mechanikai elv alkalmazásai (Hungarian) [On the applications of the mechanical principle of Fourier]. *Mathematikai és Természettudományi Értesitö*, 12:457–472, 1893–1894.

[122] C. E. Ferreira, A. Martin, and R. Weismantel. Solving multiple knapsack problems by cutting planes. *SIAM Journal on Optimization*, 6:858–877, 1996.

[123] M. Fischetti, A. Lodi, and P. Toth. Solving real-world ATSP instances by branch-and-cut. In M. Jünger, G. Reinelt, and G. Rinaldi, editors, *Combinatorial Optimization—Eureka, You Shrink!, Papers Dedicated to Jack Edmonds*, volume 2570 of *Lecture Notes in Computer Science*, pages 64–77. Springer, 2003.

[124] M. Fisher. Vehicle routing. In M. O. Ball, T. L. Magnanti, C. L. Monma, and G. L. Nemhauser, editors, *Network Routing*, volume 8 of *Handbooks in Operations Research and Management*, pages 1–79. North-Holland, 1997.

[125] M. L. Fisher. The Lagrangian relaxation method for solving integer programming problems. *Management Science*, 27:1–18, 1981.

[126] P. Flener, A. M. Frisch, B. Hnich, Z. Kiziltan, I. Miguel, J. Pearson, and T. Walsh. Breaking row and column symmetries in matrix models. In P. Van Hentenryck, editor, *Principles and Practice of Constraint Programming (CP 2002)*, volume 2470 of *Lecture Notes in Computer Science*, pages 462–476. Springer, 2002.

[127] F. Focacci, A. Lodi, and M. Milano. Cost-based domain filtering. In J. Jaffar, editor, *Principles and Practice of Constraint Programming (CP 1999)*, volume 1713 of *Lecture Notes in Computer Science*, pages 189–203. Springer, 1999.

[128] F. Focacci, A. Lodi, and M. Milano. Solving TSP with time windows with constraints. In *International Conference on Logic Programming (ICLP 1999)*, pages 515–529. MIT Press, 1999.

[129] F. Focacci, A. Lodi, and M. Milano. Cutting planes in constraint programming: An hybrid approach. In R. Dechter, editor, *Principles and Practice of Constraint Programming (CP 2000)*, volume 1894 of *Lecture Notes in Computer Science*, pages 187–201. Springer, 2000.

REFERENCES

[130] F. Focacci and W. P. M. Nuijten. A constraint propagation algorithm for scheduling with sequence dependent setup times. In U. Junker, S. E. Karisch, and S. Tschöke, editors, *Proceedings of the International Workshop on Integration of Artificial Intelligence and Operations Research Techniques in Constraint Programming for Combintaorial Optimization Problems (CPAIOR 2000)*, pages 53–55, Paderborn, Germany, 2000.

[131] L. R. Ford and D. R. Fulkerson. Maximal flow through a network. *Canadian Journal of Mathematics*, 8:399–404, 1956.

[132] L. R. Ford and D. R. Fulkerson. *Flows in Networks*. Princeton University Press, 1962.

[133] B. R. Fox. Chronological and non-chronological scheduling. In *Proceedings of the First Annual Conference on Artificial Intelligence: Simulation and Planning in High Autonomy Systems*, Tucson, Arizona, 1990.

[134] E. C. Freuder. Synthesizing constraint expressions. *Communications of the ACM*, 21:958–966, 1978.

[135] E. C. Freuder. A sufficient condition for backtrack-free search. *Communications of the ACM*, 29:24–32, 1982.

[136] A. Frisch, B. Hnich, Z. Kiziltan, I. Miguel, and T. Walsh. Global constraints for lexicographic orderings. In P. Van Hentenryck, editor, *Principles and Practice of Constraint Programming (CP 2002)*, volume 2470 of *Lecture Notes in Computer Science*, pages 93–108. Springer, 2002.

[137] A. Fügenschuh and A. Martin. Computational integer programming and cutting planes. In K. Aardal, G. L. Nemhauser, and R. Weismantel, editors, *Discrete Optimization*, Handbooks in Operations Research and Management Science, pages 69–121. Elsevier, Amsterdam, 2005.

[138] D. R. Fulkerson. Blocking and anti-blocking pairs of polyhedra. *Mathematical Programming*, 1:168–194, 1971.

[139] R. Garfinkel and G. L. Nemhauser. Optimal political districting by implicit enumeration techniques. *Management Science*, 16:B495–B508, 1970.

[140] T. Gellermann, M. Sellmann, and R. Wright. Shorter-path constraints for the resource constrained shortest path problem. In R. Barták and M. Milano, editors, *Integration of AI and OR Techniques in Constraint Programming for Combinatorial Optimization Problems (CPAIOR 2005)*, volume 3524 of *Lecture Notes in Computer Science*, pages 201–216. Springer, 2005.

[141] B. Gendron, H. Lebbah, and G. Pesant. Improving the cooperation between the master problem and the subproblem in constraint programming based column generation. In R. Barták and M. Milano, editors, *Integration of AI and OR Techniques in Constraint Programming for Combinatorial Optimization Problems (CPAIOR 2005)*, volume 3524 of *Lecture Notes in Computer Science*, pages 217–227. Springer, 2005.

[142] A. M. Geoffrion. Generalized Benders decomposition. *Journal of Optimization Theory and Applications*, 10:237–260, 1972.

[143] A. Ghouila-Houri. Caractérisation des matrices totalement unimodulaires. *Comptes rendus de l'Académie des Sciences de Paris*, 254:1192–1194, 1962.

[144] M. L. Ginsberg. Dynamic backtracking. *Journal of Artificial Intelligence Research*, 1:25–46, 1993.

[145] M. L. Ginsberg and D. A. McAllester. GSAT and dynamic backtracking. In *Principles and Practice of Constraint Programming (CP 1994)*, volume 874 of *Lecture Notes in Computer Science*, pages 216–225. Springer, 1994.

[146] F. Glover. A bound escalation method for the solution of integer linear programs. *Cahiers du Centre d'Etudes de Recherche Opérationelle*, 6:131–168, 1964.

[147] F. Glover. Maximum matching in a convex bipartite graph. *Naval Research Logistics Quarterly*, 316:313–316, 1967.

[148] F. Glover. Surrogate constraints. *Operations Research*, 16:741–749, 1968.

[149] F. Glover. Surrogate constraint duality in mathematical programming. *Operations Research*, 23:434–451, 1975.

[150] F. Glover. Tabu search—Part I. *ORSA Journal on Computing*, 1:190–206, 1989.

[151] A. Goldberg and R. Tarjan. A new approach to the maximum flow problem. *Journal of the ACM*, 35:921–940, 1988.

[152] R. E. Gomory. Outline of an algorithm for integer solutions to linear programs. *Bulletin of the American Mathematical Society*, 64:275–278, 1958.

[153] R. E. Gomory. An algorithm for the mixed integer problem. RAND technical report, RAND Corporation, 1960.

[154] R. E. Gomory. Solving linear programming problems in integers. In R. Bellman and M. Hall, editors, *Combinatorial Analysis*, volume X of *Symposia in Applied Mathematics*, pages 211–215. American Mathematical Society, 1960.

[155] R. E. Gomory. On the relation between integer and noninteger solutions to linear programs. In *Proceedings of the National Academy of Sciences*, volume 53, pages 260–265, 1965.

[156] R. E. Gomory. Some polyhedra related to combintorial problems. *Linear Algebra and its Applications*, 2:451–558, 1969.

[157] F. Granot and P. L. Hammer. On the use of boolean functions in 0-1 programming. *Methods of Operations Research*, 12:154–184, 1971.

[158] F. Granot and P. L. Hammer. On the role of generalized covering problems. *Cahiers du Centre d'Études de Recherche Opérationnelle*, 17:277–289, 1975.

[159] H. Greenberg. A branch-and-bound solution to the general scheduling problem. *Operations Research*, 8:353–361, 1968.

REFERENCES

[160] M. Grönkvist. Using constraint propagation to accelerate column generation nin aircraft scheduling. In M. Gendreau, G. Pesant, and L.-M. Rousseau, editors, *Proceedings of the International Workshop on Integration of Artificial Intelligence and Operations Research Techniques in Constraint Programming for Combintaorial Optimization Problems (CPAIOR 2003)*, Montréal, 2003.

[161] I. E. Grossmann, J. N. Hooker, R. Raman, and H. Yan. Logic cuts for processing networks with fixed charges. *Computers and Operations Research*, 21:265–279, 1994.

[162] M. Grötschel, L. Lovász, and A. Schrijver. *Geometric Algorithms and Combinatorial Optimization*. Springer, Berlin, 1988.

[163] M. Grötschel and M. W. Padberg. On the symmetric traveling salesman problem I: inequalities. *Mathematical Programming*, 16:265–280, 1979.

[164] Z. Gu, G. L. Nemhauser, and M. W. P. Savelsbergh. Sequence independent lifting of cover inequalities. In *Proceedings of the 4th International Conference on Integer Programming and Combinatorial Optimization (IPCO 1995)*, volume 920 of *Lecture Notes in Computer Science*, pages 452–461. Springer, 1995.

[165] Z. Gu, G. L. Nemhauser, and M. W. P. Savelsbergh. Sequence independent lifting in mixed integer programming. *Journal of Combinatorial Optimization*, 4:109–129, 2000.

[166] P. L. Hammer, E. L. Johnson, and U. N. Peled. Facets of regular 0-1 polytopes. *Mathematical Programming*, 8:179–206, 1975.

[167] P. Hansen. The steepest ascent mildest descent heuristic for combinatorial programming. Presentation at Congress on Numerical Methods in Combinatorial Optimization, Capri, 1986.

[168] I. Harjunkoski and I. E. Grossmann. A decomposition approach for the scheduling of a steel plant production. *Computers and Chemical Engineering*, 25:1647–1660, 2001.

[169] I. Harjunkoski and I. E. Grossmann. Decomposition techniques for multistage scheduling problems using mixed-integer and constraint programming methods. *Computers and Chemical Engineering*, 26:1533–1552, 2002.

[170] M. Held and R. M. Karp. The traveling-salesman problem and minimum spanning trees. *Operations Research*, 18:1138–1162, 1970.

[171] M. Held and R. M. Karp. The traveling-salesman problem and minimum spanning trees: Part II. *Mathematical Programming*, 1:6–25, 1971.

[172] L. Hellsten, G. Pesant, and P. van Beek. A domain consistency algorithm for the stretch constraint. In M. Wallace, editor, *Principles and Practice of Constraint Programming (CP 2004)*, volume 3258 of *Lecture Notes in Computer Science*, pages 290–304. Springer, 2004.

[173] P. Van Hentenryck. *Constraint Satisfaction in Logic Programming*. MIT Press, Cambridge, MA, 1989.

[174] P. Van Hentenryck and J.-P. Carillon. Generality versus specificity: An experience with AI and OR techniques. In *Proceedings of the American Association for Artificial Intelligence (AAAI-88)*, 1988.

[175] P. Van Hentenryck, L. Michel, L. Perron, and J.-C. Régin. Constraint programming in OPL. In *International Conference on Principles and Practice of Declarative Programming (PPDP 1999)*, Paris, 1999.

[176] J. Hiriart-Urruty and C. Lemaréchal. *Convex Analysis and Minimization Algorithms*. Springer, Berlin, 1993.

[177] W. J. Van Hoeve. A hybrid constraint programming and semidefinite programming approach for the stable set problem. In F. Rossi, editor, *Principles and Practice of Constraint Programming (CP 2003)*, volume 2833 of *Lecture Notes in Computer Science*, pages 407–421. Springer, 2003.

[178] A. J. Hoffmann and J. B. Kruskal. Integral boundary points of convex polyhedra. In H. W. Kuhn and A. W. Tucker, editors, *Linear Inequalities and Related Systems*, pages 223–246. Princeton University Press, 1956.

[179] J. N. Hooker. Generalized resolution and cutting planes. *Annals of Operations Research*, 12:217–239, 1988.

[180] J. N. Hooker. Input proofs and rank one cutting planes. *ORSA Journal on Computing*, 1:137–145, 1989.

[181] J. N. Hooker. Generalized resolution for 0-1 linear inequalities. *Annals of Mathematics and Artificial Intelligence*, 6:271–286, 1992.

[182] J. N. Hooker. Logic-based methods for optimization. In A. Borning, editor, *Principles and Practice of Constraint Programming (CP 2002)*, volume 874 of *Lecture Notes in Computer Science*, pages 336–349. Springer, 1994.

[183] J. N. Hooker. Inference duality as a basis for sensitivity analysis. In E. C. Freuder, editor, *Principles and Practice of Constraint Programming (CP 1996)*, volume 1118 of *Lecture Notes in Computer Science*, pages 224–236. Springer, 1996.

[184] J. N. Hooker. Constraint satisfaction methods for generating valid cuts. In D. L. Woodruff, editor, *Advances in Computational and Stochastic Optimization, Logic Programming and Heuristic Search*, pages 1–30. Kluwer, Dordrecht, 1997.

[185] J. N. Hooker. *Logic-Based Methods for Optimization: Combining Optimization and Constraint Satisfaction*. Wiley, New York, 2000.

[186] J. N. Hooker. Integer programming duality. In C. A. Floudas and P. M. Pardalos, editors, *Encyclopedia of Optimization, Vol. 2*, pages 533–543. Kluwer, New York, 2001.

[187] J. N. Hooker. Logic, optimization and constraint programming. *INFORMS Journal on Computing*, 14:295–321, 2002.

[188] J. N. Hooker. A framework for integrating solution methods. In H. K. Bhargava and M. Ye, editors, *Computational Modeling and Problem Solving in the Networked World (Proceedings of ICS2003)*, pages 3–30. Kluwer, 2003.

[189] J. N. Hooker. A hybrid method for planning and scheduling. In M. Wallace, editor, *Principles and Practice of Constraint Programming (CP 2004)*, volume 3258 of *Lecture Notes in Computer Science*, pages 305–316. Springer, 2004.

[190] J. N. Hooker. A hybrid method for planning and scheduling. *Constraints*, 10:385–401, 2005.

[191] J. N. Hooker. Planning and scheduling to minimize tardiness. In *Principles and Practice of Constraint Programming (CP 2005)*, volume 3709 of *Lecture Notes in Computer Science*, pages 314–327. Springer, 2005.

[192] J. N. Hooker. A sesarch-infer-and-relax framework for integrating solution methods. In R. Barták and M. Milano, editors, *Integration of AI and OR Techniques in Constraint Programming for Combinatorial Optimization Problems (CPAIOR 2005)*, volume 3524 of *Lecture Notes in Computer Science*, pages 243–257. Springer, 2005.

[193] J. N. Hooker. Unifying local and exhaustive search. In L. Villaseñor and A. I. Martinez, editors, *Avances in la Ciencia de la Computación (ENC 2005)*, pages 237–243, Puebla, Mexico, 2005.

[194] J. N. Hooker. Duality in optimization and constraint satisfaction. In J. C. Beck and B. M. Smith, editors, *Integration of AI and OR Techniques in Constraint Programming for Combinatorial Optimization Problems (CPAIOR 2006)*, volume 3990 of *Lecture Notes in Computer Science*, pages 3–15. Springer, 2006.

[195] J. N. Hooker. Logic-based modeling. In G. Appa, L. Pitsoulis, and H. P. Williams, editors, *Handbook on Modelling for Discrete Optimization*, pages 61–102. Springer, 2006.

[196] J. N. Hooker. Planning and scheduling by logic-based Benders decomposition. *Operations Research*, to appear.

[197] J. N. Hooker and C. Fedjki. Branch-and-cut solution of inference problems in propositional logic. *Annals of Mathematics and Artificial Intelligence*, 1:123–139, 1990.

[198] J. N. Hooker and M. A. Osorio. Mixed logical/linear programming. *Discrete Applied Mathematics*, 96–97:395–442, 1999.

[199] J. N. Hooker and G. Ottosson. Logic-based Benders decomposition. *Mathematical Programming*, 96:33–60, 2003.

[200] J. N. Hooker, G. Ottosson, E. S. Thorsteinsson, and H.-J. Kim. A scheme for unifying optimization and constraint satisfaction methods. *Knowledge Engineering Review*, 15:11–30, 2000.

[201] J. N. Hooker and H. Yan. Logic circuit verification by Benders decomposition. In V. Saraswat and P. Van Hentenryck, editors, *Principles and Practice of*

Constraint Programming: The Newport Papers, pages 267–288, Cambridge, MA, 1995. MIT Press.

[202] J. E. Hopcroft and R. M. Karp. A $n^{5/2}$ algorithm for maximum matchings in bipartite graphs. *SIAM Journal on Computing*, 2:225–231, 1973.

[203] V. Jain and I. E. Grossmann. Algorithms for hybrid MILP/CP models for a class of optimization problems. *INFORMS Journal on Computing*, 13:258–276, 2001.

[204] R. G. Jeroslow. Cutting plane theory: Algebraic methods. *Discrete Mathematics*, 23:121–150, 1978.

[205] R. G. Jeroslow. Representability in mixed integer programming, I: Characterization results. *Discrete Applied Mathematics*, 17:223–243, 1987.

[206] D. S. Johnson. Fast algorithms for bin packing. *Journal of Computer and Systems Sciences*, 8:272–314, 1974.

[207] E. L. Johnson. Cyclic groups, cutting planes, and shortest paths. In T. C. Hu and S. Robinson, editors, *Mathematical Programming*, pages 185–211. Academic Press, 1973.

[208] M. Jünger, G. Reinelt, and G. Rinaldi. The traveling salesman problem. In M. O. Ball, T. L. Magnanti, C. L. Monma, and G. L. Nemhauser, editors, *Network Models*, Handbooks in Operations Research and Management Science, pages 225–330. Elsevier, Amsterdam, 1995.

[209] U. Junker, S. E. Karish, N. Kohl, B. Vaaben, T. Fahle, and M. Sellmann. A framework for constraint programming based column generation. In J. Jaffar, editor, *Principles and Practice of Constraint Programming (CP 1999)*, volume 1713 of *Lecture Notes in Computer Science*, pages 261–275. Springer, 1999.

[210] I. Katriel and S. Thiel. Fast bound consistency for the global cardinality constraint. In F. Rossi, editor, *Principles and Practice of Constraint Programming (CP 2003)*, volume 2833 of *Lecture Notes in Computer Science*, pages 437–451. Springer, 2003.

[211] L. G. Kaya and J. N. Hooker. A filter for the circuit constraint. In *Principles and Practice of Constraint Programming (CP 2006)*, volume 4202 of *Lecture Notes in Computer Science*, pages 706–710. Springer, 2006.

[212] M. O. Khemmoudj, H. Bennaceur, and A. Nagih. Combining arc consistency and dual Lagrangean relaxation for filtering CSPs. In R. Barták and M. Milano, editors, *Integration of AI and OR Techniques in Constraint Programming for Combinatorial Optimization Problems (CPAIOR 2005)*, volume 3524 of *Lecture Notes in Computer Science*, pages 258–272. Springer, 2005.

[213] H.-J. Kim and J. N. Hooker. Solving fixed-charge network flow problems with a hybrid optimization and constraint programming approach. *Annals of Operations Research*, 115:95–124, 2002.

[214] Z. Kiziltan. *Symmetry Breaking Ordering Constraints*. PhD thesis, Uppsala University, 2004.

REFERENCES

[215] W. Kocjan and P. Kreuger. Filtering methods for symmetric cardinality constraint. In J. C. Régin and M. Rueher, editors, *Integration of AI and OR Techniques in Constraint Programming for Combinatorial Optimization Problems (CPAIOR 2004)*, volume 3011 of *Lecture Notes in Computer Science*, pages 200–208. Springer, 2004.

[216] N. Kohl. Application of or and cp techniques in a real world crew scheduling system. In *Proceedings of the International Workshop on Integration of Artificial Intelligence and Operations Research Techniques in Constraint Programming for Combintaorial Optimization Problems (CPAIOR 2000)*, Paderborn, Germany, 2000.

[217] H. W. Kuhn. The Hungarian method for the assignment problem. *Naval Research Logistics Quarterly*, 2:83–97, 1955.

[218] J.-L. Laurière. A language and a program for stating and solving combinatorial problems. *Artificial Intelligence*, 10:29–127, 1978.

[219] S. L. Lauritzen and D. J. Spiegelhalter. Local computations with probabilities on graphical structures and their application to expert systems. *Journal of the Royal Statistical Society B*, 50:157–224, 1988.

[220] Y. C. Law and J. H. M. Lee. Global constraints for integer and set value precedence. In M. Wallace, editor, *Principles and Practice of Constraint Programming (CP 2004)*, volume 3258 of *Lecture Notes in Computer Science*, pages 362–376. Springer, 2004.

[221] E. L. Lawler. *Combinatorial Optimization: Networks and Matroids*. Holt, Reinhart and Winston, New York, 1976.

[222] S. Lee and I. Grossmann. A global optimization algorithm for nonconvex generalized disjunctive programming and applications to process systems. *Computers and Chemical Engineering*, 25:1675–1697, 2001.

[223] S. Lee and I. Grossmann. Generalized disjunctive programming: Nonlinear convex hull relaxation and algorithms. *Computational Optimization and Applications*, 26:83–100, 2003.

[224] S. Lee and I. E. Grossmann. Global optimization of nonlinear generalized disjunctive programming with bilinear equality constraints: Applications to process networks. *Computers and Chemical Engineering*, 27:1557–1575, 2003.

[225] J. Little and K. Darby-Dowman. The significance of constraint logic programming to operational research. In M. Lawrence and C. Wilsden, editors, *Operational Research Tutorial Papers (Invited tutorial paper to the Operational Research Society Conference, 1995)*, pages 20–45, 1995.

[226] A. Lopez-Ortiz, C.-G. Quimper, J. Tromp, and P. van Beek. A fast and simple algorithm for bounds consistency of the alldifferent constraint. In *International Joint Conference on Artificial Intelligence (IJCAI 2003)*, pages 245–250, 2003.

[227] D. W. Loveland. *Automated Theorem Proving: A Logical Basis*. North-Holland, Amsterdam, 1978.

[228] A. Mackworth. Consistency in networks of relations. *Artificial Intelligence*, 8:99–118, 1977.

[229] C. T. Maravelias and I. E. Grossmann. Using MILP and CP for the scheduling of batch chemical processes. In J. C. Régin and M. Rueher, editors, *Integration of AI and OR Techniques in Constraint Programming for Combinatorial Optimization Problems (CPAIOR 2004)*, volume 3011 of *Lecture Notes in Computer Science*, pages 1–20. Springer, 2004.

[230] H. Marchand, A. Martin, R. Weismantel, and L. Wolsey. Cutting planes in integer and mixed integer programming. *Discrete Applied Mathematics*, 123:397–446, 2002.

[231] H. Marchand and L. A. Wolsey. Aggregation and mixed integer rounding to solve MIPs. *Operations Research*, 49:363–371, 2001.

[232] S. Martello and P. Toth. *Knapsack Problems: Algorithms and Computer Implementations*. Wiley, New York, 1990.

[233] S. Martello and P. Toth. Lower bounds and reduction procedures for the bin packing problem. *Discrete Applied Mathematics*, 28:59–70, 1990.

[234] A. Martin and R. Weismantel. Contribution to general mixed integer knapsack problems. Technical report SC 97-38, Konrad-Zuse-Zentrum für Informationstechnik Berlin, 1997.

[235] D. A. McAllester. Partial order backtracking. Manuscript, AI Laboratory, MIT, Cambridge, MA, 1993.

[236] G. P. McCormick. Computability of global solutions to factorable nonconvex programs: Part I—Convex underestimating problems. *Mathematical Programming*, 10:147–175, 1976.

[237] K. Mehlhorn and S. Thiel. Faster algorithms for bound-consistency of the sortedness and the alldifferent constraint. In R. Dechter, editor, *Principles and Practice of Constraint Programming (CP 2000)*, volume 1894 of *Lecture Notes in Computer Science*, pages 306–319. Springer, 2000.

[238] A. Meisels and A. Schaerf. Modelling and solving employee timetabling problems. *Annals of Mathematics and Artificial Intelligence*, 239:41–59, 2002.

[239] M. Milano and W. J. van Hoeve. Building negative reduced cost paths using constraint programming. In P. Van Hentenryck, editor, *Principles and Practice of Constraint Programming (CP 2002)*, volume 2470 of *Lecture Notes in Computer Science*, pages 1–16. Springer, 2002.

[240] R. Mohr and G. Masini. Good old discrete relaxation. In Y. Kodratoff, editor, *Proceedings of the 8th European Conference on Artificial Intelligence (ECAI 1988)*, pages 651–656. Pitman Publishers, 1988.

[241] U. Montanari. Networks of constraints: Fundamental properties and applications to picture processing. *Information Science*, 7:95–132, 1974.

REFERENCES

[242] M. W. Moskewicz, C. F. Madigan, Y. Zhao, L. Zhang, and S. Malik. Chaff: Engineering an efficient SAT solver. In *Proceedings of the 38th Design Automation Conference (DAC 2001)*, pages 530–535, 2001.

[243] M. Müller-Hannemann, W. Stille, and K. Weihe. Evaluating the bin-packing constraint, Part I: Overview of the algorithmic approach. Technical report, Technische Universität Darmstadt, 2003.

[244] M. Müller-Hannemann, W. Stille, and K. Weihe. Evaluating the bin-packing constraint, Part II: An adaptive rounding problem. Technical report, Technische Universität Darmstadt, 2003.

[245] M. Müller-Hannemann, W. Stille, and K. Weihe. Evaluating the bin-packing constraint, Part III: Joint evaluation with concave constraints. Technical report, Technische Universität Darmstadt, 2003.

[246] M. Müller-Hannemann, W. Stille, and K. Weihe. Patterns of usage for global constraints: A case study based on the bin-packing constraint. Technical report, Technische Universität Darmstadt, 2003.

[247] D. Naddef. Polyhedral theory and branch-and-cut algorithms for the symmetric TSP. In G. Gutin and A. P. Punnen, editors, *The Traveling Salesman Problem and its Variations*, pages 29–116. Kluwer, Dordrecht, 2002.

[248] D. Naddef and S. Thienel. Efficient separation routines for the symmetric traveling salesman problem I: General tools and comb separation. *Mathematical Programming*, 92:237–255, 2002.

[249] D. Naddef and S. Thienel. Efficient separation routines for the symmetric traveling salesman problem II: Separating multi handle inequalities. *Mathematical Programming*, 92:257–285, 2002.

[250] A. Nedic and D. P. Bertsekas. Incremental subgradient methods for nondifferentiable pptimization. *SIAM Journal on Optimization*, 12:109–138, 2001.

[251] G. L. Nemhauser and L. A. Wolsey. *Integer and Combinatorial Optimization*. Wiley, New York, 1999.

[252] A. Neumaier. Complete search in continuous global optimization and constraint satisfaction. In A. Iserles, editor, *Acta Numerica 2004*, pages 271–369. Cambridge University Press, 2004.

[253] P. Nobili and A. Sassano. Facets and lifting procedures for the set covering polytope. *Mathematical Programming*, 45:111–137, 1989.

[254] W. P. M. Nuijten. *Time and Resource Constrained Scheduling*. PhD thesis, Eindhoven University of Technology, 1994.

[255] W. P. M. Nuijten and E. H. L. Aarts. Constraint satisfaction for multiple capacitated job shop scheduling. In A. Cohn, editor, *Proceedings of the 11th European Conference on Artificial Intelligence (ECAI 1994)*, pages 635–639. Wiley, 1994.

[256] W. P. M. Nuijten and E. H. L. Aarts. A computational study of constraint satisfaction for multiple capacitated job shop scheduling. *European Journal of Operational Research*, 90:269–284, 1996.

[257] W. P. M. Nuijten, E. H. L. Aarts, D. A. A. van Erp Taalman Kip, and K. M. van Hee. Randomized constraint satisfaction for job-shop scheduling. In *AAAI-SIGMAN Workshop on Knowledge-Based Production Planning, Scheduling and Control*, 1993.

[258] M. Osorio and F. Glover. Logic cuts using surrogate constraint analysis in the multidimensional knapsack problem. In C. Gervet and M. Wallace, editors, *Proceedings of the International Workshop on Integration of Artificial Intelligence and Operations Research Techniques in Constraint Programming for Combintaorial Optimization Problems (CPAIOR 2001)*, Ashford, U.K., 2001.

[259] G. Ottosson, E. Thorsteinsson, and J. N. Hooker. Mixed global constraints and inference in hybrid IP-CLP solvers. In *Proceedings of CP99 Post-Conference Workshop on Large-Scale Combinatorial Optimization and Constraints*, ¡http://www.dash.co.uk/wscp99¿, pages 57–78, 1999.

[260] G. Ottosson, E. Thorsteinsson, and J. N. Hooker. Mixed global constraints and inference in hybrid CLP-IP solvers. *Annals of Mathematics and Artificial Intelligence*, 34:271–290, 2002.

[261] M. Padberg. On the facial structure of set packing polyhedra. *Mathematical Programming*, 5:199–215, 1973.

[262] M. Padberg. A note on zero-one programming. *Operations Research*, 23:833–837, 1975.

[263] M. Padberg and G. Rinaldi. An efficient algorithm for the minimum capacity cut problem. *Mathematical Programming*, 47:19–36, 1990.

[264] C. Le Pape. Implementation of resource constraints in ILOG SCHEDULE: A library for the development of constraint-based scheduling systems. *Intelligent Systems Engineering*, 3:55–66, 1994.

[265] V. T. Paschos. A survey of approximately optimal solutions to some covering and packing problems. *ACM Computing Surveys*, 29:171–209, 1997.

[266] G. Pesant. A filtering algorithm for the stretch constraint. In T. Walsh, editor, *Principles and Practice of Constraint Programming (CP 2001)*, volume 2239 of *Lecture Notes in Computer Science*, pages 183–195. Springer, 2001.

[267] T. Petit, J. C. Régin, and C. Bessiere. Specific filtering algorithms for over-constrained problems. In T. Walsh, editor, *Principles and Practice of Constraint Programming (CP 2001)*, volume 2239 of *Lecture Notes in Computer Science*, pages 451–463. Springer, 2001.

[268] Y. Pochet and R. Weismantel. The sequential knapsack polytope. *SIAM Journal on Optimization*, 8:248–264, 1998.

REFERENCES

[269] Y. Pochet and L. A. Wolsey. Integer knapsack and flow covers with divisible coefficients: Polyhedra, optimization, and separation. *Discrete Applied Mathematics*, 59:57–74, 1995.

[270] F. P. Preparata and S. J. Hong. Convex hulls of finite sets of points in two and three dimensions. *Communications of the ACM*, 20:87–93, 1977.

[271] S. Prestwich. Exploiting relaxation in local search. In *First International Workshop on Local Search Techniques in Constraint Satisfaction (LSCS 2004)*, Toronto, 2004.

[272] J.-F. Puget. A fast algorithm for the bound consistency of alldiff constraints. In *National Conference on Artificial Intelligence (AAAI 1998)*, pages 359–366. AAAI Press, 1990.

[273] C.-G. Quimper, A. López-Ortiz, P. van Beek, and A. Golynski. Improved algorithms for the global cardinality constraint. In M. Wallace, editor, *Principles and Practice of Constraint Programming (CP 2004)*, volume 3258 of *Lecture Notes in Computer Science*, pages 542–556. Springer, 2004.

[274] C.-G. Quimper, P. van Beek, A. López-Ortiz, A. Golynski, and S. B. Sadjad. An efficient bounds consistency algorithm for the global cardinality constraint. In F. Rossi, editor, *Principles and Practice of Constraint Programming (CP 2003)*, volume 2833 of *Lecture Notes in Computer Science*, pages 600–614. Springer, 2003.

[275] W. V. Quine. The problem of simplifying truth functions. *American Mathematical Monthly*, 59:521–531, 1952.

[276] W. V. Quine. A way to simplify truth functions. *American Mathematical Monthly*, 62:627–631, 1955.

[277] R. Raman and I. E. Grossmann. Modeling and computational techniques for logic based integer programming. *Computers and Chemical Engineering*, 20:563–578, 1994.

[278] R. Rasmussen and M. A. Trick. A benders approach to the constrained minimum break problem. *European Journal of Operational Research*, to appear.

[279] P. Refalo. Tight cooperation and its application in piecewise linear optimization. In J. Jaffar, editor, *Principles and Practice of Constraint Programming (CP 1999)*, volume 1713 of *Lecture Notes in Computer Science*, pages 375–389. Springer, 1999.

[280] P. Refalo. Linear formulation of constraint programming models and hybrid solvers. In R. Dechter, editor, *Principles and Practice of Constraint Programming (CP 2000)*, volume 1894 of *Lecture Notes in Computer Science*, pages 369–383. Springer, 2000.

[281] J.-C. Régin. A filtering algorithm for constraints of difference in CSP. In *National Conference on Artificial Intelligence (AAAI 1994)*, pages 362–367. AAAI Press, 1994.

[282] J.-C. Régin. Generalized arc consistency for *global cardinality* constraint. In *National Conference on Artificial Intelligence (AAAI 1996)*, pages 209–215. AAAI Press, 1996.

[283] J.-C. Régin. The symmetric alldiff constraint. In T. Dean, editor, *Proceedings of the International Joint Conference on Artificial Intelligence (IJCAI 1999)*, vol. 1, pages 420–425, Stockholm, 1996. Morgan Kaufmann Publishers.

[284] J.-C. Régin. Using constraint propagation to solve the maximum clique problem. In F. Rossi, editor, *Principles and Practice of Constraint Programming (CP 2003)*, volume 2833 of *Lecture Notes in Computer Science*, pages 634–648. Springer, 2003.

[285] J.-C. Régin. Modeling problems in constraint programming. In *Tutorial presented at conference on Principles and Practice of Constraint Programming (CP 2004)*, Toronto, 2004.

[286] C. Ribeiro and M. A. Carravilla. A global constraint for nesting problems. In J. C. Régin and M. Rueher, editors, *Integration of AI and OR Techniques in Constraint Programming for Combinatorial Optimization Problems (CPAIOR 2004)*, volume 3011 of *Lecture Notes in Computer Science*, pages 256–270. Springer, 2004.

[287] J. A. Robinson. A machine-oriented logic based on the resolution principle. *Journal of the ACM*, 12:23–41, 1965.

[288] R. Rodošek, M. Wallace, and M. Hajian. A new approach to integrating mixed integer programming and constraint logic programming. *Annals of Operations Research*, 86:63–87, 1997.

[289] B. K. Rosen. Robust linear algorithms for cutsets. *Journal of Algorithms*, 3:205–212, 1982.

[290] L.-M. Rousseau. Stabilization issues for constraint programming based column generation. In J. C. Régin and M. Rueher, editors, *Integration of AI and OR Techniques in Constraint Programming for Combinatorial Optimization Problems (CPAIOR 2004)*, volume 3011 of *Lecture Notes in Computer Science*, pages 402–408. Springer, 2004.

[291] L. M. Rousseau, M. Gendreau, and G. Pesant. Solving small VRPTWs with constraint programming based column generation. In N. Jussien and F. Laburthe, editors, *Proceedings of the International Workshop on Integration of Artificial Intelligence and Operations Research Techniques in Constraint Programming for Combintaorial Optimization Problems (CPAIOR 2002)*, Le Croisic, France, 2002.

[292] R. Sadykov. A hybrid branch-and-cut algorithm for the one-machine scheduling problem. In J. C. Régin and M. Rueher, editors, *Integration of AI and OR Techniques in Constraint Programming for Combinatorial Optimization Problems (CPAIOR 2004)*, volume 3011 of *Lecture Notes in Computer Science*, pages 409–415. Springer, 2004.

[293] A. Sassano. On the facial structure of the set covering polytope. *Mathematical Programming*, 44:181–202, 1989.

[294] N. W. Sawaya and I. E. Grossmann. A cutting plane method for solving linear generalized disjunctive programming problems. Research report, Department of Chemical Engineering, Carnegie Mellon University, 2004.

[295] N. W. Sawaya and I. E. Grossmann. Computational implementation of nonlinear convex hull reformulations. Research report, Department of Chemical Engineering, Carnegie Mellon University, 2005.

[296] L. Schrage and L. Wolsey. Sensitivity analysis for branch and bound integer programming. *Operations Research*, 33:1008–1023, 1985.

[297] A. Schrijver. *Theory of Linear and Integer Programming*. Wiley, New York, 1986.

[298] M. Sellmann. An arc-consistency algorithm for the minimum-weight all different constraint. In P. Van Hentenryck, editor, *Principles and Practice of Constraint Programming (CP 2002)*, volume 2470 of *Lecture Notes in Computer Science*, pages 744–749. Springer, 2002.

[299] M. Sellmann. Cost-based filtering for shorter path constraints. In F. Rossi, editor, *Principles and Practice of Constraint Programming (CP 2003)*, volume 2833 of *Lecture Notes in Computer Science*, pages 694–708. Springer, 2003.

[300] M. Sellmann and T. Fahle. Constraint programming based Lagrangian relaxation for a multimedia application. In C. Gervet and M. Wallace, editors, *Proceedings of the International Workshop on Integration of Artificial Intelligence and Operations Research Techniques in Constraint Programming for Combintaorial Optimization Problems (CPAIOR 2001)*, Ashford, U.K., 2001.

[301] M. Sellmann, K. Zervoudakis, P. Stamatopoulos, and T. Fahle. Crew assignment via constraint programming: Integrating column generation and heuristic tree search. *Annals of Operations Research*, 115:207–225, 2002.

[302] G. Shafer, P. P. Shenoy, and K. Mellouli. Propagating belief functions in qualitative markov trees. *International Journal of Approximate Reasoning*, 1:349–400, 1987.

[303] A. Shamir. A linear time algorithm for finding minimum cutsets in reducible graphs. *SIAM Journal on Computing*, 8:645–655, 1979.

[304] P. Shaw. A constraint for bin packing. In M. Wallace, editor, *Principles and Practice of Constraint Programming (CP 2004)*, volume 3258 of *Lecture Notes in Computer Science*, pages 648–662. Springer, 2004.

[305] P. P. Shenoy and G. Shafer. Propagating belief functions with local computation. *IEEE Expert*, 1:43–52, 1986.

[306] J. A. Shufelt and H. J. Berliner. Generating hamiltonian circuits without backtracking. *Theoretical Computer Science*, 132:347–375, 1994.

[307] J. P. M. Silva and K. A. Sakallah. GRASP—a search algorithm for propositional satisfiability. *IEEE Transactions on Computers*, 48:506–521, 1999.

[308] S. S. Skiena. *The Algorithm Design Manual*. Springer, New York, 1997.

[309] S. F. Smith. OPIS: A methodology and architecture for reactive scheduling. In M. Zweben and M. S. Fox, editors, *Intelligent Scheduling*, pages 29–66. Morgan Kaufmann, San Francisco, 1995.

[310] R. Stubbs and S. Mehrotra. A branch-and-cut method for 0-1 mixed convex programming. *Mathematical Programming*, 86:515–532, 1999.

[311] B. Sturmfels. *Gröbner Bases and Convex Polytopes*. American Mathematical Society, Providence, 1995.

[312] M. Tawarmalani and Nikolaos V. Sahinidis. *Convexification and Global Optimization in Continuous and Mixed-integer Nonlinear Programming: Theory, Algorithms, Software, and Applications*. Springer, 2002.

[313] R. R. Thomas. The structure of group relaxations. In K. Aardal, G. L. Nemhauser, and R. Weismantel, editors, *Discrete Optimization*, Handbooks in Operations Research and Management Science, pages 123–170. Elsevier, Amsterdam, 2005.

[314] E. Thorsteinsson. Branch and check: A hybrid framework integrating mixed integer programming and constraint logic programming. In T. Walsh, editor, *Principles and Practice of Constraint Programming (CP 2001)*, volume 2239 of *Lecture Notes in Computer Science*, pages 16–30. Springer, 2001.

[315] E. Thorsteinsson and G. Ottosson. Linear relaxations and reduced-cost based propagation of continuous variable subscripts. *Annals of Operations Research*, 115:15–29, 2001.

[316] C. Timpe. Solving planning and scheduling problems with combined integer and constraint programming. *OR Spectrum*, 24:431–448, 2002.

[317] P. Torres and P. Lopez. On not-first/not-last conditions in disjunctive scheduling. *European Journal of Operational Research*, 127:332–343, 2000.

[318] P. Toth and D. Vigo. Models, relaxations and exact approaches for the capacitated vehicle routing problem. *Discrete Applied Mathematics*, 123:487–512, 2002.

[319] M. Trick. A dynamic programming approach for consistency and propagation for knapsack constraints. In C. Gervet and M. Wallace, editors, *Proceedings, Integration of AI and OR Techniques in Constraint Programming for Combinatorial Optimization Problems (CPAIOR 2001)*, pages 113–124, Ashford, U.K., 2001.

[320] M. Trick. Formulations and reformulations in integer programming. In R. Barták and M. Milano, editors, *Integration of AI and OR Techniques in Constraint Programming for Combinatorial Optimization Problems (CPAIOR 2005)*, volume 3524 of *Lecture Notes in Computer Science*, pages 366–379. Springer, 2005.

REFERENCES

[321] E. Tsang. *Foundations of Constraint Satisfaction*. Academic Press, London, 1983.

[322] M. Türkay and I. E. Grossmann. Disjunctive programming techniques for the optimization of process systems with discontinuous investment costs-multiple size regions. *Industrial Engineering Chemical Research*, 35:2611–2623, 1996.

[323] M. Türkay and I. E. Grossmann. Logic-based MINLP algorithms for the optimal synthesis of process networks. *Computers and Chemical Engineering*, 20:959–978, 1996.

[324] W.-J. van Hoeve. A hyper-arc consistency algorithm for the soft alldifferent constraint. In M. Wallace, editor, *Principles and Practice of Constraint Programming (CP 2004)*, volume 3258 of *Lecture Notes in Computer Science*, pages 679–689. Springer, 2004.

[325] R. J. Vanderbei. *Linear Programming: Foundations and Extensions*. Kluwer, Boston, 1996.

[326] A. Vecchietti, S. Lee, and I. E. Grossmann. Characterization and formulation of disjunctions and their relaxations. In *Proceedings of Mercosul Congress on Process Systems Engineering (ENPROMER 2001)*, volume 1, pages 409–414, Santa Fe, Chile, 2001.

[327] A. F. Veinott and G. B. Dantzig. Integral extreme points. *SIAM Review*, 10:371–372, 1968.

[328] R. R. Vemuganti. Applications of set covering, set packing and set partitioning models: A survey. In D.-Z. Du and P. Pardalos, editors, *Handbook of Combinatorial Optimization, Vol. 1*, pages 573–746. Kluwer, Dordrecht, 1998.

[329] M. Wallace, M. S. Novello, and J. Schimpf. ECLiPSe: A platform for constraint logic programming. *ICL Systems Journal*, 12:159–200, 1997.

[330] H. P. Williams. Linear and integer programming applied to the propositional calculus. *International Journal of Systems Research and Information Science*, 2:81–100, 1987.

[331] H. P. Williams. *Model Building in Mathematical Programming, 4th Ed.* Wiley, New York, 1999.

[332] H. P. Williams. The formulation and solution of discrete optimization models. In G. Appa, L. Pitsoulis, and H. P. Williams, editors, *Handbook on Modelling for Discrete Optimization*, pages 3–38. Springer, 2006.

[333] H. P. Williams and H. Yan. Representations of the all_different predicate of constraint satisfaction in integer programming. *INFORMS Journal on Computing*, 13:96–103, 2001.

[334] P. Van Hentenryck with contributions by I. Lustig, L. Michel, and J. F. Puget. *The OPL Optimization Programming Language*. MIT Press, Cambridge, MA, 1999.

[335] L. A. Wolsey. Faces for a linear inequality in 0-1 variables. *Mathematical Programming*, 8:165–178, 1975.

[336] L. A. Wolsey. The *b*-hull of an integer program. *Discrete Applied Mathematics*, 3:193–201, 1981.

[337] L. A. Wolsey. Valid inequalities for 0-1 knapsacks and MIPs with generalized upper bound constraints. *Discrete Applied Mathematics*, 29:251–261, 1990.

[338] L. A. Wolsey. *Integer Programming*. Wiley, New York, 1998.

[339] Q. Xia, A. Eremin, and M. Wallace. Problem decomposition for traffic diversions. In J. C. Régin and M. Rueher, editors, *Integration of AI and OR Techniques in Constraint Programming for Combinatorial Optimization Problems (CPAIOR 2004)*, volume 3011 of *Lecture Notes in Computer Science*, pages 348–363. Springer, 2004.

[340] H. Yan and J. N. Hooker. Tight representations of logical constraints as cardinality rules. *Mathematical Programming*, 85:363–377, 1995.

[341] B. Yehuda, J. Geiger, J. Naor, and R. M. Roth. Approximation algorithms for the vertex feedback set problem with application in constraint satisfaction and bayesian inference. In *Proceedings of 5th Annual ACM-SIAM Symposium on Discrete algorithms*, pages 344–354, 1994.

[342] T. H. Yunes. On the sum constraint: Relaxation and applications. In P. Van Hentenryck, editor, *Principles and Practice of Constraint Programming (CP 2002)*, volume 2470 of *Lecture Notes in Computer Science*, pages 80–92. Springer, 2002.

[343] T. H. Yunes, A. V. Moura, and C. C. de Souza. Exact solutions for real world crew scheduling problems. Presentation at INFORMS national meeting, Philadelphia, 1999.

[344] T. H. Yunes, A. V. Moura, and C. C. de Souza. Hybrid column generation approaches for urban transit crew management problems. *Transportation Science*, to appear.

Index

0-1 knapsack cut, 266
0-1 knapsack problem, **159**
0-1 linear inequality, **159**, 261
  implication of, 159, 246, 417
0-1 linear metaconstraint, **416**
0-1 linear programming, **262**
  cutting planes for, 262
0-1 linear system
  Benders cuts for, 179, 417
  continuous relaxation of, 262, 417
  cutting planes for, 417
0-1 nonlinear inequality, 246
0-1 resolution, **165**, 246, 417
  completeness of, 166, 417

2-connected graph, **383**
2-satisfiability problem, 157

Absorbsion
  for inequalities, **160**
  for logical clauses, **149**, 169
Aircraft scheduling, 11
Airline crew scheduling, 10–11, 51, 53
All-different (alldiff) metaconstraint, 35, 37, **187**, 247, 362, 417
  bounds propagation for, 191, 418
  convex hull relaxation of, 363, 368, 413, 418
  domain filtering for, 37, 189, 418
  matching model for, 188, 247
  MILP model for, 368
  minimum weight, 418
  modeling with, 200, 369
  multiple alldiffs, 418
  relaxation of, 363, 366, 368, 370, 413, 418
  separating cuts for, 365
  set-valued, 418
  soft, **443**
  symmetric, **445**
  with cost variable, 370

Alternating path, **189**
Among metaconstraint, **418**
Ant colony optimization, 102
Arc consistency, 245
  generalized, **108**
Assignment problem, **200**, 418
  0-1 model for, 332, 334
  generalized, **295**
Assignment problem relaxation, 10
  of traveling salesman problem, 380
At-least metaconstraint, 421
At-most metaconstraint, 421
Augmenting path, **189**, **196**, 246–247

Backjumping, 71
BARON, 14
Basic solution, **131**, 253
  starting, 255
Basic variable, **131**, 253
Batch scheduling, 10, 12
Bayesian nmetwork, 245
Belief logic, 245
Benders cut, **77**, 140
  classical, 137
  for 0-1 linear system, 179, 417
  for integer linear system, 179, 432
  for machine scheduling, 82
  for minimum-cost planning and
    scheduling, 223, 242, 248
  for minimum-makespan planning and
    scheduling, 224, 243, 248
  for minimum-tardiness planning and
    scheduling, 226–227, 243, 248
  from branching dual, 179
  from inference dual, 120
  from linear programming dual, 137
  from subadditive dual, 179
  strong, 83
Benders decomposition, 4

as constraint-directed search, 8, **76**
classical, 104, 136, 140
for machine scheduling, 78
for planning and scheduling, 222, 224–226, 241–243, 426
for vehicle routing, 88
in local search, 95
logic-based, 10–11, 14, **77**–78, 104, 120, 222, 241
relaxation of subproblem, 80, 400–403, 405
Big-M constraint, 323
Big-M relaxation
for disjunction of linear equations, 306
for disjunction of linear inequalities, 303, 412
for disjunction of linear systems, 300, 412
for disjunction of nonlinear systems, 317
for vector-valued indexed linear element metaconstraint, 360
of disjunction of nonlinear systems, 438
of element metaconstraint, 357
of linear disjunction metaconstraint, 434
Bin packing metaconstraint, **419**
domain filtering for, 420
relaxations of, 420
Bipartite matching, **188**
Boat party scheduling, 11
Boolean variable, 30, 153–154, 158, 415
Bounds completeness, **110**, 124
Bounds consistency, **110**, 245
Bounds propagation, 4, **25**, 29, 109
for all-different metaconstraint, 191, 418
for cardinality metaconstraint, 420
for cumulative scheduling, 232, 236, 238, 240, 426
for diffn metaconstraint, 428
for disjunctive scheduling, 213–214, 217, 428
for element metaconstraint, 185
for integer programming, 432
for knapsack constraints, 25
for linear metaconstraint, 124
for linear programming, 135
for nonlinear constraints, 41
for semicontinuous piecewise linear metaconstraint, 260, 440
for set-valued variables, 57
for set sum metaconstraint, 58, 62
with dual multipliers, 135
Branch and bound, **22**
Branch and check, 86, 104
Branch and cut, 4, 14
Branch and infer, 4, **21**
Branch and price, 10–11, 14, 51, 103
Branch and relax, **22**, 28, 32, 41, 51, 175
in local search, 93
Branching, 3

as constraint-directed search, 71
by domain splitting, 40, 43, 51
by interval splitting, 41
chronological, 72
constraint-directed, 65
first-fail, 40
in GRASP, 90, 92
in local search, 91
in machine scheduling, 85
in simulated annealing, 92
inexhaustive, 91
on a constraint, 6, 22, 29
on a domain constraint, 29
on a piecewise linear metaconstraint, 260
on a set-valued variable, 58
on a variable, **21**, 65
order of, 113
Branching dual, **175**, 246
Branching heuristic, 43
Branching search, **21**, 28, 40, 43, 51, 63, 70
Bucket elimination, 245

Call center scheduling, 10
Candidate solution, **17**, 21–22, 63
Capacitated facility location problem
MILP model for, 326, 437
Capital budgeting problem
0-1 model for, 330
Cardinality clause, **163**, 246
0-1 model for, 330
implication by 0-1 linear inequality, 163, 421
implication by another cardinality clause, 164, 421
relaxation of, 421
Cardinality clause metaconstraint, **421**
Cardinality conditional, **422**
relaxation of, 422
Cardinality conditional metaconstraint, **422**
Cardinality metaconstraint, 35, **194**, 247, 375, 420
0-1 formulation for, 378
bounds propagation for, 420
convex hull relaxation of, 376, 378, 421
domain filtering for, 196, 420
flow model for, 195, 247
relaxation of, 379, 421
separating cuts for, 377
symmetric, **446**
Certificate, 117
CHAFF, 103
Change metaconstraint, **422**
domain filtering for, 423
Channeling constraint, 34, **36**, 183
propagation of, 39
Chronological backtracking, 72

INDEX   477

Chvátal-Gomory cut, 170, 246, **263**, 276,
    411–412
  as rank 1 cut, 281
Chvátal-Gomory procedure, 165, 262, 276,
    417, 433
  completeness of, 264, 277, 282
Chvátal function, **172**
Circuit metaconstraint, 88, **199**, 247, 379,
    423
  0-1 model for, 379
  comb inequalities for, 382
  domain filtering for, 201–202, 423
  flow model for, 204
  in traveling salesman problem, 379
  modeling with, 200
  relaxation of, 380, 423
  separating cuts for, 380
Circuit verification problem, 11, 14
Clause, 68, **149**, 152, 434
  cardinality, **163**, 246
  convex hull relaxation of, 336
  empty, **151**
  filter for, 69
  Horn, **152**, 246
  implication by 0-1 linear inequality, 161
  implication of, 149
  multivalent, 158
  unit, **151**
Clause learning, 4, 67, **70**, 103
Clique, **273**
Clique inequality, **273**–274, 412, 443
Clique metaconstraint, **424**
  domain filtering for, 424
  relaxation of, 424
Closed set, **299**, 311
Closure, **299**, 311, 316
CNF, **68**, 86, 149
  conversion to, 149
Co-NP, 117
Column generation, **52**, 55, 62–63, 103
Comb inequality, 382, 413
Complementary slackness, **130**, 139, 145,
    310
Complete constraint set, **108**
Complete inference method, 106, **108**
Completeness
  bounds, **110**, 124
  domain, **108**, 124, 245
  k-, **110**, 114, 125, 245
  of constraint set, **108**
  of inference method, **108**
  strong k-, **111**
Computer processor scheduling, 12
Conditional metaconstraint, 30, 59, 103,
    424, 434
  relaxation of, 425
Conditional proposition, 68

Conflict clause, 66, **70**–72, 119
Conformity, 65, **73**–74, 94
Conjunctive normal form, **68**, 86, 434
Connected component, **188**
Consistency, 107
  k-, 245
  arc, 245
  bounds, **110**, 245
  generalized arc, 245
  hyperarc, **108**, 245
  interval, 245
  k-, 112, 114
Constraint-based control, 6
Constraint-directed branching, 65
Constraint-directed search, 3, **63**, 65, 71,
    103, 245
  based on inference dual, 8
  for vehicle routing, 98
  inexhaustive, 94
Constraint, **107**
  channeling, 34, **36**, 183
  equivalent, **107**
  global, 1, 6
  implication of, **107**
  inequality, 140
  knapsack, 25, 100, 432
  knapsack covering, **24**
  knapsack packing, **24**
  precedence, 82
  satisfaction of, **107**
  valid, **16**
  violation of, **107**
Constraint programming, 1–2, 22
  integrated methods in, 13
Constraint propagation, **19**
Constraint store, **19**–20
Continuous global optimization, 1, 4, 22, 41,
    103
Continuous relaxation, 27, 30, 52
Convex combination, **298**
Convex graph, **192**, 247
Convex hull, 30, **298**
  closure of, 316
Convex hull relaxation, 11, 32, 299
  of alldiff metaconstraint, 363, 368, 413,
    418
  of cardinality metaconstraint, 376, 378,
    421
  of disjunction of linear systems, 31, 299,
    319, 412
  of disjunction of nonlinear systems, 313,
    412, 438
  of element metaconstraint, 353, 413, 430
  of fixed-charge function, 261
  of indexed linear element metaconstraint,
    356
  of linear disjunction metaconstraint, 434

of logical formulas, 336, 435
of MILP model, 320
of semicontinuous piecewise linear
metaconstraint, 259, 441
of sum metaconstraint, 445
of vector-valued element metaconstraint,
359, 430
of vector-valued indexed linear element
metaconstraint, 360
Convex set, **313**
Coupled variables, **112**
Cover, **266**
for a knapsack inequality, 59
minimal, **266**
Cover inequality, **266**, 411, 417, 432
CP-AI-OR, 14
Crew scheduling, 62
Cumulative (scheduling) metaconstraint,
230, 248, 425
bounds propagation for, 232, 236, 238,
240, 426
MILP-based relaxation of, 394, 426
MILP model for, 335, 394, 413
relaxation of, 393–394, 398, 410, 426
relaxation when a Benders subproblem,
401–403, 405, 413, 426
Cumulative scheduling, 230
Cut
0-1 knapsack, 266
Benders, 137, 140
Chvátal-Gomory, 170, 246, **263**, 276,
411–412
flow, 437
Gomory, 173, 275, **278**, 412, 417, 433
integer rounding, **263**
knapsack, **26**, 29, 50, 103, 266, 332, 417,
433
lift-and-project, 417, 436
mixed integer Gomory, 436
mixed integer rounding, **283**, 285, 412,
436
separating, 5, 22, 250, 273, **275**, 417
valid, **262**
Cutset metaconstraint, **426**
domain filtering for, 426
Cutting plane, 4, 11, 22, 26, 249, **262**
for 0-1 linear system, 417
for alldiff metaconstraint, 365
for cardinality metaconstraint, 377
for circuit constraint, 380
for circuit metaconstraint, 382
for cumulative (scheduling)
metaconstraint, 398, 410
for disjunction of linear systems, 307
for disjunctive (scheduling)
metaconstraint, 391, 393
for integer linear system, 278, 433

for knapsack inequality, 26
for logical clauses, 348
for mixed integer linear system, 285, 436
for set packing problem, 272–273, 443
for traveling salesman problem, 380, 382
for vector-valued element metaconstraint,
359
Cutting stock problem, 62
Cycle metaconstraint, **426**
domain filtering for, 427
relaxation of, 427

Davis-Putnam-Loveland algorithm, **67**, 70,
103, 119
as constraint-directed search, 8, 67
De Morgan's laws, 149
Degeneracy, **253**
Degree of inequality, **159**
Dependency graph, **112**, 115, 245
and backtracking, 113
width of, 113
Depth-first search, 29
Diagonal sum, **165**
Diffn metaconstraint, 63, **427**
bounds propagation for, 428
Digital recording problem, 10–11
Disjunction
factored relaxation for, 325, 340
of linear equations, 306, 434
of linear inequalities, 303, 434
of linear systems, 32, 298, 319, 434
of nonlinear systems, 313, 438
product relaxation for, 325
Disjunctive (scheduling) metaconstraint, 79,
212, 385, 428
bounds propagation for, 213–214, 217, 428
disjunctive relaxation of, 386
MILP-based relaxation of, 388, 429
MILP model for, 334, 388
relaxation of, 385–386, 388, 391, 393, 429
relaxation when a Benders subproblem,
80, 401–403, 405, 429
Disjunctive modeling, 103
Disjunctive scheduling, **212**, 247, 385
Distribute metaconstraint, **194**, 420
Domain, **16**
Domain completeness, **108**, 124, 245
and projection, 109
Domain filtering, 12, **19**, 25, **109**
for all-different metaconstraint, 37, 189,
418
for bin packing metaconstraint, 420
for cardinality metaconstraint, 196–197,
420
for change metaconstraint, 423
for circuit metaconstraint, 201–202, 423
for clique metaconstraint, 424

INDEX

for cutset metaconstraint, 426
for cycle metaconstraint, 427
for element metaconstraint, 39, 183, 430
for indexed linear element, 48
for indexed linear metaconstraint, 48, 185, 432
for knapsack constraint, 25
for lex-greater metaconstraint, 433
for linear programming, 124
for logical clauses, 69, 152
for min-n metaconstraint, 436
for nvalues metaconstraint, 439
for path metaconstraint, 440
for same metaconstraint, 441
for soft alldiff metaconstraint, 443
for stretch metaconstraint, 38, 210, 444
for sum metaconstraint, 445
for symmetric alldiff metaconstraint, 446
for symmetric cardinality metaconstraint, 446
for traveling salesman problem, 109, 111
for value precedence metaconstraint, 447
in dynamic programming, 247
Domain reduction, **19**, 109
by Lagrange multipliers, 44
Domain store, **19**–20, 40
as a relaxation, 40
Domination
between inequalities, 141, 143
between linear inequalities, 122
DPL algprithm, **67**
Dual multiplier, 52, 116, 176
in bounds filtering, 135
Dual variable, 10, 251
Dual
branching, 171, **175**, 246
inear programming, 411
inference, 7–8, 105, **115**, 117, 128, 245
Lagrangean, 8, **143**, 246, 293–294, 412
linear programming, 8, 127–**128**, 133, 245, 309
relaxation, 7, 9, 250–**251**, 411
subadditive, 170–**171**, 246
superadditive, 8, 170–**171**
surrogate, 8, **141**, 144, 147, 246, **292**
Duality, 7
of search and inference, 8, 105, 117
of search and relaxation, 9
strong, **116**, 128, 170, **252**
weak, 136, **144**, **252**
Duality gap, **116**, 142, 144, 147, **252**
Dualized constraint, **293**
Dynamic backtracking, 65, 158
partial-order, 65, **72**
Dynamic programming, 208, 247
domain filtering in, 247
for sequential lifting, 268

for shortest path, 212
model for stretch metaconstraint, 208, 247
nonserial, 245

ECLiPSe, 14
Edge finding, 83, 87, 104, 121, **213**, 247–248
extended, **236**, 426
for cumulative scheduling, 231, 426
for disjunctive scheduling, 213, 428
polynomial-time algorithm for, 214, 234
Element metaconstraint, 38, 61, 103, **182**, 246, 429
big-M relaxation of, 357
bounds propagation for, 185
convex hull relaxation of, 353, 413, 430
domain filtering for, 39, 183, 430
indexed linear, 47, **183**
multidimensional, **183**
relaxation of, 49, 352, 430
vector-valued, **358**, 429
Elementary matrix, **258**
Employee scheduling, 11, 33, 103, 422
Empty clause, **151**
Energetic reasoning, **239**, 248, 426
Energy, **231**
Enumerated problem, **20**
Enumerated restriction, **21**
Epigraph, **20**
Eta vector, **258**
Exhaustive restrictions, **17**
Exhaustive search, 3, **17**, 21, 64
Extended edge finding, **236**
Extreme ray, 126, 129, 137

Facet-defining inequality, **272**
Factored relaxation, 43, 61, 103, 326
Factory retrofit problem, 11
Farkas Lemma, **122**, 138, 245, 411
for inequalities, 123
Farm planning, 62
Feasible set, **16**
Feasible solution, **16**
Filtering, 4, **19**
domain, 25, **109**
First-fail branching, 40
Fixed-charge function, 261
big-M model for, 323
big-M relaxation of, 323
convex hull relaxation of, 261, 322
MILP model for, 321
Fixed-charge network flow problem, 11, 437
Fixed-charge problem, 321, 11
Fixed cost, 30
Fixed point (in propagation), 26, 42, 125
Flow cut, 437
Flow metaconstraint, **430**
Flow shop scheduling, 10

Fourier-Motzkin elimination, **125**, 127, 138, 245
Freight transfer problem, 24
　MILP model for, 330

Gcc metaconstraint, **194**, 420
Generalized arc consistency, **108**, 245
Generalized assignment problem, **295**
　Lagrangean relaxation for, 295
Generalized cardinality metaconstraint, **194**, 420
Genetic algorithm, 102
Global constraint, 1, 6
Global optimization, 1, 4, 41, 103
　continuous, 22
Global optimum, 41
Gomory cut, 173, 275, **278**, 412, 417, 433
Good charcterization, 118
Gröbner basis, 433
GRASP, 4, 89–**90**
　as branching search, 90, 92
　for vehicle routing, 96
　with relaxation, 98
Greedy algorithm
　in GRASP, 90, 96

Hamiltonian cycle, **199**, 201, 247, 423
　permissible, **203**
Hamiltonian edge, **199**, 203, 247
Hamiltonian graph, 206
Heuristic method, 1, 4, 89, 104
Homogeneous function, **171**
Horn clause, **152**, 246
Hungarian algorithm, 247, 370
Hyperarc consistency, **108**, 245

Implication, **107**
　between 0-1 linear inequalities, 159, 246
　between linear inequalities, 122
Implicit enumeration, 13
Incumbent solution, **29**
Indexed linear element metaconstraint, **183**, 430–431
　convex hull relaxation of, 356
　domain filtering for, 185
　vector-valued, 183, 360
Indexed linear metaconstraint, 47, **431**
　domain filtering for, 48, 432
　relaxation of, 49, 432
Induced subgraph, **188**
Induced width, 245
Inequality constraint, 140
Inequality
　0-1 linear, **159**
　clique, **273**–274, 412, 443
　comb, 413
　cover, **266**, 411, 417, 432
　degree of, **159**
　domination of, 141, 143
　facet-defining, **272**
　feasible, 141
　integer linear, 275, 432
　mixed integer linear, **282**, 284
　odd cycle, **272**, 411, 443
　orthonormal representation, 443
Infeasible problem, **16**
Inference, 4, **18**, 105
Inference dual, 7–8, 105, **115**, 117, 245
　for cumulative scheduling, 242
　for disjunctive scheduling, 120, 223
　for inequality constraints, 141, 143
　for integer programming, 171, 175
　for linear programming, 127–128, 256
　for propositional satisfiability, 119, 435
　in constraint-directed search, 8, 119
　in sensitivity analysis, 8, 118, 130
Input refutation, **343**
Input resolution, **341**, 435
　and linear relaxation, 343
　and rank 1 cuts, 341, 344, 348, 412, 435
Integer linear inequality, 275, 432
Integer linear metaconstraint, 25, **432**
Integer linear system
　Benders cuts for, 179, 432
　bounds propagation for, 432
　relaxation of, 432
Integer programming, 12, 22, 51, 54, **171**, 246, 282, 432
　inference dual for, 142, 171, 175
　Lagrangean dual for, 144
　sensitivity analysis for, 173, 175
　separating cuts for, 278
Integer rounding cut, **263**
Integral polyhedron, **286**
Integrated methods
　advantages of, 9
　applications of, 11
　computational performance, 9
　history of, 13
　software, 14
Interior point method, 255, 411
Intersection graph, **272**
Interval consistency, 245
Interval matrix, **291**

Jackson preemptive schedule, **214**, 428
Job shop scheduling problem, 11
JPS, **214**

k-completeness, **110**, 114, 125, 245
　and backtracking, 111
　for 0-1 linear inequalities, 167
　for logical clauses, 152
　of linear system, 125

INDEX 481

k-consistency, **112**, 114, 245
  and backtracking, 112
  for 0-1 linear inequalities, 168, 417
  for logical clauses, 154
k-resolution, 154, 435
k-tree, 245
Knapsack constraint, **24**–25, 100, 103, 114, 432
  0-1, 103
  bounds propagation for, 25
Knapsack covering constraint, **24**, 329
Knapsack cut, **26**, 29, 103, 266, 332, 417, 433
  for nonmaximal packing, 59
Knapsack packing constraint, **24**, 329

Lagrange multiplier, 52, 130, **143**, **293**
  in domain reduction, 44, 52, 62, 146, 432
Lagrangean dual, 8, **143**, 246, 293–294, 412
  as inference dual, 143
  complementary slackness for, 145
  concavity of, 145
  for integer programming, 144
  solving, 296
Lagrangean function, **293**
Lagrangean relaxation, 10, 12, 148, 291, **293**, 412
  for linear programming, 293
Left-shifting, **239**
Lesson timetabling, 9, 11
Lex-greater metaconstraint, **433**
  domain filtering for, 433
Lift-and-project cut, 417, 436
Lifting coefficient, 267, 270
Lifting
  recursive algorithm for, 268
  sequence-independent, 269, 411, 417
  sequential, 267, 411, 417
  superadditive function for, 271
Linear disjunction metaconstraint, 11, 32, 60–61, 63, **433**
  big-M relaxation for, 412
  big-M relaxation of, 300, 434
  convex hull relaxation for, 299, 412
  convex hull relaxation of, 434
  separating cuts for, 307, 434
Linear inequality, 122
  domination of, 122
  feasible, 122
Linear metaconstraint, 30, 37, 60
  filtering for, 124
Linear programming, 28, 52, 121, **127**, 131, 246, 252, 411
  bounds propagation for, 135
  inference dual for, 127, 256
  interior point method for, 255, 411
  Lagrangean relaxation for, 293

  relaxation duality for, 256
  sensitivity analysis for, 130, 133
  simplex algorithm for, **254**
  surrogate relaxation for, 256
Linear programming dual, 8, 127–**128**, 133, 245, 256, 309, 411
Literal, **68**
Local optimum, 41
Local search, 4, **88**
  as branching, 91
  with relaxation, 96
Location theory, 245
Logic-based Benders decomposition, 10–11, 14, **77**–78, 222, 241
Logic metaconstraint, 30, **434**
  domain filtering for, 152
Lot sizing problem, 60, 437
  MILP model for, 327

Machine scheduling, 11, 78, 104, 120–121
  bounds propagation for, 83
Makespan, **78**
Markovian property, 208
Master problem
  in Benders decomposition, **77**, 80
  restricted, 52
Master scheduling, 88
Matching, **188**
  maximum cardinality, 189
Material conditional, 68, 149
Mathematical programming, 1–2
Max-n metaconstraint, 436
Max back heuristic, 381, 384, 413
Maximum cardinality matching, 189, 246
  on convex graph, 192
Maximum clique problem, 424
Maximum flow, 196, 247
  and all-different metaconstraint, 38
Medication problem, 86
Metaconstraint, **6**, 9, 18, 106, 415
  all-different (alldiff), 35, 37, 187, 247
  among, **418**
  at-least, 421
  at-most, 421
  bin packing, **419**
  cardinality, 35, **194**, 247, 420
  change, **422**
  circuit, 88, **199**, 247
  clique, **424**
  conditional, 30, 59, **424**, 434
  cumulative (scheduling), **230**, 425
  cutset, **426**
  cycle, **426**
  diffn, 63, **427**
  disjunctive (scheduling), 79, **212**, 428
  distribute, **194**, 420
  element, 38, 61, 103, **182**, 246, 429

482

flow, **430**
gcc, **194**
generalized cardinality, **194**, 420
indexed linear, 47, **431**
indexed linear element, 430–431
integer linear, 25, **432**
lex-greater, **433**
linear, 30, 37, 60
linear disjunction, 11, 32, 60–61, 63, **433**
logic, 30, **434**
max-n, 436
MILP, **435**
min-n, **436**
network design, **437**
nonlinear disjunction, 11, **438**
nvalues, 35, 62, **198**, 247, 438
path, 57, **439**
pattern, 444
piecewise linear, **440**
same, **441**
set covering, **441**
set packing, **442**
set sum, 57, 62
soft alldiff, **443**
sort, 441
stretch-cycle, 36, 60–61, **208**, 444
stretch, 36, 60, **207**, 247, 444
sum, **444**
symmetric alldiff, **445**
symmetric cardinality, **446**
usedby, 441
value precedence, **446**
Metaheuristic, 89
MILP-based relaxation
  of cumulative (scheduling) metaconstraint, 394
  of disjunctive (scheduling) metaconstraint, 388, 429
MILP metaconstraint, **435**
MILP model, 249, 412, 435
  based on disjunctive modeling, 324
  based on knapsack modeling, 329
  convex hull relaxation of, 320
  relaxation of, 249, 283, 318, 320, 436
  representability as, 319
  separating cuts for, 285
MILP representability, **319**, 412
Min-n metaconstraint, **436**
  domain filtering for, 436
Minimum cutset problem, 426
Minimum flow problem, 58
Mixed integer Gomory cuts, 436
Mixed integer linear inequality, **282**, 284
Mixed integer linear programming, 80, 245, 249, 275
Mixed integer rounding cut, **283**, 285, 412, 436

separating, 285
Modeling, 9
  with metaconstraints, 6–7
Monotone variable, **347**
Mosel, 14
Multidimensional knapsack problem, 11

Neighborhood, **88**
Network design metaconstraint, **437**
  relaxation of, 437
Network design problem, 11, 437
Network flow model, 430
  capacitated, 291, 430
  fixed charge, 11, 437
  for alldiff metaconstraint, 198
  for alldiff relaxation, 366
  for cardinality metaconstraint, 195, 197, 431
  for circuit constraint, 204
  for circuit metaconstraint, 431
  for nvalues metaconstraint, 198–199
  for same metaconstraint, 441
  max back heuristic for, 381
  maximum flow in, 196
  minimum-cost flow in, 366, 431
  minimum capacity cut for, 381
  multi-commodity, 431
  total unimodularity of, 289, 291
  with gains, 431
Nogood, 3, **63**–64, 66, 71
  as relaxation, 65, 75
  obtained from inference dual, 119
Nogood processing, 65, 72
Nonbasic variable, **131**
Nonlinear disjunction metaconstraint, 11, **438**
  big-M relaxation for, 317
  big-M relaxation of, 438
  convex hull relaxation for, 313, 412, 438
Nonlinear programming, 41, 313
Nonpreemptive scheduling, **212**, 230
Nonserial dynamic programming, 245
Not-first not-last rule, **217**, 248
  for cumulative scheduling, 238, 426
  for disjunctive scheduling, 217, 428
  polynomial-time algorithm for, 218
NP, 117, 128
Nurse scheduling, 33
Nvalues metaconstraint, 35, 62, **198**, 247, 438
  domain filtering for, 439
  flow model for, 199

Odd cycle, **272**
Odd cycle inequality, **272**, 411, 443
Odd hole, **272**
Operations research

INDEX 483

integrated methods in, 14
OPL Studio, 14
Optimal solution, **16**
Optimization problem, **16**
Orthogonal Latin square problem, 11
Orthonormal representation inequality, 443

Packing
    for knapsack inequality, **26**
    maximal, **27**
Parallel resolution, 65, 73, 94, 103, 435
Partial-order dynamic backtracking, 65, **72**, 74, 86, 103, 121
    in local search, 94
Partial order dynamic backtracking, 4
Particle swarm optimization, 102
Path in a graph, **188**
Path metaconstraint, 57, **439**
    domain filtering for, 440
Pattern in stretch metaconstraint, 36, 207
Pattern metaconstraint, 444
Penultimate variable, **72**–74
Permissible edge, 204
Permissible hamiltonian cycle, **203**
Phase I problem, 255
Phase II problem, 255
Physician scheduling, 11
Piecewise linear function, 9, 11, **259**, 411, 440
Piecewise linear metaconstraint, 259, **440**
    bounds propagation for, 260, 440
    convex hull relaxation for, 259, 441
Pigeon hole problem, 157
Planning and scheduling, 9–12, 78, 104, 222, 241
    Benders decomposition for, 222, 224–226, 241–243, 426
    to minimize cost, 222, 241
    to minimize makespan, 224, 242, 402
    to minimize number of late jobs, 225, 243, 403
    to minimize total tardiness, 226, 243, 404
Polyhedral cone, 126, 138
Polyhedron, 4, 30, 131
    integral, **286**
Power plant scheduling, 61
Precedence constraint, 82
Preemptive scheduling, **212**
Preprocessing, 4
Pricing a variable, 51–52, 55
    directed graph model, 55
Primal graph, **112**
Primal problem, **116**
Prime implication, 32, **151**, 157, 434
Probing, 40
Problem restriction, **17**, 29
Processing network design problem, 11

Product configuration, 9, 11, 45, 103
Production planning, 11, 30
Programming
    constraint, 1–2, 22
    integer, 12, 22, 51, 54
    linear, 28, 52
    mathematical, 1–2
    mixed integer linear, 249, 275
    nonlinear, 41
Projection, 20, **108**, 110, 124
    polyhedral, 125
Propagation, **19**
    bounds, **25**, 29, 109
Propositional logic, 67, **148**, 335
    relaxation of, 335
Propositional satisfiability problem, 12, 66–**67**, 74
    relaxation of, 335, 435
Pruning of a search tree, 29, 66

Quadratic assignment problem, **372**
    0-1 model for, 334
    relaxation of, 373

Recession cone, **319**
Reduced cost, 52, 55, **132**, 134–135, 253
Reduced-cost variable fixing, 9, 11, 135
Reduction for inequalities, **160**
Relaxation, 4, **16**, 19
    and nogoods, 65
    assignment problem, 9
    continuous, 27, 30, 52
    convex hull, 11, 32
    domain store, 40
    factored, 43, 61, 103, **325**, 340
    for semicontinuous piecewise linear metaconstraint, 259
    in GRASP, 98
    in local search, 96
    Lagrangean, 10, 12, 148, 291, **293**, 412
    of 0-1 linear system, 262, 417
    of alldiff constraint, 370
    of alldiff metaconstraint, 363, 366, 368, 413, 418
    of Benders subproblem, 80, 400–403, 405
    of bin packing metaconstraint, 420
    of cardinality clause, 421
    of cardinality conditional, 422
    of cardinality metaconstraint, 376, 378–379, 421
    of clique metaconstraint, 424
    of conditional metaconstraint, 425
    of cumulative (scheduling) metaconstraint, 393–394, 398, 426
    of cycle metaconstraint, 427
    of disjujnctive (scheduiling) metaconstraint, 386

of disjunction of linear equations, 303, 306
of disjunction of linear inequalities, 412
of disjunction of linear systems, 299–300, 412
of disjunction of nonlinear systems, 313, 317, 319, 412, 438
of disjunctive (scheduling) metaconstraint, 385, 388, 391, 393, 429
of element metaconstraint, 352, 357, 413, 430
of generalized assignment problem, 295
of indexed linear element, 49
of indexed linear element metaconstraint, 356
of indexed linear metaconstraint, 49, 432
of integer linear system, 432
of linear programming problem, 293
of logical clause, 336
of MILP model, 283
of network design metaconstraint, 437
of nonlinear constraints, 43
of propositional formulas, 335
of quadratic assignment problem, 373
of semicontinuous piecewise linear metaconstraint, 441
of set covering metaconstraint, 442
of set packing metaconstraint, 443
of sum metaconstraint, 445
of traveling salesman problem, 380
of vector-valued element metaconstraint, 359
of vector-valued indexed linear element metaconstraint, 360
product, **325**
semidefinite programming, 10
surrogate, 256, 291–**292**, 411
tight, 249
time window, 80
Relaxation dual, 7, 9, 250–**251**, 411
for inequality constraints, 292–293
for linear programming, 256
Renamable Horn clause set, **152**, 158
Residual graph, **196**
Resolution, 33, **150**, 246, 264, 435
k-, 154, 435
0-1, **165**, 246, 417
as generator of rank 1 cuts, 340, 412
completeness of, 151, 435
for generation of separating cuts, 348, 413, 435
for multivalent clauses, 158
for projection, 153
for tighter relaxation, 340
input, 435
parallel, 65, 73, 94, 103, 435
unit, **152**, 246, 435

Resource-constrained scheduling, 11, 426, 428
Resource-constrained sequencing, 419
Restriction, 3, **17**, 75
in Benders decomposition, 77
in branching search, 29
in constraint-directed search, 64
in local search, 88, 90
Right-shifting, **239**
Roof point, 169
Root node, 29
Roster for airline crew, 53

Saddle function, 293
Safe methods, 45
Same metaconstraint, **441**
domain filtering for, 441
network flow model for, 441
Satisfiability problem, 12, 66–**67**, 74
2-, 157
Scheduling
aircraft, 11
airline crew, 11, 51, **53**
batch, 10, 12
boat party, 11
call center, 10
computer processor, 12
cumulative, 230, 248
disjunctive, **212**, 247, 385
employee, 11, 33, 103, 422
flow shop, 10
machine, 11, 78, 104, 120
master, 88
nonpreemptiove, 230
nonpreemptive, **212**
nurse, 33
physician, 11
preeemptive, **212**
resource-constrained, 11, 426
sports, 10
transit crew, 11
SCIP, 14
Search-infer-and-relax, **3**, 12, 17, 65, 103, 117
Search, 3, **17**
branching, **21**, 28, 40, 43, 51, 63
constraint-directed, 3, **63**, 65, 103
depth-first, 29
exhaustive, 3, **17**, 21, 64
local, 4, **88**
steepest ascent, 145
tabu, 4, 89–**90**, 101, 104
Search tree, 28
Search variable, 77, 80
Selection function, **65**–66, 72
Semicontinuous function, 11, 259, 411, 440
Semidefinite programming relaxation, 10

INDEX

Sensitivity analysis, 8, **118**
  for integer programming, 173, 175, 182
  for linear programming, 130, 133
  with inference dual, 8
Separable function, 259, 440
Separating cut, 5, 22, 250, 273, **275**
  for alldiff metaconstraint, 365
  for cardinality metaconstraint, 377
  for circuit metaconstraint, 383
  for disjunction of linear systems, 307
  for integer linear system, 278, 417
  for logical clauses, 348, 413, 435
  for mixed integer linear system, 285
  for traveling salesman problem, 380, 383, 413
  for vector-valued element metaconstraint, 359
Separating mixed integer rounding cut, 285
Separating resolvent, **348**
Separator, **202**
Separator graph, **202**
Separator
  heuristic method for finding, 204
Sequence-independent lifting, 269
  superadditive function for, 271
Sequential lifting, 267
  recursive algorithm for, 268
Set-valued variable, 56
  bounds propagation for, 57
  branching on, 58
Set covering metaconstraint, **441**
  relaxation of, 442
Set covering problem
  0-1 model for, 329, 442
Set packing metaconstraint, **442**
  relaxation of, 443
Set packing problem, 272
  0-1 model for, 330, 442
  cutting planes for, 272, 443
Set partitioning problem
  0-1 model for, 330
Set sum metaconstraint, 57
  bounds propagation for, 58, 62
Setup cost, 61
Shadow price, **130**, 134
Shortest path problem, 57–58, 62, 439
  dynamic programming recursion for, 212
  resource-constrained, 11
SIMPL, 14
Simplex algorithm, **254**, 411
Simulated annealing, **89**, 91
  as branching search, 92
Soft alldiff metaconstraint, **443**
  domain filtering for, 443
Solution, **16**
  candidate, **17**, 21–22, 63
  feasible, **16**

incumbent, **29**
  optimal, **16**
Sort metaconstraint, 441
Space packing problem, 427
Specially ordered set, 259
Sports scheduling, 10
State transition, **208**
Steepest ascent search, 145
Stretch-cycle metaconstraint, 36, 60–61, **208**, 444
Stretch metaconstraint, 36, 60, **207**, 247, 444
  domain filtering for, 38, 210, 444
  dynamic programming model, 208, 247
Strip packing problem, 11
Strong k-completeness, **111**
  and backtracking, 111
Strong k-consistency
  and backtracking, 112
Strong duality, **116**, **252**
  for branching dual, 171
  for linear programming, 128–129
  for subadditive dual, 170, 172
Strongly connected component, **188**, 191
Structural design problem, 10–11
Subadditive dual, 170–**171**, 246
Subadditive function, **171**
Subgradient, **298**
Subgradient optimization, 246, 252, 296, 412
Subproblem
  in Benders decomposition, 77
Subtour-elimination constraints, 380
Sum metaconstraint, **444**
  convex hull relaxation of, 445
  domain filtering for, 445
Superadditive dual, 8, 170–**171**
Superadditive function, **171**, 270
  for sequence-independent lifting, 271
Supply chain management, 78
Supporting inequality, **304**
  for disjunction of linear inequalities, 304
Surrogate, **122**, 124, 126
Surrogate dual, 8, **141**, 144, 147, 246, **292**
Surrogate relaxation, 291–**292**, 411
  for linear programming, 256
Symmetric alldiff metaconstraint, **445**
  domain filtering for, 446
Symmetric cardinality metaconstraint, **446**
  domain filtering for, 446
Symmetry, 40
Symmetry breaking, 40, 433, 446

Tabu list, 89–**90**, 94
Tabu search, 4, 89–**90**, 104
  as constraint-directed search, 90, 101
Time-window relaxation, 80, 401, 413
Total unimodularity, **287**, 412

Traffic diversion problem, 12
Transit crew scheduling, 10–11
Transportation problem, 11
Traveling salesman problem, **200**, 379, 413, 423
   0-1 model for, 332, 379, 413
   comb inequalities for, 382
   filtering for, 109, 111
   relaxation of, 380
   separating cuts for, 380, 413
   symmetric, **379**
   with time windows, 11, 88, 95
Traveling tournament problem, 10–11
Truncation errors, 45
Truth function, 149

Unary resource constraint, 212
Unbounded problem, **16**, 254
Uncapacitated facility location problem
   MILP model for, 327
Unit clause, **151**
Unit clause rule, 69, 74, 121
Unit refutation, **343**
Unit resolution, **152**, 246
   and linear relaxation, 343, 412
   and relaxation, 341
   for Horn clauses, 152, 435
Usedby metaconstraint, 441

Valid constraint, **16**
Valid cut, **262**
Valid inequality, 26, **262**
Value precedence metaconstraint, **446**
   domain filtering for, 447
Variable index, 37–38, 46, 60–61, 88
   and element constraint, 38, 429
Variable subscript, 37
Vector-valued element metaconstraint, **358**, 429
   convex hull relaxation of, 359, 430
   separating cuts for, 359
Vector-valued indexed linear element metaconstraint, 183
   big-M relaxation of, 360
   convex hull relaxation of, 360
   filtering for, 187
Vehicle routing, 11
Vehicle routing problem, 88, 95, 427
   generalized GRASP for, 96

Warehouse shelving problem, 375
Weak duality, 136, **144**, **252**
Weierstrass Theorem, 123
Width of dependency graph, 113, 115, 245

Zero-step lookahead, 111, 170

Early Titles in the
INTERNATIONAL SERIES IN
OPERATIONS RESEARCH & MANAGEMENT SCIENCE
Frederick S. Hillier, Series Editor, *Stanford University*

Saigal/ *A MODERN APPROACH TO LINEAR PROGRAMMING*
Nagurney/ *PROJECTED DYNAMICAL SYSTEMS & VARIATIONAL INEQUALITIES WITH APPLICATIONS*
Padberg & Rijal/ *LOCATION, SCHEDULING, DESIGN AND INTEGER PROGRAMMING*
Vanderbei/ *LINEAR PROGRAMMING*
Jaiswal/ *MILITARY OPERATIONS RESEARCH*
Gal & Greenberg/ *ADVANCES IN SENSITIVITY ANALYSIS & PARAMETRIC PROGRAMMING*
Prabhu/ *FOUNDATIONS OF QUEUEING THEORY*
Fang, Rajasekera & Tsao/ *ENTROPY OPTIMIZATION & MATHEMATICAL PROGRAMMING*
Yu/ *OR IN THE AIRLINE INDUSTRY*
Ho & Tang/ *PRODUCT VARIETY MANAGEMENT*
El-Taha & Stidham/ *SAMPLE-PATH ANALYSIS OF QUEUEING SYSTEMS*
Miettinen/ *NONLINEAR MULTIOBJECTIVE OPTIMIZATION*
Chao & Huntington/ *DESIGNING COMPETITIVE ELECTRICITY MARKETS*
Weglarz/ *PROJECT SCHEDULING: RECENT TRENDS & RESULTS*
Sahin & Polatoglu/ *QUALITY, WARRANTY AND PREVENTIVE MAINTENANCE*
Tavares/ *ADVANCES MODELS FOR PROJECT MANAGEMENT*
Tayur, Ganeshan & Magazine/ *QUANTITATIVE MODELS FOR SUPPLY CHAIN MANAGEMENT*
Weyant, J./ *ENERGY AND ENVIRONMENTAL POLICY MODELING*
Shanthikumar, J.G. & Sumita, U./ *APPLIED PROBABILITY AND STOCHASTIC PROCESSES*
Liu, B. & Esogbue, A.O./ *DECISION CRITERIA AND OPTIMAL INVENTORY PROCESSES*
Gal, T., Stewart, T.J., Hanne, T. / *MULTICRITERIA DECISION MAKING: Advances in MCDM Models, Algorithms, Theory, and Applications*
Fox, B.L. / *STRATEGIES FOR QUASI-MONTE CARLO*
Hall, R.W. / *HANDBOOK OF TRANSPORTATION SCIENCE*
Grassman, W.K. / *COMPUTATIONAL PROBABILITY*
Pomerol, J-C. & Barba-Romero, S. / *MULTICRITERION DECISION IN MANAGEMENT*
Axsäter, S. / *INVENTORY CONTROL*
Wolkowicz, H., Saigal, R., & Vandenberghe, L. / *HANDBOOK OF SEMI-DEFINITE PROGRAMMING: Theory, Algorithms, and Applications*
Hobbs, B.F. & Meier, P. / *ENERGY DECISIONS AND THE ENVIRONMENT: A Guide to the Use of Multicriteria Methods*
Dar-El, E. / *HUMAN LEARNING: From Learning Curves to Learning Organizations*
Armstrong, J.S. / *PRINCIPLES OF FORECASTING: A Handbook for Researchers and Practitioners*
Balsamo, S., Personé, V., & Onvural, R./ *ANALYSIS OF QUEUEING NETWORKS WITH BLOCKING*
Bouyssou, D. et al. / *EVALUATION AND DECISION MODELS: A Critical Perspective*
Hanne, T. / *INTELLIGENT STRATEGIES FOR META MULTIPLE CRITERIA DECISION MAKING*
Saaty, T. & Vargas, L. / *MODELS, METHODS, CONCEPTS and APPLICATIONS OF THE ANALYTIC HIERARCHY PROCESS*
Chatterjee, K. & Samuelson, W. / *GAME THEORY AND BUSINESS APPLICATIONS*
Hobbs, B. et al. / *THE NEXT GENERATION OF ELECTRIC POWER UNIT COMMITMENT MODELS*
Vanderbei, R.J. / *LINEAR PROGRAMMING: Foundations and Extensions, 2nd Ed.*
Kimms, A. / *MATHEMATICAL PROGRAMMING AND FINANCIAL OBJECTIVES FOR SCHEDULING PROJECTS*
Baptiste, P., Le Pape, C. & Nuijten, W. / *CONSTRAINT-BASED SCHEDULING*
Feinberg, E. & Shwartz, A. / *HANDBOOK OF MARKOV DECISION PROCESSES: Methods and Applications*
Ramík, J. & Vlach, M. / *GENERALIZED CONCAVITY IN FUZZY OPTIMIZATION AND DECISION ANALYSIS*
Song, J. & Yao, D. / *SUPPLY CHAIN STRUCTURES: Coordination, Information and Optimization*
Kozan, E. & Ohuchi, A. / *OPERATIONS RESEARCH/ MANAGEMENT SCIENCE AT WORK*
Bouyssou et al. / *AIDING DECISIONS WITH MULTIPLE CRITERIA: Essays in Honor of Bernard Roy*

Early Titles in the
INTERNATIONAL SERIES IN
OPERATIONS RESEARCH & MANAGEMENT SCIENCE
(Continued)

Cox, Louis Anthony, Jr./ *RISK ANALYSIS: Foundations, Models and Methods*
Dror, M., L'Ecuyer, P. & Szidarovszky, F./ *MODELING UNCERTAINTY: An Examination of Stochastic Theory, Methods, and Applications*
Dokuchaev, N./ *DYNAMIC PORTFOLIO STRATEGIES: Quantitative Methods and Empirical Rules for Incomplete Information*
Sarker, R., Mohammadian, M. & Yao, X./ *EVOLUTIONARY OPTIMIZATION*
Demeulemeester, R. & Herroelen, W./ *PROJECT SCHEDULING: A Research Handbook*
Gazis, D.C./ *TRAFFIC THEORY*
Zhu/ *QUANTITATIVE MODELS FOR PERFORMANCE EVALUATION AND BENCHMARKING*
Ehrgott & Gandibleux/ *MULTIPLE CRITERIA OPTIMIZATION: State of the Art Annotated Bibliographical Surveys*
Bienstock/ *Potential Function Methods for Approx. Solving Linear Programming Problems*
Matsatsinis & Siskos/ *INTELLIGENT SUPPORT SYSTEMS FOR MARKETING DECISIONS*
Alpern & Gal/ *THE THEORY OF SEARCH GAMES AND RENDEZVOUS*
Hall/*HANDBOOK OF TRANSPORTATION SCIENCE - 2nd Ed.*
Glover & Kochenberger/ *HANDBOOK OF METAHEURISTICS*
Graves & Ringuest/ *MODELS AND METHODS FOR PROJECT SELECTION: Concepts from Management Science, Finance and Information Technology*
Hassin & Haviv/ *TO QUEUE OR NOT TO QUEUE: Equilibrium Behavior in Queueing Systems*
Gershwin et al/ *ANALYSIS & MODELING OF MANUFACTURING SYSTEMS*
Maros/ *COMPUTATIONAL TECHNIQUES OF THE SIMPLEX METHOD*
Harrison, Lee & Neale/ *THE PRACTICE OF SUPPLY CHAIN MANAGEMENT: Where Theory and Application Converge*
Shanthikumar, Yao & Zijm/ *STOCHASTIC MODELING AND OPTIMIZATION OF MANUFACTURING SYSTEMS AND SUPPLY CHAINS*
Nabrzyski, Schopf & Węglarz/ *GRID RESOURCE MANAGEMENT: State of the Art and Future Trends*
Thissen & Herder/ *CRITICAL INFRASTRUCTURES: State of the Art in Research and Application*
Carlsson, Fedrizzi, & Fullér/ *FUZZY LOGIC IN MANAGEMENT*
Soyer, Mazzuchi & Singpurwalla/ *MATHEMATICAL RELIABILITY: An Expository Perspective*
Chakravarty & Eliashberg/ *MANAGING BUSINESS INTERFACES: Marketing, Engineering, and Manufacturing Perspectives*
Talluri & van Ryzin/ *THE THEORY AND PRACTICE OF REVENUE MANAGEMENT*
Kavadias & Loch/*PROJECT SELECTION UNDER UNCERTAINTY: Dynamically Allocating Resources to Maximize Value*
Brandeau, Sainfort & Pierskalla/ *OPERATIONS RESEARCH AND HEALTH CARE: A Handbook of Methods and Applications*
Cooper, Seiford & Zhu/ *HANDBOOK OF DATA ENVELOPMENT ANALYSIS: Models and Methods*
Luenberger/ *LINEAR AND NONLINEAR PROGRAMMING, 2nd Ed.*
Sherbrooke/ *OPTIMAL INVENTORY MODELING OF SYSTEMS: Multi-Echelon Techniques, Second Edition*
Chu, Leung, Hui & Cheung/ *4th PARTY CYBER LOGISTICS FOR AIR CARGO*
Simchi-Levi, Wu & Shen/ *HANDBOOK OF QUANTITATIVE SUPPLY CHAIN ANALYSIS: Modeling in the E-Business Era*

* A list of the more recent publications in the series is at the front of the book *